HUMAN–INSECT INTERACTIONS

HUMAN–INSECT INTERACTIONS

Sergey Govorushko

Chief Research Scholar
Pacific Geographical Institute of Russian Academy of Sciences
Professor Far Eastern Federal University
Vladivostok
Russia

CRC Press
Taylor & Francis Group
Boca Raton   London   New York

CRC Press is an imprint of the
Taylor & Francis Group, an **informa** business

A SCIENCE PUBLISHERS BOOK

Cover Illustrations:

Top left. Plate with fried insects in Thailand.
Photo credit: I. Arzamastcev, 20 April 2017
Top right. Cocoons of the silkworm *Bombyx mori* and silk thread, which obtained after unwinding.
Photo credit: M. Lochynska, 03 of April 2017
Bottom left. Worm-infested fruit. Prepared by graphic designer, A. Parvez.
Bottom right. Mosquitos in the lower part of the Indigirka River (north-eastern Yakutia, Russia).
Photo credit: S.M. Govorushko, July 1975

CRC Press
Taylor & Francis Group
6000 Broken Sound Parkway NW, Suite 300
Boca Raton, FL 33487-2742

First issued in paperback 2020

© 2018 by Taylor & Francis Group, LLC
CRC Press is an imprint of Taylor & Francis Group, an Informa business

No claim to original U.S. Government works

ISBN-13: 978-1-4987-1949-0 (hbk)
ISBN-13: 978-0-367-78158-3 (pbk)

Library of Congress Cataloging-in-Publication Data

Names: Govorushko, Sergey, 1955- author.
Title: Human-insect interactions / Sergey Govorushko, chief research scholar,
   Pacific Geographical Institute, of Russian Academy of Sciences,
   Vladivostok, Russia.
Other titles: Human, insect interactions
Description: Boca Raton, FL : CRC Press, 2017. | "A science publishers book."
   | Includes bibliographical references and index.
Identifiers: LCCN 2017040117 | ISBN 9781498719490 (hardback : alk. paper)
Subjects: LCSH: Insects. | Beneficial insects. | Insect pests.
Classification: LCC QL463 .G67 2017 | DDC 595.7--dc23
LC record available at https://lccn.loc.gov/2017040117

Visit the Taylor & Francis Web site at
http://www.taylorandfrancis.com

and the CRC Press Web site at
http://www.crcpress.com

Preface

Insects are the **most numerous class** of living organisms on Earth. You cannot find a human who is not in contact with them, yet everyone has his or her own personal experience of interaction with insects. The book offered here for your attention is an attempt to generalize the experiences of all people in order to provide a general consideration of the interaction between human and insects. I am not an entomologist but a geomorphologist by education; however, all of my books, in one way or another, illustrate different aspects of the interaction between society and the environment. This book is also focused primarily on the *interaction*; all data on the biology, ecology, and behavior of insects are aimed at the best understanding of the **character of interaction** between humans and insects.

Strictly speaking, it is **not only insects** that are discussed in the book. Its content was expanded to include members of the closely related group of **arachnids** (spiders, scorpions, and ticks). If a section is completely or to a great degree devoted to arachnids, this is mentioned in its title.

Naturally, the preparation of this book would be impossible without the **assistance of entomologists**. These entomologists gave advice as to which subjects to include in the book, provided published material (their own and those by other authors) and photos, aided with various problems (e.g., with commonly used and scientific [Latin] names of species), and solved a number of organizational issues.

I wish to express my gratitude to the Academician P.Ya. Baklanov, Scientific leader of the Pacific Geographical Institute, Far Eastern Branch of the Russian Academy of Sciences (FEB of RAS), for every possible support over many years in the abovementioned studies.

The support and assistance of current colleagues at the Pacific Geographical Institute are also gratefully acknowledged, including Mr. V.B. Primak, Dr. S.M. Krasnopeev, Prof. S.V. Osipov, Dr. V.V. Yermoshin, and Mr. A.V. Vlasov.

Numerous colleagues from Russia have also contributed indirectly to this book, including Mr. A.G. Blummer (All-Russian Plant Quarantine Center, Moscow), Dr. E.S. Koshkin (Institute of Water and Ecology Problems, FEB of RAS, Khabarovsk), Prof. L.P. Esipenko (All-Russian Research Institute of Biological Plant Protection, Krasnodar), Dr. I.A. Kerchev (Institute of Systematics and Ecology of Animals, Novosibirsk), Prof. D.I. Berman (Institute of Biological Problems of North, FEB of RAS, Magadan), Dr. A.R. Manukyan (Kaliningrad Amber Museum), Prof. S.I. Chernysh (St. Petersburg State University, St. Petersburg), Dr. Y.M. Yakovlev (A.V. Zhirmunsky Institute of Marine Biology, FEB of RAS, Vladivostok), and others.

I have had the unique opportunity of becoming acquainted and collaborating with numerous international colleagues. Of particular note are Prof. C. Tisdell (University of Queensland, Brisbane, Australia), Dr. V. Haddad, Jr. (São Paulo State University, Botucatu, Brazil), Dr. M. Lochynska (Institute of Natural Fibres and Medicinal Plants, Poznan, Poland), Prof. J. White (Women's and Children's Hospital, Adelaide, Australia), Dr. E.P. Cherniack (University of Miami Miller School of Medicine, Florida, USA), Dr. P. Querner (University of Natural Resources and Life Sciences, Vienna, Austria), Dr. B.P. Zakharov (American Museum of Natural History, New York, USA), Dr. J. Amendt (Institute of Forensic Medicine, University of Frankfurt am Main, Germany), and Dr. S. Fattorini (University of Aquila, Italy).

Special thanks to Prof. Yu.M. Marusik (Institute of Biological Problems of North, FEB of RAS, Magadan, Russia), for checking the entomological aspects of the Russian text, and to Ms. D.M. Miller (Boulder, Colorado, USA) and Ms. A. Houska (Princeton, New Jersey, USA) for closely editing the manuscript and persisting in overcoming differences between the Russian and English wording, which I hope has allowed the book to become clearer for the Western reader. The author would like to thank the translator, V.M. Karpets (Pacific Geographical Institute, FEB of RAS, Vladivostok, Russia) and Dr. V. Fet (Marshall University, Huntington, West Virginia, USA), who checked the zoological aspects of the text (including the common and scientific names of species and other taxa) and suggested a number of important corrections.

Of a major importance to the author was the positive attitude and helpfulness of Ms. L.P. Slavinskaya, for assistance in preparation of the manuscript for publication, and Ms. E.V. Oleinikova, for preparation of the electronic versions of the maps.

When reading a book, questions caused by the **different meaning** of terms in different countries are unavoidable. For example, notably, "sand flies" appears to be a very imprecise term. It may indicate different species in different countries. In the United States, "sand fly" may refer to certain horse flies of the family Tabanidae (e.g., "greenheads") or to members of the family Ceratopogonidae (e.g., sand gnat, sand flea, no-see-em); outside the United States, "sand fly" may refer to members of the subfamily Phlebotominae within the family Psychodidae. Similarly, in Russian literature, a single word is used for ticks and mites.

Different usage of terminology by different authors is also possible. Note that species of the insects discussed in this book are known by **more than one scientific name**, reflecting different opinions on generic placement or precise status. Often, the text of the book includes the names used by the authors, but for full retrieval of information, it is also necessary to examine the names that follow.

Therefore, the author would be very grateful to all who, after discovering an error or uncertainty in the book, send pertinent information to me at **sgovor@tig.dvo.ru** or **sgovor@inbox.ru**. In addition, e-mailed photos illustrating interaction between humans and insects would be appreciated.

Contents

Part 2: Negative Aspects

Appendixes

Introduction

At the present time, more than one million species of insects have been described (Schowalter 2013). This number is more than half the animal species inhabiting the Earth. The overwhelming majority of insects live on land. Among the insects adapting to aquatic environments, the freshwater species predominate; very few of them live at sea. For example, the water strider (*Halobates micans*) lives among the floating algae in the Sargasso Sea 2,400 km (1,500 miles) from the nearest land (Komarova 2008).

The entomofauna of tropical regions is much more diverse than the fauna of insects in temperate and especially cold zones. This difference is related mainly to the fact that the insects are animals with variable body temperatures. Therefore, in zones with temperate and cold climates, almost all insects fall into a long hibernation, to which only a few species have been able to adapt.

The insects have taken the position of ecological preeminence over the past 400 million years. Their **competitive advantages** are as follows (http://www.cals.ncsu.edu/course/ent425/text01/success.html; Komarova 2008):

1. Small size, which allows them to (a) use minimal resources for survival and reproduction; and (b) to avoid predation because they can hide in the smallest shelters, such as cracks in rocks, beneath the bark of the trees, and under blades of grass.

2. High reproductive potential, which is associated with (a) high fecundity (females often produce large numbers of eggs); (b) high fertility (most of the eggs hatch); (c) and short life cycles (in most cases, 2–4 weeks).

3. High adaptation capacities, such as the following: (a) adaptation to new kinds of food (they have adjusted to such non-characteristic kinds of foods as glue, tanned leather products, the corks of wine bottles, chocolate, tobacco, etc.); (b) unique tolerance of difficult environments (they inhabit environments such as mountaintops with heights up to 6,000 m, the surfaces of glaciers, hot springs with temperatures up to 50°C, caves, and oil puddles); and (c) acquisition of resistance to insecticides (now, more than 500 species of insects are known to be immune to insecticides, and a number of insects are resistant to compounds from more than one chemical family).

Clearly, insects have effects on humans. Accordingly, human activities also have impacts on insect populations. In this book, the different aspects of these interactions will be considered. In addition, along with insects, the book considers certain other arthropods, including arachnids (spiders, scorpions, ticks) and myriapods.

PART 1

Positive Aspects

From a human perspective, insects can be divided into **beneficial** and **destructive** species. From this point of view, the overwhelming majority of insect species do not fall within one or the other category. It is evident that such an anthropocentric approach is too limited. In addition, some insects can be both "right" and "wrong" from different perspectives. From the standpoint of impacts on plant cultivation, the locust is an absolute evil. However, one can also consider it from a gastronomic perspective. In a number of regions, the locust is used for food, which allows us to consider it as a protein source.

Now, the concept of usefulness is less and less used to evaluate the role of different species of living things in nature. In the interrelated living world, there are no "useful" or "destructive" species of insects. Each species occupies its ecological niche in the natural ecosystem and its position in the cycle of matter and is a holder of unique genetic information.

Nevertheless, it is difficult to completely stop using the criteria of "right" and "wrong." An attempt has been made in this book to expand their meaning. For example, when insects fall into nutrient-poor lakes and headwater streams, they bring in considerable amounts of carbon, nitrogen, and phosphorus (Mehner et al. 2005). From an anthropocentric viewpoint, these benefits are not that great; however, they are very good for the ecosystem.

The science of **ethnoentomology** deals with the interactions between insects and people ("ethno," study of humans, and "entomology," study of insects). Ethnoentomology has revealed how insects were used in the past and how they are used at present by people for food, rituals, medicine, and other purposes (https://en.wikipedia.org/wiki/Ethnoentomology).

However, it is evident that this approach is **one-sided**. Besides the positive aspects of human interactions with insects, there are also many examples of the negative effects of insects on human activities. In connection with the foregoing, the materials will be nonetheless considered on the principle of "good/bad." This book will start with **the good things**.

1.1

Ecosystem Services

Ecosystem services are benefits that are obtained by people from ecosystems. They are divisible into four categories: (1) *provisioning* (food and water supply); (2) *supporting* (nutrient cycles, pollination, and formation of soils); (3) *regulating* (contributing to a consistent supply of other ecosystem services through density-dependent feedback); and (4) *cultural* (spiritual and recreational values) (Millennium Ecosystem Assessment 2005; Jorgensen 2010). Insects, being a part of the ecosystems, contribute significantly to these services.

The importance of insects is due to the **huge number** of insects. On the average, the numbers of insects per square meter are as follows: 1,000–1,500 in tropical forests; 500–600 in temperate forests; 300–400 in savanna, steppe, and forest-tundra; and 200 in desert, tundra, and short-grass Alpine meadow (Panfilov 1977). The maximum values significantly exceed the above-noted numbers. For example, the numbers of termites per square meter can reach 15,000 individuals (Jouquet 2011). The annual productivity of insects per hectare of land is 500–1,000 kg (Panfilov 1977).

When it comes to the **economic costs** of ecosystem services, their annual average for the United States alone is estimated at $57 billion (Losey and Vaughan 2006).

1.1.1 Pollination

During the pollination of plants, the pollen is transferred from anthers of stamens to stigmae, and it is possible only when the plants are flowering. The transfer of pollen from stamens to stigmae can be facilitated by the wind, water, insects, birds, and other animals.

About 80% of all plant pollination is biotic (https://en.wikipedia.org/wiki/Pollination). For agricultural plants, Klein et al. (2007) found, on the basis of data analysis for 115 key food crops from 200 countries, that 87 of them depend on pollination by animals, whereas 28 do not. However, **global production volumes** provide a different picture: only 35% of global agricultural production is provided by about 800 cultivated plants requiring pollination by *animals* (Nicholls and Altieri 2013). This difference is because the crops include mainly fruits and vegetables, while the basic volume of output is provided by cereal crops pollinated by *wind*.

Nevertheless, the contribution of pollinators to food production is great. It is estimated that, in their absence, **losses** of global agricultural production would reach about 3–8% (Bretagnolle and Gaba 2015). Different economic valuations of the biotic pollination services are given in Table 1.1-1 below.

The role of insects in biotic pollination is crucial. For example, they pollinate more than 80% of plants in Europe (Ellison and Gotelli 2009). Not all insects visiting flowers for the sake of nectar are good for cross-pollination. Such insects as beetles, true bugs, and aphids, even if they enjoy the nectar, **do not efficiently pollinate** the flowers. More often, they do harm to the plants. The burrowing wasps, cuckoo wasps, gallflies, parasitic wasps, and sawflies are **practically ineffective**. Some benefit is provided by root-eating flies–syrphid flies and some butterflies.

Table 1.1-1. An economic valuation of biotic pollination services.

Annual economic value	Region	Source	Notes
$153 billion	Globe	Melathopoulos, Cutler, Tyedmers (2015)	
more than [sic] $150 billion	Globe	Patiny et al. (2009)	
$120 billion	Globe	Costanza et al. (1997)	
EUR14.6 [+/–3.3] billion (1991–2009)	Europe	Leonhardt et al. (2013)	Pollination by insects; 12 (+/–0.8)% of the total economic value of annual crop production
$14.29 billion in 1996 (directly dependent crops) + (indirectly dependent crops)	United States	Calderone (2012)	(2009 USD)
$10.69 billion in 2001 (directly dependent crops)	United States	Calderone (2012)	(2009 USD)
$15.12 billion by 2009 (directly dependent crops)	United States	Calderone (2012)	(2009 USD)
$15 billion	United States	Irwin and Kampmeier (2003)	Pollination by domesticated honey bees
$20 billion	United States	http://www.si.edu/encyclopedia_si/nmnh/buginfo/benefits.htm	Pollination by honey bees
US$12 billion	Brazil	Giannini et al. (2015)	They analyzed 141 crops and found that 85 depend on pollinators
EUR 825.1 million (2012)	Poland	Majewski (2014)	Losses resulting from insufficient numbers of pollinators are estimated at EUR 728.5 million
36.7 million pounds	United Kingdom	Garratt et al. (2014)	Two apple varieties: Gala and Cox
$367 million	State of Georgia, United States	Barfield et al. (2015)	13% of the total production value of crops studied and 3% of the total production value

The bumblebees (Photo 1.1-1) and certain species of social wasps are important. However, **bees** play a critical part in the pollination of agricultural entomophilous cultures. There are about 20,000 species of bees in the world (Winfree 2010). The high importance of bees, particularly honey bees (Photo 1.1-2), is determined by the **following factors**:

- High volume of pollinating work. A strong bee family visits over a billion flowers during a season for collecting nectar and pollen.
- A large number during spring. At temperate latitudes, the numbers of wild insects/pollinators is very low at the beginning of flowering (for example, a family of bumblebees retains only a queen at this time). The honey bees hibernate as the extended families and about 10,000 old bees in each bee family take part in collection of nectar and pollen.
- Wide spectrum of the plants used. Many species of insects specialize in pollination of particular species of plants. A honey bee collects nectar and pollen from all accessible plants belonging to different families, genera, and species. In this case, the worker bees quickly switch to visiting plants at the peak of flowering; that is, when the need for pollinators is greatest.
- Body structure, adapted in the best way to perform pollination. For the bee body with its hair covering, thousands of pollen grains are transferred to styles as a result of repeated visits to the flowers.

Photo 1.1-1. Bumblebees work on flowers three times faster than domestic bees and prefer flowers with a deep corolla which are, often inaccessible to bees. Bumblebee pollinating European columbine *Aquilegia vulgaris* is shown.
Photo credit: By Roo72 (Own work) [CC BY-SA 3.0 (http://creativecommons.org/licenses/by-sa/3.0) or GFDL (http://www.gnu.org/copyleft/fdl.html)], via Wikimedia Commons, 26 May 2008

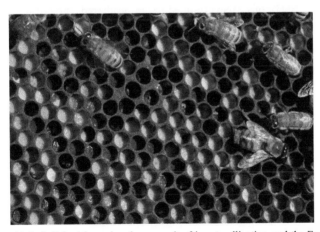

Photo 1.1-2. About one third of all food is produced as a result of insect pollination, and the European honey bee, *Apis mellifera*), is responsible for about 80% of this. The photo shows honey bee comb showing cells filled with different colored pollen collected from different plants.
Photo credit: Nick Pitsas, Commonwealth Scientific and Industrial Research Organisation (CSIRO)

For example, in the southern Non-Chernozem belt of Russia, up to 83 species of insects from five orders participate in the pollination of buckwheat; however, up to 80–90% of the pollinating work falls on honey bees (Vazhov 2013).

Nevertheless, for pollination of particular species of plants, insects specialized to pollinate just them are often preferable. For example, **bumblebees** pollinate mainly legumes. They work on flowers three times faster than domestic bees and prefer flowers with a deep corolla, often inaccessible for bees. The seed harvest of fodder crops such as red clover depends on them completely (Marikovsky 2012). For example, when this plant was introduced to New Zealand, plantings did not provide seeds due to that absence of insects that are able to pollinate them. The issue was dealt with by *introducing* bumblebees (Mamayev and Bordakova 1985).

The individual representatives of **solitary bees** are well adapted to the openings of flowers and pollination of alfalfa. The syrphid flies pollinate carrots very effectively.

1.1.2 Cleaning services

Cleaning services are provided by insects that eat the decomposing remains of animals and plants. One can identify **three categories of such insects**: (1) insects feeding on droppings; (2) insects feeding on spoils; and (3) insects feeding primarily on decaying plant bodies.

1.1.2.1 Removal of dung

The most important cleaning service rendered by insects is utilization of **animal dung**, which is often a source of pathogens. Such insects include dung beetles (Photo 1.1-3); beetles of the Hydrophilidae family; termites; and larvae of dung flies, houseflies, some syrphid flies, soldier flies, and tachinid flies. Animal dung may be harmful for health due to pathogens and recycling of dung mitigates health problems. However there are other advantages of utilisation of dung. For example, in Australia the excrement of cows has led to a constant decrease of pasture productivity and this required the introduction of appropriate insect species.

A problem that occurred during the period of colonization of Australia is well known. Initially, there was no cattle stock, sheep, and wild ruminant animals, and the beetles feeding on their dung were absent. The local dung beetles consumed only the dung of marsupial mammals. When the British brought cows to Australia, this had a **disastrous impact**. The vast herds of domestic animals left more than 200 million dung pies every day (Marikovsky 2012). The dung of one cow occupied an area of 365 m^2 each year (Zakhvatkin 2003); this excess resulted in the steady reduction of the pastures' productivity. Decomposition of manure lasted 3–5 years; the grass under it rotted, and these points were later occupied by weeds.

At the end of the 19th century, the government of Great Britain purchased more than a quarter of a million **dung beetles** (Mirzoyan et al. 1982) from the countries of Europe and Asia. Of 57 species of imported beetles, about 20 species did well. The dung beetles adapted to the new conditions and, within three years, resolved the problem (http://www.psciences.net/main/sciences/biology/articles/shuky.html).

This is hardly the only case of the deliberate introduction of dung beetles. In 2011, 10 species were exported from Australia to New Zealand. The **economic effects** from this measure reach $150 million per year in the reduction of atmospheric greenhouse gas (GHG) emissions (http://biotechlearn. org.nz/news_and_events/news/2011_archive/dung_beetles_imported_to_aid_soil_fertility). The Afro-Asian dung beetle, *Digitonthophagus gazella,* was introduced to the south of the United States (to South Carolina) from California (Thomas 2001). The farmers on Hawaii have also imported dung beetles from East Africa to accelerate the decomposition of manure (http://www.cals.ncsu.edu/course/ent425/text01/impact1.html).

Photo 1.1-3. The most important cleaning service rendered by insects is utilization of animal dung. The photo shows a dung beetle rolling a ball of dung.
Photo credit: CSIRO

Each geographic zone has its own characteristic set of dung-removing insect. Within the forest zone, the **dung beetles** of the genera *Geotrupes* and *Aphodius* and larvae of many dung flies predominate, while in the steppe, beetles of the genera *Copris, Scarabaeus, Gymnopleurus*, and others predominate. In the tropical savanna, **termites** are of great importance. At least 126 species of termites feeding on the dung of 18 species of mammals are known (Freymann et al. 2008). The termites process the manure fully within three months while, in their absence, 20–30 years are required for manure to disappear (Whitford 1982).

Most **coprophagous beetles** feed on fresh manure of mammalian herbivores (Holter and Scholtz 2007). Sometimes, greater specialization is observed: for example, the beetles *Aphodius fossor* and *Onthophagus taurus* feed only on cow manure. At the same time, the flies *Cypsela equina* and *Polietes albolineata* are typical of the dung of horses and do not visit the dung of cattle. The large dung beetle *Heliocopris colossus* makes use solely of elephant manure (Marikovsky 2012), and other beetles feed only on the manure of yaks. Investigations related to the preference of certain beetles for the dung of certain animals were carried out in Brazil (Bogoni and Hernandez 2014). The authors collected 426 individuals of 17 species of dung beetles preferring the manure of carnivores (cougar, *Puma concolor*); omnivores (crab-eating fox, *Cerdocyon thous*); black capuchin (*Sapajus nigritus*); and herbivores (South American tapir, *Tapirus terrestris*). The researchers found that the dung of the crab-eating fox (*Cerdocyon thous)* was in great demand with all 17 species (Bogoni and Hernandez 2014).

The rate of manure processing depends on **many factors**: species of coprophagous beetles, temperature, humidity, and others. For example, in the tropical grassland in Veracruz, Mexico, loss of 80% of manure bulk takes place after three days during the rainy season and after 17 days in the dry season. The major dung beetles there are *Euoniticellus intermedius, Digitonthophagus gazella,* and *Copris lugubris* (Cruz et al. 2012).

The colonization of dung by different species of insects occurs in **specific sequences**. In the case of cow manure, stable flies (*Lyperosia, Haematobia*) appear at first and lay their eggs alongside fresh excrement. At times, the flies emerge in as little as a few minutes. After this, the first beetles, Hydrophilidae (*Sphaeridium, Cercyon*), appear, which may float in the still semiliquid substrate.

When the cow manure cakes dry slightly, **dung beetles** settle in them. The system of passages they create in the excrement improves air exchange. Later, **flies** (*Scatophaga, Mesembrina, Morellia, Hylemya*) appear that need more oxygen for the development of eggs. Apart from them, in excrement coated with a crust, the **predatory beetles** of the Histeridae and Staphylinidae families are found. When most of the flies have finished laying eggs, the **syrphid flies** *Rhingia* and Stratiomyidae (*Geosargus, Microchrysa*) emerge (Galperin 2007; Tixier et al. 2015).

It is interesting to note that problems with processing dung exist now and again. In the steppes and deserts, the day cake is roasted from below when it comes in contact with heated ground and dries up quickly from above. As a result, the dung becomes impassable for insects, so they process only dung deposited during the night and early morning. In this case, the day cakes are utilized by people who use them as a fuel.

1.1.2.2 Decomposition of carcasses

The decomposition of carcasses is another service that insects provide. Why are the carcasses of small animals (mice, field mice, moles, birds, etc.), which are killed every day in great numbers, so rarely found in fields and forests? Where do they go? Who takes them away from pathways, grass clearings, and the fringes of the forest? The insects that destroy a great deal of carrion include, the carrion beetles (Silphidae), skin beetles, larvae of flies (Diptera), including flesh flies (Sarcophagidae), and blowflies (Calliphoridae), as well as some small ticks.

The **necrobionts**, the destroyers, are a major component of any ecosystem; they execute the complete decomposition of dead organic matter. If they were absent, then all the nutrients would be bound in dead bodies and new life might not emerge (Odum 1986). In the wild, animals die continually, and their carcasses become a resource for necrophagous organisms. In terrestrial ecosystems, most dead tissues are utilized by insects; namely, flies (Diptera) and beetles (Coleoptera) (Liabzina 2011a; Dekeirsschieter et al. 2011; Pushkin 2004, 2012; Liabzina and Uzenbaev 2013).

Kocarek (2003) carried out detailed field investigations of the decomposition of rat carcasses in Opava, Czech Republic. He studied different stages of the decomposition process (fresh, bloated, decayed, and dry stages) in different habitats (meadows, forests) and during different seasons (spring, summer, and fall). Kocarek recorded a total of 145 Coleoptera species belonging to 22 families, on the carcasses. The numbers of beetle species on the carcasses initially increased, reaching a maximum in the decay stage and afterwards decreasing gradually. The maximum numbers of beetles were observed in the **spring**, while the populations were reduced consistently in the summer and fall. Within **forests**, the numbers of necrophagous beetles were higher as compared to the **meadow habitats** (Kocarek 2003).

Animal carcasses also occur in water bodies, appearing there both as a result of natural death and under the action of external factors. In the seas and oceans, insects are practically **absent** and are not involved in the disposal of carcasses; this function is generally performed by echinoderms (brittle stars), crustaceans: (shrimps, crabs, amphipods), and mollusks (gastropods). In **freshwater bodies**, insects constitute up to 70–90% of benthic communities (Fenoglio et al. 2014); however, there are no specialized necrophagous organisms among them. The decomposition of animal carcasses is accomplished by omnivorous invertebrates, while insects play a key role.

For example, in the lakes of Karelia (Russia), 47 species of invertebrates have been found on animal carcasses; insects were prevalent among them. Leading roles in the decomposition of carcasses were played by larvae of the **chironomids** *Polypedilum convictum* and *Pentapedilum exectum*, and larvae of the **deerfly** *Chrysops relictus*, as well as adult dytiscid beetles (genera *Acilius* and *Dytiscus*) (Liabzina 2013).

As with manure decomposition, colonization of carcasses by different species of invertebrates is carried out in specific sequences. For example, during the decomposition of kangaroo carcasses in grassy eucalypt woodlands, **beetles** are most abundant in species composition and numbers during the first six weeks. These insects are gradually replaced by **mites** during the next six weeks (Barton et al. 2014). In the central part of the Ural Mountains (Russia), flies dominate in the early stages of the cadaveric microsuccession, while beetles are present at the advanced stages (Ermakov 2013).

It is essential to give special consideration to **necrophagous ants**; there is extensive literature on them (Campobasso et al. 2009; Chin Heo et al. 2009; Devinder and Bharti 2001). They are always present on carcasses: on meadows and agrocoenoses of Eurasia, they generally include *Lasius niger* and *Myrmica rubra* (Liabzina 2011b), while in coniferous forests they include *Formica polyctena* and *Formica aquilonia* (Berezhnova and Thsurikov 2013). In tropical areas, **red imported fire ants** (*Solenopsis invicta*) have been observed feeding on the carcasses of vertebrate animals (Clark and Blom 1991).

Where carcasses are present at high concentrations, they are disposed of at a great pace. For example, a mouse carcass present at the dome of an ant hill is fully processed by the ants within 24 hours, while a carcass on an ant path (at an air temperature of 18°C) is disposed of within five days (Liabzina 2011b). However, in most cases, the appearance of ants results in a two- or threefold delay in the decomposition process. This is because they exterminate many other necrophagous insects, sharply depleting their diversity (Chin Heo et al. 2009).

The role of necrophages is so important that it is the basis of an aphorism by Carl von Linné: three flies can dispose of the corpse of a horse in less time than it would take a lion.

1.1.2.3 Breakdown and recycling of organic matter

The breakdown and recycling of organic matter is accomplished by both terrestrial and aquatic insects. Terrestrial organic matter comprises mainly **plant litter** (leaves, bark, needles, and twigs that have fallen to the ground).

The insects and related arthropods (for example, ticks, millipedes, and crustaceans duch as woodlice) mechanically crush the plant tissues, **increasing** the total surface area of the destroyed parts of the plants. This process then contributes to the faster penetration of different microorganisms into the dead plant material. For example, as a result of fir-needle processing by oribatid mites (moss mites), the surface area of the needles increases 10,000 times (Berg and McClaugherty 2008).

In different natural zones, the **composition of insects** that disintegrate plant litter is varied. For example, in the **tropics**, hydrophilic termites process all dying wood materials and plant remains. In the **tropical savanna** of west Africa, they consume 60% of the annual woodfall and 3% of the annual leaffall, representing 24% of total litter production (Schowalter 2013). In the **steppes** and **deserts**, darkling beetles play a large part in the disintegration of plant litter.

In the **forests** of the temperate belt, the insect destructors of plant litter are most numerous. They include ship-timber beetles (Lymexylidae), larvae of long-horned beetles (Cerambycidae), jewel beetles (Buprestidae), horntails (Siricidae), crane flies (Tipulidae), and carpenter ants (*Camponotus*). The contribution of springtails (Collembola) is also essential. Flowing through the intestinal tract of the springtails, a mass of plant and animal tissue is crushed and, thereby, the surface area of the plant matter accessible to microorganisms is repeatedly increased along with the surface exposure to air and water. In some places, they produce 175 cm^3 of humus per square meter (Akimushkin 1975).

The **decomposition rates** for different components of litter are different. For example, in the taiga of northern European Russia, the rates of decomposition are as follows, in descending order: birch leaves, cowberry leaves, pine needles, spruce needles, moss, branches, and bark (Likhanova 2014). In the southern Far East of Russia, the needles of the Korean pine (*Pinus koraiensis*) decompose more slowly; then the foliage of Mongolian oak (*Quercus mongolica*) follows, and the fastest decay is a characteristic of the foliage of the lime tree (*Tilia mandshurica*) (Ganin 2011).

In addition, many representatives of the soil fauna produce enzymes that decompose plant remains and make the nutritional substances **available again** for plant growth. For example, the larvae of dipterous insects (different flies, gnats, cecidomyids, etc.) decompose plant litter, improving the humic status of soil, increasing humic acids, increasing the content of nitrogen from ammonium compounds, and yielding a high humus content (Ross et al. 1985). For example, the larvae of fungus gnats can transform leaves into black mold humus within three days (Zhdanova 2012). The ability to transform minerals chemically is also characteristic of termites (Sako et al. 2009).

The **chemical transformation** of plant litter also takes place in the intestinal tracts of springtails, plant remains are partially mineralized during digestion (Akimushkin 1975). The contribution of microarthropods (among which the insects are the inescapable link) to the decomposition of plant litter can reach 80% (Schowalter 2013). Where litter destructors are present in small numbers, such as in the sparse larch forests, fibrous peatlike unproductive layers accumulate.

There are also **freshwater insects** that feed on decaying organic substances that are suspended in or are gravitating to the bottom of water bodies. This organic matter consists of plant remains (both leaf litter caught in the water bodies from the terrestrial ecosystems and the remains of aquatic plants), animal excrement, and animal carcasses.

The aquatic insects feeding on organic matter lead **different lives**. Several of them crawl freely at the bottom, for example, larvae of stone flies (Plecoptera) and caddis flies (Trichoptera), water slaters (Isopoda), and some aquatic mites (Hydrachnidiae). Others dig, in part or in whole, into the bottom, such as larvae of some dragonflies, mayflies (Ephemeroptera), and nonbiting midges (Chironomidae). The third kind are attached, temporarily or permanently, or even accrete, to the substrate (to the aquatic plants and other bottom objects); these include the larvae of nonbiting midges and caddis flies dwelling in the fixed little houses and pupae of many insects.

The rate of organic matter breakdown depends on many **factors**: composition of insect destructors, their abundance, water temperature, oxygen content, water mixing, and others (Halvorson et al. 2015). As with the processing of plant litter on land, dispersion and eating are of prime importance. The insects divide the plant litter into minor particles, repeatedly increasing the surface area, which contributes to accelerated microbial treatment (Patrick 2013). The consumption of organic matter by shredding insects can reach 50% (Hieber and Gessner 2002).

The intensity of organic matter processing by aquatic insects is highest in the **middle latitudes** due to an abundance of the shredding insects and seasonal restriction of the arrival of the largest amount of organic matter (Raposeiro et al. 2014). The specific contribution of their processing is largest in the high latitudes (Macadam and Stockan 2015).

The **chironomids** are the most significant organisms in transforming organic matter into mineral materials in freshwater ecosystems. Under favorable living conditions, their larvae dominate in biomass over that of other benthic invertebrates in rivers and lakes. At the present time, no fewer than 5,000 species of chironomids are known. Numerous papers confirm the important role of these insects in the processing of organic matter (Creed et al. 2009; Poulsen 2014; Holker et al. 2015).

1.1.3 Soil formation and soil conditioning

Soil formation is a natural process forming soil from rocks, while **soil reclamation (conditioning)** increases the fertility of a soil.

The major **factors** in soil formation include climate, underlying strata, organisms, relief, and time. The springtails (Collembola) play an essential part in soil formation. Traditionally, they were considered to be apterous insects; however, in the modern classification, they are considered to be entognathes, a separate primitive group of arthropods related to insects.

Initially, open rocky surfaces become inhabited by crustaceous lichens, which grow extremely slowly. As the rocks are overgrown, **oribatid mites** and **springtails** begin feeding on the lichens. They live and reproduce under lichens for generations. A layer of their excrement is gradually accumulated along with mineral particles of the rock, forming a soil. Thereafter, other arthropods (including insects) arrive and improve the soil.

A primary role of insects in soil amendment involves the following **factors**: (1) soil loosening; (2) decomposition of plant and animal remains (plant litter, carcasses, and animal excrement); (3) removal of stones from the surface to the deeper layers; and (4) soil fertilization.

1. **Soil loosening.** Many insects inhabit the ground in the same manner as other arthropods (crustaceans, arachnids, and millipedes). Some of them remain for only a few days while they pass through stages of eggs, larvae, or pupae, whereas others dwell there for a prolonged period. For example, the Japanese beetle (*Popillia japonica*) resides in the ground in the egg, larva, and pupa stages for almost **eleven months** a year. The adult insects emerge from the ground in order to feed and copulate; then they digs new underground dwellings and lay their eggs (https://en.wikipedia.org/wiki/Japanese_beetle).

   The larvae of beetles of the subfamily Melolonthinae remain in the ground from **two to three years**, while the larvae of wireworms (Elateridae) remain there from two to six years. The larval stage of the 17-year locust lasts from **13 to 16 years**; the insect is located in the ground all this time. Mole crickets (Gryllotalpidae) remain in the soil the best part of their lives, mating there and laying eggs, and some ground beetles (Carabidae) never emerge on the surface.

   The loosening process is accomplished in **two ways**: (1) use of soil as a refuge; and (2) mechanical and masticatory work of insects. Many insects use the soil as a temporary refuge (hibernation, protection against high temperatures, pupation). The depths to which different species of insects penetrate the soil depend on many factors (species of insect, soil type, season, etc.). For example, some ground beetles penetrate the soil to a depth of only 2.5 cm below the surface. The holes of cicada killers are situated at depths of 30–35 cm. The wintering holes of earth-boring dung beetles in sandy soil reach horizons of 100–110 cm and more (http://ours-nature.ru/b/book/15/page/5-navozniki-i-bronzovki/102-nora-i-navoznaya-kolbasa). Tiger beetles dig holes more than 180 cm deep, and the 17-year periodical cicadas penetrate the soil to depths of up to 5.5 m (http://www.tinlib.ru/biologija/priklyuchenija_s_nasekomymi/p37.php). Termite nests in sandy areas have been found at depths of up to 12 m (www.floranimal.ru/orders/2709.html).

   Vast numbers of various insects carry out activities in soil, tunneling through it, mixing it, or throwing it onto the surface. Colossal work carried out by insects increases the number of cavities in soil and, therefore, lowers its density and facilitates the penetration of air and water to the soil.

2. **Decomposition of plant and animal remains.** Key indicators of the fertility of soils include the presence of different chemical elements: phosphorus, potassium, nitrogen, sulfur, and others. These

elements are in the soil in the solution state and are washed out of it by atmospheric precipitation. These elements are replenished when different parts of plants fall to the ground. These plant remains form a layer of litter on the soil surface. Animal dung and carcasses also contribute to this process. Processing by insects of different types of organic matter into humus is considered in Sections 1.1.2.1.–1.1.2.3.

3. **Removal of stones from the surface to the deeper layers.** This factor of soil formation and conditioning is characteristic, to a greater degree, of the stony deserts. Generally, ants settle under stones, which serve as the roofs of heating chambers. The ants grub under stones, and they take out the ground from under them in building their anthills; this process contributes to the embedding of stones into the ground (Berman and Zhigulskaya 1996; Marikovsky 2012). In the absence of ants, stones would cover the ground surface with a continuous mantle. Note that the stones are obstacles for humans but not for soil.

4. **Soil fertilization.** Insects help fertilize soils in various ways. The **most common case** is the storage of food reserves. For example, ants store nectar, pollen, seeds, and parts of plants in their nests. Some wasps burrow holes that are filled with larvae and spiders, while the cicada killer fills burrows with cicadas (http://www.tinlib.ru/biologija/priklyuchenija_s_nasekomymi/p37.php). In the areas where cattle are raised, dung beetles fertilize the soil by dragging manure to it. For example, some earth-boring dung beetles dig about 250 g of manure into the ground during their short life spans (Marikovsky 2012). Manure itself in its pure form is useless for soil; it improves soil fertility only after processing by the larvae of insects. In addition, fertilization occurs when insects eat litter and dung.

Another way insects fertilize soil is through surface fertilization by their bodies, that is to say, when insects die and fall to the ground, their bodies are utilized by other arthropods, fungi, or microorganisms. Additional number of microelements due to dead bodies increases soil fertility. The total biomass of the terrestrial insects is vast, and they form a vitally important factor, ecologically. For example, about 500 kg of gnats (tiny flying insects belonging to the suborder Nematocera of the order Diptera) fall on 1 km² of forests and meadows adjacent to water bodies. In terms of the equivalent amount of microelements, this equates to approximately 16 kg of nitrogen, 9 kg of phosphorus, and 6 kg of calcium (Volovnik 1990). The average weight of larvae per one hectare of broad-leaved forests is 200–300 kg, which produce more than 200 kg of excrement over a season. They are also excellent fertilizers (Zhdanova 2012).

Therefore, insects have essential roles in soil formation and conditioning.

1.1.4 Insects as a food resource for animals

The insects are the key element of the food pyramid. The specific weight of insects in the diet of different animals varies considerably. Some animals feed *solely* on insects; for other animals, they are one of the major sources of food, while the remainder eat insects *from time to time*. How often they are used for food also depends on many other **factors**: nature of habitat (relief, vegetation mantle, and other characteristics), season, and **climatic characteristics** (temperature, humidity, and other parameters).

1.1.4.1 Insects as a food resource for invertebrates

Invertebrates that eat insects include scorpions, camel spiders, harvestmen, spiders, centipedes, crustaceans, flatworms, stone centipedes, and a formidable army of entomophagous insects (carnivores and parasites). The feature uniting them is carnivory; they differ considerably from each other in other ways.

The spectrum of insects eaten by invertebrates varies enormously. **Extremely focused specialization** is inherent in some primitive ants. For example, in the humid savanna of Nigeria, the ants *Megaponera foetens* consume only termites of the Macrotermitinae subfamily; more often, those are *Macrotermes bellicosus* (72%) and *Odontotermes* (22%) (http://antclub.ru/lib/brian-m-v/obshchestvennye-nasekomye/

pishcha/osy-i-muravi-kak-khishchniki). *Centromyrmex gigas*, which inhabitsArgentina and Brazil, eats only termites (http://www.antwiki.org/wiki/Centromyrmex_gigas). The ants of the *Hypoponera* genus (subfamily Ponerinae) feed on springtails (Collembola) (http://www.antwiki.org/wiki/Hypoponera).

Narrow nutritional specialization also occurs in other taxa. Ruby-tailed wasps prefer exclusively bark beetles. Many spiders feed primarily on ants, in particular, some species of the Salticidae (jumping spiders) and Oecobiidae families (Cushing 2012). Some tarantulas eat only ants. The larvae of ladybirds *Diomus thoracicus* in French Guiana feed only on *Wasmannia auropunctata* ants (Vantaux et al. 2010).

On the other hand, there are **omnivorous invertebrates** that eat a wide variety of animals they can handle. For example, mantids would eat any insect or spider (http://spidersworld.ru/informacija-o-paukakh/vragi-paukov/). The scorpions feed on ants, beetles, bugs, grasshoppers, locusts, crickets, flies, millipedes, and spiders. The camel spiders (Solifugae) primarily consume insects (termites, bees, darkling beetles) as well as millipedes, spiders, scorpions, woodlice, and others (http://zoogalaktika.ru/photos/invertebrata/arthropoda/arachnida/solifugae).

Adult dragonflies catch various creatures that they can handle: flies, mayflies, stone flies, caddis flies, lacewings, beetles, butterflies, and others. However, flies, which generally represent the basic mass of airborne insects (http://www.odonata.su/content-view-152.html), are the basis of their nutrition. Ground beetles that eat the larvae and pupae of different butterflies serve as an example of moderate specialization.

These insects also **differ in the environments** where their hunts for insects take place. For example, dragonflies procure food predominantly in the air. Fairy flower wasps hunt for the larvae of chafers, fiddler beetles, and rhinoceros beetles below the ground. Some ground beetles run fast on the ground feeding on aphids, insect larvae, and slugs. Centipedes also procure their food (dormant flies, wasps, and other insects) on top of the ground. Damselflies feed primarily on small insects collected from vegetation. The forest caterpillar hunter (*Calosoma sycophanta*) and lesser searcher beetle (*Calosoma inquisitor*), as well as bronze carabid (*Carabus nemoralis*), forage on trees for the caterpillars of different butterflies. Water beetles (Dyticsidae) inhabit stagnant water bodies and procure their food (wigglers and larvae of mayflies, dragonflies, and caddis flies) in water (http://dom-sad-og.ru/nasekomye-protiv-nasekomyx-vreditelej/).

Sometimes, hunting takes place at the **interface of air and water**. Many aquatic insects are an essential food for terrestrial species. For example, in northern Sweden, the contributions of aquatic insects to the diet of littoral wolf spiders (Lycosidae), riparian spiders (Linyphiidae), and ground beetles (Carabidae) were estimated to be 44%, 60%, and 43%, respectively (Stenroth et al. 2015).

The **times of day** when the hunting for insects occurs also differ. First of all, some predators hunt in *daylight hours*; however, many species feed predominantly at night. For example, some ground beetles hunt for their prey (aphids, slugs, beetles, earthworms, and snails) at *night*.

Cannibalism is characteristic of many invertebrates that prey on insects. Cannibalism is an everyday occurrence among dragonflies, mantids, spiders, mosquitos, and other invertebrates. For example, dragonflies eat their smaller relatives—damselflies or even weaker dragonflies—while raider ants (*Cerapachys biroi*) and New World army ants (*Nomamyrmex esenbeckii*) feed on other ants (https://en.wikipedia.org/wiki/Myrmecophagy).

Invertebrates that prey on insects vary considerably in their **mobility**. Dragonflies and the larvae of antlions show extremes in mobility. Dragonflies are very quick predatory insects that catch their prey in the air. Meanwhile, antlion larvae usually sit at the bottom of a sand crater, digging slightly into it, and await their prey. Most often, ants are caught in these traps (http://nacekomoe.ru/publ/hiwnye-nasekomye/).

The differences in the **metabolic rates** of invertebrates are enormous. Some species are characterized by extremely *high metabolism*. For example, the assassin fly (*Satanas gigas*) is so ravenous that it is able to eat all day long. This characteristic is also inherent in the seven-spot ladybird (*Coccinella septempunctata*). The larva devours 100 adult aphids or 300 of their larvae every day. Meanwhile, scorpions have a *stunted metabolism* that, in addition, they can regulate. Under unfavorable conditions, they can do without food throughout the year (http://os-nauka.ru/scorpions/).

It is evident that the insects are a key food resource for invertebrates including other insects.

1.1.4.2 Insects as a food resource for amphibians

The amphibians are largely represented by frogs, salamanders, and newts. There are a total of about 7,000 amphibian species, 90% of which are frogs and toads (https://en.wikipedia.org/wiki/Amphibian). Insects are of paramount importance in the nutrition of all amphibians (both adult individuals and their larvae). Subsisting on insects is especially characteristic of **frogs**, because insects account for up to 95% of the latter's diet.

In the **temperate latitudes**, the feeding of frogs is nonselective; that is, they consume everything they can catch in the course of hunting (Ruchin 2014). This depends on both the abundance of different insects where the particular individuals live and features of the biology of the frogs. Generally, frogs and toads specialize in catching moving prey that travel across their field of vision (Browne 2009). Toads feed mainly on non-flying insects (beetles, ants, larvae, etc.). Toads do not jump well, so flying insects are practically absent in their diet (although they can also catch a fly or mosquito that flies nearby from time to time). Frogs, which keep their heads up and jump high, are able to catch food higher up and in the air (http://zooschool.ru/amfib/33_2.shtml).

Many papers have been devoted to the **nutrition of frogs**. In particular, a number of articles have been concerned with the nutrition of different species in different regions of Russia (Ruchin and Alekseev 2012; Ruchin 2014; Timkina and Odintsev 2014). In the Tyumen region (Russia), for example, insects form the basis of the diet of the swamp frog (*Rana arvalis*) (92.5%), while for the common toad (*Bufo bufo*), this index is 78% (Timkina and Odintsev 2014).

The spectrum of food of the grass frog (*Rana temporaria*) in Mordovia was studied by A.B. Ruchin (2014). He found that the major portion of prey organisms for this frog included beetles (first of all, ground beetles [43.95%], mollusks [11.63%], and caterpillars; i.e., larvae of owl moths, cankerworms, etc. [10.35%]). Considerable numbers of Diptera (flies, crane flies, mosquitoes) and small ticks were also recorded.

In the **tropics** where the humidity and temperature conditions are maximally favorable for amphibians, there are forms that specialize in different types of nutrition. For example, the Mexican toads (*Rhinophrynus*), Malagasy toads (*Rhombophryne*), Siamese toads (*Glyphoglossus*), and African toads (*Hyperolius* and *Rhynobatrachus*), like a number of other tropical amphibians, tend to feed on ants and termites (http://myreptile.ru/forum/index.php?topic=5237.0).

Salamanders also feed on insects, but their value as a food source compared to frogs is much lower. The larger the individual salamanders, the less important is the role of insects. So larger species such as the Japanese giant salamander (*Andrias japonicus*), is able to eat crabs, fish, small mammals and amphibians, aquatic insects. Aquatic insects in the food structure of large species play a very minor role. Meanwhile, the specific weight of insects and other invertebrates in the diet of the smaller dusky salamanders (*Desmognathus*) in the Appalachian Mountains is much higher. Their diet includes flies, beetles and beetle larvae, leafhoppers, springtails, moths, spiders, grasshoppers, and mites (https://en.wikipedia.org/wiki/Salamander).

The role of insects in the diet of **newts** is fairly large. In the aquatic stage of life, they eat the larvae of insects, most often nonbiting midges. For example, caddis worms play a large role in the diet of the Central Asian salamander (*Ranodon sibiricus*) (74% of total weight of gastric content). The larvae of Diptera comprise 55.7% of the total weight of the gastric content in the common newt (*Triturus vulgaris*) during the reproduction period (i.e., when it does not emerge on land) (http://zooschool.ru/amfib/33.shtml). On land, beetles, larvae of butterflies, millipedes, oribatid mites, spiders, and earthworms become the major components of the diet of this species.

1.1.4.3 Insects as a food resource for reptiles

The reptiles that ingest insects include turtles, crocodilians, lizards, tuataras, worm lizards (Amphisbaenidae), and snakes. Most lizards, as well as some turtles and snakes, feed on insects. Mainly the terrestrial insects are used as food; however, nine species of reptiles eat aquatic insects (Macadam and Stockan 2015).

The insect-eating **lizards** include, for example, the following species (http://amccorona.com/wp-content/uploads/2014/11/Feeding-Insect-eating-reptiles.pdf): (1) agamas (Agamidae); (2) American chameleon (*Anolis* spp.); (3) basilisk lizards (*Basiliscus* spp.); (4) Old World chameleons (Chamaeleonidae); (5) chuckwallas (*Sauromalus* spp.); (6) fence lizards (*Sceloporus undulatus*); (7) geckos; (8) monitor lizards; (9) most skinks; (10) spiny-tailed lizards (*Uromastyx* spp.); (11) tegus (*Tupinambis* spp.); and (12) water dragons (*Physignathus cocincinus*).

For the most part, lizards feed on a wide spectrum of insects (crickets, locusts, grasshoppers, cockroaches, flies, ants, wasps, bees, beetles, butterflies and their larvae, etc.) as well as spiders; however, some species are stenophagous creatures, meaning they specialize in eating specific types of foods. For example, the thorny dragon (*Moloch horridus*) feeds only on ants, especially *Ochetellus flavipes* and other species of *Iridomyrmex* or *Ochetellus* (https://en.wikipedia.org/wiki/Thorny_dragon#Diet). One individual can often eat thousands of ants in a day (Browne-Cooper 2007). The horned lizards also feed nearly exclusively on ants (https://en.wikipedia.org/wiki/Myrmecophagy). The share of ants in the diet of agamas and skinks is also high.

The **tuatara**, a rare species of reptiles that inhabits a few rocky islands off the coast of New Zealand, eats medium-sized insects such as beetles and crickets, as well as spiders (http://animals.about.com/od/reptiles/a/what-reptiles-eat.htm). Insects also constitute a large share of the diet in worm lizards (Amphisbaenidae). Since these reptiles are geobionts, they consume predominantly soil invertebrates, including insects. However, some of them prefer terrestrial insects. For example, the Iberian worm lizard (*Blanus cinereus*) visits anthills after rain and feeds on the inhabitants (https://en.wikipedia.org/wiki/Iberian_worm_lizard).

Some **turtles** and **tortoises** also feed on insects. Their diet includes dragonfly larvae, mayflies, caddis flies, beetle larvae, crickets, grasshoppers, mealworms (beetle larvae, not a species of worm), roaches, wax worms (a kind of moth caterpillar), and others; however, in most cases they are not a major food source.

What turtles in the **wild eat** depends greatly on the type of turtle and its natural habitat. For example, sea turtles do not eat insects because the latter are practically absent at sea. As for the freshwater turtles, they feed on insects that are most available to them. For instance, the red-eared slider eats mosquito larvae (http://www.orkin.com/other/mosquitoes/mosquito-predators/), while painted turtles feed on crickets and other insects. Terrestrial tortoises consume, in particular, grubs, beetles, and caterpillars (http://www.livescience.com/45539-what-do-turtles-eat.html).

The share of insects in the diet of **snakes** is imperceptible. Insects become the prey of only the smallest snakes. For example, dwar racers (*Eirenis*) eat small insects with soft coverings, stone centipedes, spiders, small scorpions, and camel spiders. And blind snakes (*Scolecophidia*) feed on termites and ants. The Peringuey's adder, which inhabits the Kalahari Desert, also consumes insects (ants, ground beetles, and termites).

1.1.4.4 Insects as a food resource for birds

Based on their diet, birds are conditionally subdivided into three **groups**: phytophagous, carnivorous (including also the insect-eating birds), and those with mixed feeding (birds that consume both plants and animals) (Photo 1.1-4). Birds that specialize in feeding on insects represent one-third of the world avifauna (http://onbird.ru/terminy-i-opredelenija-ornitologii/e/entomofagi).

The relation between insects and other foods of birds varies within **wide limits**. A few birds feed nearly **exclusively** on insects. This class can include swifts, swallows, nighthawks, nutcrackers, chiffchaffs, redstarts, kinglets, tree creepers, wagtails, and others. There are birds for which insects are the **predominant food,** but there are also birds that eat them quite rarely. Few birds **never eat any insects at all**. During the summer season, even plant-eating birds not only eat insects but also feed them to their nestlings. For example, insects (sharpshooters, flies, larvae of butterflies) and spiders are, for the first few days, the predominant food of flappers (partridge chicks) (http://www.egir.ru/bird/46.html).

Photo 1.1-4. Southern Red-billed Hornbill *Tockus rufirostris* consume both plants and insects. A female Southern Red-billed Hornbill eating a dragonfly on the verge of a road in the Kruger National Park is shown.
Photo credit: By Taejo - Own work, CC BY-SA 3.0, https://commons.wikimedia.org/w/index.php?curid=303791, July 2004

Different birds feed on a wide variety of insects. Some species **eat all insects they can catch**; for example, redstarts, starlings, and crows. The wryneck, on the other hand, has a restricted diet; it eats and feeds its nestlings using the ants of predominantly one species—the common black ant (*Lasius niger*) (http://www.duhzemli.ru/animal/bird-protection/insect.html).

If you follow the direction of lowering of the universality of feed, **the intermediate position** is occupied by the following birds. Antbirds feed on grasshoppers, crickets, cockroaches, praying mantises, stick insects, larvae of butterflies and moths, spiders, scorpions, and centipedes (https://en.wikipedia.org/wiki/Antbird). The golden-crested kinglet eats representatives of all basic groups of insects encountered within its habitat; however, it prefers beetles, mosquitos, ants, and sawfly larvae but avoids ants (http://www.ecosystema.ru/08nature/birds/morf/morf2.htm).

All warblers feed on beetles (weevils, leaf beetles), flies, true bugs, and caterpillars (http://www.sad-sevzap.ru/vse-o-zashchite-rastenii/ptitsy-pomogayut-sadu). Chickadees eat predominantly sawflies, caterpillars, and aphids (http://www.ecosystema.ru/08nature/birds/morf/morf2.htm). Woodpeckers feed on bark beetles, while the spotted flycatcher (*Muscicapa striata*) predominantly catches flies (http://www.duhzemli.ru/animal/bird-protection/insect.html).

There are also considerable differences in the **ways** birds catch insects (http://www.ecosystema.ru/08nature/birds/168s.php): (1) picking up insects from the ground with the beak (characteristic of doves, starlings, larks, pipits); (2) penetrating soft soil with the beak (waders); (3) digging into the soil (gallinaceous birds); (4) catching insects in the air through watching and jumping (flycatchers); (5) catching insects in the air after a long pursuit (swallows, swifts, falcons); (6) pecking trees in order to extract insects from the wood (woodpeckers); (7) drawing by the beak insects hiding in cracks in bark (tree creepers, nuthatches, titmice); and (8) pecking from the water surface or water column (tattlers, storks, herons, lapwings).

Naturally, the methods birds use to catch insects depend on the environment where the hunting takes place. Many birds catch insects only while they are **in flight**; these account for 313 of the world's species (http://baob.wikidot.com/aerial-insectivores). Some species specialize in procuring insects **in trees** (titmice, tree creepers, nuthatches, orioles, cuckoos). Some birds catch insects at the water **surface and below** (plovers, storks, herons, lapwings). Thirty species of birds feed on aquatic insects (Macadam and Stockan 2015).

Birds pursue insects day and night. The **daytime hunters** include swifts (100 species), tree swifts (4 species), swallows and martins (86 species), wood swallows (11 species), and others. At **night**, insects are eaten by pratincoles (8 species), nightjars, nighthawks, potoos, owlet-nightjars, and other species.

Therefore, insects are a vitally important source of food for birds. Many birds depend on insects during the former's whole life cycle, whereas for other birds, insects are important in the diet during specific periods of their life (rearing of nestlings, molt). This is because insects are the key source of protein for growing nestlings; therefore, even the phytophagous birds (partridges, blackcocks, great grouses, hazel hens, pheasants) feed their rising generation exclusively with insects.

1.1.4.5 Insects as a food resource for fish

Fish feed on both terrestrial insects that have fallen into the water and those which live in water throughout life or during some part of it. Some fish, such as striped Archer fish banded archerfish (*Toxotes jaculatrix*) shot down by jets of water insects and spiders in the water. Insects are unimportant in the diet of **marine fish** because marine insects are few and insects are rarely carried out to sea from the shore. In contrast, insects play a large role in the feeding of the **freshwater fish** (Giller and Greenberg 2015; Suter and Cormier 2015; Macadam and Stockan 2015). About 40% of freshwater fish, e.g., CYPRINID FISHES **eat only insects** (Marikovsky 2012).

The **proportion** of the terrestrial and aquatic insect species in the diet of ichthyofauna depends on many **factors**, such as the size of a water body, fish species and their size, geographical zone, water flow regime, and time of the year. One hundred and fifty-seven species of fish feed predominantly on aquatic insects (Macadam and Stockan 2015). However, when some insects fly en masse (for example, Bullhead (Cottus gobio), common minnow (Phoxinus phoxinus) stone flies or mayflies, etc.), even those species of fish that do not normally eat these insects will feed on them. Such seasonality is characteristic primarily of the middle and high latitudes; in the tropics, seasonality in the feeding of fish on flying insects is less expressed.

A number of papers have been devoted to the **role of terrestrial insects** in the nutrition of fish. For example, according to data obtained by Baxter et al. (2005), insects that fall into streams comprise 30–80% of the diets of young salmon. In some North American lakes, the share of terrestrial insects in the diet of trout (*Oncorhynchus* spp.) averages 50% (Francis and Schindler 2009) while it is also 50% for *Priapichthys annectens,* a fish species in Costa Rica (Small et al. 2013). Terrestrial insects accounted for half of the diet of this dominant species of fish. According to these authors, the average input rates of terrestrial insects ranged from 5 to 41 mg dry mass/m²/day.

Detailed studies on the importance of terrestrial and aquatic invertebrates to **salmonid diets** were carried out on 25 streams in Finland (Syrjanen et al. 2011). In autumn, black flies and caddis fly larvae turned out to be the most important food items in all 25 streams. The highest values of the terrestrial prey were recorded for streams flowing through deciduous forests. The terrestrial invertebrates in the diet of fish reached, on average, 17%; however, this share increased with the size of the individual.

A similar relationship between the sizes of fish and shares of terrestrial and aquatic insects in their diet was also recorded for the **Taiwan salmon** (*Oncorhynchus masou formosanus*). The authors found out that although the aquatic invertebrates were the most important prey for this fish species, the terrestrial invertebrates became more and more important sources of food as the size of the fish increased (Liao et al. 2012).

Many publications have also been devoted to the **contribution** of aquatic insects to the diet of fish. About 100,000 species of insects from 12 orders are known to pass one or several life stages in freshwater bodies (Dijkstra et al. 2014). Aquatic insects hold a prominent place in the diet of some species of fish. For example, according to an index considering abundance and weight, mayflies constitute 98% of the diet for such species as the oscar (*Astronotus ocellatus*) (Reis and Santos 2014).

In the stomachs of the **lake sturgeon** (*Acipenser fulvescens*) in the Winnipeg River, Manitoba, Canada, the number of Diptera comprises 78% of the total amount of invertebrates consumed (Barth et al. 2013). According to the index of relative importance, mayflies constitute 80% of the diet of the **stingray** (*Potamotrygon signata*) inhabiting the Parnaíba River basin (northeastern Brazil) (Moro et al. 2012). In the stomachs of the **freshwater drum** (*Aplodinotus grunniens*) in the Wabash River (the biggest right-bank tributary of the Ohio River) in the midwestern United States, caddis flies comprise 69% of the contents (Jacquemin et al. 2014). In the food bolus of young **Siberian salmon** in the Kadi River (the lower Amur River basin), the larvae of chironomids account for up to 100% of the contents by frequency, and 99% by weight (Yavorskaya 2008).

By comparing of the **importance** of different species of insects to fish, it is clear that the larvae of **chironomids** are the most important food item for fish. They are of great importance in the tropics, temperate latitudes, and the Arctic. Their role is most important in stagnant water bodies or slow-flowing streams with soft bottoms (Worischka et al. 2015). They are of the greatest significance to bottom-feeding fish such as sturgeon (Hamidoghli et al. 2014).

Other **important insects** in the diet of fish include larvae of caddisflies and stream mayflies, mainly in rivers with rock bedding, fast current, and clean water.

1.1.4.6 Insects as a food resource for mammals

Feeding on insects is characteristic of many species of mammals. For some species, insects are the **major food**, and for others, insects **supplement** the major food source. The distinction between these two categories of mammals is fairly conventional. The concept of "major feed" is also fairly fuzzy because the insects can compose 99% of the diet for some species, while for others, insects will only exceed half of the diet. It is evident that because the insects are small, the animals that feed predominantly on them must not be too large as well.

The **ten groups of mammals** for which insects play a key role in the diet include the following: shrews, moles, bats, anteaters, armadillos, echidnas, numbats, aardvarks, pangolins, and aardwolves.

The eating of insects appears to be the only characteristic these animals have in common. These mammals differ greatly in size. For example, the **Etruscan shrew** (*Suncus etruscus*) weighs around 2 g (https://en.wikipedia.org/wiki/Shrew), whereas the **giant armadillo** (*Priodontes maximus*) can weigh as much as 54 kg (https://en.wikipedia.org/wiki/Armadillo). Similarly, there are major differences in physiology, living habits, feeding mechanisms, and other characteristics.

There are appreciable differences in the **metabolic rates** of these mammals. Generally, the smaller the animal, the higher the metabolic rate. However, some animals are characterized by significant deviations from this relationship. For example, all the anteaters have a relatively low metabolism. At the same time, shrews, moles, and bats have very high metabolic rates. For example, shrews should eat 1.5–2 times their own weights every day (https://en.wikipedia.org/wiki/Shrew). One case was recorded when a great bat (*Nyctalus noctula*) ate in succession 115 larvae of mealworms (*Tenebrio molitor*) (Marikovsky 2012).

Mammals also eat a variety of insects; some animals eat a **large variety** of insects. For example, shrews eat practically all soil-inhabiting insects as well as spiders, woodlice, worms, and other organisms. Armadillos also have a widely varying diet; their menu includes almost 500 different foods, most of which are insects and invertebrates such as beetles, cockroaches, wasps, fire ants, scorpions, spiders, and snails (https://www.nwf.org/Wildlife/Wildlife-Library/Mammals/Nine-Banded-Armadillo.aspx).

Meanwhile, numbats, anteaters, aardvarks, and pangolins feed almost **exclusively** on ants and termites (https://en.wikipedia.org/wiki/Numbat; http://www.rumbur.ru/nature/1182-trubkozub; http://www.animalsglobe.ru/muravyedi/; https://bioweb.uwlax.edu/bio203/s2012/grosshue_crai/diet.htm). The rest of the animals in this category fall somewhere in **between**. In the diet of moles, such insects as mole crickets, click beetles, larvae of scarab beetles (Scarabaeidae), larvae of may beetles, millipedes, and earthworms predominate. Meanwhile, moths, mosquitoes, and beetles are important food sources for bats (http://animals.mom.me/list-insecteating-mammals-8065.html).

There are also a vast number of species of **omnivorous mammals** that, to one extent or another, feed on insects. Such animals include hedgehogs, desmans, many rodents—pikas, ground squirrels, dormice and hazel mice, mice and voles—and such predators as least weasels, ferrets, mongooses, weasels, otters, raccoon dogs, badgers, and even foxes, wolves, and many species of bears. Occasionally, some kangaroos also eat insects.

The **variety of insects** eaten can vary considerably even within one family. For example, the sloth bear (*Melursus ursinus*) eats predominantly termites (https://en.wikipedia.org/wiki/Sloth_bear). The diet of the sun bear (*Helarctos malayanus*) includes bees in addition to termites (https://en.wikipedia.org/wiki/Sun_bear). The variety of species eaten by the Asian black bear (*Ursus thibetanus*) is more diverse:

besides termites and bees, it feeds on different beetles and their larvae (https://en.wikipedia.org/wiki/Asian_black_bear). The American black bear (*Ursus americanus*) consumes ants, termites, wasps, bees, and yellow jackets (https://en.wikipedia.org/wiki/American_black_bear).

Some **primates**, such as marmosets, tamarins, tarsiers, galagos, and aye-ayes, also eat insects. To a greater degree, small-bodied frugivorous primates eat insects. M.A.H. Bryer et al. (2015) studied the feeding of red-tailed monkeys (*Cercopithecus ascanius*) in Kibale National Park, Uganda. They observed that, among all the eaten insects, 74% were cicadas (suborder Homoptera), 14% caterpillars (order Lepidoptera), and 7% long-horned grasshoppers (order Orthoptera). Even though red-tailed monkeys expended less than 10% of their feeding time catching insects, they obtained 24% of their daily protein intake and 14% of their energy from insects (Bryer et al. 2015).

Therefore, feeding on insects among mammals is fairly widespread. From the viewpoint of calories consumed, this is profitable for big animals when insects are available in large quantities. However, even when small quantities of insects are consumed, they are often an important source of vitamins and fatty acids.

1.1.5 Insects as a source of nutrients for plants

Another ecosystem service rendered by insects is the **provision of nutrients to plants.** About 630 species of carnivorous plants from 19 families are known that have adapted to catching and digesting small animals, mainly insects (https://en.wikipedia.org/wiki/Carnivorous_plant). These generally include plants growing near water and perennial herbs (sometimes, half-shrubs and small shrubs. They occur in all parts of the world (http://www.animalsandenglish.com/animal--insect-eating-plants.html).

Carnivorous plants use five basic **types of traps** for catching prey: (1) pitfall traps (pitcher plants) trap prey in a rolled leaf that contains a pool of digestive enzymes or bacteria; (2) snap traps utilize rapid leaf movements; (3) flypaper traps use a sticky mucilage; (4) bladder traps suck prey with a bladder that generates an internal vacuum; (5) lobster pots, also known as eel traps, force prey to move towards a digestive organ with inward-pointing hairs. These traps may be active or passive, depending on whether movement aids the capture of the prey (https://en.wikipedia.org/wiki/Carnivorous_plant).

Most carnivorous plants eat flying, foraging, or crawling insects. Generally, those are small aquatic insects (wigglers), etc. as well as crustaceans such as water fleas. Flies, mosquitos, and different small beetles and ants are often caught. More rarely, spiders will see an insect captured by a plant and dive directly into its digestive fluid. They begin to suck out the prey captured by the plant, thereby easing digestion. Soon thereafter, the spiders themselves are digested (https://en.wikipedia.org/wiki/Carnivorous_plant).

Surely, the significance of this ecosystem service is **much lower** than that of the kinds of services previously discussed. Nevertheless, it is also of importance. The carnivorous plants are adapted to grow in places with plenty of light where soils are shallow or are deficient in nutrients, especially nitrogen (acidic bogs, rock outcroppings, etc.). Such way the plants receive the nutrients they need to synthesize their own proteins from insects.

Use of Insects in Science

Different species of insects are used in a variety of scientific studies. Owing to the diversity of insects, there is always a possibility to find one that will satisfy the necessary requirements of experiments in the most diverse academic fields. Some examples of such **academic fields** include the social behavior of animals and their capacity for learning and elementary brain activity (ants, termites, bees, cockroaches, etc.); development of new engineering units in bionics (dragonflies, butterflies, beetles, bumblebees); and genetic research (*Drosophila*).

The study of insects helps researchers investigate nature and contributes to such **fundamental sciences** as systematics, physiology, biochemistry, genetics, animal ecology, hydrobiology, biophysics, and bionics, as well as to a variety of the applied disciplines (e.g., agriculture, forestry, veterinary entomology, and medical parasitology) and economic branches (e.g., beekeeping, silkworm breeding, seed raising, vegetable production, grassland management, berry growing, and forestry).

1.2.1 Insects as an object of genetic research

For more than 100 years, the most important subject of genetic research has been **fruit flies** (*Drosophila*). These are small flies (in most cases, with a body length of 2–4 mm) that are found around the world; a total of more than 1,600 species are known. In genetic research, more than 10 species (*Drosophila melanogaster*, *D. simulans*, *D. mercatorum*, *D. yakuba*, *D. erecta*, *D. ananassae*, and others) are used; however, the common fruit fly (*Drosophila melanogaster*) is by far the **most significant** (Photo 1.2-1). In the present-day literature, the terms common fruit fly, fruit fly, *Drosophila*, and *Drosophila melanogaster* are in many cases **synonymous** (https://en.wikipedia.org/wiki/Drosophila).

The following **features** have contributed to the extensive use of this species in genetic research (Kozak 2007): (1) short period of development (10–14 days); (2) high breeding performance (about 100 to 175 descendants can be obtained from one pair of insects); (3) small number of chromosomes (four pairs); (4) convenience of breeding under laboratory conditions; (5) great number of easily distinguishable characteristics; and (6) cells of the salivary glands of the fly's larvae contain giant chromosomes, which are particularly useful for research.

The American biologist **Thomas Hunt Morgan** began his experiments with *Drosophila* in 1906. Initially, he captured flies in grocery and fruit shops and raised them in the laboratory. Morgan's fly room at Columbia University became legendary. Its area was 35 m^2, and it held eight work stations. Food for the flies was also cooked in this room. Large numbers of cans and bottles were used to raise the flies from larvae and use them in experiments.

Among numerous *Drosophila*, Thomas Hunt Morgan and his colleagues (Calvin Bridges, Alfred H. Sturtevant, and H.J. Muller) observed **noticeable differences** among the flies and, in addition, determined that these differences are inherited. For example, common flies have gray-yellow bodies, but sometimes

Photo 1.2-1. Fruit flies (Drosophila) are widely used for genetic research. The common fruit fly (*Drosophila melanogaster*) is most significant species among them. The photo shows *D. melanogaster* types (clockwise): brown eyes with black body, cinnabar eyes, sepia eyes with ebony body, vermilion eyes, white eyes, and wild-type eyes with yellow body. COLOR
Photo credit: https://en.wikipedia.org/wiki/Drosophila_melanogaster

flies with black bodies were encountered. Common flies have red eyes, but rarely flies with white eyes were found. The researchers looked intently at the antennas of the flies, bristles in different parts of their bodies, and other characteristics. During the first year of investigations in Morgan's laboratory, 14 mutants were discovered, while by 1914 their number reached 168. When enough mutants were identified and the varieties of flies characterized by one or another **mutation** were obtained, experiments on their interbreeding were carried out.

In the course of these experiments, the researchers observed a vast number of **features**: eye color (for example, in addition to common red-eyed flies, white-, yellow-, and pink-eyed flies were found), coloration of bodies, different numbers of bristles, and various shapes and sizes of wings. The **first major discovery** was related to the fact that the attribute of white eyes is linked with sex. This mutation was not among the first detected, but it was essential. In 1910, the first genetic work by T. Morgan, "Sex-limited inheritance in *Drosophila*", was devoted to the mutation related to white-colored eyes. By 1913, T.H. Morgan revealed a variety of *Drosophila* mutations resulting in modified eye coloration and wing morphology. The subsequent **five mutations** he discovered also turned out to be sex-linked (Yurchenko et al. 2015).

The first publication was followed by a series of papers by T.H. Morgan and his colleagues that reported numerous experiments with *Drosophila* which proved that genes are the material units responsible for hereditary variance. The chromosomes of the cell nucleus were identified as the **carriers** of these genes. The truly monumental work of Morgan and his colleagues allowed the integration of cell biology and genetics into a single whole and resulted in the creation of the chromosome theory of heredity. The fundamental works by Thomas Hunt Morgan "The Physical Basis of heredity" (1919), "The Theory of the Gene" (1926), "The Scientific Basics of Evolution" (1932), and others mark the progressive advance of the science of genetics.

The **chromosomal theory** of heredity confirmed and reinforced the principles of segregation and independent inheritance of characteristics demonstrated by Gregor Mendel with pea plants. In 1933, T.H. Morgan received the Nobel Prize in physiology and medicine "for identifying chromosomes as the vector of inheritance for genes" (Koryakov and Zhimulev 2009).

The fruit fly *Drosophila* is one of the most widely used organisms in the biological studies. From the beginning of the 20th century up to the present day, *Drosophila* has been a model in the majority of genetic laboratories around the world. Scientists around the world study *Drosophila* not with a goal of investigating the insects as such but because the results of the investigations can be extended to all living organisms on Earth—including humans, as 70% of the fruit fly genome is similar to that of humans (https://en.wikipedia.org/wiki/Insect). Using this fly, the field of genetics has made many discoveries. The popularity of *Drosophila* is so great that a **yearbook** devoted to it is published that contains an abundance of diverse information on it.

1.2.2 Insects as an object of gerontology research

The insects also are of considerable importance for gerontological investigations. Naturally, mammals would be more appropriate to study the aging process in humans; however, their use is frequently inexpedient due to their long lifetimes and high maintenance costs. In this connection, insects are acceptable as **alternative model objects** for investigation of aging because they have relatively short lifetimes and expenses for their upkeep are much lower.

It should be noted that, in a number of cases, data on study subjects obtained under laboratory conditions do not correlate with some parameters (lifetime, mortality, etc.) observed for them in the wild (Roach and Carey 2014). Nevertheless, the regularities of the **aging process** that are revealed remain the same.

Many species of insects are used as models in gerontological investigations. **In most cases**, these are *Drosophila*, silkmoth (*Bombyx mori*), Eastern lubber grasshopper (*Romalea microptera*), and some social insects (ants and bees) (Lee et al. 2015).

The crucial (although not as exclusively as in genetics) part is played by *Drosophila*. Besides their advantages considered in the previous section, the **important** upside is that 60% of human disease-related genes are present in flies (Matthews et al. 2005). Another **advantage** is the small space required for raising them. For example, breeding about 100 pairs of flies requires only 500 cm^3 of space (Lee et al. 2015).

Drosophila is widely used for investigating the aging mechanism. In particular, the relationship between longevity and calories was established. For feeding of *Drosophila* in the laboratory, a nutrient medium is used that consists largely of a solution of distilled water, yeast, and sugar. Studies found that when the yeast concentration was decreased from 16% to 2%, the lifetime of flies increased by 30%. An increase in life span owing to food calorie restriction by 30–40% (without undernutrition) was also established for nematodes, rodents, and monkeys (Lee and Min 2013). For females and males of Australian cactus fly species (*Telostylinus angusticollis*, family Neriidae), dietary restriction extends life span by 65% (Adler et al. 2013).

Numerous investigations have been carried out on the effects of different materials on life span. For example, studies found that a diet low in phosphate **increases** the life span of *Drosophila*. This dependence also proved to be true for mammals, including humans: the lower the phosphate content in the blood of mammals, the **longer** the lifetime (http://www.gersociety.ru/information/magg2009/). Other chemicals that have been studied include insulin, rapamycin, adenosine monophosphate, and sirtuin (Rogina 2011; Stenesen et al. 2013).

Fruit flies have frequently been used for investigating such **age-related diseases** as sarcopenia (loss of muscle mass specific to aging), cancer, neurodegenerative diseases, cardiovascular disease, immunosenescence, and metabolic diseases (Lee et al. 2015; Demontis et al. 2013; Hoffmann et al. 2013; Seong et al. 2013).

The attractiveness of the **silkmoth** as a model for gerontological investigations is associated with the following **features** of this insect: (1) large-sized body, which makes it a suitable model for examination of tissue-specific aging (for example, the midgut and fat body can be easily isolated from silkworms for pharmacological experiments to discover anti-aging interventions); (2) inability of the butterflies to fly simplifies working with them; (3) similarity of pharmacokinetic mechanisms in silkmoth and mammals; and (4) more than 5,000 genes in silkmoth have been found to be orthologous to human disease-related genes, mainly related to skeletal, neurologic, and growth diseases (Zhang et al. 2014; Feyereisen and Jindra 2012; Imamura et al. 2013).

The Eastern lubber grasshopper (*Romalea microptera*) is also one of the most popular models for these studies, determined by the following **factors**: (1) it does not fly; (2) it is characterized by relatively large sizes (up to 7–8 cm); (3) it can be acquired in great numbers and at a low price; (4) its lifetime is 120 days, so it is useful for studying aging since an organism with a longer life span might be too sensitive to other putative interventions; and (5) researchers can easily control the quantity of food that is eaten, which is impossible for fruit flies (Lee et al. 2015).

Such social insects as ants and bees stoke the attention of gerontologists due to the great **differences** in the lifetimes of different castes. Ants have a caste system composed of a single queen, numerous female workers, soldiers, reproducing females, and males. The lifetimes of ants belonging to different castes differ greatly. In particular, the average lifetime of queens is much longer than that of individual female workers, even those that have common parents. As the genotypes of queens and female workers are identical, the differences in the life span can be caused by aging determinant genes. As a result, ants are a **promising model** for identification of such genes (Lee et al. 2015).

Similar to those in ants, **honey bee castes** are composed of queens, female workers, and males known as drones. A bee can become a queen or a female worker depending on what it ingests during the larval stage. Feeding with royal jelly results in the emergence of a queen, while feeding on honey and pollen causes female workers to develop. Although the genotypes of queens and female workers are the same, the queens are able to live about 2 years, while the female workers live only about 20–40 days in summer and 140 days during winter (Rueppell et al. 2007). Researchers have proposed the hypothesis that differences in membrane fatty acid composition are responsible for the differences between the lifetimes of queens and worker bees (Haddad et al. 2007). A higher content of vitellogenin in the yolk protein of queens is thought to be a factor determining a longer life span of queens (Corona et al. 2007).

In addition to the insects that have already been discussed, monarch butterflies (*Danaus plexippus*), some beetles, house flies, crickets, and mosquitoes have also been used in gerontological investigations (Lyn et al. 2011; Kelly and Tawes 2013; Lee et al. 2015).

1.2.3 Insects as an object of other medical and biological research

Besides genetics and gerontology, insects are objects of research in other medical and medico-biological fields. They played an important (although indirect) part as early as the late 17th century in studies of bacteria and microbes by the Dutch researcher Antonie van Leeuwenhoek. He used the **eye of a louse** as his standard unit of measurement for microscopic observations (http://www.ucmp.berkeley.edu/history/leeuwenhoek.html).

Yet another example of indirect participation of insects in medico-biological studies is related to the silkmoth. In the mid-19th century, an outbreak of **silkmoth disease** was registered in France (Vaucluse Province). The disease quickly spread to other provinces of the country and, after a time, reached Spain, Italy, Turkey, Syria, Moldavia, Russia, and China. The disease resulted in vast losses in the silkworm breeding of these countries.

The French Ministry of Agriculture sent Louis Pasteur to one of the centers of silkworm breeding, the town of Alais. Exploring grounds where of silkworm moth deaths had occurred, he found in dead larvae microscopic corpuscles that, in his judgment, caused the illness. In 1870, Pasteur published the paper "Investigations of diseases of silkworms", in which he showed the **infective nature** of diseases transmitted through grain (eggs of silkmoths) upon contact of healthy insects with diseased ones and through infected food. Later on, generalizing his observations, Pasteur formulated the **bacterial theory** of disease; however, the first association between pathogens and disease grew out of Louis Pasteur's studies of silkworm diseases (1865–1870) (http://www.cals.ncsu.edu/course/ent425/text01/impact2.html).

The **grasshopper** is another insect used in medico-biological investigations. Grasshoppers are susceptible to all major taxonomic groups of **pathogens**, including viruses, bacteria, protozoa, and nematodes. In this connection, they can be used as a model for **testing antibiotics**.

For example, the **Eastern lubber grasshopper** (*Romalea microptera*) was used to test 11 commercial antibiotics against a pathogenic *Encephalitozoon* species (Microsporidia). The grasshoppers were infected with Microsporidia, and then the antibiotics were orally administered. The use of two types of antibiotics (fumagillin and thiabendazole) turned out to be **relatively efficient** and resulted in a marked decline in the quantity of pathogenic spores. The application of quinine was **less efficient** and resulted in noticeable but insufficient decreases in the numbers of spores. Feeding with such antibiotics as albendazole, ampicillin, chloramphenicol, griseofulvin, metronidazole, sulfadimethoxine, and tetracycline proved **unsuccessful**: the treated grasshoppers were not significantly different from the infected controls. The use of streptomycin led to a **nonsignificant increase** in the number of spores (Johny et al. 2007).

However, fruit flies (*Drosophila*) are the **most important** insects in this field as well. They are used in the genetic modeling of some human diseases, including such neurodegenerative disorders as Parkinson's, Huntington's, spinocerebellar ataxia, and Alzheimer's diseases. The fruit fly is frequently used to study mechanisms forming the basis of oxidative stress, immunity, diabetes, and cancer, as well as drug abuse (https://en.wikipedia.org/wiki/Drosophila_melanogaster).

Drosophila is widely used to identify the extent of mutagenic activity of different medicinal preparations, depending on the schedule of administration, dosage, time of exposure (Chshiyeva 2006). They are successfully used to examine the **effects of drugs** designed to treat the human heart, and *Drosophila* has been used to study dilated cardiomyopathy (Wolf et al. 2006).

Drosophila serves as the closest natural model for the investigation of the anopheline mosquito (*Anopheles gambiae*), yellow fever mosquito (*Aedes aegypti*), and common house mosquito (*Culex pipiens*), which are vectors of dangerous **infectious diseases** of humans: malaria, dengue fever, yellow fever, and West Nile encephalitis. Examination of *Drosophila* also offers a clue to understanding genetic processes identified as important for agricultural insects such as bees and silkworms, and insect pests such as locusts and many species of beetles and lice (Yurchenko et al. 2015).

Drosophila has also made a substantial contribution to **developmental biology**. For example, the Nobel Prize for physiology and medicine in 1995 was presented for discoveries concerning the genetic control of early embryonic development. The award was presented to Edward B. Lewis, Christiane Nüsslein-Volhard, and Eric F. Wieschaus. All of them conducted investigations on the *Drosophila melanogaster*.

These researchers investigated the effects of different genes on **embryogenesis**. They checked about 20,000 genes and selected 150 ones, without which the normal course of ontogeny is impossible. The methods they used included individual gene silencing and breeding of species without this gene. Therefore, they produced a generation without some genes and could evaluate its effects on the development of the embryo. This field of study is of great importance for humanity because it may help researchers understand how some serious diseases arise (http://kodomo.fbb.msu.ru/~partyhard/term2/pr6/nobelprize1995.html).

1.2.4 Insects as bioindicators

Bioindicators (biological indicators) are organisms that respond to environmental changes. Accordingly, **bioindication** is the assessment of environment quality according to the condition of a location's biota, based on observations of indicator species. For instance, the appearance or disappearance of certain insect species in nearby water bodies is a indication of quality of water. For example, larvae of harlequin flies Chironomus riparius, drone fly Eristalis tenax, etc. are indicators of low quality of water. Many species of insects are good indicators of the condition of different natural components due to their short life cycle, high rate of reproduction, and high sensitivity to environmental changes. Therefore, they can warn against such changes before they become visible (Schowalter 2013).

Numerous papers have been devoted to the theme of bioindication. There is no one established opinion on a number of questions. We will consider **three categories** of bioindicators: (1) environmental indicators (representing the effect of economic activity on the natural components); (2) ecological indicators (demonstrating the direction and character of changes in ecosystems); and (3) biodiversity indicators (characterizing the complexity of biological systems and the different qualities of its components).

The initial stage of any type of bioindication study is the selection of the biological indicator species, including insects. It is important to chose most appropriate species of insects as biological indicator species. Ideally, the species should meet the **following criteria**: (1) sufficient population; (2) widespread occurrence; (3) capability to reflect the desired parameter of the environment (accumulation of controlled pollution, ecosystem changes, etc.); (4) clear correlation (for example, between the levels of pollution and the bioindicator and natural component); (5) low mobility; (6) easy accessibility; (7) high specificity (at low specificity, a bioindicator responds to different factors, while at high specificity it responds to only one factor); and (8) high sensitivity (at low sensitivity, a bioindicator responds only to strong deviations of a factor from the standard, while at high sensitivity it responds to slight deviations).

1.2.4.1 Insects as environmental indicators

These indicator species reflect, first of all, **pollution** of the environment they predominantly inhabit (aquatic, soil, etc.). The response of insects to environmental pollution can be evidenced as **follows**: (1) emergence or disappearance; (2) variation in population size; (3) change in appearance of (a) sizes and proportions, (b) body cover, (c) coloration, and (d) deformities; (4) accumulation of pollutants in the organism; (5) changes in behavior; (6) reduction in breeding performance; (7) total change in ontogenesis, that is, individual development of the organism, **such as** (a) distortion of the molting process, (b) reduction of developmental stages, and (c) change in life span.

A dying-out of bees when the environment is polluted with pesticides serves as an example of **extinction** of insects due to pollution. An **opposite example** is the emergence of bloodworm larvae of a buzzer midge (*Chironomus plumosus*), when water bodies are contaminated with sewage waters. An example of **population variations** within a pollution zone is a sharp **increase** in the number of sugescent phytophages, especially aphids, and bark and stem insects. For example, sharp **decrease** in the numbers of some ground beetles, rove beetles, and a number of book lice have been observed near cement and metallurgical plants where salts of heavy metals are released (Chernyshev 1996).

Changes in **appearance** are manifested in many ways. *Sizes and proportions* of bodies change quite frequently. For example, for a number of aphids, differences in head breadth, length of femur, tibia, and siphon are reliably fixed at contaminated and uncontaminated sites. When insects fed on contaminated food, the sizes of their larvae usually decreased. The *body covers* differed too. For example, in the descendants of black bean aphid (*Aphis fabae*), cuticle polygons and granulation changed markedly after sulfite ions were added to their food (Grigoriev et al. 2008).

The classic example of *color change* is the phenomenon of industrial melanism detected for about 100 species of butterflies in habitats with intensive industrial development. This phenomenon was first observed in Manchester (England) in the mid-19th century. Previously, all peppered moths (*Biston betularia*) had white-grayish wings with dark dots which provided the assimilative coloration on the tree trunks. However, in the woods around industrial centers, the stems of trees are often found to be smoked (Photos 1.2-2, 1.2-3). In such areas, the black coloration is an advantage, reducing the risk of the butterflies being eaten by birds. Industrial melanism is also observed among beetles (two-spotted lady beetle *Adalia bipunctata*), firebug *Pyrrhocoris apterus* (Mukhina 2007), some aphids and cicadas, which also exhibit, together with color changes, variation in body sizes and microstructure of the surface (Barabanshchikova 2013).

Emergence of disfigurements in appearance of insects start to emerge. Xenobiotics such as diesel fuel, DDT (dichlorodiphenyltrichloroethane), and other compounds cause disruptions of the morphogenetic processes in the ontogenesis of insects. Experiments have shown that the proportion of abnormal pyralid moths increased from 5 to 35% when PbO was added to their food (Grigoriev et al. 2008). **The buildup of pollutants** in food chains is well-known. Insects, like other living organisms, are able to accumulate pollutants in their tissues. Phytophagous insects receive them predominantly from the plants they eat, while predacious insects ingest them from their prey. Because carnivorous insects are on a higher level of the food pyramid, they are characterized by higher concentrations of pollutants.

Changes in behavior in polluted areas are quite typical. For example, braconid wasps cease to lay eggs into the bodies of aphids and execute cleaning movements under the action of insecticides (Marikovsky 2012). When the content of compounds of copper, zinc, and chromium, as well as salts of mercury, cadmium, and lead, are increased, the communication behavior of ants and bees is adversely affected. Even with low doses of pollutants, the construction of caddisworm cases changes (Chernyshev 1996). Where electromagnetic field pollution exists, increased aggressiveness and anxiety, and decreased performance of bees are observed (Orlov 1990), while changes in behavior are observed

Photo 1.2-2. Insects are environmental indicators. A well-known phenomenon is the phenomenon of industrial melanism—color change due to industrial pollution. A classic example is the peppered moth evolution where the number of dark-colored moths is increased. Photo shows *Biston betularia f. carbonaria*, the black-bodied peppered moth in Dresden, Pohrsdorfer Weg (Saxony, Germany).
Photo credit: by Olaf Leillinger [CC-BY-SA-2.5 and GNU FDL], via Wikimedia Commons at 2006-06-13

Photo 1.2-3. In a clean environment, a number of light-colored peppered moths is increased. The white-bodied peppered moth *Biston betularia f. typica*, in Dresden, Pohrsdorfer Weg (Saxony, Germany) is shown.
Photo credit: by Olaf Leillinger [CC-BY-SA-2.5 and GNU FDL], via Wikimedia Commons at 2006-06-13

in beetles, mosquitoes, and butterflies (Presman 2003). Under the influence of super-high frequencies, disorganization of ants' behavior occurs in which they lose the capability to inform each other about food sources (Kozlov 1990).

Reductions in breeding performance were registered, in particular, for springtails (*Onychiurus armatus*, *Orchesella cincta*) at sites contaminated with heavy metals, as well as for aphids and gypsy moths when areas were fumigated with sulfur vapor. Impacts on **ontogenesis** consist of disturbances of *molting processes*. For example, the larval stage was extended for owl moths (*Scotia segetum*) in cases of copper poisoning, and for gypsy moths exposed to fumigation with fluorine hydride (HF) and methyl mercaptan. As for butterflies, numbers of pupating larvae and the percent of imago emergence decreased in polluted environments.

Terms of development were reduced for springtails (*Isotoma notabilis*, *Onychiurus armatus*) in areas with heavy metals pollution. For owl moths (*Scotia segetum*), they decreased by 4–7 days on exposure of the organism to cadmium chloride ($CdCl_2$). Generally, the *life span* also is reduced. This was observed for larvae of the gypsy moth, silkmoth, pine tree lappet, and pine looper (*Bupalus piniarius*) feeding on polluted food as well as for the burrowing grasshopper (*Acrotylus patruelis*) and larvae of the brown blowfly (*Calliphora vicina*) subjected to industrial fumes (Grigoryev et al. 2008).

The number of species of insects used to indicate environmental conditions is very high. The selection of species depends on many factors (goals of the indication, main pollutants, geographic position, etc.). Bioindication is often carried out in the composition of beekeeping products (honey, wax, pollen, propolis). All the pollutants (heavy metals, radioactive agents, pesticides, etc.) accumulate in these materials, providing information on environmental pollution in areas where they were collected.

Information has been obtained through studies of the contents of heavy metals and radioactive substances in products of beekeeping close to highways in the Krasnodar Krai of Russia (Moreva et al. 2010), and pesticide pollution of agricultural areas of Greece (Balayiannis and Balayiannis 2008), Poland (Barganska et al. 2016), Italy (Porrini et al. 2014), and Australia (Sanchez-Bayo and Goka 2014).

Everything in nature is interconnected, and pollution of one of the components of an ecosystem, as a rule, affects the others. For example, atmospheric pollutants precipitating on the land surface pollute soil and vegetation. Then, as a result of washoff by rain and melting snow, these pollutants reach watercourses and partially penetrate into the groundwater. Later on, vegetation receives pollutants from the soil, and the processes continue. Although insects are most often used to evaluate pollution levels in soils and surface waters, these levels indicate pollution of other environmental components.

For example, in Murmansk Oblast (Russia), soil pollution levels were evaluated with the use of the millipede *Monotarsobius curtipes* (Petrashova 2010). However, it was evident that the soil pollution reflected the atmospheric pollution caused by the operation of the Kandalaksha aluminum smelter and Severonikel plant within an affected zone of waste release where the studied area was located.

Firebugs (*Pyrrhocoris apterus*) are sometimes used to estimate overall levels of anthropogenic impacts. This species of insects forms colonies on the ground at the feet of trees; it sucks plant sap and feeds on fallen seeds of plants. More rarely, it feeds on small insects. Conclusions about the state of a habitat can be drawn based on observations of patterns on its **front wings** (elytrae): the higher the anthropogenic load, the greater the phenotypic variability (Mukhina 2007; Barabanshchikova 2013).

However, insects are most often used to evaluate the state of specific natural components. For example, in the case of **surface waters**, different caddis flies and larvae of dragonflies are most popular. Publications devoted to their use as indicators of the state of the environment include, for instance, papers by Whatley and colleagues (2014) for the Netherlands; Barman and Gupta (2015) for northeastern India; Barbosa de Oliveira Jr. et al. (2015) for Para State, northern Brazil; Taylor (2013) for all 50 states of the United States; and Azam et al. (2015) for Pakistan.

For evaluation of **soils**, ground beetles (Carabidae) are often used. For example, Kiselev (2005) studied ground beetles in the industrial zones of Tula city (Russia). Of 36 species of ground beetles encountered there, he selected the strawberry ground beetle (*Pterostichus melanarius*) as a model. This selection is explained by the fact that in this species, a high level of correlation was found between the contents of metals in soil and indices of the fluctuating asymmetry of the elytrae patterns.

An example of studies evaluating **atmospheric** pollution is Selikhovkin's (2013) investigations of the reaction of insects to the industrial air pollution in the Irkutsk, Novosibirsk, and Leningrad Oblasts of Russia. This researcher used the gypsy moth (*Lymantria dispar*) and sallow leafroller moth (*Anacampsis populella*) as models. Sulfur dioxide was found to have the strongest impact directly through the air, while fluorine affected the insects through their food.

Beetles are most often used to indicate the state of **vegetation** in an area (Cano and Schuster 2008; Viegas et al. 2014; Bicknell et al. 2014). Viegas et al. (2014) observed a negative response in dung beetle communities to increasing forest fragmentation in southern Brazil. Bicknell and colleagues (2014) explored the effects of timber cutting activity in French Guiana on dung beetle communities. Species were classified as indicators of logged (*Hansreia affinis* and *Eurysternus caribaeus*) and unlogged forest (*Canthidium aff. centrale* and *Deltochilum carinatum*). The key factors affecting dung beetle assemblages were tree extraction intensity, bare ground cover, and ground cover by leaf material.

Methods of estimating the state of **surface waters** have been developed to the **greatest extent**. Special tables allow one to determine water contamination levels according to the presence or absence of various species of insects. If water contamination is estimated from zero (purest) to 10 (most dirty), then intervals within which major taxons of aquatic insects can live will be as **follows** (http://www.wpwa.org/documents/education/Biological%20sampling.pdf): 0–7 for Ephemeroptera (mayflies); 0–4 for Plecoptera (stoneflies); 0–8 for Trichoptera (caddis flies); 1–3 for Odonata (dragonflies); 5–9 for damselflies; 0–4 for Megaloptera (dobsonflies); 2–10 for Diptera (true flies); and 2–5 for Coleoptera (beetles).

1.2.4.2 Insects as ecological and biodiversity indicators

Ecological indicators allow us to obtain information on ecosystems—associations of living organisms and their habitats. Changes take place in ecosystems not only due to human activities. Frequently, they are caused by natural **development** of ecosystems. There are numerous natural rhythms: rhythm of solar activity (11-year and "secular" 80- to 90-year cycles); long-period lunar tide (49 years); variations of positions of the Earth's rotation poles (7-year cycle); and others. Periodic variation in atmospheric circulation and, consequently, distributions of air temperatures on earth and water temperatures in the ocean, territory moistening levels, water content in lakes, and recession and expansion of mountain glaciers are related to these phenomena in relatively complicated ways (Govorushko 2009).

All of these factors cause changes in ecosystems, and insects frequently help to reveal the presence of these changes. Information on ecosystem changes can be obtained by comparing changes in different areas (spatial comparison) and by analyzing changes within the same area (comparison in time; Gerlach 2013). It is important that the insects are able to respond very fast to any changes in the environment and are far more sensitive indicators of such changes, especially climatic ones, as compared to vegetation.

Soil insects are often studied to analyze the soil development. There are a variety of **elementary soil processes** (gleyification, prairification, formation of leaf litter, steppification, salinification, etc.). One can use soil insects as indicators of these processes. For example, Mordkovich (1997) has identified eight ecological groups of ground beetles (Carabidae) for the Baraba Steppe (southern West Siberia, Russia): valley bog, bog, solonchak, forest, woodland meadow, solonetz, meadow, and steppe groups.

Each of these groups clearly indicates an **elementary soil process**: the gley formation process in the upper part of soil is characterized by the valley bog ecological group of ground beetles; peat formation by the bog group; solonchak processes (halobionts) by the solonchak group; the process of leaf-litter formation by the forest group; solodization by the woodland meadow group; alkalization by the solonetz group (small flat ground beetles inhabiting the cracks); meadow humus accumulation by the meadow group; and steppe soil-forming processes by the steppe group (Mordkovich 1997).

One can determine certain **soil characteristics** (mechanical composition, humus type, degree of humification of organic remains, acidity [pH], calcium content, and hydrothermal soil regime) according to the presence or absence of different invertebrates. For example, coarse humus is diagnosed by soil centipedes (Geophilidae), while soft humus is diagnosed by the larvae of crane flies (Tipulidae). The degree of compost maturity can be determined by collembolans (in mature composts, their white soil forms predominate). A high calcium content is indicated according to abundance in the soil of calciphilic diplopods (Slepchenko 2006).

Woodlice are indicators of **heavy earth** (in sandy soils, their burrows collapse). The vertical distribution of small soil arthropods correlates to total soil space. An important indicator is frequently the presence of white grubs (may beetle larvae). For example, in East Siberia, their occurrence in soil is related to the fact that the permafrost occurs not higher than 2.2–3 m from surface and that, in winter, there is no closure of the seasonal and perennial freezing layers. In the European part of Russia, their presence in soil indicates the presence of deep **groundwater** (Mukhamedzhanova and Konshin 2015).

Mass mortality of trees in dark coniferous forests serves as an indicator of natural cyclicity (Kozin 2013). This is a normal occurrence as the trees grow older. Xylophagous insects characteristically show a high sensitivity in responding to the least variations of a tree's condition; they occupy trees during the stage of physiological weakening, and they form, as a result, so-called ecological groupings of species (Lindeman 1993). Initially, trees are attacked by the most active insect species that are able to reduce the tree's resistance and develop fairly successfully on viable food objects. As the resistance of trees drops, they are occupied by new species of xylophagous insects. The appearance of each new species of insects on the food object gives a qualitative indication of the stage of decrease in the tree's resistance (Alekseev and Demakov 2000).

Insects can also provide information on a **water body's development**. The larvae of aquatic insects have chloride cells, which are able to actively absorb anions, especially chloride ions, in order to stabilize their concentration in the hemolymph. Generally, these cells are located on the gills (in mayfly larvae) or on the abdomen (larvae of caddis flies). The quantity of these cells is in inverse proportion to the degree of salinity, and after each molt their quantity is brought into line with the environmental salinity. Therefore, from molt to molt, one can determine changes in water body salinity (Grigoryev et al. 2008).

Many insects are **temperature indicators**. Some species of insects are not encountered if winters are extremely cold, while others, on the contrary, require cold temperatures, for instance, –27–30 degrees Celsium. For many insects, accumulated warm temperatures are of major importance. The presence of particular species of insects by itself means that cumulative temperatures are higher, while its absence suggests that the temperatures do not reach the required value. Presence of certain insect species within an area also allows one to infer the presence of particular animals and plants, which the insects feed on.

As for **biodiversity**, there are several meanings of this term. **Most often**, it means the abundance of species determined as the number of species present on a designated territory (from a small habitat to a country or a biogeographic region) or the ratio of the number of species to the total number of individuals. The principles of object selection to characterize the biodiversity of a territory should take into account many factors (Yashina 2011).

For example, Zhigalski (2011) chose representatives of three insect orders referring to **different trophic levels** to estimate the biological diversity of the Ural forest ecosystems: carnivores (ants), phytophagous insects (lepidopterans), and saprophagous insects (crane flies). The distribution of the integrated index of the species diversity of all listed groups of organisms in the zonal and azonal aspects reflects the overall picture of biodiversity of different taxonomic groups: fishes, amphibians, reptiles, birds, and mammals.

1.2.5 Use of insects for paleoecological reconstructions

The study of organisms that existed in the past geological periods is referred to as paleontology. Correspondingly, it contains **two disciplines** related, directly or indirectly, to insects: (1) paleoentomology, the study of entomolites (fossil insects); and (2) paleoecology, the study of environmental conditions, and living organisms, and interrelationships of organisms of the geological past.

Because the insects emerged about 400 million years ago (https://en.wikipedia.org/wiki/Timeline_of_the_evolutionary_history_of_life), they are of some significance for reconstructing past environments (i.e., **paleoenvironment**). Each species of insects lives within a specific range of climatic and physiographic conditions. These ranges vary for different species; however, in comparing them, one can always identify the relatively narrow interval where the fossil insects are found. Knowing their age, one can judge many environmental parameters existing there at that time.

These parameters can **include** temperatures (their minimum or maximum values, accumulated temperatures above 0°C, etc.), humidity, and even the salinity of water. For example, based on spiders (their imprints), one can judge what **kind of water body** existed: fresh (extended legs) or saline (pulled-up legs).

There are two basic **types** of insect burials (Zherikhin et al. 2008): (1) direct burial in sedimentary deposits (when the dead or alive insect is covered by bottom sediments); and (2) indirect burial in fossil containers (when an insect is found in another fossil object; resins are the most important type of fossil container).

The **first type** is predominant. The highest probability of such disposal is characteristic of the inhabitants of stagnant water. It is also high for semiaquatic insects, especially those inhabiting riparian vegetation. The lowest chances for disposal are characteristic of inhabitants of rapid water courses, soil insects, and desert and mountain insects (Kalugina 1980). In order that a dead insect be present in bottom deposits, **three conditions** must be fulfilled: (1) its dead body should fall into a water body; (2) it needs to overcome water surface tension forces; and (3) the dead body should be covered with the sediment (Zherikhin et al. 2008).

Any paleoecologic and paleogeographic **reconstructions** should consider the features of insect burial. They vary for different taxa, different stages of development, and even for different populations of the same species. In most cases, preservation of fossil insects is poor. Their chitinous covers, generally being rigid and resistant to chemicals, have a low mechanical strength. The corpses break up quickly into separate fragments in water. In most cases, fossil insects contain a disproportionally large number of the hardest exoskeletal parts (for example, elytrae of beetles, cockroaches, and orthopteran and homopteran insects); in these cases, the legs, antennae, and other fragile fragments disintegrate. The larvae of insects and imagoes with thin coats and delicate wings (mayflies, flies, etc.) are rarely found (Kalugina 1980).

Fossil insects in a good state of preservation are characteristically found in the following **cases**: (1) burial in fossil resins; (2) burial in natural asphalts; (3) burial in volcanic ash; (4) burial in perennially frozen rocks; and (5) burial in peat. Burial in the hydrothermal deposits, burrows of animals, as well as burial in karst cracks and funnels are also of some importance. The remains of animals that have fallen into them also contain the remains of necrophagous insects. However, similar burials are rarely discovered. Besides the fossil insects themselves, signs of their activity are also observed; for example, excrements, constructions (more often, cases of caddis flies), and damage to plants.

Fossil resins are buried resins of plants. Resinous plants occur in 26 families, or about 9% of the total number of families of higher plants (http://www.woodyman.ru/publ/173-1-0-4640). The maximum amount of resin is produced by trees, and the content in coniferous trees is vastly greater than that in deciduous trees. Because insects buried in fossil resins are of primary importance for paleoecological reconstructions, they are considered separately in Section 1.2.5.1.

Points of natural oil ingress are frequently observed on the earth's surface. This oil accumulates in topographic depressions and results in the formation of oil lakes. After the light ends evaporate from such lakes, they turn into **pitch lakes**. Insects fall on their surfaces, get stuck there, and perish, with no chance to escape. Owing to the conservation properties of natural asphalts, fossil insects found in them are well preserved. The best known example is the La Brea Tar Pits in Los Angeles, California. The arthropods found there include flies, beetles (Photo 1.2-4), grasshoppers, leaf-cutter bees, pill bugs, scorpions, termites, and water fleas. Similar lakes are also known in Peru, Trinidad, Iran, Russia, Venezuela, and Poland (https://en.wikipedia.org/wiki/La_Brea_Tar_Pits).

The burial of insects related to **volcanic activity** is confined to tuffites (mixture of ash and sedimentary materials at the bottom of water bodies) and to vitric tuffs (fossil volcanic ashes). Ash burial of insects is related to the fall of ash directly into a water body and transport of fallen ash down slopes. An abundance of flies is characteristic of these burials (Rasnitsyn and Quicke 2002).

Because **perennially frozen rocks**, on a geologic scale, were formed most recently (according to different data, their ages are estimated at 300,000 to 1 million years), only insects existing at that time are preserved in them. Their preservation is frequently very good. While mechanical damage to the insect skeleton does takes place, while the scales and hairs are often preserved and chitin does not lose its strength (Kuzmina 2003).

Photo 1.2-4. Paleoentomology is the study of entomolites (fossil insects). Burial in natural asphalts allows for insects to be preserved well. The best known example is the La Brea Tar Pits in Los Angeles, California. The photo shows a water scavenger beetle (*Hydrophilus* sp.) having size 12 cm and age about 35,000 years from La Brea Tar Pits.
Photo credit: By Didier Descouens (Own work) [CC BY-SA 4.0 (http://creativecommons.org/licenses/by-sa/4.0)], via Wikimedia Commons, 22 January 2011

The remains of insects also keep quite well in **peat**. Some insects exist in peat from the cradle to the grave. For example, the peat cockroach (*Panesthia* sp.), widespread in Vietnam, feeds on residues of wood and leaves of linden and oak. Sometimes, the northern mole cricket (*Neocurtilla hexadactyla*) inhabits peat. Most often, insects from adjoining areas and water areas are buried in peat. For example, in northern Sweden, diving beetles, *Hydroporus*, rove beetles, bark beetles, weevils, and ground beetles are observed in peat samples. In samples of Gorbunovsky peatland (north Ural), the stag beetle (*Platycerus caraboides*), click beetle (*Selatosomus melancholicus*), ground beetles (*Carabus menetriesi, Chlaenius quadrisulcatus*), chrysomelid beetles (Alticinae), and rose chafer (*Potosia metallica*) have been found (Zinovyev 2010).

Numerous **examples** of paleoecological reconstructions based on investigations of fossil insects are known. For example, in Mongolia, fossil insects have been found in more than 100 locations, with almost all burials in lacustrine deposits. Fossil insects were detected in deposits of all systems from Permian to the Paleogene, and records for the Jurassic and Cretaceous deposits were especially complete. There, the ancient entomofauna is numerous and diverse, which allowed researchers to reconstruct quite confidently the living environment of insects for different geological periods and different regions of the country (Ponomarenko and Prokin 2014).

Nevertheless, fossil insects are, to a greater degree, used to acquire information on the environmental parameters existing in the relatively recent past. For almost 60 years, fossil insects have been used to determine **paleotemperatures**. The pioneer works in this regard were publications by G.R. Coope in the early 1960s. He analyzed the present-day distribution of different species of ground beetles and, using detailed climatic maps, determined suitable temperatures for all ground beetles of the fossil complex (Kuzmina 2003).

Later on, the mathematical method of paleoclimate reconstruction, called the **mutual climatic range method** (MCR), was developed. It was based on the comparison of maps of fossil insect occurrence with the temperature ranges of recent species, but paleotemperatures are determined by using computer technologies (Elias 1994; Elias et al. 1999; Elias 2000). A host of publications concern the reconstruction of temperatures of the Late Pleistocene and Holocene for a number of European countries, the United States, and Canada, which are discussed in a review by Kuzmina (2003).

For example, Khorasani (2013) identified environmental variations in northern Sweden over the last 8,000 years by analyzing fossil insects. She determined the averaged summer and winter temperatures of air, variations in the boundaries of distribution of different wood species, dynamics of bog formation, and other environmental variations. Changes in the **numbers of dung beetles** allowed her to draw conclusions about the development of cattle breeding in different parts of the territory under study.

Yet another example is a study of soil hydrologic conditions for separate sections of the Arkhangelsk Region (Russia) based on morphoecological features of fossil oribatids from peat deposits. This study found representatives of 46 species, 34 genera, and 25 families of these mites (Krivolutsky and Sidorchuk 2005; Sidorchuk 2004).

Quaternary insects are also of interest to **archaeologists** because they help them to reconstruct the environmental conditions under which ancient humans existed. Numerous archeological excavations at various locations have been involved. To a greater degree, they were investigated in England and, to a lesser extent, in continental Europe and North America. Fossil insects were found in some archeological locations in Greenland, South America, and Egypt (Elias 1994).

In 1970, for example, the **oldest known man-made footpath** in Europe was found, the 1,800 m long Sweet Track across the Somerset Levels in southwest England. The study of the chitinous external skeletons of beetles found there helped researchers to revise concepts of the Neolithic climate in southern Great Britain. Some species of beetles living here at that time also exist now. The remaining species, including those that could not adapt to climate variations, have disappeared. In particular, a ground beetle *Oodes gracilis* now occurs only in those places of continental Europe where the winters are colder than in the Neolithic period by 2–3°C and the summers are warmer by 2–3°C, which suggests climate variation (Coles and Coles 1988). In addition, based on the remains of insects, one can judge where ancient humans **lived** and what kind of **food reserve** they kept for long periods (Lemdahl 1990; Hellqvist 2004).

Photo 1.2-5. Insects have some significance in the reconstruction of a paleoenvironment. Each species of insects lives within a specific range of climatic and physiographic conditions. These ranges vary for different species; however, in comparing them, one can always identify the relatively narrow interval where the fossil insects are found. Knowing their age, one can judge many environmental parameters existing there at that time. Thephoto shows an ant about 7 mm long in Baltic amber. Photo credit: By Anders L. Damgaard—www.amber-inclusions.dk—Baltic-amber-beetle [CC BY-SA 3.0 (http:// creativecommons.org/licenses/by-sa/3.0)], via Wikimedia Commons, 3 December 2010

Fossil insects also serve as one of the indicators of the age of **sedimentary rocks** and provide considerable help to geologists **searching for minerals**.

1.2.5.1 Use of fossil resin inclusions for paleoecological reconstructions

Resins are products of plants that have solidified in the air. After it falls to the ground, a drop of resin dries, acidifies, and, after several thousand years, nothing is left of it. In order to pass into the **fossil state**, hardened resin should be in water or be buried under the ground, usually in humid clay or sandy sedimentary rock (Lyukhin 2014). The remains of a living organism (most often, insects) are found as **inclusions** in the fossil resin (Photo 1.2-5).

Most fossil resins are known as **amber**. The legitimacy of the use of this term is frequently in doubt. There is a point of view that this term is applicable only for a highly polymerized variety of fossil resin widespread mainly on the southwestern coast of the Baltic Sea and in the Dnieper River basin. According to this point of view, all the other varieties should be referred to as fossil amber-like resins (Bogdasarov 2006). Within the expanded meaning, any fossilized resin of natural origin aged over 1 million years old can be called amber (http://amber-trade.com/mining/). We will use the term "amber" in this particular sense.

Many publications have considered the different questions related to amber. So, there are the widely differing hypotheses on amber production and formation of its deposits (Matsui 2010; Yushkin et al. 2006; Lyukhin 2014). In the 19th century, **six types** of amber had been identified (http://finesell.ru/vsjo-pro-jantarj/raznovidnosti-jantarja.html). Now, different sources suggest the existence of **more than 100** (Vavra 2009), or even more than 250 types of amber (Lyukhin 2014).

By now we know a vast number of amber deposits; they have been found on all continents except Antarctica (Map 1-1). However, from the standpoint of inclusions of insects in ambers, the four **most important regions** are as follows: (1) Baltic region (Kaliningrad Oblast of Russia, Lithuania, Poland, Germany, and Denmark); (2) Ukraine (Dnieper River basin); (3) Dominican Republic; and (4) India (Gujarat state).

The age of **Baltic amber** is about 40–54 million years (Coleman 2006). It was produced by the coniferous trees of the family Sciadopityaceae. It has been estimated that over 10^5 tons of Baltic amber were produced (Wolfe et al. 2009). Currently, it is extracted with high intensity. In the year 2000 alone, 510 tons of amber were extracted in this region; the amber contained, according to rough estimates, about 2 million inclusions (Clark 2010).

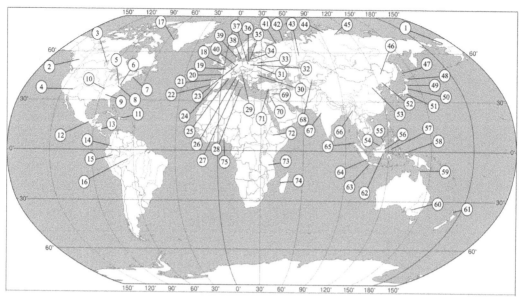

Map 1.1. Global distribution of amber locations. Prepared by author with use next sources: http://andy321.proboards. com/thread/46825; Ragazzi 2008; Bogdasarov 2008; http://academic.emporia.edu/abersusa/geograph.htm; http://schools-wikipedia.org/wp/a/Amber.htm; http://gems.minsoc.ru/eng/articles/amber; http://schools-wikipedia.org/wp/a/Amber.htm, https://www.mindat.org/show.php?id=188&ld=1#themap; http://www.brost.se/eng/education/find.html; http://academic. emporia.edu/abersusa/go340/students/miles/).

Legend: 1. Avalik, Alaska; 2. Grassy Lake, Alberta, Canada; 3. Cedar Lake, Manitoba, Canada; 4. Sheridan, Wyoming, USA; 5. Cape Fear River, North Carolina, USA; 6. Neuse River, North Carolina, USA; 7. Sayreville, New Jersey, USA; 8. Lee Creek Mine, Aurora, North Carolina, USA; 9. eastern Alabama, USA; 10. Acme Clay Pits, Malvern, Arkansas, USA; 11. numerous sites in Dominican Republic (La Bucara amber mine, La Toca amber mine, La Cumbre amber mine, Palo Quemado, Los Cacaos, Bayaguana area); 12. Chiapas, Mexico; 13. Caribbean sea-shore of Nicaragua; 14. Pena Blanca, Colombia; 15. Santiago river in far northern Peru; 16. Amazonas, Brazil; 17. Disko Bay, West Greenland; 18. Hastings, England; 19. Oise, France; 20. Charente-Maritime, France; 21. Alava, Spain; 22. Estoril parish, Portugal; 23. Bitterfeld, Germany; 24. Utrillas Teruel, Aragon, Spain; 25. Marseilles, France; 26. numerous sites in North-Eastern Italy (Redagno, Pietralba, Cortina d'Ampezzo, Val Badia, Ra Stua, Vernasso, Monte di Malo, Bolca, Salcedo, Sedico, Riva del Garda); 27. Northern Apennines, Italy; 28. Sicily, Italy; 29. Vogelberg, Salzburg, Austria; 30. Colti, Buzau, Romania; 31. Klesovo-Dubrovitskoe deposit, Rivne region, Ukraine; 32. Konin, Poland; 33. Polesye, south of Belarus; 34. Lithuania; 35. Jantarny (Palmnicken), Kaliningrad, Russia; 36. Gdansk, Poland; 37. Belchatow coal mine, Lodz Voivodeship, Poland; 38. Falsterbo peninsula, Sweden; 39. Denmark; 40. Hannover, Germany; 41. Mouth of Mezen River; 42. Mouth of Pechora River; 43. Kamensky District, Sverdlovsk Oblast; 44. Sea coast between Pyasina River and Khantanga River; 45. Mouth of Khantanga River, Russia; 46. Nehe, Heilongjiang, China; 47. Dolinsk, Sakhalin, Russia; 48. Monbetsu, Hokkaido, Japan; 49. Kuji, Iwate prefecture, Japan; 50. Chōshi, Chiba prefecture, Japan; 51. Mizunami, Gifu Prefecture and Kizu, Kobe, Japan; 52. Fushun, province Liaoning, China; 53. Nanyang, province Henan, China; 54. numerous sites (Merit-Pila Coal Mine, Daro, Kuala Matu beach) in Sarawak, Borneo; 55. Pinangah, state Sabah, island of Borneo, Malaysia; 56. Tarakan Basin, Tarakan, Indonesia; 57. Sulawesi, Indonesia; 58. Papua, Indonesia; 59. Cape York, Queensland, Australia; 60. Allendale, Creswick, Hepburn Shire, Victoria, Australia; 61. Kauri, Northland, New Zealand; 62. Banjarmasin, Kalimantan, Indonesia; 63. Jampang, Java, Indonesia; 64. Lampung, Sumatra, Indonesia; 65. Banda Aceh, Sumatra, Indonesia; 66. Tanai amber mine, Hukawng Valley, Kachin State, Myanmar (Burma); 67. Cambay, north-west coast of India, Gujarat Province; 68. Southern Kazakhstan Province (Yuzhno-Kazakhstanskaya Oblast), Kazakhstan; 69. Wadi Zerka north of Amman, Jordan; 70. numerous sites in Lebanon and Israel (Jezzine, Hammana, western slopes of Mount Lebanon, Golan Heights); 71. South-Eastern Anatolia Region, Turkey; 73. Amhara Province, Ethiopia; 74. Tanzania; 75. western coastal area of Madagascar; 76. Nigeria

The remaining deposits of amber are not nearly as large as those in the Baltic. By the early 1980s, about 3,000 species of arthropods had been identified thanks to the Baltic amber (Sredobolsky 1984). Thirty years later, their number had reached 3,500 species (Penney and Preziosi 2014). However, discovery of new species in inclusions continues with considerable intensity (Tolkanitz and Perkovsky 2015; Bukejs and Alekseev 2015; Soszynska-Maj and Krzeminski 2015).

The age of **Ukrainian (Rivne) amber** is about the same as that of the Baltic deposits. Early on, it was thought that deposits containing amber were transported from the Baltic Sea coast to the territory of the present Ukraine by a glacier (Srebrodolsky 1984). However, this assumption has not been proved. Later on, many insects absent in the Baltic amber were found in inclusions in the Ukrainian amber. For example, 9 of 37 species of ants identified in Ukrainian amber were not found in Baltic amber (Perkovsky et al. 2007). The rates of extraction of the Ukrainian (Rivne) amber are much lower than those for the Baltic. Over 16 years (1993–2009), 32,025 kg of amber were extracted there (Perkovsky et al. 2010).

Dominican amber was produced by the extinct trees *Hymenaea protera* (family Fabaceae) rather than the coniferous plants (https://ru.wikipedia.org/wiki/Hymenaea_protera). Its age is estimated at 15–40 million years (https://en.wikipedia.org/wiki/Dominican_amber). In the Dominican amber, as compared to the Baltic amber, inclusions of insects are found many times over: the proportion of stones with inclusions is 50–60%, while that for Baltic amber is not more than 6%. The inclusions in Dominican amber number 505 species of insects (Manukyan and Vaishat 2010). New species are also discovered quite regularly (Poinar and Legalov 2015; Poinar 2015; Peris et al. 2015).

Indian amber was formed by resin of trees belonging to the Dipterocarpaceae family. Its age is 50–52 million years, and the amber is extracted from the Cambay deposit in western India. This deposit is fairly ordinary, and only 150 kg of amber has been extracted from it (Rust et al. 2010). However, its major advantage is the high solubility in different solvents (toluene, xylene, dichloromethane, and chloroform; Mazur et al. 2014). This characteristic simplifies the identification of insects. In total, 700 arthropods assigned to 14 orders and more than 55 families and 100 species of arthropod inclusions have been discovered in the Cambay amber (Rust et al. 2010).

In most cases, in amber from all of the regions, inclusions of insects have been confined to the multilayered resin sinters. The lower layer to which an insect was stuck is more viscous than the upper layer. The insect touching it cost it its life. The following burial was carried out as a result of smooth bedding of resin portions forming the upper sinter rather than by way of its gradual bogging down in the resin (Manukyan 2011). It usually consists of many thin laminas. In most cases, the viscosity of the upper resin sinter was so insignificant that it could not deform even the softest tissues of organism, as is evidenced by the integrity of the finest hairs, fibrils in wings, open-worked coloration of eyes, and other small structural details. For the most part, the entomofauna of inclusions consists of small forest insects.

The **age** of ambers found in different regions varies from less than one million to 125–130 million years. Inclusions of insects of the Neogene (Dominican amber) and Paleogene (Baltic amber) ages have been explored to the greatest extent. Therefore, analysis of the samples of arthropods contained in the Baltic amber demonstrated that the dipterans amount to more than half of all insects. Among them, Nematocera, combining mosquitoes and other mosquito-shaped dipterans, predominate.

One of the **most numerous groups** of amber arthropods includes the spiders, representing 41 families. The pseudoscorpions were represented by 12 genera and 9 families, while harvestmen belonging to 8 families were observed. In Baltic amber, 13 genera of cockroaches and 51 species of butterflies were observed (Sredobolsky 1984).

An analysis of entomofauna allowed researchers to reconstruct the **climatic characteristics** of that epoch. The climate in northern Europe was humid and warm with clear differentiation into the dry and wet periods of the year. Generally, it resembled the present-day climate of southern Europe and the subtropics. The average annual temperature did not fall below 18°C (Sredobolsky 1984). Based on investigations of insect frass in Baltic amber, Nuorteva and Kinnunen (2008) concluded that the succiniferous forests grew in the area of present-day southern Finland.

Considerably fewer paleoecological reconstructions have been based on investigations of inclusions in ambers of the Cretaceous period (65–135 million years ago). Resins from the Cretaceous period containing insects have been discovered in Lebanon (Choufani et al. 2015), Morocco (Engel et al. 2015), France (Perrichot 2015), Canada (McKellar and Engel 2014), Burma (Myanmar; Caterino et al. 2015), the United States (New Jersey), Russia (Taimyr Peninsula and Yakutia), Transcaucasia, and other locations. However, the number of such findings is much lower, and they have been explored to a far lesser extent than inclusions in the Dominican and Baltic ambers. Therefore, findings in Ethiopia of amber aged 93–95 million years with perfectly preserved fossil remains of arthropods aroused much interest (Schmidt et al. 2010).

In aleuritic sediments of the Debre Libanos Sandstone Unit in the northwest part of the Ethiopian upland, several hundred ambers were discovered, and 62 of them were subjected to detailed study. Generally, they had sizes of 1–5 cm, while the largest of them reached about 25 cm in diameter. These ambers registered a small fragment of a **complete paleoecosystem** with remains of insects, plants, bacteria, and fungi. The chemical composition of amber looks much like the composition of the resin of metasperms—trees of the pea family (genus *Hymenaea*).

In 9 tested pieces of amber, 30 arthropod individuals were found, which belonged to the arachnids (mites and spiders) and 12 families of hexapods (Hexapoda), belonging to collembolans (Collembola), book lice (Psocoptera), bugs (Hemiptera), thrips (Thysanoptera), zorapterans (Zoraptera), butterflies (Lepidoptera), beetles (Coleoptera), flies (Diptera), and hymenopterans (Hymenoptera), etc. (Schmidt et al. 2010). An analysis of entomofauna that was found allowed researchers **to reconstruct** in great measure the ecosystems existing on the territory of the present-day Ethiopia in those times.

1.2.6 Forensic entomology

Forensic entomology is a science and type of practical activity related to the application of data on insects and other arthropods in forensic science, mainly for criminal trials. In addition, it is used in civil actions related to insurance and inheritances.

1.2.6.1 History and present state

The **earliest documentation** of a criminal investigation that used insects is the book *Washing Away of Wrongs*, published in 1247. It was written by the judicial intendant Sung Tz'u, who lived in China in AD 1188–1251. The book describes the following case, which happened in 1235. During harvesting operations, a Chinese farm worker was slaughtered. The investigation showed that the wounds were given by a sickle used for harvesting rice, so suspicion fell on the farm workers using such sickles in the field.

The local authorities convened the farmers on the village square with their sickles and told them to put their sickles on the ground and get out of the way. After a while, flies began to gather to one of the sickles. Its detailed examination demonstrated the presence of traces of blood and tissues, which was evidence that this particular sickle was the assassination weapon. The owner of the sickle acknowledged the crime (Haskell 2006).

The year 1668, when the work of Italian physician and naturalist Francesco Redi (1626–1697) was published, is considered to be the **next milestone** in forensic entomology. He proved that the "meat worms" present in molding meat were maggots hatched from eggs laid by flies rather than the result of spontaneous generation. In an experiment, he used fabric to cover pots containing rotten meat; the fabric prevented flies from laying eggs in the meat (https://en.wikipedia.org/wiki/Francesco_Redi).

In the 18th through 19th centuries, mass exhumation of burials from previous wars were carried out in France and Germany. In the course of these operations, different degrees of corruption of bodies were found, suggesting the bodies were buried at different times. The French physicians M.J.B. Orfila and C.A. Lesueur, examining the exhumed bodies, concluded that maggots played a major role in their disintegration. Since 1850, medicolegists began using insects for determine the time of burial of the corpses (Benecke 2001).

In that year, one more **landmark event** in forensic entomology happened. The French investigator Dr. Marcel Bergeret was performing repair work in a boarding house near Paris, in 1850, when he found a mummified body of a newborn child behind the heating flue of a fireplace. When he exhumed the body, Dr. Bergeret discovered larvae of the common flesh fly (*Sarcophaga carnaria*) and some moths. He ascertained that the body was occupied by flies in the summer of 1848, while the moths had gained access in 1849. These finding implicated the previous tenants in the murder and cleared the present residents of the crime (Gennard 2007; Sathe et al. 2013).

The **next major figure** in this field was an army veterinarian and entomologist, Jean Pierre Megnin (1828–1905). His fundamental books were entitled, *Fauna of the Tombs* (*Faune des Tombeaux*, 1887) and *Fauna of Corpses* (*La Faune des Cadavres: Application de l'Entomologie*, 1894). He generalized the experience of 15 years of medicolegal experience with corpses concerning the development of necrophagous insects. In particular, he identified **eight stages** of human decomposition related to the succession of insects colonizing the body after death. These books are the founding works of modern forensic entomology (Benecke 2001).

Beginning in the 1980s, forensic entomology began to use different modern biochemical and genetic methods (gene engineering, DNA analysis, etc.). In analyzing the gut contents of particular insects, determining the species of the object of blood sucking or eating, and its genetic identification blood samples of humans have appeared. Similar investigations have been carried out for a **wide spectrum of arthropods**: mosquitoes, blackflies, tsetse flies, and mites (Chaika 2003).

At the present time, forensic entomology is the fastest growing scientific field having practical application. Studies in this field are coordinated by the North American Forensic Entomology Association (http://www.nafea.net/) and the European Association for Forensic Entomology (http://www.eafe.org/). Among the **countries** with the greatest development in this field are Canada, the United States, Belgium, Germany, and France. However, numerous countries study different aspects of forensic entomology: Argentina (Mariani et al. 2014), Brazil (Barros-Souza et al. 2008, 2012), the Netherlands (Krikken and Huijbregts 2001), Iran (Tüzün et al. 2010), and others.

International conferences on forensic entomology are held regularly. Several **journals** publish articles on these themes on a regular basis. Among them are the *Journal of Forensic Sciences*, *Journal of Forensic Dental Sciences*, *Journal of Forensic Medicine and Toxicology*, *Journal of Medical Entomology*, *Canadian Society of Forensic Science Journal*, *Forensic Science International*, and *The American Journal of Forensic Medicine and Pathology*. A number of standards and practical **guidelines** have been developed (Smith 1986; Marchenko and Kononenko 1991; Amendt et al. 2007).

1.2.6.2 Insects of forensic importance

There are different classifications of the entomofauna of corpses (Marchenko 1991; Smith 1986) According to the classification of Smith (1986), it is divided into **four groups**:

1. Necrophagous species feeding on the carrion (Diptera: Calliphoridae; Coleoptera: Silphidae (partially), Dermestidae).
2. Predators and parasites feeding on the necrophagous species (Coleoptera: Silphidae (partially), Staphylinidae; Diptera (larvae [maggots] of some necrophagous species are predators at the late stages of development, for example, *Chrysomya* [Calliphoridae], *Hydrotaea* [Muscidae]).
3. Omnivorous species (wasps, ants, and some beetles feed both on the corpse and around it).
4. Accidental species (springtails, spiders).

Representatives of the first two groups are the **most important** for forensic entomology. Predominantly, these are representatives of the orders Diptera (flies) and Coleoptera (beetles) (Joseph et al. 2011). The Diptera (flies) first discover and occupy a corpse, and they are the primary group of insects causing its corruption and feeding on it. They are first because of the high level of development of their olfactory system and organs of sight, as well as their fast and agile flight. Beetles occupy a corpse well after the flies (Chaika 2003).

In total, the **fauna of corpses** includes 522 species of organisms belonging to three phyla, nine classes, 31 orders, 151 families, and 359 genera. Four arthropod orders (Coleoptera, Diptera, Hymenoptera, and Araneida) account for 78% of the species, while, two families of beetles (hister beetles [Histeridae] and rove beetles [Staphylinidae]) and three families of flies (flesh flies [Sarcophagidae], blowflies [Calliphoridae], and houseflies [Muscidae]) account for 26% of the total fauna (http://molbiol.ru/forums/index.php?act=Attach&type=post&id=71555).

The entomofauna, both in a corpse and around it, changes in accordance with the **degree of decomposition** (fresh, inflated, corrupted, and dry). The flesh flies and blowflies can fly to a corpse within a few minutes after death, while the house-flies colonize the corpse until the body reaches the bloat stage of decomposition (Joseph et al. 2011). Beetles eliminate the dry corpse: bacon beetles, rove beetles, hister beetles, and carrion beetles. The coats of animals, hairs, wool, horns, and feathers are also eaten by the larvae of moths (Chaika 2003). A detailed description of the **occupation sequence** of a loose-lying corpse by groups of insects is given in a monograph by K.G.V. Smith (1986).

Ants (Formicidae) are found in corpses at all stages of decomposition. Some species of ants detect a corpse during the first 3 to 6 hours. For example, in the forests and meadows of Karelia (Russia), the red wood ant (*Formica polyctena*), common black ant (*Lasius niger*), and red myrmicine ant (*Myrmica rubra*) are the first insects to find corpses. The early emergence of the ants in a corpse can markedly affect the disposition process. As the first colonizers, they destroy a large number of eggs and larvae of necrobionts. This destruction increases the term of the corpse corruption, so the determination of death time on the basis of blowflies developing in a corpse can be erroneous (Liabzina 2011b).

Using the **species composition** of ants, one can sometimes determine the biotope and time of year when an individual died. The time of death is known to have been determined on the basis of the development of colonies of yellow crazy ants (*Anoplolepis gracilipes*). Sometimes, ants can contaminate material that could have been of value as evidence. For example, there are known cases when extractions of the Argentine ant (*Linepithema humile*; formerly *Iridomyrmex humilis*) resulted in the false definition of human blood collected from the body of a sacrifice (Chaika 2003).

Mites are frequently found in corpses. In the literature, 212 mite species associated with carcasses have been described (Perotti and Braig 2009; Perotti et al. 2009). Many mites are accidentally delivered to corpses by flies or beetles (the phoresy phenomenon). However, mites are often indicators of the time elapsed from the moment of death. Detailed information concerning the use of different species of mites to determine the postmortem interval is given in a work by Chaika (2003).

Fleas and **lice** are of an insignificant interest to judicial entomologists because they leave corpses as soon as their temperature begins to drop. Among scorpions and spiders, there are no specialized inhabitants of corpses and, therefore, their importance in forensic examination is also insignificant. In corpses found in water, the larvae of some species of midges, mayflies, caddis flies settle most often (Merritt and Wallace 2009). Aquatic insects are often critical to forensic entomology, which will be discussed in the next section.

1.2.6.3 Solving forensic problems

Entomological investigations allow workers to solve the following **problems** (Marchenko and Kononenko 1991; Sathe et al. 2013): (1) determination of the time of death; (2) determination of how long a corpse has been present in the place where it was found; (3) determination of whether a corpse has been moved; (4) determination of the initial location of a corpse and the route of its movement; (5) determination of environmental conditions under which the death happened; (6) detection of the potential cause of death in the case of suicide; and (7) determination of the place of origin of narcotic plant raw material.

Determination of the time of death is the major task of forensic entomology. This is accomplished based on the rate of necrobiotic insect development, which is determined from a detailed examination of the macro- and microfauna of the corpse and its environment. When the postmortem interval (i.e., time death and detection of the corpse) is short, blowflies are most often used. Different **stages** of their development on the corpse clearly show its age: less than 4 hours, there are no eggs and larvae; from 4–6 to 10–12 hours, eggs are laid; in 24 hours, the larvae emerge; during 1 week, their sizes increase; and, in 2 weeks, pupation begins (Sang Eon Shin et al. 2015).

Different **ages** of corpses and their locations and conditions (at the surface, buried belowground, submerged, mummified, burned, etc.), show their own sets of insects; there is a large number of corpses. All of them are in different conditions, so there is a large combinations of conditions in which different insects living on them. So, there are own sets of insects. These questions are extensive and have been studied in detail. Many publications consider such things as ways of determining the postmortem interval, possible errors in its determination, procedures for collecting insects, and necessary equipment (Marchenko and Kononenko 1991; Joseph et al. 2011; *Forensic Entomology* 2007; Singh et al. 2008; Gennard 2007; Krikken and Huijbregts 2001, etc.).

Determination of how long a corpse has been present where it is found is also a common task of forensic entomology. For example, knowledge of the species and age of midge larvae collected from a corpse can provide useful information on how long it has been present in water. Similar information can be acquired by analyzing the quantity of periphyton and number of larvae of mayflies in a corpse (Merritt and Wallace 2009).

Determination of whether a corpse has been moved. Forensic entomology often allows one to establish the true location of a person's death. Different insects have their own habitats, and their presence in corpses beyond their limits suggests a corpse has been moved. For example, the brown blowfly (*Calliphora vicina*) is a synanthropic species. If its larvae are found in a corpse discovered away from a populated area, then this can suggest that death was in the town and then the corpse was moved (Chaika 2003). In some cases, establishing the place of death is possible with the use of other arthropods. For example, using cases (indusia) of caddis flies found in a corpse, one can determine if the corpse has been moved from the specific habitat of larvae (Merritt and Wallace 2009).

Determination of the initial location of a corpse and the route of its movement. Intact insects or insect fragments discovered in any means of transport can provide useful information on the path of its travel, provided it passes through areas with specific entomofauna. Small insects often adhere to the front of a vehicle, including the headlights, because they are attracted by the light. Some insects—for example, dipterans (midges)—can even lay eggs on a vehicle, mistaking its dew-drenched glossy surface for the surface of a water body.

Different small arthropods, including soil dwellers, can be present on shoes. Discovery of larvae or chrysalids of necrophilous flies in automobile car suggests that it could be used to transport a corpse. If the eggs of blowflies are found on a drowned man, then one can assume that the deceased was killed elsewhere, days prior to its discovery, and then later transported and drowned to suppress traces of the crime.

Determination of environmental conditions under which the death happened. Different species of insects prefer different environmental conditions (illumination intensity, temperature, humidity, etc.). Analyzing the character of the entomofauna of corpse can sometimes lead to conclusions in this regard. The places eggs are laid by light-loving and shade-loving species of flies can serve as an example. Some species of blowflies (for example, green blowflies [*Lucilia*]) are heliophilic and lay eggs on surfaces heated by the sun, while blue bottle flies [*Calliphora*] prefer shaded surfaces; Krikken and Huijbregts 2001; Chaika 2003).

Detection of the potential cause of death in the case of suicide is also possible through analysis of corpse entomofauna. In cases of intoxication with poisons or narcotics, they enter the blood, urine, intestinal tract, hair, and nails of humans. However, for a decomposed corpse, it is impossible to take tissue samples and to detect the existence of poisons. Therefore, analysis of larvae or pupae of flies feeding on a corpse is of considerable importance, as traces of poisons can be preserved in them. In this way, one can detect heroin, cocaine, malation, mercury, triazolam, oxazepam, chloropyraminum, phenobarbital, and other substances (http://molbiol.ru/forums/index.php?act=Attach&type=post&id=71555).

Determination of the place of origin of narcotic plant raw material. When narcotic plants are confiscated, it is desirable to determine the ways of their illegal transport. For this purpose, their habitat must be ascertained. Insects are widely found in narcotic plants. If they belong to prevailing species, then the benefit of their discovery is not great. However, if endemic species are found among these insects, then the origin of the narcotic plant origin may be identified. For example, 272 species of insects are connected with common hemp (*Cannabis sativa*).

Dried flowers, seeds, and leaves of this plant are called marijuana. A related form of narcotic (hashish) is produced from the laticifer of hemp. Once, a lot of marijuana was confiscated in New Zealand. An analysis of insects found in it showed the presence of species endemic to the Tanintharyi Region (formerly Tenasserim), Myanmar. These findings allowed investigators to determine where the raw material originated (Chaika 2003).

Numerous particular examples of criminal investigation using the methods of forensic entomology are given in publications by Smith (1986), Krikken and Huijbregts (2001), Gennard (2007), and Sathe et al. (2013).

Use of Insects in Medicine and the Cosmetics Industry

The medical use of insects and insect-derived products is known as **entomotherapy**. Although entomotherapy has been practiced for many millennia, it is still little known in the scientific world. The insects used in medicine can be live, cooked, ground, in infusions, in plasters, and as ointments; they are used in both curative and preventive medicine.

Although the use of insects in modern medicine is fairly uncommon, their applications are quite diverse. An analysis of 371 species of medicinal insects was conducted, and they were found to be used for treatment of disorders in the **following human systems**: digestive (58 species); skin (57); respiratory (34); reproductive (31); circulatory (31); nervous (28); kidney (22); eyes (21); bones (19); neuromuscular (13); immunological (6); hearing (4); endocrine (4); and others (43) (Costa-Neto 2005).

Among other applications, the red dye carmine, derived from the cochineal, is one of the most heat-resistant natural dyes, resisting oxidation and discoloration and is far more stable than many synthetic food dyes. In medical-biological research, it is used to stain histological material (Vtorova 2012). In addition, the insects can be used to identify plants with pharmaceutically active compounds (Helson et al. 2009).

1.3.1 Use of insects in folk medicine

Folk medicine is a complex of empirical data accumulated by people concerning therapeutic agents as well as therapeutic and hygienic skills for diagnosis, prevention, and treatment of diseases. People have always experienced **wounds** and **diseases**, so folk medicine exists just as long as humankind itself exists. Medical treatment was always in demand.

People attempted to apply different plants to the wounds and to use plant-based teas and other recipes in cases of discomfort. Of course, these attempts and experiments were not always successful, but they provided knowledge and experience. Gradually, useful recipes were accumulated that were transferred from generation to generation. Naturally, there were holders of all this knowledge; they were called **different names** in different places: healers, powwow, voodoo, charlatans, quacksalvers, fetishers, peai, saltimbanco, witch doctors, shamans, bonesetters, magicians, etc.

It is hard to say when and how insects were first used in medicine. The Chinese have used insects in medicine for almost 3,000 years (Feng et al. 2009). Up to a point, medicine was integrated and there was no concept of folk medicine. Folk medicine **was distinguished** from conventional medicine in the later part of the 18th century (https://ru.wikipedia.org/wiki/Народная_медицина).

The earliest of the known **books** of standards regulating the quality of medicinal agents (pharmacopeia) was written in ancient **China** about 2,000 years ago. The book was called *Sheng nongbencaojing* (Sheng

Nong's Herbal Classic) and contained the recipes for more than 20 medicinal drugs made of eight species of insects. Later, more than 50 similar reference books were issued. Among them was the monumental trait *Bencaoganmu* (Compendium of Materia Medica). It was written in 1578 by the preeminent Chinese physician and pharmacologist Li Shizhen and contained the recipes for more than 1,800 pharmaceutical drugs (https://en.wikipedia.org/wiki/Li_Shizhen), including **73 drugs** derived from 60 species of insects (Feng et al. 2009).

Today, folk medicine is used to different degrees in different parts of the world. It is **most popular** in China, India, some regions of Africa, and Central and South America. For example, in China there are now about 300 species of **medicinal insects** belonging to 70 genera, 63 families, and 14 orders. About 1,700 Chinese medical prescriptions include insects or insect-derived crude drugs. Diverse **parts** of the insects are used: total insects' bodies, insect exuvia, eggs and egg shells, secretions of insects, etc. (Feng et al. 2009).

For example, the extract from Chinese black mountain ants, *Polyrhachis vicina*, is highly valued in Tibetan and Chinese medicine as an agent of prolongation of life and promotion of sexual vigor (https://en.wikipedia.org/wiki/Ethnoentomology). Silkworm larvae are used for treating impotence (Ahn et al. 2008).

In **India**, the larvae of leaf-miner flies (Agromyzidae), feeding mainly at jatropha (bushes and trees of the genus *Jatropha*, milkweed family), are boiled and ground into a paste and then ingested to induce lactation and to soothe the gastrointestinal tract (Srivastava et al. 2009). Also, this paste is used for reducing fever (http://www.livingrichlyonabudget.com/insects-used-as-medicine-around-the-world). Water striders (Hemiptera, Gerridae) are used in the rural locality of Bagbahera (Chhattisgarh State, central India) for treating dog bites (Srivastava et al. 2009).

In **Africa**, some species of grasshoppers are collected, dried in the sun, and then crushed into a powder that is mixed with water and ash; this paste is applied to the forehead to relieve headaches (https://en.wikipedia.org/wiki/Ethnoentomology). On the coast of Lake Victoria, *Chironomus* midges are crushed to powder and given to sick and weak children. In the Democratic Republic of the Congo, some true bugs (Hemipterans) are used to calm insanity. The bugs are mixed with mud of the river by the same name, and the ill person is instructed to eat this mixture (Macadam and Stockan 2015).

The practice of using insects for medical purposes is also characteristic of **South America**. For example, at least 42 species of insects are used in the folk medicine of Bahia, in northeastern Brazil. These insects are from 9 different orders and the Hymenopterans, including 22 species that are used the most (Costa-Neto 2002).

In **Central America**, the silkworm (*Bombyx mori*), imported there by the Spanish and Portuguese, is a popular insect used for medicinal purposes. The cooked pupae are used for the treatment of apoplexy, bronchitis, pneumonia, convulsions, hemorrhages, frequent urination, and aphthae. In addition, the dung produced by the larvae is eaten to improve circulation and alleviate the symptoms of cholera (intense vomiting and diarrhea) (Ramos-Elorduy de Concini and Pino Moreno 1988).

In some parts of **Mexico**, chapulines (grasshoppers of the genus *Sphenarium*) serve as a diuretic agent for treatment of kidney diseases. Eating them reduces edema and alleviates peptic disorders. However, there is a risk of infection of the patient with nematodes, which these grasshoppers often harbor. Also in Mexico, red harvester ants (*Pogonomyrmex barbatus*) are used for treatment of rheumatism, arthritis, and poliomyelitis through the immunological responses to their stings (https://en.wikipedia.org/wiki/Ethnoentomology).

Insects are not used not only as components of medicinal agents. At times, they help **to diagnose** a case. For example, the Amazonian Indians detect diabetes by watching to see whether ants swarm over urine, which in diabetics contains high levels of sugar that attract the ants (Ratcliffe et al. 2014). Sometimes, insects are applied as a **medical device**. If the subcutaneous administration of a drug is necessary, it is laid on the patient's skin and then a termite is agitated and placed on the skin of the patient. In cases of termite stings, their mandibles are the effective injection device (Srivastava et al. 2009).

Some indigenous peoples of Central America (in particular, the Mayans) use the soldier cast of the army ant for **suturing wounds**. Usually, an agitated ant is placed into the affected spot, its mandibles are held up to the wound edges. Then it is pressed, and the ant stings the patient. Then the head with pressed wound edges is left, while the thorax and abdomen are torn off. Furthermore, the salivary gland secretions of ants have antibiotic properties (https://en.wikipedia.org/wiki/Ethnoentomology). In a like manner, bulldog ants (*Myrmecia* sp.) are used by the healers of New Guinea (Dunayev 1997).

Sometimes, the **dwellings of insects** are used for medicinal purposes. For example, in some areas of India and Africa, the material composing termite mounds is cooked and ground to a paste. Then this paste is applied on a wound to prevent the introduction of infection; it is also used to treat internal hemorrhages (Srivastava et al. 2009). This remedy is also prescribed for treatment of ulcers, rheumatic diseases, and anemia (Chakravorty et al. 2011).

One application of live insects for medicinal purposes in the treatment of syphilis, is no longer practiced because new, more acceptable treatment modalities have emerged. The causative organism, a spirochete bacterium (*Treponema pallidum*), can reproduce only at 36.8–37.2°C. Increases in temperature quickly results in its death. In order to increase the body temperature to 39–40°C, patients were specially infected with other diseases; for example, malaria. The high fever resulting from this infection hampered the progression of syphilis and sometimes resulted in the complete destruction of *Treponema pallidum* (Irwin and Kampmeier 2003).

It is possible to see that the use of insects in folk medicine was and remains **fairly extensive**. The problem is that, together with really effective remedies, some treatments are totally unacceptable. So far, only a few of the remedies applied in folk medicine have passed extensive clinical trials to prove their efficacy. However, folk medicine is still one of the most important sources of new natural medicinal treatments.

1.3.2   Creation of new medicines

Insects are an important source of medicinal agents for modern medicine, as they possess immunological, anesthetic, bactericidal, diuretic, and antiarthritic properties. Chemical analysis of 14 species of insects revealed the presence of proteins, terpenoids, sugars, polyols (high-molecular-weight alcohols) and mucilages, saponins, polyphenolic glycosides, quinones, anthraquinone glycosides, cyanogenic glycosides, and alkaloids (Costa-Neto 2002). Below, we list examples of actual and potential medicines based on insects.

The **blowflies** (Calliphoridae) are transmitters of agents of some infectious diseases. However, they can also be a very useful insect from a medical point of view. From the hemolymph of maggots of a dipteran (*Calliphora vicina*), a protein called **alloferon** was isolated that has been successfully applied to treat herpes infections and viral hepatitis B (Chernysh and Gordia 2011; Chernysh et al. 2002). The **allostatins**—another group of peptides from the maggots of the blowfly (*C. vicina*)—could possibly be used in cancer immunotherapy (Chernysh and Kozuharova 2013).

It is believed that humankind is now experiencing the **end of the era of antibiotics**. Disease agents are now evolving dramatic resistance to antibiotics, resulting in the return of diseases such as tuberculosis, which was almost eradicated, even in industrialized countries (Bode 2009). In 2008, a gene that codes for the NDM-1 (New Delhi metallo-beta-lactamase) enzyme was discovered, which makes bacteria resistant to the overwhelming majority of antibiotics (https://en.wikipedia.org/wiki/New_Delhi_metallo-beta-lactamase_1). It was described in detail by Yong et al. (2009). This gene is able to convert common pathogenic bacteria into superbugs that can give rise to diseases of the genitourinary system and abdominal cavity as well as pneumonia, sepsis, meningitis, and other conditions.

If the effectiveness of modern antibiotics is impaired, then another way to control bacteria is the improvement of **human resistance** to them. It is hoped that in developing medicines to combat superbugs, the same old brown blowfly (*Calliphora vicina*) will help. These flies inhabit a very infective environment. In response to infection, organs are activated that discharge to their blood, in one stroke, dozens of substances with antibacterial activity. As a result, cocktails of these substances block bacterial development in that environment. In a paper by Chernysh et al. (2015), a new approach to solving the key problem of modern medicine—overcoming the resistance of pathogenic bacteria to antibiotics—was suggested. It is based on the use of natural complexes of antimicrobial peptides of blowflies (Calliphoridae).

The simplest and most **common way** of administering medicinal agents is peroral (by mouth). However, **transdermal** drug delivery using patches is sometimes more expedient. It decreases the number of side effects and relieves the gastric irritation and drug metabolism. Tsaia and Chang (2013) manufactured a biologically inspired microcapillary patch that is self-adhesive and reusable, and can sustain a controlled drug release. The patch demonstrates strong adhesive characteristics, but it is easily removed and its different antiinflammatory drug unguents can be released calculably and regularly for several days.

Sericin is a protein forming part of silk released by silkworm caterpillars (*Bombyx mori*) as well as other arthropods (caddis flies, spiders, etc.). Long time ago, it was considered to be a waste material of silk production. Every year, up to 50,000 tons of sericin were discharged with wastewaters into the global environment. Today, it is seen under a different angle. Over the last few years, its numerous positive features (antioxidative, antibacterial, anti-coagulative, and promotion of cell growth and differentiation, etc.) have been revealed (Wang et al. 2014). Its moderate antiprotozoal properties have also been registered (Macadam and Stockan 2015).

It has long been known that the salivary glands of blood-sucking insects contain **anticoagulants**. For example, extracts of the salivary glands of horseflies have been applied in Oriental medicine through the ages to prevent blood coagulation. Today, the blood-sucking insects such as true bugs (Hemiptera), flies and mosquitoes (Diptera), fleas (Siphonaptera), and lice (Anoplura) and ticks are recognized to have a **huge potential** for creating new anticoagulants (Ratcliffe et al. 2011). The mechanism of anticoagulant formation in the salivary glands of *Anopheles* mosquitoes has been studied in detail (Figueiredo et al. 2012).

One more promising component for the creation of new drugs is **cantharidin**. It is a poison without color and odor contained in the hemolymph of representatives of blister beetles (Meloidae) (Photo 1.3-1). They are widely present in the genus Epicauta. Insects producing cantharidin also include the Spanish fly (*Lytta vesicatoria*), false blister beetles (Oedemeridae), cardinal beetles (Pyrochroidae), and soldier beetles (Cantharidae, or Cantharididae). The bodies of these insects contain up to 5% cantharidin (https://en.wikipedia.org/wiki/Cantharidin).

Preparations containing cantharidin (in general, powders, tinctures, and ointments made of Spanish fly, *Lytta vesicatoria*), were widely **used from ancient times** till the 20th century as a means of strengthening sexual vigor. They were also applied for wart removal. However, all of the preparations containing cantharidin are on the list of "problem drugs," because, in case of overtreatment, they adversely affect the liver, kidneys, central nervous system, and digestive tract. A drug based on cantharidin was created for treatment for warts and other skin problems and was approved by the U.S. Food and Drug Administration in 2004.

Photo 1.3-1. Cantharidin is a promising component for the creation of new drugs. It is a poison without color and odor contained in the hemolymph of representatives of blister beetles (Meloidae). Blister beetle *Hycleus lugens* in Tanzania is shown.
Photo credit: By Muhammad Mahdi Karim (Own work) [GFDL 1.2 (http://www.gnu.org/licenses/old-licenses/fdl-1.2.html)], via Wikimedia Commons, 2010

1.3.3 Wound healing

There are **two ways** to heal wounds by using insects: (1) maggot therapy; and (2) healing with honey.

A method of **maggot therapy** is based on feeding of maggots of blowflies (Calliphoridae) on necrotic tissues. The maggots are placed on a wound and left there for some time, and the maggots consume all of the necrotic tissues, leaving the wound clean; they secrete digestive juice that digests the necrotic cells (Photo 1.3-2). **Most often**, the larvae of the common green bottle blowfly (*Lucilia sericata*) are used for this purpose, although the larvae of such species as the Australian sheep blowfly (*Lucilia cuprina*), black blowfly (*Phormia regina*), and brown blowfly (*Calliphora vicina*) are also used (Sherman 2014).

Larvae were used for **healing wounds** thousands of years ago by the Maya Indians, Australian Aborigines, and ancient Burmese (Ratcliffe et al. 2014). In human history, there have been several periods when this method was newly disclosed and began to be actively applied, and then periods of unawareness followed.

Western literature records the use of maggots as long ago as the **16th century** in France (Jurga and Morrison 2009). The French physician Ambroise Paré, who was the court surgeon of four French kings, focussed on the remarkable effects of fly larvae on wound healing (Whitaker et al. 2007). Afterwards, the phenomenon of debridement with larvae was observed during Napoleon's Egyptian campaign in Syria, 1798–1801, by the army's general surgeon Dominique-Jean Larrey. Later, this mode of treatment was actively applied during the Napoleonic Wars (1799–1815) and American Civil War (1861–1865) (Sherman 2014).

Several decades later, during World War I, this phenomenon was **again** observed. Two seriously wounded soldiers of the German army were found seven days after a battle, and wounds of each of them were infested with maggots of blowflies. After the wounds were cleaned, they turned out to be in such good condition that this fact engaged the attention of surgeons, especially since such wounds often ended with death (Kruglikova and Chernysh 2013).

Between World War I and World War II, maggot therapy received especially **wide acceptance**; however, after the discovery of **antibiotics** in the mid-1940s, it lost popularity again. Nevertheless, increased resistance of pathogens to antibiotics in the late 1980s gave rise to the **next explosion** of interest in maggot therapy (Irwin and Kampmeier 2003).

Now, maggot therapy is going through another **peak of popularity**. In the United States, this method was named "maggot debridement therapy", while in Great Britain, it is called "biosurgery". In 2004, the U.S. Food and Drug Administration (FDA) and the British National Health Service (NHS) issued permits to use maggots for surgical resection of necrotic tissues of humans and animals (https://en.wikipedia.org/

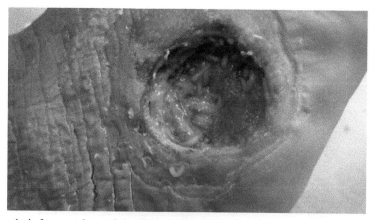

Photo 1.3-2. A method of maggot therapy is based on allowing maggots of blowflies (Calliphoridae) to feed on necrotic tissues. The maggots are placed on a wound and left there for some time, and the maggots consume all of the necrotic tissues, leaving the wound clean; they secrete digestive juice that digests the necrotic cells. Maggot debridement therapy on a wound on a diabetic foot is shown.
Photo credit: By Alexsey Nosenko/Maggot Medicine [CC BY 3.0 (http://creativecommons.org/licenses/by/3.0)], via Wikimedia Commons

wiki/Maggot_therapy). Now, this method is used in more than 1,500 health care centers in Europe and the United States.

Maggot therapy can be divided into **three processes**: (1) debridement of wounds from necrotic tissue, (2) disinfection (killing of microbes); and (3) hastened wound healing (growth stimulation). The maggots, eating in the decomposing tissues of wounds, not only remove these tissues and small fragments of bones but also prevent the reproduction of pathogenic bacteria with their secretions. In addition, the maggots release allantoin, which is a consolidant (Kruglikova and Chernysh 2013).

However, flies taken from the natural environment can carry tetanus or gangrene bacteria. Therefore, specially cultivated, sterile maggots are used for maggot therapy. This method of healing wounds is of prime importance for people with diabetes. Their wounds heal over extremely slowly, bacteria often enter them, and open ulcers develop that often force medical workers to amputate the affected extremities or to conduct major surgeries (Ratcliffe et al. 2014).

Over the last few years, intensive research into different aspects of the maggots' effect on wound healing has taken place. Comprehensive experimental data on the antibacterial action of "surgical maggots" exosecretions has been accumulated. It has been shown that the antibacterial agents of the exosecretions inhibit the growth of the harmful pathogen, methicillin-resistant *Staphylococcus aureus* (MRSA). Methods of treating long-lasting wounds without the direct use of maggots also are being developed. Researchers are planning to use dressings with a gelatinous substance to which the enzymes released by maggots will be added (Kruglikova and Chernysh 2013).

Using **honey** for healing wounds resembles in many respects of the above-mentioned use of maggots. The healing influence of honey on wounds has been known for a long time. It has been used for this purpose over **four millennia**. In 1894, the antimicrobial characteristics of honey were recorded; however, with the appearance of antibiotics, active application of honey was suspended, and within several decades it was considered only as a folk remedy.

At present, interest in honey for wound care has started to **revive**. It is used for therapy of both **acute wounds** (for example, burn, surgical, and traumatic wounds), and **chronic wounds** (for example, in cases of surgical infection, pressure injuries, etc.). The management of wounds with honey prevents the appearance of scars and cicatrices. This is because the honey attracts moisture from the air, which reduces the risk of maceration and accelerates the process of wound healing with the formation of new epithelial cells—new skin (Cherniack 2010).

1.3.4 Venom therapy

One more important aspect of the medical use of insects is venom therapy; that is, treatment of diseases with the aid of insect venoms. In this segment of medicine, the venoms of bees, wasps, ants, scorpions, spiders, larvae, and other arthropods are used. Some of the drugs based on these venoms have **already shown** good results and are mass-produced, while the others are at the stage of **clinical trials**. The venoms of some insects have been considered for design of such drugs, and research is being conducted in this field.

The treatment modes using insect venoms can be most **diverse**: by means of direct stings and also through agents containing the venom of an insect (rubbing of an ointment into the skin; administration through skin and mucous membranes using the direct electric current; and intake in the form of tablets, aerosols, steam inhalations, etc.).

Bee venom therapy (BVT) is also called **apitherapy** or beesting therapy. Apitherapy has been known for more than 5,000 years as an effective mode of therapy for diseases and traumas (http://www.bterfoundation.org/bvt). Bee venom therapy was practiced by ancient Egyptian, Chinese, Indian, Babylonian, and Greek therapists, including Hippocrates (Choudhary et al. 2014).

The practice of bee venom therapy **varies** widely in our time. As a rule, in Europe and the United States, the venom of the honey bee (*Apis mellifera*) is applied, while other countries (China, Japan, India), the venom of the Indian bee (*Apis dorsata*) and Asian honey bee (*Apis cerana*) is used (Dragomirescu and Sood 2009). Many drugs based on these venoms have been mass-produced for a long time. These drugs are described by Simics (1999), Khismatullina (2005), Omarov (2009), and others. Venom therapy is **widely used** in China, Korea, Japan, Russia, Romania, Bulgaria, Brazil, and the United States (Lee et al. 2005; Dragomirescu and Sood 2009).

Differing views exist as to diseases that are cured with the use of apitherapy. Such diseases listed by some authors are **extremely diverse**. For example, a work by Choudhary et al. (2014) includes more than thirty diseases, including the **following**: arthritis, frozen shoulder, diseases of the central and peripheral nervous system (multiple sclerosis, dementia, post-stroke paralysis, polyneuritis, ganglion nerve inflammation, cerebellar ataxia, syringomyelia, myopathy, Parkinson's disease, trigeminal neuralgia), diseases of the heart and blood system (hypertension, arteriosclerosis, endarteritis, angina pectoris, arrhythmia), skin diseases (eczema, dermatitis, psoriasis, baldness), other diseases (cancer, colitis, ulcers, asthma, bronchitis, pharyngitis, tonsillitis, ear nerve neuritis), and HIV.

A book by Khismatullina (2005) lists more than 70 diseases. In the 1960s, in the period of the greatest popularity of therapy with bee venom, physicians in some countries tried to apply it for treatment of practically **all diseases**. However, later apitherapy returned to the methods that were evaluated positively in the early 20th century and where apitherapy is far more efficient than other health aids. These include treatment of musculoskeletal system disorders, diseases of the nervous system (Omarov 2009), and some immune-related diseases (Oršolić 2012).

However, investigation of bee venom properties continues, and effective remedies based on bee venom for other diseases may appear in the future. In particular, numerous investigations aimed at the use of bee venom for treating cancer are being carried out now. It has been revealed that **melittin** (which comprises 30–50% of the dried venom mass) has a great antitumor activity and can potentially be used for treatment of leukemia, and renal, lung, liver, prostate, bladder, and breast cancer (Heinen and Gorini da Veiga 2011).

The medicinal use of **wasp venom** (Photo 1.3-3) is also under study. It has been revealed that the venom of the Brazilian social wasp, *Polybia paulista,* is capable of destroying cancer cells while leaving healthy tissues unaffected. Experiments have established that it is able to fight cancer of the blood, prostate gland, and urinary bladder (Leite et al. 2015).

Ant venom has anti-inflammatory, antiseptic, bactericidal, analgesic, and cleansing effects on the human organism. It is applied to relieve radiculitis, back pain, varicose veins, arthrthis, gout, strained ligaments, bruises, fractures, and dislocations (http://www.ayzdorov.ru/lechenie_ykys_myravi.php).

Dried and living red ants (*Formica rufa*) are used to prepare medical drugs—rubs and tinctures for treating the joints and neuralgia (https://en.wikipedia.org/wiki/Formica_rufa). The venom of devil-tree ants (*Myrmelachista schumanni*) is applied to treat rheumatism (Dunn 2007).

Photo 1.3-3. Bee venom therapy or beesting therapy has been known for more than 5,000 years as an effective mode of therapy for diseases and traumas. The medicinal use of wasp venom is under study yet. Wasp stinger, with droplet of venom is shown.
Photo credit: By Pollinator [CC BY 3.0 (http://creativecommons.org/licenses/by/3.0)], via Wikimedia Commons

Scorpion toxin is also actively used in medicine. For example, in March 2011, the medicinal agent VIDATOX, based on blue scorpion (*Rhopalurus junceus*) venom, was registered in Cuba. Produced by the company Labiofam, this drug is designed for adjunctive therapy in the treatment of different types of cancer. VIDATOX is recommended to be applied in the course of chemotherapy and X-ray therapy as a remedy supporting immunity, cleaning the organism from products of treatment, and restoring the weight and appetite (Díaz-García et al. 2013).

The venom of the **Asian scorpion** (*Mesobuthus martensii*) and the **Indian black scorpion** (*Heterometrus bengalensis*) is used in traditional Chinese medicine for pain relief and treatment of meningitis, epilepsy, apoplectic attacks, and rheumatic diseases (Cherniack 2011). In 1993, the potassium channel blocker margatoxin, which prevents neointimal hyperplasia, a common cause of bypass graft failure, was synthesized from the venom of the Central American bark scorpion (*Centruroides margaritatus*) (Costa-Neto 2005).

Spiders have been used in traditional Chinese medicine over the millennia; however, the mechanism of their venom action still has **not been adequately investigated**. In this regard, intensive research on rats and mice is being carried out. Experiments on rats have shown that the peptides of Chilean rose tarantula (*Grammostolas spatulata*) venom block pain reactions (Park et al. 2008).

The pain-alleviating effect of Trinidad chevron tarantula (*Psalmopoeus cambridgei*) venom has been established in experiments on mice (Cherniack 2011). It also has been stated that the venom of spiders negatively affects the proliferation of cancer cells. For example, theranekron, a remedy based on venom of the spider *Tarantula cubensis*, showed effective antineoplastic action in cases of breast cancer in dogs (http://onkologia.maxbb.ru/topic239.html).

Several investigations of the antitumor potential of insect larva venoms have been performed. For example, it has been established that peptides in the hemolymph of giant silk moth (*Hyalophora cecropia*) larvae possess antimicrobial activity and are a potent anticancer agent against a variety of tumor cell lines (Heinen and Gorini daVeiga 2011).

1.3.5 Other kinds of human therapy

A number of other methods of using insects to treat humans are also known; Products produced by bees play a critical role in insect-based therapy. For example, royal jelly is applied in many countries to treat different chronic skin diseases (such as eczema, atopic dermatitis, microbial and seborrheic dermatitis, lupus erythematosus, psoriasis, alopecia, warts, and ulcers).

Propolis is a resinous, waxy mixture collected by honey bees from the spring buttons of trees (such as poplar, alder, and birch) and used as a hive insulator and sealant; it is used to treat anemia, gastrointestinal ulcers, arteriosclerosis, and hypertonia (Ramos-Elorduy de Concini and Pino Moreno 1988). Propolis is also used as an auxiliary remedy to heighten the effect of medicaments, including antibiotics; to stimulate immunity; and as prophylaxis against mycotic complications. In addition, it is used to suppress inflammation, to stimulate reparative processes, and as a general strengthening agent (Horn and Leybold 2006).

A **pollen** collected by bees is applied in the medical field for treatment of enteropathy; for improvement of appetite, of one's physical strength; for stimulation of intellectual activity; and for other purposes (https://en.wikipedia.org/wiki/Ethnoentomology). Many papers have been written on the subject of using honey for medical purposes. In some sources, citation of its medicinal properties alone occupies more than one page. However, scientifically, the described properties of honey sometimes have been found to be **unconfirmed** or only partially confirmed and should be considered with a certain degree of skepticism. Nevertheless, the applicability of honey in cases of some catarrhal diseases and heart and digestive tract diseases is, as a rule, not contested (Omarov 2009; Feng et al. 2009).

The **wart-biter** grasshopper (*Decticus verrucivorus*) was used to remove warts. It would bite the wart, releasing onto it a droplet of digestive juice. In China and Japan, the yellow "juice" eructed by the grasshoppers used to be collected and sold in pharmacies. A tincture of **black** and **scarlet cicada** (*Huechys sanguinea*) is used in China in external application to fight hepatitis. **Mole crickets** are also used there as a diuretic agent (Marikovsky 2012).

Experience using **butterflies** in homeopathic medicine is described in a book by the French pediatrician Patricia Le Roux (2009). She has successfully applied them for treatment of hyperactive (ADHD) children. The author has found that these remedies are also very effective in treating various skin problems such as urticaria or eczema. In her work, Patricia Le Roux has used 13 butterfly and moth remedies: California sister (*Adelpha californica*), tailed jay (*Graphium agamemnon*), lilac beauty (*Apeira syringaria*), processionary caterpillar (*Thaumetopoea pityocampa*), brown-tail moth (*Euproctis chrysorrhoea*), brimstone butterfly (*Gonepteryx rhamni*), death's head hawkmoth (*Acherontia atropos*), blue morpho butterfly (*Morpho menelaus*), European peacock butterfly (*Aglais io*), marsh fritillary butterfly (*Euphydryas aurinia*), cabbage white (*Pieris brassicae*), small tortoiseshell butterfly (*Aglais urticae*), and fox moth (*Macrothylacia rubi*) (Le Roux 2009).

1.3.6 Use of insects in the cosmetics industry

The bee is also considered to be the most important insect in the cosmetics industry. **Honey** and other products of beekeeping were applied as cosmetics as early as the Stone Age. As a natural cosmetics product, honey still has not lost its popularity. It is included in many creams, curative-cosmetic emulsions, cosmetic packs, balms, lotions, sticks, and other products. "Honey" production lines are present in many cosmetics companies.

The **effects** of honey on human skin are manifested in its softening, improvement of turgor, and recovery of the elasticity of muscle fibers. Honey penetrates easily into cutaneous pores, feeds the skin and regulates fluid balance, activates the metabolic processes in tissues, keeps the skin healthy, and prevents premature wrinkling. Beeswax also occupies a prominent place in cosmetics. It is included without fail in the composition of high-quality lipsticks and lip balms, reducing dryness, and preventing exfoliation. The wax, which is white, serves as a base in facial nutritional, astringent, cleansing, and whitening creams and packs. The wax forms a superfine barrier film on the skin and preserves it against moisture loss.

Favorable effects on the facial skin are also produced by another product of beekeeping—**royal jelly**. Royal jelly has a rejuvenating effect and stimulates the metabolic processes in the skin. Creams containing royal jelly are applied for aging and withering skin.

In cosmetology, **chitosan**—an organic compound extracted from chitin that, in turn, forms a part of exoskeleton of all insects—is widely used. The maximum quantities (3–10% of their body dry weight) are contained in beetles (Coleoptera), flies (Diptera), mayflies (Ephemeroptera), dragonflies and damselflies (Odonata), stone flies (Plecoptera), and caddis flies (Trichoptera) (Cauchie 2002).

Chitosan is largely applied as a moistener that, when applied to the skin, contributes to recovery of the natural moistening factor. Along with moistening, chitosan protects the skin against aggressive environmental factors and increases its resistance to ultraviolet irradiation. At the present time, the **principal source** of chitin are freshwater crustaceans; however, insects may well replace them (Macadam and Stockan 2015).

In the state of Bahia (Brazil), **common houseflies** (*Musca domestica*) are used for cosmetic purposes. There, baldness is treated by rubbing the head with a mass of crushed common houseflies (*M. domestica*), while acne is controlled by washing the face with an infusion made from their scutellum (Costa-Neto and Oliveira 2000). Carmine, a dye, which is made of cochineal (*Coccus cacti*)—is used for painting tooth powders, some lipsticks, shampoos, and rouge.

1.4

Use of Insects in Technology (Insect Bionics)

Bionics is a science that uses biological processes and methods to solve engineering problems. It can be also described as a theory of creating engineering systems with characteristics that approach those of living organisms. In other words, bionics is a mixture of biology and engineering. The science of bionics is related to biology, physics, and chemistry, and it is used in electronics, navigation, communication, and many other branches of science, engineering, and high technology. The year 1960 is commonly considered to mark the birth of bionics (http://bio-nica.narod.ru/). The results of research in this field are published in many journals. In particular, in Great Britain, the journal *Bioinspiration & Biomimetics* publishes research involving the study and distillation of principles and functions found in biological systems.

The features of biological systems are of interest for engineering due to following **reasons**: (1) the technical feasibility of new principles and units of information processing, and creation of new elements of automation and computing systems; (2) the possibility of assimilating the structural and mechanical characteristics of biological systems for development of new types of technical devices; and (3) the possibility of applying chemical and energetic processes occurring in these systems with high efficiency for development of new technological processes (*Great Medical Encyclopaedia*, v. 3, 1976).

Bionics comprises the following **fields** (Skurlatova 2015): (1) biological bionics, which examines the processes occurring in biological systems; (2) theoretical bionics, which constructs mathematical models of these processes; and (3) engineering bionics, which uses the models of theoretical bionics for solving engineering problems.

The insects, which possess a wide range of uncanny abilities, are an essential and integral part of biological systems. Among them are the **special abilities** of olfactory sensation, vision, auditory perception, flying, jumping, navigation, chemical synthesis, exquisite structure, and others. These qualities are of strong interest for solving engineering problems. In view of this, insects are the focus of the modern science aimed at the comprehensive ascertainment of the mechanisms of these capabilities and their introduction into the technologies.

The bionics of insects is focused on the solution of engineering problems and the development of engineering systems with characteristics similar to those of living organisms and their parts. In most cases, a vast superiority of the "technical" characteristics of insects over those of humans is observed at all levels. Thus problems already solved by nature may also be solved by humans with the help of insects. The **most promising insects** within this framework are butterflies, moths, flies, bees, beetles, dragonflies, and fleas (Liu et al. 2014).

Investigation of using insects for solving engineering problems can be carried out according to different **principles**. One can use an approach applied by Liu et al. (2014), who have sequentially analyzed the bionic applications of motor (walking, jumping, flight), visual, olfactory, auditory, and other abilities. Our approach is based on the uses of insects in different **engineering fields**, and in this case the principle of "from the general to the specific" is observed.

1.4.1 Insect bionics applications for creating technical systems

A technical system is an assemblage of interrelated material elements designed to improve human efficiency. A special case is an **aircraft**—a technical system designed for the transportation of man and/ or cargo in the atmosphere. Three decades after the emergence of aircraft (1903), the aircraft industry met with the phenomenon of **flutter**; that is, the vibration of wings arising suddenly and rapidly at specific speeds (http://aviation.cours-de-math.eu/ATPL-081-POF/flutter.php). Due to these vibrations, planes were falling apart in the air in just a few seconds.

The solution offered by scientists corresponds structurally to the geometry of the **wing of a dragonfly**, which has chitinous bosses (pterostigma) at the ends of the wing leading edges. These structures quench the nuisance vibrations of the wing (http://bio-nica.narod.ru/). Pterostigma are also present in other insect groups, such as snakeflies, hymenopterans, and megalopterans (https://en.wikipedia.org/wiki/Pterostigma).

Bombardier beetles are ground beetles (Carabidae) belonging to two subfamilies (Brachininae and Paussinae) that were so named due to a peculiar defence mechanism. They are able to accurately shoot a self-heating mixture of chemical agents from the glands in the rear of their abdomen. The temperature of the mixture at the instant of firing reaches 100°C, and its ejection is accompanied by a heavy sound (https://en.wikipedia.org/wiki/Bombardier_beetle). At present, biologists together with specialists in the field of thermodynamics aspire to copy this mechanism of bombardier beetles in order to solve the problem of the gas supply of **aircraft turbines** (Guillot and Meyer 2013).

It should be noted that many flying capabilities of insects look extremely attractive for use in **aviation**. For example, dragonflies easily lift objects of more than 15 times their own weight. This ability is explained by specific kinematics (Photo 1.4-1): the fore and rear wings of dragonflies of almost all families act in phase opposition (phase shift is equal to about a half of the beating cycle), while other insects are "single-engined": their fore and rear wings are "linked" (Grodnitsky 1999).

Photo 1.4-1. Dragonflies easily lift objects of more than 15 times their own weight. This ability is explained by specific kinematics: the fore and rear wings of dragonflies of almost all families act in phase opposition (phase shift is equal to about a half of the beating cycle), while other insects are "single-engined": their fore and rear wings are "linked". Photo shows Brown Hawker Dragonfly *Aeshna grandis* in flight.
Photo credit: By Tony Hisgett from Birmingham, UK—Brown Hawker Dragonfly in flight 7 Uploaded by Magnus Manske, CC BY 2.0, https://commons.wikimedia.org/w/index.php?curid=21110069

Excellent **maneuverability** is characteristic of the flight of insects. For example, dragonflies of some species perform with ease such aerial stunts as loop-the-loop and barrel roll. Flies, dragonflies, butterflies, wasps, and bees of some species are able to promptly change the direction of their flight and move in any direction, including backwards. The nectarivorous hawkmoths (Sphingidae) can hover over flowers for a long time (Eberle et al. 2015; http://www.f-mx.ru/biologiya/nasekomye_v_zhizni_cheloveka.html). The tobacco hornworm (*Manduca sexta*) is able to accelerate sharply from one position and then to slow down rapidly (George et al. 2013).

The flights of insects have always presented many mysteries. Gradually, some of them were solved, but then new ones emerged. For example, in the early 20th century, the lift of bumblebee wings was calculated. The bumblebee was chosen because it has small wings with reference to the mass and size of its own body. The paradoxical conclusion that was reached was that the bumblebee is **not able to fly**—but it does fly, breaching physical laws (https://en.wikipedia.org/wiki/Bumblebee).

It was found relatively recently that the error was due to the fact that, in the case of bumblebee flight, the formulas for calculating aircraft lift were applied. However, an aircraft, in contrast to the bumblebee, does not beat the air with its wings, and the mechanics of its flight are entirely different. The flight of a bumblebee and movement of its wings were caught on camera and reviewed in the slow motion. It was found that the bumblebee wings are not rigid. They bend and twist, and the stroke angles change (Peterson 2004).

It has also been discovered that at very high frequencies of the movement of bumblebee wings, leading-edge vortices arise that are removed after the wing beat is over. These vortices are characterized by differences in air pressure that create lift. Therefore, the principles of the flight of the bumblebee cannot be used for the construction of conventional aircraft. Possibly, they can be used for designing helicopters with flexible blades.

The **wings of insects** are complex structures having many bosses and cavities, as well as different microscopic veins and wrinkles. However, roles of all of these wing features were unclear for a long time. In order to clarify these roles, **locust** movement was recorded by high-speed digital video cameras in an air tunnel. Based on the detailed kinematics of a wing, a three-dimensional model of hydrodynamic processes was developed and deformation of wings during flight was analyzed. At the second stage of the study, the researchers dealt with a "virtual locust".

They successively removed the small details of relief from the wings, and then also smoothed the unusual bending of the wings. It was found that the small details do not introduce noticeable changes in the airflow, while the wing shape itself just makes the flight of insect so effective. If the locust as altered in the computer were real, then its flight would be much more energy-consuming and slow (Young et al. 2009).

Turbulence is one of the most dangerous natural processes to aviation. In nature, flying animals deal with gusts of wind and different eddies. How do they stay on course in such cases? In order to answer this question, Combes and Dudley (2009) performed experiments with 10 species of wild bees in the jungle of Panama. In front of a high-speed camera recording the flight of bees, biologists intentionally produced strong erratic airflow to see how they would overcome these obstacles.

Aromatic scents were used in order to stimulate insect flight. Maximum flying speeds were recorded depending on the wind force and direction, and differences in the behavior of bees under different conditions were also identified. Researchers have found that with increases in "atmospheric perturbances", bees change their **aerodynamic configuration**—namely, they straddle their massive hind legs. Although this action results in a 30% increase in aerodynamic resistance and a maneuverability penalty, it also improves the stability of insects in vortex flows (Combes and Dudley 2009).

Many studies of different aspects of **insect flight** have been carried out (Sane 2003). Studies by the members of the Entomology Department at the Leningrad University, USSR, starting in the early 1980s, were the first extensive and detailed works in this field. For the first time, in order to observe insect flight, a stroboscope was used instead of a high-speed camera. This allowed researchers to visualize air flows (vortices) that create lift and to obtain data on the speed of these vortices.

Because the technological equipment at that time was inferior to that available now, a mirror was installed at a 45° angle to obtain three-dimensional images with the use of a single camera to determine the patterns of wing motion. The flight of almost all orders of insects were studied, resulting in a number of publications (Brodsky and Ivanov 1983; Brodsky and Grodnitsky 1986; Brodsky 1991, 1994; Grodnitsky and Morozov 1993; Grodnitsky 1996, 1999).

The composition and structure of the **wing tissue** of different insects have been studied in detail: dragonflies (Chen et al. 2013; Appel et al. 2015); honey bees (*Apis mellifera*) (Vance and Roberts 2014; Ma et al. 2015); house crickets (*Acheta domesticus*) (Sample et al. 2015); blowflies (*Calliphora*) (Ganguli et al. 2010); morpho butterflies (Niu et al. 2015), and others. The effects of wing deformation on their load-lifting capacity have been investigated (Bao et al. 2006; Kovalev 2008; Mountcastle and Combes 2013; Qi and Gordnier 2015).

Different aspects of insects' **wing beats** have been studied. For example, Sponberg et al. (2015) studied the motor features of insect flight muscles in the tobacco hornworm (or Carolina sphinx; *Manduca sexta*). Mukherjee and Ganguli (2012) modeled the wing beats of three different dragonfly species (*Aeshna multicolor, Anax parthenope julius*, and *Sympetrum frequens*). Ha et al. (2013) experimentally studied the relationship between wing beat frequency and resonant frequency of 30 individuals of eight insect species from five orders: Odonata (*Sympetrum flaveolum*), Lepidoptera (*Pieris rapae, Plusia gamma*, and *Ochlodes*), Hymenoptera (*Xylocopa pubescens* and *Bombus rupestris*), Hemiptera (*Tibicen linnei*), and Coleoptera (*Allomyrina dichotoma*).

The **different aspects of insect flight** have been examined; for example, aerodynamic efficiency in flying dung beetles (*Heliocopris hamadryas*) (Johansson et al. 2012) and maneuverability in flight of different insects (Eberle et al. 2015; Beatus and Cohen 2015). Up to now, researchers and engineers have failed to use many secrets of insects in the aircraft industry; however, humans still try to come nearer to them in their engineering solutions.

1.4.2 Insect bionics applications for creating technical installations

A **technical installation** is a component part of a technical system and a unit of industrial product. Insect bionics is most commonly used for the engineering of robots and other automatic devices that act based on a predetermined program. Robots, which obtain information from the outside world through the use of sensors (analogs of sense organs of living organisms), independently perform production and other operations commonly performed by humans.

Robotechnics is an extensively developing scientific and engineering discipline. Robots can be considered based on their spatial motion (wheeled, walking, running, rolling, hopping, metachronal motion, swimming, brachiating, hybrid, spinning) (https://en.wikipedia.org/wiki/Robot_locomotion). They can be specialized according to the **environment** where the locomotion occurs (on the ground and underground, on water and under water, in the atmosphere and in the cosmos), the **types of tasks** performed, etc. To a greater degree, the bionics of insects is used to develop the methods of motion.

Locomotion modes of insects include running, crawling, jumping, climbing, and flying (Combes and Dudley 2009). Surface texture (roughness, geometry, etc.) has marked effects on their motion on solid media.

The **walking motion** of insects is of interest from the viewpoint of developing alternative locomotion methods for robots that can traverse surfaces impassable for wheeled vehicles. For example, the RHex robot created by Boston Dynamics is widely known. It is a hexapod robot that walks on six legs and is characterized by good mobility over a wide range of terrain types at speeds exceeding five body lengths per second (2.7 m/s) (Saranli et al. 2001), and abilities to climb slopes exceeding 45 degrees, to swim, and to climb stairs (https://en.wikipedia.org/wiki/Rhex).

The **discoid cockroach** (*Blaberus discoidalis*), which inhabits the countries of Central America, Venezuela, and Colombia, as well as the United States (Florida), can serve as a prototype for this robot. The structure of its footsteps allows it to overcome with ease obstacles when it moves on rough surfaces and to travel over unordinary surfaces with traditional speed. In addition, the tarsi of the cockroach are covered with small spinules that bend easily in one direction. Thus the insect can extract tarsi that become entrapped in surface irregularities, and because the spinules don't bend in the opposite direction, the tarsi do not bend and the insects are sure footed when running (et al. 2001).

The **desert cockroach** (*Arenivaga investigata*) is another cockroach contributing to robotechnics. It is abundant in sand dunes of the Colorado Desert and southwestern California in the United States. Researchers found that the displacement speed of its extremities is very irregular: as long as the tarsus touches on sand, it moves inch by inch, but as soon as the tarsus lifts, the speed rises sharply. Scientists at the Georgia Institute of Technology have constructed a model of the robot SandBot for motion on sand. Inspired by the six-legged cockroach, it has a length of 30 cm and a mass of 2.3 kg (Li et al. 2011).

The robot has adopted from the cockroach the ability to walk on **three legs**. Because a tripod is a stable structure, it stands steadily on three legs. The "tarsi" themselves are C-shaped arcs of a circle that rotate around the "knob" at the upper end of the "C" character contour rather than around its center of curvature. At any time, the robot rests on three "tarsi"; for example, the fore and hind ones from the left side and the middle one from the right side. Then positions of the tarsi change: due to the eccentric position of the rotation center, the standing arcs rise, and arcs that are in the air descend. Now two supports turn out on the right, while one support is on the left. Due to the eccentric position of arcs, the speed of tarsus motion changes. In the air, they move faster, while when they touch the surface, their speed is slower. The SandBot is able to move through deep sand at a speed of up to 0.3 m/s (Li et al. 2011).

Israeli engineers have created a miniature robot called TAUB (Tel Aviv University and Ort Braude College) that is **able to hop** like a grasshopper. It has a mass of less than 30 g and copies the kinematics of the legs of a grasshopper. The legs are bent and then locked in a bent position and abruptly released. The fast release of energy fires the insect as from a catapult. The robot TAUB is not a perfect copy of a grasshopper, but it uses the same principles for jumping. The legs of the robot are rigid steel wire springs. It is able to high-jump (up to 3.3 m) and long-jump (up to 1.4 m) (Grasshopper—new chassis for robots 2015).

The possibility of using the principles of the functioning of a grasshopper's back legs for robot engineering attracted attention as early as 1983 (http://www.cals.ncsu.edu/course/ent425/text01/impact2. html). In addition, the superior kinematics of the hind legs of the locust was noted earlier, and the use of the flexible-rigid hopping mechanism of the locust was proposed for the development of a jumping robot (Chen et al. 2011).

Robots based on **water striders** have been constructed; they are not only able to move over a water surface but also able to jump over it. This development is based on the use of porous nickel foam, a material with extremely hydrophobic (water-repellent) properties. The foam is applied to the six "legs" of the robot, allowing it to not sink into water. Four legs are used for support, while two are used for jumping. The mass of the microrobot is only 11 g, and it is able to high-jump (14 cm) and long-jump (34 cm) (Zhao et al. 2012).

Several **aircraft drones** also have been created. In a number of cases, **butterflies** were used as prototypes. For example, the robot Entomopter was developed at the Georgia Tech Research Institute (GTRI). Liquid fuel is burned to move its wings, which resemble those of the night butterfly and move at a frequency of 10 Hz. The spread of its wings reaches 18 cm (https://en.wikipedia.org/wiki/Entomopter). It is expected that robots of this type, with a wingspread of about 1 m, will fly over the Martian surface in the foreseeable future. The light flying units with beating wings are optimal for flights in a rare atmosphere (Jaroszewicz et al. 2013).

The German company Festo has created ultralight bionic motion butterflies. They weigh only 32 g, and their elastic wings have a span of 50 cm (http://www.robotspacebrain.com/robotic-butterfly-by-festo/). These wings are attached to the almost weightless frame of thin graphite rods. The wings beat 1 to 2 times a second and may fly about 3 minutes. Maximum flying speeds reach 2.5 m/s. The body frame of every butterfly is equipped with infrared sensors that allow two-way communication between robots and the host computer that serves as a central control system (http://www.engineering.com/DesignerEdge/DesignerEdgeArticles/ArticleID/9870/eMotion-Butterflies-Demonstrate-Coordinated-Robotic-Flight.aspx).

In the development of robots, locomotor functions are not the only ones taken from insects. Navigation abilities, behavior, etc. are also used. For example, the **desert ants** *Cataglyphis*, which inhabit deserts and steppes, survey the local environment in a zigzag fashion in search for food but then return to their niduses (nests) practically in a straight line. Their record is 592 m of zigzag searches and only 140 m for the return trip. In estimating the distance they have covered, the ants use information on the distance they have covered from its receptors; while in navigation they depend on the **solar compass**, optic cells that help determine the direction in which the light is polarized. This information allows them to determine their location with regard to the sun even in cloudy weather (Guillot and Meyer 2013). Scholars at the University of Zurich have copied the method of ants' navigation and applied it in the mobile robot Sahabot 2 instead of conventional navigational instruments such as GPS (Lambrinos et al. 2000).

In another case, Festo has copied the **behavior of ants**, which make individual decisions as well as work as a team while they are in constant touch with their counterparts. The robot BionicANT is 13.5 cm long, and its electric supply is provided by two 7.2 W batteries. Piezoceramic transducers are built into the foot drives of each ant, while two cameras controlling the movements are attached to the head. Optic sensors mounted in the thoracic segment of the robot are responsible for tracking the device's location. The management of electronic insects is based on an electro-adaptive intellectual program forcing individual robots to work for a shared objective and to conform to the centralized control system. The robot-ants will be used in the manufacture of different articles (Griffiths 2015).

1.4.3 Insect bionics applications for creating tools

Insects frequently provide ideas for the development of **tools** (auxiliary technical devices). A widely known example is the **chipper-type chain**, invented by Joseph Buford Cox for chain saws. In the 1930s Cox, a woodcutter in the state of Oregon (USA), turned his attention to larva of the longhorn beetle (*Ergates spiculatus*), furrowing out its way in the hard wood with an uncommon easiness (https://en.wikipedia.org/wiki/Joseph_Buford_Cox).

The C-shaped jaws of the larva of the cerambycid beetle (*Ergates spiculatus*) allow it to easily gnaw out a way both along and across wood fibers. This mechanism led to the creation of a saw chain that makes it possible to saw wood more productively and with more handy sharpening. Cox's inspiration was the alternating cuts made by the curved mandibles of the cerambycid beetle (*Ergates spiculatus*) (Lucia, 1981); while one is cutting, the other acts as a depth gauge. As was stated above, he invented what is now known as the chipper-type chain for chain saws. In 1947, he and his wife, Alice, set up The Oregon Saw Chain Co. and began to manufacture the saw chains. Most saw chains produced in the world are still based on this invention of Joseph Buford Cox (Green 2012).

The **ovipositors** of some insects serve as biological models of certain technical installations. Roughly speaking, the ovipositor is a chitinized tube situated at the rear end of the body of a female insect and is designed for laying eggs. Due to their extreme attractiveness for technical applications, ovipositors have often been subjected to detailed study.

For example, Vincent and King (1996) used a scanning electron microscope to examine the ovipositors of horntail sawflies (*Sirex noctilio*) and their parasitoid wasps—western giant ichneumons (*Megarhyssa nortoni nortoni*). Ovipositors of the wasp *Ephialtes* (Hymenoptera: Ichneumonidae), which are parasites of longhorn and jewel beetles (Korzunovich 2005), have also been studied.

In India, the ovipositors of parasitoid fig wasps (*Apocryta westwoodi*), which lay their eggs into larvae of pollinator fig wasps (*Ceratosolen fusciceps*), were studied. Researchers found that the ends of the ovipositors are zinc-tipped, giving them hardness that corresponds to that of cement used in dental prosthetics (Kundanati and Gundiah 2014). Earlier, researchers found manganese and zinc in the ovipositors and mandibles of hymenopterous insects (Quicke et al. 1998).

The mechanism of the ovipositor penetrating into wood proved to be as **follows** (Vincent and King 1996). The ovipositor consists of two sections, each covered with denticles facing the direction opposite to the movement of the ovipositor. When the denticles of one section go into the wood mass, the other section goes forward. Due to such fast motion, in which the sections alternately go ahead and are fastened into the wood, the ovipositor saws up in the liber of pine, drilling a straight channel, without bends or fractures, up to 20 mm deep (http://wol.jw.org/ru/wol/d/r2/lp-u-ru/102011093#h=1).

A neurosurgical **flexible catheter** functioning on the principle of ovipositor motion has been created based on the structure of the wasp's ovipositor. The silicone tip of this flexible catheter consists of two moving parts with microscopic denticles. This structure allows it to penetrate deeply into brain tissues while causing minimum trauma. In contrast to conventional rigid surgical probes, the flexible catheter with this type of tip is able to penetrate through tissues in a way that is safest for the patient, avoiding areas of risk—as in brain surgery, for example (Quicke and Fitton 1995).

This principle was used to construct a **boring assembly**. This assembly is equipped with a unique needle consisting of two parts functioning alternately. One needle has denticles that perforate a substrate (soil) under pressure in order to preclude compaction, while the other has pockets to lift fragments to the surface. This bioinspired unit weights less than 1 kg, consumes only 3 W of electric power, and is able to reach depths of 1 to 2 m (Gouache et al. 2010; Guillot and Meyer 2013).

The complexity and expediency of the "operating device" of **sand wasps** (Ammophilinae) have been given attention. The wasps are provided with powerful "sledgehammers" with which they burrow holes for their nests. This convenient device includes a small air pocket located in the sternum to create a pneumatic effect. First, the wasp actively beats air with its wings. Their oscillations, produced by alternate compacting of the pocket, send portions of air through channels—"hoses"—to the base of the insect's jaws. The jaws begin to vibrate intensively. In this case, once the wasp's jaws get in touch with stone firmly soldered by clay, the latter flies apart (Zhdanova 2012).

Reversible interlocking is inspired by the wing-locking device of a beetle where densely populated microhairs (termed microtrichia) on the cuticular surface form numerous hair-to-hair contacts to maximize lateral shear adhesion. The beetle has a sensitive touch sensor that is able to distinguish pressure, shear, and torsion. When the beetles rest, the setae on their wings are attracted to the hair covering their body due to static electricity. In a sensor based on this principle, two sheets are connected to each other by hairy sides face-to-face and are connected to a resistance sensor. The sheets are made of polymer fibers that are coated with metal to make them electrically conductive (Pang et al. 2012).

Electric current can easily pass from one sheet to the other, and the resistance depends on the total area of contact between nanohairs. When the positions of the nanohairs are changed, the resistance changes too, which is the basis of sensor operation. When the sheets are sandwiched together, the nanohairs are attracted to one another and lock together, just like the beetle hairs. The device is then wired up so that an electrical current can be applied, and it is covered in a layer of soft, protective polymer (Bourzac 2012).

A prototype constructed at the Seoul National University was able to record the pressure corresponding to a body weighing 510 g on an area of 1 m². By analyzing the resistance change, one can distinguish between three types of mechanical strain: pressure, which comes straight down on the sensor; shear, a frictional slide along the surface; and torsion, a twisting motion (http://www.russianelectronics.ru/leader-r/news/9318/doc/60430/).

1.4.4 Insect bionics applications for creating devices

Devices are instruments designed for machine and installation control, process monitoring, computations, and other purposes. Widely **used types** include sensors (sensitive elements, transducers that transform environmental parameters into mostly electrical signals suitable for technical application) and analyzers (units for determining the composition and properties of some substances).

Interactions of some living organisms, including insects, with the outside world occur with the aid of **sensory organs**. For example, the human has five such sensory organs: eyes (organ of sight), ears (organ of hearing), nose (organ of smell), skin (organ of touch), and tongue (organ of taste). Each of these sense organs responds to particular environmental stimuli. Some of them can respond to remote stimulation (for example, organs of sight, hearing, and smell), while the others (organs of taste and touch) respond only upon direct contact.

Many insects have different sensors/analyzers that transform the energy of external stimuli (heat, light, and mechanical) into the energy of **nervous impulses**. For now, these analyzers greatly surpass their engineering analogs in diminutiveness and sensitivity. In such cases, it is reasonable to adopt the abilities of insects using the methods of fixation they use in the construction of various devices.

The species diversity of insects is very large, and therefore the development of their sense organs is very different from that in humans. As for the sense of sight, for example, the structure of the **compound eyes** of insects is most remarkable. They consist of many (sometimes thousands) of individual photoreceptor

Photo 1.4.-2. The structure of the compound eyes of insects is very remarkable. They consist of many (sometimes thousands) of individual photoreceptor units, ommatidia, appearing as narrow, strongly elongated cones converging by their vertexes deep in the eye while the cone bases form its reticulate surface. Modern researchers try to use compound eyes as the basis for inventing diverse sensors. Eyes of dragonfly are shown.
Photo credit: By Dustin Iskandar from Kuching, Malaysia (Big Eyes) [CC BY 2.0 (http://creativecommons.org/licenses/by/2.0)], via Wikimedia Commons, 19 May 2013

units, ommatidia, appearing as narrow, strongly elongated cones converging by their vertexes deep in the eye while the cone bases form its reticulate surface (https://en.wikipedia.org/wiki/Eye#Types_of_eye).

Each ommatidium has an extremely limited angle of sight and "sees" only a minute section of an object in front of the eyes, that section toward which the axis of this ommatidium is directed (Photo 1.4-2). But because the ommatidia are closely adjacent to each other and their axes diverge in a beam-like manner, the compound eye senses the thing as a whole (Chen et al. 2011). Doubtless the advantage of compound eyes is their wide **angle of view** (reaching almost 360° for some species, except for the dead point just behind the body). Modern researchers try to use compound eyes as the basis for inventing diverse sensors (Sergeev and Blagodatsky 2015).

For example, a device for determining **aircraft speed** was created based on investigations of the compound eye of the common housefly (*Musca domestica*). The eyes of the fly allow it to see many images of an object. When this object moves, it passes from one image to the other, which in turn makes it possible to determine the speed of its movement with high accuracy. Here, movement-detecting neurons play the key role. They evaluate the relative velocity of a moving object that moves across the retina (Guillot and Meyer 2013). A device working on this principle was named a "fly's eye".

It must be noted that, over the last twenty years, several insect-based **visual motion sensors** have been designed and implemented. For example, at the Aix-Marseille University, France, the elementary motion detector (EMD) was created for use with microaerial vehicles (Roubieu et al. 2012). A biomimetic vision sensor for detecting aircraft wing deformation was developed by members of the US National Aeronautics and Space Administration (NASA) and the University of Wyoming (Frost et al. 2013). Chen et al. (2011) present a review of different devices designed on the principles of insect sight.

The characteristic sound of a flying fly arises due to vibrating halteres, which are reduced rear winds. It appears that, while the fly is in flight, the halteres determine the fly's deviation from the horizontal position. The planes in which two halteres vibrate are perpendicular to each other and make an angle of about 45° with the insect axis. When the fly rotates in flight, the Coriolis force affects the vibrating halteres. This force is recorded by sensilla, and therefore the fly is provided with information on the orientation of its body in space (https://en.wikipedia.org/wiki/Halteres). Inspired by these halteres, a biomimetic **gimbal-suspended gyroscope** has been developed. It is applied in high-speed aircrafts and rockets to sense rotation angle or rotating speed (Droogendijk et al. 2014).

The mechanism of the sonic and hearing apparatus of aquatic insects is of the utmost interest. Study of this mechanism made it possible to develop an innovative method of **communication between vessels** in water. The approach makes it possible to communicate without signal output to the atmosphere, in order to avoid information capture and decoding (Zhdanova 2012); 2.2% of insect species use this communication method. Among these are some beetles, caddis flies, and heteropterans (Cocroft and Rodríguez 2005).

Much remains unknown in the field of **information exchange** among insects. Whereas 3.6 million motorcar accidents take place in the world every year, 80 million locusts moving within 1 km^2 never collide with each other. Researchers at the University of Newcastle upon Tyne in England have ascertained that a specific large neuron helps the locusts to avoid collisions. Now they are working to introduce the principles of operation of this neuron into car safety systems. The model for this effort is the Lobula Giant **Movement Detector** (LGMD) cell of the locust (Badia and Verschure 2004).

Many insects have an extremely acute sense of smell. Time and time again, researchers have attempted to create an "**electronic nose**"; a review of these attempts is given in the works by Sankarana et al. (2012) and Glatz and Bailey-Hill (2010). For example, the silkmoth female excretes a pheromone called bombycol to attract the male.

It was shown that the male responds to even very weak concentrations of bombycol in the air (only 200 molecules/m^3) and may get wind of a female at a distance of up to 10 km (Guillot and Meyer 2013). The task of designing a detector with such colossal sensitivity is unlikely to be accomplished for a long time and, up to now, it is inaccessible for human technologies (Marques and de Almeida 2000).

Insects are able to detect many stimuli that cannot be perceived by humans. For example, migrating insects respond to variations in **magnetic fields**, and the majority of insects perceive infra- and ultrasonic oscillations. The bee's eyes respond to **ultraviolet light**, while the eyes of a cockroach respond to **infrared radiation**. The grasshopper perceives infrared radiation owing to a cone on the 12th joint of its cirri (antennae). Ants and termites sense emissions from **radioactive substances** (Skurlatova 2015).

Termites (*Heterotermes indicola*) are able to produce and perceive **electric and magnetic fields** (Manzoor et al. 2015). Termites placed in closed glass vessels exercised an "influence" on individuals outside the vessels by creating an alternating electric field. Grasshoppers respond to oscillations an amplitude equal to half the hydrogen nucleus diameter. If we could understand the mechanism with which they do it, it would be possible to develop highly precise seismic instruments (Zlotin 1989).

1.4.5 Insect bionics applications for creating new materials

Insects frequently give ideas for creating new **materials** (matter or mixtures of matter of which the products are made). A problem of postconsumer plastics is well known. About 14 billion pounds (6,300,000 tons) of garbage in which plastic comprises the major part are dumped every year into the World Ocean (http://dnevniki.ykt.ru/_Emerald_/522481). According to data of the International Union for Conservation of Nature, postconsumer plastics cause the deaths of 1 million birds, 100,000 marine mammals, and a vast number of fish every year. Processing postconsumer plastics is necessary because incineration results in the release of toxic agents and the period of plastics decay reaches 100–200 years.

Inexpensive biodegradable materials that might replace plastics must be developed. The **wings of insects** became the standard for creating such materials. The insect cuticle is known to be among the hardest structures in the world. The cuticle provides physical and chemical protection without added weight or volume. In the course of re-creating this structure, researchers from the Wyss Institute for Biologically Inspired Engineering, at Harvard University, managed to create a film having a structure and composition similar to those of the insect cuticle (Mowatt 2011). This new material is called **Shrilk**. Its hardness and strength are identical with those of aluminum, but it is two times lighter than the metal. This material is a biological product and can be easily reprocessed without adverse consequences to the environment (https://en.wikipedia.org/wiki/Shrilk).

The well-known **spider silk** is a very complex product (Photo 1.4-3). Several types of glands of the spider take part in its production. The thread itself is not similar to a hair—it is an analogue of a girl's braid, made of threads of different thickness that are connected together by spider's spinnerets. Another unusual property of a spider web is its **internal hinge mechanism**: an item suspended on the web thread can be rotated without restriction in the same direction, and, in this case, the thread will not kink and create any noticeable counteracting force (https://en.wikipedia.org/wiki/Spider_web).

Photo 1.4-3. Spider silk is 40 times thinner than a human hair, four times sturdier than steel, and much more elastic than the synthetic fiber Kevlar used to make bulletproof vests. A cable made of filaments of the spider silk with a thickness of a human thumb is able to withstand the weight of 10 buses. The spinneret of an Australian garden orb weaver spider (*Eriophora transmarina*) is shown.

Photo credit: By Jason7825 (Own work) [CC BY-SA 4.0 (http://creativecommons.org/licenses/by-sa/4.0)], via Wikimedia Commons, 16 March 2014

Spider silk possesses many other **extraordinary properties**. Biological fiber, of which it consists, is 40 times thinner than a human hair, four times sturdier than steel, and much more elastic than the synthetic fiber Kevlar used to make bulletproof vests. A cable made of filaments of the spider silk with a thickness of a human thumb is able to withstand the weight of 10 buses (Guillot and Meyer 2013). The spider silk is very resilient and breaks only under a tension of 200–400% (https://en.wikipedia.org/wiki/Spider_silk). Other natural materials, including what is produced by silkworms, cannot be compared with it in strength (Xu and Lewis 1990).

Naturally, the possibility of producing spider silk is of high interest because its **potential applications** are numerous; for example, production of bulletproof vests, parachutes, sports clothes, and fishing nets; construction of cable-stayed bridges; and the making of artificial tendons and surgical sutures. However, the collection of spider silk is a laborious process. It is also complicated by the fact that the maintenance of a high number of spiders in a single location is practically impossible because they kill each other in the competition for territory.

Genetic engineering has provided an advance toward the resolution of this problem. There are seven types of spider silk, and dragline silk is the strongest (Saravanan 2006). Professor Randy Lewis from the University of Wyoming has isolated a gene responsible for the production of dragline silk and cloned it into the DNA of a goat. It is possible the silk can be produced using the goat milk (Jones et al. 2010; Zhu et al. 2015).

Investigations concerning the **different aspects** of spider silk have been widely carried out and are reflected in a large number of papers. For example, a search on the website of the journal *Chemical & Engineering News* reveals 110 related publications (http://pubs.acs.org/action/doSearch?text1=spider+silk&field1=Title). A similar search in the database of WEB of Science gives 408 such publications (http://apps.webofknowledge.com/Search.do?product=UA&SID=W2b5PfIcABguZr22AZD&search_mode=GeneralSearch&prID=4c22e3fd-83e9-4440-86f0-c9b1dcefd14a).

Resilin is another remarkable natural material. Its extraordinary elasticity allows a flea to jump very high (the height of jumps exceeds 150 times the flea's size) and allows a fly to make 500 million wing beats throughout its life. This ultra-elastic product isolated from locust tendon returns 97% of the energy applied to it, and only 3% of the stored energy is degraded into heat (https://en.wikipedia.org/wiki/Resilin).

Studies of resilin concern its properties (Su et al. 2014; Qin et al. 2012), applicability in different fields (Li and Kiick 2013; Renner et al. 2012; Balu et al. 2014), and its distinctive features among the different species of insects: honey bees (Ma et al. 2015), dragonflies (Appel et al. 2015), and locusts (Burrows and Sutton 2012), etc.

A vast number of insects are able to climb on vertical walls or to walk upside down on ceilings. For a long time, these abilities were poorly understood. A group of scientists from Germany has studied the nature of **adhesive surfaces** on the legs of ladybird beetles (*Coccinella septempunctata*). They ascertained that the insects have thin adhesive setae on each leg, allowing the ladybird beetles to move on surfaces of any type. It turned out that different parts of each seta are characterized by their own composition and properties. The setae themselves are quite hard and rigid, and their ends are soft and flexible. It is assumed that this structure allows it to attain the best adhesion with surfaces (Peisker et al. 2013).

However, the composition and structure of adhesive setae on the tarsi of ladybird beetles are so complex that, at the present time, it is impossible to reproduce this microstructure. But the researchers hope that the best is yet to come (http://insectalib.ru/news/item/f00/s02/n0000224/index.shtml). The ladybird beetle is not the only arthropod whose tarsi have been subjected to thorough study. Similar investigations also have been carried out for dung beetles (Coprinae) (Dai et al. 2007) and other insects (Gorb and Filippov 2014) as well as tarantulas (Peattie 2009).

The **water spider** (*Argyroneta aquatica*) is the only spider that has a fully aquatic mode of life. However, the lungs of this spider do not allow it to breathe oxygen underwater. The diving-bell spider builds a specific "diving-bell web" with a down-directed funnel and to fill it with air (Neumann and Kureck 2013). In order to gather the air, it turns upside down and uses its legs to capture small air bubbles, which are held on millions of setae. It transfers a small amount of air to its house, which also keeps the spider dry (Seymour and Hetz 2011).

The ability of the material to **stay dry** after long time in water is very promising. A group of researchers directed by Thomas Stegmaier at the University of Bonn, in cooperation with the Institute of Textile Technology and Process Engineering (ITV) in Denkendorf (Germany), has created a fabric that stays dry for four days in water (which is tenfold longer than other existing fabrics). It is able to entrap the surrounding air passively (without any energetic interventions). A principle of its creation was adopted from the water spider *Argyroneta aquatica* (Stegmaier et al. 2008).

The potential applications of this material are quite diverse, ranging from the production of dry swimming suits that prevent hypothermia to the coating of ships or pipelines for the purpose of reducing the friction of liquid circulating along the ships' hulls or in the pipes (Guillot and Meyer 2013).

Yet another example of the capability to keep from becoming wet in water is provided by a mosquito. Researchers at the Center for Molecular Science, Institute of Chemistry, Chinese Academy of Sciences, turned their attention to the fact that the surfaces of mosquito eyes in a fog **remain dry** while the adjacent beard is covered with microscopic droplets of water. The reason for this is that the air becomes trapped in spaces between nanonipples and microhemispheres to form a stable air cushion that acts as an effective water barrier (Gao et al. 2007).

The ability to move over **smooth surfaces** (for example, glass) even under water is characteristic of the green dock leaf beetle (*Gastrophysa viridula*). Their legs have setae that are covered with a hydrophobic oil; these setae cling to the surface due to capillary contacts (in the same way that a wet piece of paper adheres to a table). The capillary contacts should break in water, but with the beetles this is not the case.

The studies showed that this ability is related to the fact that, under water, **air bubbles** protecting the place of contact from water intrusion are always present on the legs of the beetle (Hosoda and Gorb 2012). Based on this mechanism, researchers have created artificial fasteners that can adhere to the glass under water without the use of glue. They were made of modified silicone and covered with a great number of "hairs" that held the air and at the same time made them adhesive (http://insectalib.ru/news/item/f00/s02/n0000209/index.shtml).

Head-stander beetle (*Onymacris unguicularis*) and fogstand beetle *Stenocara gracilipes* belonging to the Tenebrionidae family inhabit the Namib Desert in South Africa. They have the ability to **obtain moisture** from the morning fog blown by the wind from the sea into the desert. The beetles rest on the ridges of the high dunes, turn their abdomen upward against the wind, and lower their heads. The fog precipitates on the elytrae and flows along the central groove of the elytrae to the mouthparts (Photo 1.4-4). The moisture received in this manner amounts to up to 40% of the body weight (Parker and Lawrence 2001).

Photo 1.4-4. Fogstand beetle *Stenocara gracilipes* is native to the Namib Desert of southern Africa. This is one of the most arid areas of the world, receiving only 1.4 centimeters (0.55 in) of rain per year. The beetle is able to survive by collecting water on its bumpy back surface from early morning fogs. Researchers at the Massachusetts Institute of Technology have emulated this capability by creating a textured surface that combines alternating hydrophobic and hydrophilic materials. A Racing Stripe Darkling Beetle at Epupa Falls, Namibia is shown.
Photo credit: Hans Hillewaert/, via Wikimedia Commons, 25 June 2007

Why do the beetle's elytrae entrap the water droplets and allow them to flow quickly rather than to evaporate or be swept away by wind? Investigations using an electron microscope showed that the rough beetle's elytrae are covered with small flat-top cones that are coated with a certain hydrophilic substance. The slopes of these cones abound with microscopic grooves covered with repellent wax. So the finest water droplets falling on the cones due to **electrostatic attraction** to the hydrophilic substance do not evaporate and are not swept away by the wind. When droplets reach a certain size, they run down (due to gravitation) to the beetle's mouth along the repellent slopes of the cones (Guillot and Meyer 2013).

Researchers wondered if they could create a material resembling this beetle's elytrae. The problem was resolved as follows. A glass surface was covered with very small glass bubbles covered with beeswax, while alcohol providing hydrophilic properties was spread on the tops of the resulting cones. As a result, independent of the temperature of the surface, the dew was easily and very efficiently captured by it. Later on, a **textile fabric** with similar properties was developed (Sarsour et al. 2010). The amount of water separated by the meshed fabric was 300 ml/mm^2 fog event.

1.4.6 Insect bionics applications for creating structures

One more application of the bionics of insects is using the ideas adopted from them for the **construction of facilities**; that is, objects resulted from the purposeful construction activity of people. One example of such a structure is the houses of the Pueblo peoples in Mexico and the southwestern United States (New Mexico and Arizona). These are multistory complexes made of clay with straw and characterized by excellent thermal regulation. It is thought that they imitate the architecture of the clay nests of solitary wasps (Guillot and Meyer 2013).

The **honeycomb-inspired apartments** in Izola, Slovenia, can serve as another example. In a competition for the design of two social housing units with an ecological design, this project received first prize for its combination of low cost, functionality, and creation of a favorable climate. The design won the contest due to its handling of the economic, rational, and functional issues; its ratio between gross versus saleable surface area; and the flexibility of its plans. The hexagonal grids of wax cells on either side of the honey bee nest are slightly offset from each other. The hexagon is considered one of the most efficient designs in nature, holding the highest volume of liquid in the strongest structure with the least amount of wax needed. This structure provides effective shadowing and ventilation for apartments (http://architectuul.com/architecture/honeycomb-apartments).

Termites are also skillful builders. The aboveground parts of their mounds (Photo 1.4-5) are made of sand, clay, processed cellulose, and other substances fastened with the saliva of the termite workers. The termite mound is constantly overbuilt as long as the colony of termites inhabiting it is living. The largest known termite mound has a height of 12.8 m. The exterior walls of a termite mound are watertight to prevent flooding by tropical showers. They reach thicknesses of 20–30 cm but have small holes for ventilation. The interior space of a cone is pierced with a great number of passageways that provide the water supply, water discharge, and ventilation. The air is moistened and cooled here because the termites—especially the egg-laying termite female—are very sensitive to the humidity of the air (https://en.wikipedia.org/wiki/Mound-building_termites).

The termites clearly **orient the locations** of their constructed "skyscrapers" with respect to the cardinal directions. The east and west "faces" of their "buildings" are extended, and weak streams of the morning and evening sun only slightly warm the construction. The narrow surface of the termite mound is always oriented toward the hot sun at noon. The vertical and side ventilation shafts and passages are designed to catch in the best way the refreshing winds and to dissipate excess heat. The passages into which the scorching wind begins to blow are temporarily blocked by termites (http://www.priroda.su/item/1376).

The **Eastgate Centre building** (Photo 1.4-6) in Harare, Zimbabwe, can be called the first building based on the construction technologies of termites. The passive ventilation system provides comfortable conditions in the building with minimum electricity consumption (Thomas 2006).

The **efficiency** of these technologies is so high that the Eastgate complex does not need to use air-conditioning. Thus it was possible to avoid the expenses of US$3.5 million that could have been required to buy air-conditioning equipment. Accordingly, this decreased electric energy consumption. The Eastgate

Photo 1.4-5. Termites are skillful builders. The exterior walls of a termite mound (which can have height of 12.8 m) are watertight to prevent flooding by tropical showers. They reach a thicknesses of 20–30 cm but have small holes for ventilation. The interior space of a cone is pierced with a great number of passageways that provide the water supply, water discharge, and ventilation. Termite mounds constructed by grass feeding termites are shown.
Photo credit: CSIRO

Photo 1.4-6. The Eastgate Centre building in Harare, Zimbabwe, can be called the first building based on the construction technologies of termites. The passive ventilation system provides comfortable conditions in the building with minimum electricity consumption. Eastgate Centre (foreground, the building with the large number of chimneys on top) is shown. Photo credit: By David Brazier (Contact us/Photo submission) [CC BY-SA 3.0 (http://creativecommons.org/licenses/by-sa/3.0)], via Wikimedia Commons, 11 January 2008

complex uses only 35% of the energy required for temperature regulation as compared to other buildings of similar size. Rents are lower by 20% than they are in neighboring buildings (Pappagallo 2012).

1.4.7 Other insect bionics applications

The insects are sources of ideas for other fields as well. The creation of new kinds of **glue** are one example. Larvae of many caddis flies make their small houses by gluing sand grains, small pebbles, bits of leaves and twiglets, etc. Researchers at the University of Arizona (USA), under the direction of Jeff Yarger, investigated the adhesive web of these insects and established its chemical composition (Addison et al. 2013). Work is being conducted to create such a glue. The following is a possible important application: patching a pipe leak in normal conditions requires draining the liquid, which is physically impossible in quite a few cases; for example, if heating pipes must be withdrawn from service in the winter (http://insectalib.ru/news/item/f00/s02/n0000221/index.shtml).

Apart from the caddis fly larvae, **golden-eyed flies** (Chrysopidae) possess a secret of producing an ideal glue. During its reproduction cycle, the golden-eyed fly female lays around a thousand eggs, which are placed on the stems or leaves of plants. Each egg is fixed on a leg 1 cm in length which serves as a fastener to the support and, at the same time, removes moisture. This leg is "riveted" to the base with a droplet of liquid glue that solidifies instantly upon contact with the air. For now, the secret of this effective product has not been discovered (Guillot and Meyer 2013).

In aviation and astronautics, **antigravity suits** designed to improve resistance of the wearer to acceleration exposure are of paramount importance. Such suits have been required since the Second World War. Visual problems were observed among fighter pilots immediately after acceleration during pursuit of enemy aircraft. Defects in the visual field arose because the blistering transference of blood to the lower body resulted in hypoxemia of the brain (https://en.wikipedia.org/wiki/G-suit). The first antigravity suit with rubber and water-filled pads was developed by Wilbur R. Franks and his colleagues at the Banting and Best Medical Research Institute at the University of Toronto (https://en.wikipedia.org/wiki/Wilbur_R._Franks).

However, such a suit is useless in modern supersonic aviation with its sharply increased accelerations. An alternative was suggested by studying **dragonflies**, insects capable of accepting overloads of up to 30 g. This is because the organs of the dragonfly float in the hemolymph, which circulates without veins and arteries. The heart of a dragonfly is shaped like a tube and constantly sets in motion the liquid surrounding it and other organs, and together they withstand the lightning-fast changes in speed (https://en.wikipedia.org/wiki/Dragonfly).

In 1987, this phenomenon attracted the attention of Swiss air force fighter pilot Andreas Reinhard. He decided to create an antigravity suit functioning on the **same principle**. However, the practical implementation of the idea was delayed for many years because it was necessary to find suitable materials. In 1996, he established the company Life Support Systems, and in the early 2000s, this company, in cooperation with the Germany company Autoflug, produced the pilot protection system Libelle G-Multiplus® (http://campfiresmusic.com/Helvetica%20Condensed/?p=libelle-g-suit).

Small strips **filled with water** are placed on the nondeforming suit from shoulders to ankles. At the moment of high acceleration, the water in these strips is displaced towards the feet, countering the pressure on all parts of the body and redirecting the blood up to the brain. In addition, this suit (system) with a water reserve of 2 liters can provide drinking water in case of urgent ejection. The name *Libelle* derives from the German word for "dragonfly". The model was successfully tested at loads of up to 12 g (http://www.wisegeek.com/what-are-g-suits.htm).

In the mid-20th century, the command of the US Army aspired to furnish its soldiers with any device that would allow the synchronization of military operations. By then, the wrist alarm clock already existed, but it had two major **disadvantages**: the sound of the alarm was faint and, in addition, the vibrations produced by the alarm misaligned the watch mechanism. Attempts to improve the watch construction were unsuccessful.

The French physicist Paul Langevin, visiting the Swiss factory Vulcain producing the watches, declared to his companions that these problems were completely solvable: if such a small insect as the grasshopper is able to emit a sound propagating over distances of tens of meters, then one can also achieve the same effect in watches (Guillot and Meyer 2013). The stridulation apparatus of the grasshoppers (as well as locusts and crickets) is located on their leathery front wings. These insects produce acoustic signals by rubbing a vein of one wing against the other.

This assertion led to the idea of placing a **special membrane,** on which the hammers pounded, between the rear housing cover and the alarm device, which was patterned on the stridulation apparatus of the grasshopper. On the one hand, the sealed chamber between the cover and the membrane provided sufficient resonance for the sound, while on the other hand, it created the effect of the "double cover plate" protecting the mechanism against moisture (http://mywatch.ru/watch-art/art_1109.html). Before the appearance of electronic devices, this Cricket Watch was very popular. It became especially famous in the United States, where it was used by four presidents: Harry Truman, Dwight Eisenhower, Richard Nixon, and Lyndon Johnson (http://montre24.com/brand/Vulcain/Vulcain/).

The **coloration of the wings** of butterflies can be pigmental (due to pigments), optical (due to light refraction), and combining these two types. The vivid coloring of the wings of some tropical butterflies is controlled by two **basic mechanisms**: diffuse reflection of the incident light at multilayers and spectral filtration of the internal pigment (Wilts et al. 2015). These plays of the light due to phenomena of light interference and diffraction are called iridescence. Thanks to this feature, the wings of butterflies change colors in cases of the slightest play of light.

The concept of a **new type of display** using reflected sunlight and requiring no additional illumination is based on this property of the butterfly. A display was developed based on interferometric modulation (IMOD)—the technology of image formation from Qualcomm called Mirasol. It is based on the idea of multicolor imaging with the use of the interference of light waves. Such a screen exactly renders colors, and image resolution is higher than that on standard screens. In addition, it consumes a tenth of the energy required for operation of a liquid crystal screen, as the mirror reflects the ambient light rather than the artificial white light emitted by the source behind the screen. That is why this technology is widely used in mobile telephones, as standard screens consume about 30% of the total energy provided by the telephone batteries (Guillot and Meyer 2013).

The same property of butterfly wings can be found useful in the **graphic arts and printing and publishing industries**. Researchers at the Cambridge University have studied the green swallowtail butterfly (*Papilio blumei*), the wings of which consist of complex microscopic structures that resemble the inside of an egg carton. These researchers ascertained that pigments are by no means the cause of their vivid coloring, because the colors are created by the streams of light that are reflected by the microscopic scales on the wings of the insects. Using a combination of nanofabrication procedures—including self-assembly and atomic layer deposition—they created structurally identical copies of the scales of the butterflies; these structures reproduced the same bright colors that can be observed on the butterfly wings (Kolle et al. 2010).

The research results described above can be applied in the **security printing industry**. These artificial structures can be used for data encoding the optical signs on bank notes, credit cards, and other valuable articles. It is interesting that the butterflies can change coloring for their own protection in the case of approaching predators. As a result, predators see only green patches in a green tropical environment, while counterparts detect the butterfly by its blue patches (http://www.sciencedaily.com/releases/2010/05/100530144025.htm).

Thus the use of principles of the functioning of insects for development of different technologies and their applications in human activities is quite extensive. Here, it must be noted that sometimes humans invent **something and only afterwards find a natural analogue** of this invention; for example, solution of the flutter problem in the aircraft industry. Sometimes in studying nature, humans make things that bear no relation to a prototype but are superficially **similar** to it (e.g., flying mechanical robots).

1.5

Commercial Use of Insects

Insects have various commercial uses. Often, the same insect is applied to a wide variety of human activities. For example, the larvae of silkworm (*Bombyx mori*) are of importance not only for the production of silk but also as a valuable foodstuff. The galls formed by the gall wasps are used for making pharmaceutical products (astringent remedies) and tanning as well as the preparation of textile dyes. The shellac produced from lac scales is also used as a colorant. Nevertheless, in most cases, one can identify the key application areas for each insect.

1.5.1 Sericulture

Sericulture is the breeding of silkworms; that is, caterpillars (larvae) of silk moths. In fact, this branch of agriculture is considered to be animal production. Because silk moths are mainly bred indoors, this industry can be referred to as "stall feed". Some silk moths are bred in an open environment (unrestricted grazing), which could be considered to be grassland farming.

The goal of sericulture is to obtain silk cocoons, which in turn are a raw material for fabricating natural silk. Basically, silk is used to manufacture clothes, although there are other applications, which will be discussed in Section 1.5.5. Therefore, sericulture is a peculiar kind of animal production supplying raw material for textile industry.

1.5.1.1 History of sericulture

Information for this section was taken from various sources, which often present conflicting data, especially with regard to dates. In most publications, the twenty-seventh century BCE is indicated to be the time sericulture originated, and the year 2640 BC is often given (Meyer 2007; Ganie et al. 2012). According to Confucius, a silkworm cocoon fell into a cup of tea of the empress Xilingji (Hsi-ling-chi), the 16-year-old wife of China's third emperor, Huangdi (Huang-Ti), in a palace garden. Trying to extract the cocoon from the drink, she found that a thread extended from the cocoon, and then she thought of weaving a fabric from the threads.

However, this story is nothing more than a **legend**. Owing to archaeological evidence, it is reliably known that, long before the described event, both raw silk and various silk fabrics were already being produced in China. In an excavated layer belonging to the Yangshao era (4000–3000 BC) in Xia County, Shanxi, a silk cocoon cut in half by a sharp knife was found.

The species to which this cocoon belonged was identified as *Bombyx mori*, the domesticated silkworm. Fragments of a primitive loom (about 4000 BC) were found out in Yuyao, Zhejiang Province. The **most ancient** silk fabric, dated at 3630 BC, was discovered during excavations at a Yangshao site in Qingtaicun at Rongyang, Henan. The fabric was wrapped around a child's remains (https://en.wikipedia.org/wiki/History_of_silk).

The evolution of sericulture and the production of silk were **widespread** in ancient China. Silk was used to pay taxes. For example, in 107 BC, the government received 24 million m² of silk fabric from people as tribute (Demidenko 2012). Silk was the main Chinese export. In 139 BC, the Great Silk Road connecting eastern China with the Mediterranean was opened (Cherry 1987).

Revealing the **secret** of manufacturing silk was punishable by death. This secret was kept for a long time, and silk production remained exclusive to the Chinese throughout three millennia (https://en.wikipedia.org/wiki/History_of_silk). Nevertheless, the secret was finally revealed to other nations; opinions vary as to where silk production was next practiced. No one knows when and how China lost the secret of silk production, and opinions vary as to where sericulture emerged next.

According to some sources, sericulture spread immediately from China to **India**, where was widespread in the Brahmaputra River valley, as well as between the Brahmaputra and Ganges Rivers (Demidenko 2012; http://insectalib.ru/books/item/f00/s00/z0000013/st012.shtml). According to other information, **Japan** was the next country where sericulture appeared, in about AD 300 (https://ru.wikipedia.org/wiki/История_шёлка). Another source argues that sericulture reached Japan through Korea (Cherry 1987).

Many countries have tried to possess the secret of silk production. Two **successful efforts** of industrial intelligence activities are widely known. In Khotan (present-day Xinjiang Uygur Autonomous Region of China), sericulture emerged when the local ruler, on the recommendation of his minister, asked to marry the Chinese princess. When the offer of marriage was accepted, the oratorical and astute emissary assured the young lady that the future for her beauty was bleak because there were no silks in the home of her fiancé. So he said she should take along silkworm eggs and mulberry seeds.

Whatever was required for silk production, the adventurous fiancée brought to Khotan. The silkworm eggs were concealed in her elaborate hairdo, which could not be checked by the border guard, while the mulberry seeds were in packs with her dowry. It is interesting that she acted much more farsightedly than her fiancé and took specialists in raising silkworms and mulberries and in silk weaving along in the guise of indoor servants (Akhmetshin 2002).

A **second successful effort** was related to the wife of the Byzantine emperor Justinian. During his reign, raw silk was delivered by Persians who continually raised prices. Supposedly, the spouse of the emperor employed the services of two mendicant friars, who managed to carry the concealed larvae of silkworms in their hollow bamboo canes (Demidenko 2012). Information differs as to when this event occurred. The following years have been mentioned: 522 (Cherry 1987; https://en.wikipedia.org/wiki/History_of_silk), 550 (Akhmetshin 2002; Meyer 2007), and 554 (https://en.wikipedia.org/wiki/Byzantine_silk). Soon after, factories for producing silk were established in Constantinople, Beirut, Antioch, Tyre, and Thebes (https://en.wikipedia.org/wiki/Smuggling_of_silkworm_eggs_into_the_Byzantine_Empire).

The **further spread** of silk production across the globe was very rapid. In the seventh century, the Arabs conquering Persia gained knowledge of silk production, and in the 800–900s, the production of silk spread along with Islam into North Africa, Spain, and Sicily (Demidenko 2012). By the twelfth century, Italy had become the silk capital of the West (Meyer 2007). From Italy, sericulture and silk weaving reached France, where several generations of kings pursued sensible protectionist policies. At the end of the seventeenth century, France had become the main supplier of silk fabrics in the Western world (Demidenko 2012).

Step-by-step, sericulture and silk production developed in many **other countries**. In Mexico they emerged in 1522 (https://en.wikipedia.org/wiki/History_of_silk); in Russia, in 1652 (http://www.ryadi.ru/tekstil/436.html); in Poland, in 1659 (Lochynska 2010); in the United States, in 1660 (Cherry 1987); in England, in the late 1680s (http://www.smith.edu/hsc/silk/papers/baird.html); and in Kenya, in 1904 (Wairimu 2010). However, in most cases, these industries were relatively small.

1.5.1.2 Insects used in sericulture

The mulberry silkworm (*Bombyx mori*, Bombycidae) is the major insect used in sericulture (Photo 1.5-1). This species can be called truly domestic because it is unable to live independently and perishes in the wild. Through many millennia of farming, the silkworm has changed to such a degree that one cannot now positively identify its wild foregoer, although it is assumed that it is the wild silk moth *Bombyx mandarina* (with which the mulberry silkworm is able to breed) (Arunkumar et al. 2006).

Photo 1.5-1. A silk moth hatched from the cocoon, mated with a male, and then it begins to lay eggs. In 4–6 days, she lays up to 800 eggs. It not eat, because its mouthparts are underdeveloped. After finish this activity it dies. Adult moth *Bombyx mori* is shown.
Photo credit: Malgorzata Lochynska, Institute of Natural Fibres and Medicinal Plants, Poznan, Poland, 12 of July 2016

The life cycle of *Bombyx mori* lasts 6 to 8 weeks, depending upon a variety of features and climatic conditions. The wing span of the moth reaches 4 cm but it has no ability to fly. Thanks to their voracity, the caterpillars grow from 3 mm to 10 cm in 30 to 32 days. This species is monophagus: larvae eat only leaves of the white mulberry (*Morus alba*). Two silk glands (spinnerets) inside caterpillars produce the silk fiber, which may be 3 to 4 km long. Silk fiber spun into cocoons are in the form of a bave, that is, a pair of fibroin fibryllas with a sericin (silk glue) covering. This wet or moist covering binds the fibers together (Chen et al. 2012).

After copulation, the female lays 300 to 500 eggs of 1.5 mm diameter. Multivoltine varieties bred in tropical regions may produce seven to eight generations of eggs annually. Univoltine varieties produce one generation of eggs in the spring and a second generation that will hibernate until the next spring. Bivoltine varieties produce two generations of eggs annually and a third generation that will hibernate until the next spring (Lochynska 2015).

Another domesticated silkworm is the **Ailanthus silk moth** (*Samia cynthia*, Eri silk moth). It is found in China, India, Malaysia, Indochina, Java, Japan, and the Philippines. In the 19th century, the Ailanthus silk moth was delivered to Europe for the production of silk, which proved to be very expensive. However, several moths flew away from captivity and established populations in such European cities as Vienna, Paris, and Strasbourg where the *Ailanthus* was cultivated as an ornamental plant. This species was also introduced in Australia, Canada, the United States, Venezuela, Uruguay, Brazil, and Tunisia (https://en.wikipedia.org/wiki/Samia_cynthia).

The caterpillars feed on leaves of such **trees** as ailanthus (*Ailanthus glandulosa*), tree of heaven (*Ailanthus altissima*), elder (*Sambucus*), walnut (*Juglans*), plum tree (*Prunus*), privet (*Ligustrum*), forsythia (*Forsythia*), castor-oil plant (*Ricinus*), golden chain (*Laburnum*), mountain ash (*Sorbus*), lilac (*Syringa*), magnolia (*Magnolia*), and bay tree (*Laurus*). Larvae will feed on other trees and shrubs, but all eggs are laid on tree of heaven (*Ailanthus altissima*) and growth is best on it. This tree is commonly grown as an ornamental plant in cities. The subspecies *Samia cynthia ricini* feeds upon castor bean. In 1999–2007, 7.72% on average of the raw silk made in India was obtained from the Ailanthus silk moth *S. cynthia* (Eri silk moth) (Anitha 2011). The silk is extremely durable, but it cannot be easily reeled off the cocoon and therefore is thus spun like cotton or wool (https://en.wikipedia.org/wiki/Samia_cynthia).

More than **500 species** of wild silkworms are also known (https://en.wikipedia.org/wiki/Wild_silk). However, the larvae of only some of them are used for weaving fabrics. The silk produced by them is usually more hard and coarse than that of the mulberry silkworm, and dyeing it is more difficult. Another difference is that wild silk is usually extracted after the moths have left cocoons, so floss is obtained rather than one long thread as in the case of the domesticated silkworm.

Wild silkworms used for fabric weaving include the following:

Assam silk moth (*Antheraea assamensis*): Its larva (and the silk produced by it) is called the muga silk worm. This species inhabits India (Assam state), Myanmar, Malay Peninsula, and the islands of

Kalimantan, Java, and Sumatra. The larvae feed on leaves of cinnamon, bay tree, magnolia, and other trees. Muga is a very rare, good-quality, fine, and durable kind of silk, and its price exceeds several times that of traditional silk. In the translation, *muga* means "amber-golden", referring to the color of the *A. assamensis* cocoon. For 600 years, this kind of silk was used solely by members of the royal family on the territory of today's Assam state (https://en.wikipedia.org/wiki/Antheraea_assamensis). The volumes of its production are not too large. In India, 0.62% on average of the produced raw silk was this type during 1999–2007 (Anitha 2011).

Tropical tasar silkworm (*Antheraea mylitta*): This species inhabits areas with tropical climates in the forests of China, Korea, and India. Efforts to domesticate this species of silkworm were unsuccessful. The cocoons of this silkworm are much larger than those of the ordinary silkworm. They contain a long double thread (up to 1,400 m in length) that easily unwinds because the glue impregnating the cocoon consists mainly of an acid sodium salt of uric acid and easy dissolves in alkaline solutions. The natural color of the silk is brown due to tannin contained in the plants on which this species of silkworm feeds. Tasar silk is mainly used for making furniture upholstery and other interior fabrics (https://accessuar.ru/articles/wild_silk/). The volumes of its production are also not too large: in 1999–2007, it accounted for 1.71% on average of the raw silk produced in India (Anitha 2011).

Chinese tussah moth *Antheraea pernyi* is found in Russia (Amur River region and south of Primorsky Krai), China, Korea, and Japan, and it became acclimated to southeastern Spain and Majorca Island. The larvae feed on leaves of different species of oak (*Quercus* spp.), chestnut, and hornbeam. Silk from the Chinese tussah moth began to be produced more than 250 years ago in China. This silk is used for producing shantung. The thread extracted from the cocoon is coarse, has high tensile strength, and yields beautiful fabrics. The color of the silk depends on the soil on which the trees grow (https://en.wikipedia.org/wiki/Antheraea_pernyi).

The **polyphemus moth** (*Antheraea polyphemus*) is widely distributed in the continental part of North America. The larvae feed on leaves of birch (*Betula*), willow (*Salix*), oak (*Quercus*), maple (*Acer*), hickory (*Carya*), beech (*Fagus*), different fruit trees (plum, peach, apricot, and cherry), and citrus trees (lemon, grapefruit, orange, and lime). The larva is extremely ravenous and eats, in less than two months, an amount of food that exceeds its weight by 86,000 times (http://ru.encydia.com/en/Antheraea_polyphemus).

The **Japanese silk moth, or Japanese oak silk moth** (*Antheraea yamamai*): This species initially inhabited Japan and the southern Russian Far East, but now it is found in China and Korea. The Japanese silk moth also has been introduced into southeastern Europe. This species has been reared in Japan for more than 1,000 years. The cocoon is easily reeled. The silk thread obtained from it is soft, thick, and shiny. It is comparable with thread of the mulberry silkworm in strength and exceeds it in elasticity (https://ru.wikipedia.org/wiki/Японская_дубовая_павлиноглазка).

The **orange-tipped oakworm moth** (*Anisota senatoria*): This species is widespread throughout the territory of the United States and southern Canada. It feeds on leaves of different oaks (*Quercus*) and possibly dwarf chestnut (*Castanea pumila*) (http://www.butterfliesandmoths.org/species/Anisota-senatoria).

The **Io moth** (*Automeris io*): The distribution of this species is slightly wider than that for previously mentioned species. The larvae feed on leaves of bird cherry, willow, balsam fir, red maple, bastard indigo, American hornbeam, beech, ash, oak, and other trees (https://en.wikipedia.org/wiki/Automeris_io).

The **wild silk moth** (*Bombyx mandarina*): This species inhabits eastern China, Taiwan, the Korean Peninsula, Japan, and Russia (southern Primorsky Krai). Wild silk moth larvae feed solely on leaves of the white mulberry (*Morus alba*). As has been said, it is possibly the ancestor of the mulberry silkworm (*Bombyx mori*).

Silk moth (*Bombyx sinensis*): This species is found in China. It is prolific but has small cocoons (http://www.liquisearch.com/wild_silk/list_of_some_wild_silk_moths_and_their_silk).

The **Promethea silk moth** (*Callosamia promethea*): This species is found in the eastern half of the United States and southern parts of eastern Canada. Larvae feed on leaves of maple, persimmon, white ash, laurel, spicebush, tulip tree, apple, poplar, peach, pear, oak, willow, and other trees (https://en.wikipedia.org/wiki/Callosamia_promethea).

The **African wild silk moth** (*Gonometa postica*): The cocoons of this silkworm are harvested commercially in Namibia, Botswana, Kenya, and South Africa; apart from these countries, the species also inhabits Zimbabwe and Mozambique. Larvae feed on *Acacia*, *Brachystegia*, *Elephantorrhiza*, *Pinus radiata*, and *Julbernardia* (https://en.wikipedia.org/wiki/Gonometa_postica).

The **African mopane silkworm** (*Gonometa rufobrunnea*) occurs in Zimbabwe, Mozambique, Botswana, Zambia, Namibia, Angola, and Malawi. Larvae feed on leaves of the mopane (aka mopani) tree (*Colophospermum mopane*) as well as some acacia. Cocoons are harvested from the ground after the silkworms have left and the cocoons fall from the tree. This is a time-consuming operation that lowers the profitability of production. The silk is close in color to that of silk produced by the tropical tasar silkworm (*Antheraea mylitta*) (http://www.treehugger.com/culture/qa-is-silk-green.html).

African lunar moth (*Argema mimosae*): This species inhabits central and south Africa (Kenya, Zimbabwe, Zaire, Tanzania, Zambia, and Malawi). Larvae feed on corkwood (*Commiphora*), marula (*Sclerocarya birrea*), and tamboti (*Spirostachys africana*) (https://en.wikipedia.org/wiki/Argema_mimosae).

The **cecropia moth** (*Hyalophora cecropia*) inhabits the Rocky Mountains and north in the majority of Canadian provinces (https://en.wikipedia.org/wiki/Hyalophora_cecropia).

The **economic significance** differs among species of silkworms. Some of them are of importance in separate regions or segments of the market. Among the most important wild silkworms from an economic viewpoint are the Assam silk moth (*Antheraea assamensis*), tropical tasar silkworm (*Antheraea mylitta*), Chinese tussah moth (*Antheraea pernyi*), Japanese silk moth or Japanese oak silk moth (*Antheraea yamamai*), African wild silk moth (*Gonometa postica*), and African mopane silkworm (*Gonometa rufobrunnea*). The global distribution of some wild silkworms is shown in Map 1-2.

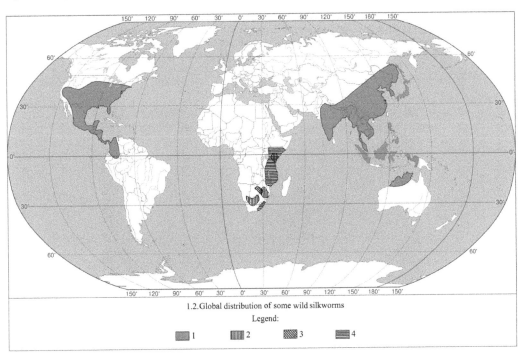

1.2. Global distribution of some wild silkworms

Legend:

1   2   3   4

Map 1.2. Global distribution of some wild silkworms (http://www.fao.org/docrep/k1100e/k1100e03.htm; Veldtman 2005; http://www.discoverlife.org/mp/20q?search=Gonometa; https://www.newikis.com/en/wiki/Gonometa_postica).
Legend: (1) species of the genus *Antheraea* (Saturniidae) secreted Tasar silk; (2) African wild silk moth *Gonometa postica*; (3) African mopane silkworm *Gonometa rufobrunnea*; (4) African moon moth *Argema mimosae*

1.5.1.3 Process of silk production

The process of true silk production consists of **three stages**: (1) sericulture proper—farming of silkworms; (2) silk spinning—production of raw silk; and (3) weaving.

Sericulture begins with the development of the mulberry silkworm, including **four stages**: eggs, larvae, pupae, and moths. After copulation with the male, the moth lays up to 800 eggs and dies. In a week's time, a larva 2–3 mm in size emerges from each egg. Within 4–6 weeks, the larvae feed on mulberry leaves (Photo 1.5-2). All this time, the larvae are in large trays with leaves put on top of one another in a special room with constant temperature and humidity. During rearing, they grow in size by 25 times and in weight hundreds of times (Photo 1.5-3).

Before pupation, the silkworm breeders move larvae to floorings of leaves and branches, wood gratings, or special fagots for building cocoons. Attaching to a stick or other base, the larvae create fuzzy

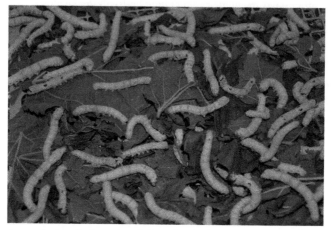

Photo 1.5-2. Each egg in a week gives the larva about 2–3 mm long. It has unimaginable cravings. The larvae need to feed a month all day and night on the leaves of mulberry tree. The appetite of caterpillars is constantly growing. From continuous operation of a huge number of jaws, silkworms in the room should rumble similar to a knocking of the rain on the roof. Picture shows the feeding caterpillars of the silkworm *Bombyx mori*.
Photo credit: Malgorzata Lochynska, Institute of Natural Fibres and Medicinal Plants, Poznan, Poland, 09 of June 2016

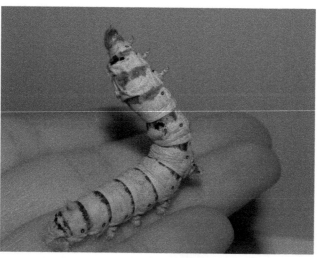

Photo 1.5-3. On the fifth day of life, the larva became motionless. It sleeps through the night, tightly clutching the leaves. Then straightens up, and close the old skin bursts, releasing the grown caterpillar. During the period of feeding larvae 4 times changing the skin and re-taken for food. Caterpillar of the moth before a molt is shown.
Photo credit: Malgorzata Lochynska, Institute of Natural Fibres and Medicinal Plants, Poznan, Poland, 14 of June 2010

netting-carcass and then twin cocoons inside it (Photo 1.5-4). They begin to secrete a jellylike substance that solidifies in the air with the formation of **silk thread**. Then the larvae wind this thread as a figure eight using rotating motions.

In 8–9 days, the cocoons are ready for **unwinding** (Photo 1.5-5). Together with pupae, they weigh 1–4 g and spin from 350 to 1000 m of continuous silk thread (Akhmetshin 2002). The cocoons of the mulberry silkworm can have different shapes (round, oval, pointed, or constricted in the middle), sizes (from 1.5 to 6 cm), and colors (white, golden, lemon-yellow, dark-yellow). They are detached and sent to procurement centers.

Within the cocoons, the living pupae will pass into the moth stage after two weeks. Because their mouthparts are undeveloped, the moths do not gnaw through the cocoon but rather secrete a special caustic agent that dissolves the upper part of the cocoon. Such cocoons cannot be unwound, and the thread will break. In order to counteract this, the pupae are killed. Often, this is done by way of **processing with hot steam** for 2–2.5 hours at temperatures not higher than 110°C.

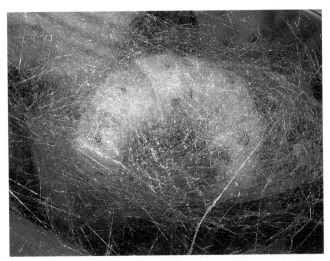

Photo 1.5-4. Caterpillar first creates a fluffy mesh-frame, and then inside it builds a cocoon. It begins to secrete a gelatinous substance which hardens in air to form a silk thread, and rotational movements wrapped this thread in the form of eight. Initial phase of the cocoon creating is shown.

Photo credit: Malgorzata Lochynska, Institute of Natural Fibres and Medicinal Plants, Poznan, Poland, 23 of June 2010

Photo 1.5-5. A silk thread consists of 75–90% of the protein fibroin and sticky substance sericin that holds the threads and gives them to disintegrate. Completely cocoon the caterpillar create during 3–4 days. Picture shows cocoons and silk thread, which obtained after unwinding. To unwind the thread, the cocoons are washed alternately in hot and cold water to dissolve the sericin. A thread of 600 silk cocoons is required for one female blouse.

Photo credit: Malgorzata Lochynska, Institute of Natural Fibres and Medicinal Plants, Poznan, Poland, 03 of April 2017

Then the cocoons are delivered to silk factories for **rewinding** on special automatic devices. In order to facilitate this process, the cocoons are usually soaked in hot water, which dissolves cericin (material that tightens the fibers into a cocoon). Because the thread of a single cocoon is too fine, the threads of five to eight cocoons are combined in the process of unwinding in order to obtain a yarn of adequate strength and thickness. In this manner, the raw silk is produced. For the production of 1 kg of raw silk, about 45,000 cocoons (18 kg) are necessary, and the total length of threads in the cocoons reaches 300–900 km.

Thereafter, **silk spinning** (processing of the raw silk into twisted threads) is performed. The skeins of raw silk are soaked in warm, soapy water, and then they are straightened and dried. After drying, the skeins are unwound on special machines. Here, the continuity and homogeneity of the threads in thickness and structure are controlled. Then doubling winding (parallel winding of threads on one spool) and twinning proper are carried out. To keep the threads from unwinding, they are treated with water vapor and afterwards are packed and delivered to the weaving factories.

Then the **weaving** starts. The silk is weaved by way of interweaving of lateral and transverse threads. In order to produce silk fabrics, different types of interweaving are applied: simple (plain, twill) or fancy (jacquard, zigzag twill, etc.). Using the different techniques of interweaving, different types of silk fabrics (atlas, crepe de chine, toile, chiffon, gauze, tissue cloth, organza, jacquarette, etc.) are produced at the weaving factories.

Just as in ancient times, the production of silk requires mostly **hand labor**. At the present time, the rewinding of thread is the only mechanized process. However, one cannot assert that the production of silk will not develop. Research studies are being carried out in this field. For example, researchers at an experimental sericultural station in the Gunma Prefecture (Japan) have been able to induce silkworms to produce threads of different colors which opens up fresh opportunities for producers of silk goods.

The secret of creating **colored cocoons** lies in what the silkworms are fed. It is known that the silkworm is very sensitive to toxic substances contained in most dyes. To the present day, one can obtain only cocoons that are white, yellow, or lime green in color. Researchers have succeeded in including nontoxic dyes in new artificial food (Sadowsky and Nesmelov 2012).

1.5.1.4 *State of world sericulture*

Sericulture exists as an economic industry more than 40 countries (Ganta and Pallamparthi 2010). The basic index of this industry is the volume of harvested silkworm cocoons. Ten countries that lead in this index are given in Table 1.5-1.

After harvesting, the cocoons are delivered to silk factories for unwinding and production of **silk threads** (raw silk). The shares of processed cocoons in different countries vary. Some countries fully process the collected cocoons, while others export them to a considerable extent. There is a large

Table 1.5-1. Countries leading in production of silkworm cocoons, reelable in 2012.[1]

Rank	Country	Production (MT)[2]
1	China	370,000
2	India	151,000
3	Uzbekistan	25,500
4	Thailand	4,600
5	Iran	3,300
6	Brazil	3,219
7	Vietnam	2,500
8	North Korea	900
9	Romania	790
10	Afghanistan	700

[1] http://faostat.fao.org/site/339/default.aspx.
[2] Metric tons.

difference between production of cocoons, raw silk, and so on in different countries. Countries that lead in volumes of raw silk production are given in Table 1.5-2.

One can see that China (80%) and India (15%) stand out sharply among the countries listed (Anitha 2011). The **remaining countries** account for only 5%. Among the countries not shown in Table 1.5-2, South Korea, Kenya, Botswana, Nigeria, Zambia, Zimbabwe, Bangladesh, Colombia, Egypt, Nepal, Bulgaria, Turkey, Uganda, Malaysia, Romania, Bolivia, Madagascar, Tunisia, Syria, and the Philippines are significant.

As for countries presented in Table 1.5-2, it is worth giving consideration to three of them: China, India, and Japan. About one million workers are involved in sericulture in **China** (Popescu 2013). Throughout the millennia, this country has remained the world leader in this industry. The mulberry silkworm is reared in a majority of Chinese provinces. Optimum conditions for the industry are characteristic of Zhejiang and Jiangsu Provinces, where sericulture provides more than 40% of all income for farmers (http://lib7.com/aziatyy/244-shelkovodstvo-kitai.html). The other sericultural regions include the Xinjiang Uygur Autonomous Region, and the Shandong, Hebei, Shanxi, Shaanxi, Anhui, Hubei, Hunan, Sichuan, Guangdong, Guangxi, Yunnan, and Guizhou Provinces (http://www.fao.org/docrep/005/ x9895e/x9895e03.htm). The key zones of oaken silkworm rearing are Liaoning Province (more than 50% of harvested cocoons), as well as the Shandong and Henan Provinces (http://lib7.com/aziatyy/244-shelkovodstvo-kitai.html).

India, as well as being the second-ranking country in the world based on sericulture development, is distinguished by the species diversity of bred silkworms and, accordingly, the types of produced silk. The rearing of mulberry silkworms prevails in India; for example, 88.7% of the silk produced in India came from its cocoons in 2007–08 (Ganta and Pallamparthi 2010). However, a number of other species of silkworms are also cultivated, producing tasar, muga, and eri silks. Mulberry silk is produced mainly in the states of Karnataka, Andhra Pradesh, Tamil Nadu, Jammu and Kashmir, and West Bengal, while non-mulberry silks are produced in Jharkhand, Chhattisgarh, Orissa, and northeastern states of India (Anitha 2011).

For some time, **Japan** has dominated in the production of raw silk, as is seen from Table 1.5-2. In the period of occupation of China (1940–45), the Japanese, occupying the key areas of sericulture, cut down about 2 billion mulberry trees in order to eliminate the competition (http://lib7.com/aziatyy/244-shelkovodstvo-kitai.html). However, the sericulture in Japan fell into disarray later on, and plantations

Table 1.5-2. Raw silk production (tons).[1]

Producer	1938	1958	1966	2001	2007	2010	2013
China	65,520	79,370	84,000	58,600	130,000	115,000	130,000
India	9,404	19,581	21,100	15,857	18,320	20,410	26,538
Japan	282,211	216,724	114,000	–	150	54	30
Brazil	403	442	1,073	1,484	1,220	770	550
Thailand	–	–	–	1,500	760	655	680
Iran	3,000	1,800	2,000	–	253	75	123
Turkey	2,348	2,934	800	–	20	18	25
Indonesia	–	–	–	–	65	20	16
USSR	23,343	28,900	34,483	–	–	–	–
North Korea	–	–	–	–	–	300 (2011)	300
Uzbekistan	–	–	–	–	950	940	980
Vietnam	–	–	–	2,000	750	550	475
World	**438,832**	**271,000**	**277,936**	**80,774**	**119,603.58 (2008)**	**136,918.42**	**159,776**

[1] Sericulture in the countries of the world (1968); Popescu (2013); Anitha (2011); The Global Silk Industry (2011); http://inserco.org/en/?q=statistics.

of mulberry trees were replaced by other agricultural plants. Now, Japan's share in world sericulture is insignificant. However, Japan, along with the United States, Italy, India, France, China, United Kingdom, Switzerland, Germany, UAE, Korea, and Vietnam, is a leading country in the use of silk fabrics.

Today, the share of natural silk in the world production of textile fibers is very small. According to different estimates, it is about 0.2% (http://inserco.org/en/?q=statistics) to 0.5% (Demidenko 2012).

1.5.1.5　Use of silk

The main purpose of silk fabrics is the **fabrication of clothes**. In addition, silk is often applied in building interiors for finishing walls and ceilings. A considerable amount of silk is used for manufacturing draperies. This fabric is also used in the production of carpets. Embroidery in silk on clothes is characterized by long-standing traditions. The history of this type of ornament numbers more than 20 centuries. Pictures embroidered in silk are also in great demand (http://anysite.ru/publication/silk).

In the **medieval period**, Mongolian rulers used silk as a currency, diplomatic gifts, and rewards to their subjects. Tribute and rewards for services were paid in silk, and silk was used for making idols, camp tents and tentage, bedding items, and other things. Silk bedding items were even used in Mongolian post houses (guesthouses); the fabric is considered quite hygienic—lice and other parasites transmitting diseases across medieval Europe are known to be unable to inhabit creases in silk blankets and linens. And for a long time, silk was used in the Far East as a paper for letters (Demidenko 2012).

The history of application of silk in the **military arts** is quite long. Mongolian fighters in the twelfth and thirteenth centuries were provided with silk shirts, as arrows rarely penetrated this fabric and more often squeezed into the wounds, so the arrows could be extracted by pulling the silk (https://en.wikipedia.org/wiki/Mongol_military_tactics_and_organization). In addition, arrows that went into silk rarely broke and could be reused (http://www.ru-expo.ru/novost-kozha_prochnee_bronezhileta).

Silk was widely used for making **flags** (banners). For example, in February 1880, instructions for making flags and banners approved by the Minister of War of the Russian Empire prescribed that cloth used for making banners made of "silk material (top-quality ripple cloth) weaved in a special manner" (Zaitseva 2007). In the 19th century, exchange of silk for military purposes was established between Japan and France (Takeda 2014).

Silk has also been used to manufacture **sieves and filters**. Under the conditions of modern intensive production with high specific loads, silk sieves in production equipment have been replaced by sieves made of synthetic materials, which are much cheaper and wear-resistant. However, silk sieves are still recommended for use in the flour-milling industry as control sieves in laboratory analyses of grinding products and flour (http://tinref.ru/000_uchebniki/04800selskoe_hozaistvo/002_tehhnolog_muku_krupi_kombikorm_1_3/038.htm). Silk is also used in fine mesh filters for fluid filtration.

The canopies of the first **parachutes** were made of the silk. However, with the invention of nylon in 1938, nylon came into use more often as a stronger and more durable material (Meyer 2007). Sometimes, silk fabrics are applied for manufacturing bag filters used for removal of dust in gases for smelting furnaces in metallurgy of ferrous and nonferrous metals, in the glass and ceramic industries, for treating gas of roasting furnaces, for different units for burning (garbage, tires, sludge), for boiler plants for industrial and community facilities, for units for the production of soot for carbon black pigment, and for coffee roasting units (Aliev 1986).

Silk was once used in the manufacture of **tires** for high-class racing bicycles. With the use of silk, the mass of these bicycles was considerable reduced (Ragulin 1981). Silk threads have been used since olden times as a **surgical suture material** applied to connect tissues with a view to reducing cicatrization. In the human organism, silk dissolves within 6–12 months. Silk is used in surgeries on the digestive tract and mucous membranes, as well as in cosmetic and plastic surgery, ophthalmology, and neurosurgery (http://www.catgut.ru/info/16.html).

1.5.2 Production of dyes

Some insects of the order Hemiptera are used for the production of dyes. Among them are, first of all, are several species of **coccids, or scales** (Coccoidea); the material to produce the red dye carmine is obtained from the females of these insects. It was once used for dyeing fabrics and yarn from which carpets were weaved and, in addition, this dye was used in ancient times for painting miniatures on parchment.

Gall wasps (Cynipidae) are also used for this purpose. Their larvae produce gall nuts, which were once used for making inks; that is, liquid color suitable for writing and creation of pictures using writing tools and dies.

1.5.2.1 Insects used in production of carmine

Female coccids (Coccoidea), which suck the moisture from plants, are used for the production of red dyes. These species are called scales. They are widespread in the wild; however, they are difficult to observe. On the outside, they resemble swellings or small scales on the roots, stems, or leaves of plants. Often, the females can be protected by wax secretions that cover them with a flat or roundish shield (hence one of the names of these insects, scales) (Zotova 2003).

Three species of coccids are most important for the production of carmine. Geographically, they are widely separated and belong to different genera and families; however, their ability to produce the carminic acid used for making dyes is a common unifying factor. Sometimes, the word *cochineal* is used to indicate all species of insects whose females yield carminic acid. These species include the following:

1. Armenian, or Ararat, cochineal (Porphyrophora hamelii) is endemic to the salt desert located along the middle reaches of the Araks River, Armenia. This species leads predominantly an underground life and inhabit the roots of two species of herbs growing on alkaline soils: the grass *Aeluropus littoralis* and the common reed *Phragmites australis*. The red dye extracted from the Ararat cochineal is known in Armenia as *vordan karmir*. It has a brilliant, saturated magenta color and earlier was widely used for coloration of yarn and threads, dyes, and inks (https://en.wikipedia.org/wiki/Armenian_cochineal).

   The *adult* female is a wingless, slow-moving insect having an oval body shape and dark-cherry blossom. The length of the female is 2 to 12 mm, the width is 1 to 6 mm, and the weight is 2 to 100 mg (on average 27 mg). The body is covered with wax secreted by special glands on the body surface. The males differ markedly from the females. They are much smaller (body length 2–4.5 mm, weight 0.6–3.4 mg), have long legs, and are capable of moving quickly.

   The **habitat** of the Ararat cochineal is very small. In the past, its range reached 11,000 ha (http://noev-kovcheg.ru/mag/2011-20/2880.html#ixzz424QwfykC); now the range has contracted to 4,000 ha in Armenia and a small patch in Azerbaijan. In the late 1980s, the total reserves of the Ararat cochineal were estimated at approximately 100 t (Zotova 2003); now they are likely much less. This decrease is related to reductions in the range of its forage plants, the grass *Aeluropus littoralis* and the common reed *Phragmites australis*. In 1987, the state wildlife sanctuary Vordan Karmir, with an area of 190 ha, was established in the semidesert of the Ararat plain for preservation of the Ararat cochineal (http://noev-kovcheg.ru/mag/2011-20/2880.html#ixzz424QwfykC). In 2008, the area of the wildlife sanctuary was increased to 220 ha (https://en.wikipedia.org/wiki/List_of_national_parks_of_Armenia).

2. The **Polish cochineal** (*Porphyrophora polonica*). Initially, its range covered a territory from France and England to China; however, the present-day distribution is many times less and touches some regions of Western and Eastern Europe and European Russia. The Polish cochineal feeds on the roots of 20 genera of plants; however, the perennial knawel (*Scleranthus perennis*) is the most important. Other important plants include mouse-ear hawkweed (*Hieracium pilosella*), bladder campion (*Silene vulgaris*), velvet bent (*Agrostis canina*), smooth ruptureworth (*Herniaria glabra*), cinquefoil (*Potentilla*), and strawberry (*Fragaria*).

The red or cherry-violet bodies of females are wide-oval, 1.5–6.6 mm in length. The eyes are well developed. The legs are large, especially the front, burrowing ones, which have large claws. The males have a length of 2.25–3.5 mm, and they are bluish-violet in color. The front legs are short, while the back ones are long and the tarsi have claws. As with the Ararat cochineal, a new generation is possible every year. On each individual plant they feed on, one can find only about 40 insects; so in order to harvest the necessary quantity of insects, thousands of plants are necessary (Łagowska and Golan 2009; https://en.wikipedia.org/wiki/Polish_cochineal).

3. The **Mexican cochineal**, or cochineal (*Dactylopius coccus*), belongs to a genus and family different than those of the Ararat and Polish cochineals. The original distribution of this species remains unknown, and it is assumed that there were two isolated ranges in the highlands of Mexico (centered in Oaxaca State, but also found in Puebla, Tlaxcala, and the Valley of Mexico) and southern Peru (Campana et al. 2015). It is believed that in the beginning, the Mexican cochineal inhabited the north of the present-day territory of Chile (Chávez-Moreno et al. 2009) as well as Bolivia (Cabrera 2005). In the 19th through 20th centuries, it was introduced to a number of other regions of the world.

The main food plants for the Mexican cochineal (*Dactylopius coccus*) are cactuses, predominantly of the genus *Opuntia*. The females are 5 mm long, while the males are 10–12 mm long. The females of cochineal *D. coccus* lead completely immobile lives underground. In their early days, they stick by the snout to a cactus stem or spine and suck the sap; they never move from this place and conceive and lay their eggs there.

In order to obtain one troy pound (373 g) of dried insects, from 18,000 to 23,000 females of the Ararat cochineal, 100,000 to 130,000 of the Polish cochineal, and 20,000 to 25,000 thousand of the Mexican cochineal are needed (https://en.wikipedia.org/wiki/Armenian_cochineal). The global present distribution of scale insects produced carmine is shown in Map 1-3.

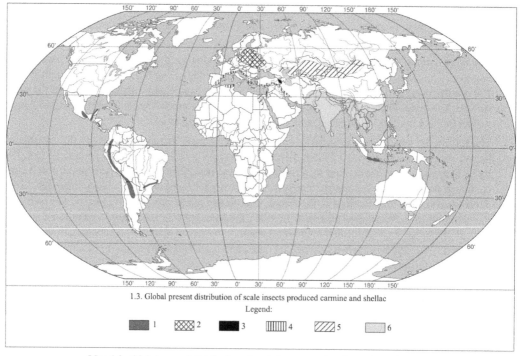

1.3. Global present distribution of scale insects produced carmine and shellac
Legend:
1   2   3   4   5   6

Map 1.3. Global present distribution of scale insects produced carmine and shellac
(http://www.anandarooproy.com/portfolio/project/76; https://en.wikipedia.org/wiki/Polish_cochineal).
Legend: (1) Mexican cochineal (*Dactylopius coccus*); (2) Polish cochineal *Porphyrophora polonica*; (3) Armenian cochineal
(*Porphyrophora hamelii*); (4) scale *Kermes vermilio* (*Coccus ilicis*): (5) various red-dye scale insects of Genus *Porpyrophora*;
(6) true lac scale *Laccifer lacca*

1.5.2.2 History of production of carmine

Carmine is one of the oldest organic pigments. For a long time, it was used in both the Old and New World. Opinion differs as to the authorship and time of origin of producing this dye. Some sources assert that the **Phoenicians** were the first to produce it (residents of the ancient state situated on the eastern coast of the Mediterranean Sea with the center in the present-day Lebanon) about 3,000 years ago (http://0lenka.dreamwidth.org/181770.html).

The scale (*Coccus ilicis*), which inhabited the leaves of the kermes oak (*Quercus coccifera*), was used for its production (https://en.wikipedia.org/wiki/Crimson). Other sources declare that long before, carmine was used by the **ancient Egyptians, Greeks, and Persians,** who mainly used the kermes scale (*Kermes vermilio*) for its manufacture (http://insectalib.ru/books/item/f00/s00/z0000013/st012.shtml).

The time when the use of carmine began in South America is also open to question. According to some data, the **Aztecs** extracted carmine from the cochineal (*Dactylopius coccus*) more than 2,000 years ago (Schowalter 2013). Others maintain that the earliest known fabrics colored with carmine were found out in **Paracas, Peru** (tenth to twelfth centuries AD). The first evidence of cochineal rearing goes back to the tenth century AD, associated with the Toltecs (people who resided in central Mexico) (Rodriguez et al. 2001).

Later on, the **Ararat cochineal** (*Porphyrophora hamelii*) came into use for the production of carmine. It is not inconceivable that this dye was manufactured as early as 714 BC. In fact, in chronicles it is stated that the Neo-Assyrian king Sargon II received a red fabric as a trophy of war from the kingdom of Urartu (the geographic predecessor of Armenia) (https://en.wikipedia.org/wiki/Armenian_cochineal). It is known that in the third century AD, the Persian king, Shapur 1 (http://penelope.uchicago.edu/Thayer/e/roman/texts/historia_augusta/aurelian/2*.html) presented a woolen fabric dyed purple to the Roman emperor Aurelian.

Rome was full of rumors of the fantabulous color of the fabric, the paints for which were produced from a certain "worm", called "vordan karmir", farmed in far-off Armenia. The paint was also used to color the graphic arts in ancient books (Zotova 2003). Carmine was **extremely expensive**. For example, in 1437–38, one gram of dried Ararat cochineal could be purchased in Constantinople for 5.3–9.8 g of gold (Łagowska and Golan 2009).

The twelfth century AD is considered to mark the beginning of production of carmine from the **Polish cochineal** (*Porphyrophora polonica*). The insects were harvested between the Volga and Rhine Rivers. Harvesting began on June 24 and lasted 2–3 weeks. The usual harvest from 1 ha was about 1 pound of females, while a day's harvest of one worker reached about a quarter of a pound, a little more than 100 g (Łagowska and Golan 2009). The dye itself was made in Germany and Poland (http://insectalib.ru/books/item/f00/s00/z0000013/st012.shtml). Later on, production began in the territory of present-day Lithuania. In the fifteenth to sixteenth centuries, the cochineal was one of the most important exports of Poland (along with grain, timber, and salt).

Trade in the Polish cochineal was primarily conducted by Jewish and Armenian merchants who bought up the cochineal from farmers in different regions of Poland and exported it to other countries. Main shipping points included the Polish towns of Krakow and Poznan, and the Prussian town of Danzig (Gdansk), while Breslau (Wrocław), Nuremberg, Frankfurt, Augsburg, and Venice were main destination points. The production of carmine from the Polish cochineal reached its peak in the 1530s. In 1534, about 30 metric tons of the dye was sold in Poznan alone (https://en.wikipedia.org/wiki/Polish_cochineal).

However, since the sixteenth century, the harvesting of both Ararat and Polish cochineal has **receded**. The Mexican cochineal, the trade in which started in 1526, entered the world market (http://www.zooeco.com/0-plant/0-plant22-8-5.html). In conquering Mexico, the Spanish conquistadors gained access to cochineal, among other natural resources. For the first time, the magenta derived from this insect was brought to Europe by Hernán Cortés as gift to his king Charles V (Zotova 2003).

The **advantages** of the Mexican cochineal over the European and Asiatic species were as follows:

The dye extracted from it was far **brighter**. For example, as compared with carmine produced from the scale (*Kermes vermilio*), its concentration was 10–12 times higher (Serrano et al. 2015).

The vital rhythm of this insect is **shorter**, and in Mexico, five generations a year were obtained; that is, "productivity" was much higher (https://en.wikipedia.org/wiki/Cochineal).

The **fat** hindering the extraction of the paint from the Ararat cochineal, where its content reaches 30%, is practically absent in the dried bodies of the Mexican cochineal (http://www.azerbaijanrugs.com/arfp-natural_dyes_insect_dyes.htm).

The cochineal was one of the most significant exports of New Spain (after gold and silver), and the base of the economy in the mountainous areas of Mexico. In view of its huge monetary value, the technology of carmine production was a Spanish state secret (Chávez-Moreno et al. 2009). In Europe, the dried bodies of the cochineal were taken for seeds for a long time. In Russia, the "grains" of cochineal were known as "clerical seed" (Zotova 2003). Only in 1703 was it realized that it is an insect (http://insectalib.ru/books/item/f00/s00/z0000013/st012.shtml).

Spain's **secret** of carmine production from the Mexican cochineal became known to others. From the late 1700s, attempts were made to acclimatize the cochineal (as well as cactuses on which the insects feed) beyond their natural range. In most of Europe (except for Spain), the scale died; however, the Mexican cochineal was successfully introduced to the Canary Islands, Argentina, Guatemala, and South Africa (Campana et al. 2015), as well as Madagascar (Łagowska and Golan 2009), Honduras, Indonesia (Java Island), and Algeria (http://www.zooeco.com/0-plant/0-plant22-8-5.html).

The **production** of Mexican cochineal (both in Mexico and in the areas where it was transplanted) increased steadily until the end of the 1870s, when it reached 7 million pounds (3,178 t) of dried insects (http://www.pinebrookhills.org/Press/Articles/cochineal_insect.htm). With the invention of aniline dyes, cochineal rearing began to decline. For example, 2,557,000 kg of it was exported from the Canary Islands in 1880–81 and only 482,000 kg was exported in 1888 (http://www.brocgaus.ru/text/054/838.htm). For some time, the cultivation of cochineal was on the decline. However, in the 1970s, after the discovery of carcinogenic and hazardous properties of many synthesized dyes, carmine production was renewed (Chavez-Moreno et al. 2009).

At the **present time**, Peru accounts for about 85% of the world production of carmine. Large harvests of Mexican cochineal also take place in the Canary Islands, southern Spain, Algeria, and countries of Central and South America (Tsatsenko 2015). However, the present-day production volumes are less than they used to be. In 2005, only 200 t of carmine were produced in Peru and 20 t in the Canary Islands (https://en.wikipedia.org/wiki/Cochineal).

1.5.2.3 Production and use of carmine

In the production of carmine, **diligent handwork** is used. For the manufacture of 1 kg of carmine, a large number of insects is needed. According to different data, 88,000 (http://www.vegparadise.com/news13.html), 100,000–150,000 (Łagowska and Golan 2009), or 150,000 (Irwin and Kampmeier 2003) are needed. The process itself is as follows. During the period when the females are laying eggs (when the females contain the most red pigment in the organism), the insects are removed from the cactuses by using hard brushes or blades. Then they are killed in acetic acid or under high-temperature stress.

Thereafter, the shells of the insects are **dried** and **milled**. The dyestuff of carmine is the carminic acid, which commonly accounts for 17–24% of the mass of the dried insects (https://en.wikipedia.org/wiki/Cochineal). Its coloring depends on environmental acidity. At pH 3 (acidic medium) the color is orange, at a slightly acidic pH of 5.5 it is red, while at pH 7 the color is purple (http://4108.ru/u/karminovaya_kislota). Therefore, to change the initial color and attain the necessary tint, mordants and adjuncts that change the acidity are added to the powder of milled cochineals (Campana et al. 2015). These substances can be lime, solutions of ammonia or sodium carbonate, salts of aluminum and calcium, and other chemicals.

During the **ancient** and **medieval periods**, dyeing fabrics was the primary purpose of carmine. In **our time**, it is mainly used in the production of foodstuffs, cosmetics, and medicinal preparations (Borges et al. 2012); it is also to some extent used in the textile industry, polygraphy, production of plastics, medical investigations, or pictorial art (manufacture of art paints).

In the **foodstuffs industry**, carmine is known as "carminic acid", "carmine", cochineal, cochineal extract, crimson lake, C.I. 75470, E120, and natural red (http://www.innovateus.net/innopedia/how-carmine-obtained). The **list of products** in which carmine is used for coloration is vast. It includes some types of cheeses; dry breakfast cereals; bitter soda beverages; alcoholic and aromatized alcohol-free beverages; jams, jellies, marmalades, and other similar products of fruit processing; glazed vegetables and fruits; canned fruits; sugar confectionery products and decorative coatings; cake bakery and pasta; ice cream; fruit ice; desserts; meat products; fish products; different appetizers and soups; sauces; dressings; and mustards.

In the cosmetics industry, carmine is used as a dye in the production of the **following goods**: lipstick, lip gloss, eye and lip pencils, eye shadows, cheek colors, eyeliners, lash mascara, face and body creams, soap, toothpastes, and liquids for oral hygiene. The pharmaceutical industry uses carmine for coloration of ointments and tablets.

Carmine is considered to be a fairly **nonhazardous additive**. Its use is permitted in Russia, the United States, and a number of European countries. However, there are the restrictions concerning its application in some countries. For example, it used only for alcoholic products in Denmark; in Sweden the confectionery products were added to alcoholic drinks, while it was completely prohibited in Finland (Łagowska and Golan 2009).

It is widely accepted that, at concentrations used in the foodstuffs industry, **side effects** are not manifested. Nevertheless, allergic responses to carmine are observed in a small percent of people in the world. The carmine present in some foodstuffs can result in anaphylactic shock. In rare instances, carmine can also cause allergic responses upon contact with skin. Therefore, in many countries there are mandates to show information on its presence in products. For example, a similar standard was accepted in the United States by the U.S. Food and Drug Administration on January 5, 2011 (http://www.fda.gov/ForIndustry/ColorAdditives/GuidanceComplianceRegulatoryInformation/ucm153038.htm).

1.5.2.4 Production of ink

Scales are not the only insects used for making dyes. Dyes also have been produced by using the larvae of other insects (gall wasps and aphids). These dyes are black in color, and they also have been used for dyeing cloth; however, they first were applied for making **iron gall ink**.

In many cases, the galls—spherical or oblong burrs with diameters of 1.5–2 cm and more—can be seen on the leaves of some trees (oak, sumac, pistachio). They are a result of the activity of parasitizing gall wasps. These wasps are usually about 5 mm long. In early spring, the female of this insect lays eggs into the leaves, the hatched larvae grow, and burrs build up around them. Such burrs are called the "**gall nuts**", due to the fact that they were used for making writing inks until quite recently (http://www.si.edu/encyclopedia_si/nmnh/buginfo/benefits.htm). Because they contain tanning agents (tannic acid; up to 70% of dry mass), gall nuts were also used for production of pharmaceutical preparations (astringent remedies) and tanning hides (http://www.pitersad.ru/galli.html).

It is believed that inks for writing were manufactured from gall nuts over a period of **not less than 1,800 years** (https://en.wikipedia.org/wiki/Andricus_kollari). Traces of the ink were detected on the Dead Sea Scrolls, which were found during 1946–56 in the caves of the Judean Desert near the Dead Sea in the West Bank of the Jordan River. The scrolls dated back to 408 BCE to 318 CE (https://en.wikipedia.org/wiki/Dead_Sea_Scrolls).

The **earliest surviving recipe** for making iron gall inks is presented in the largest encyclopedic work *Naturalis Historia* written by the ancient Roman author Pliny the Elder (living in 22–24 to 79 CE). However, there are numerous recipes for making these inks. For example, the U.S. Postal Service had its own official recipe for making inks, which were produced for all post offices of the country for use by its clients (https://en.wikipedia.org/wiki/Iron_gall_ink).

One of the recipes is as follows: Take 3 parts gall nuts (galls), 2 parts sulfate of iron, 2 parts arabic gum (resin of acacia), and 60 parts water. The nuts are powdered and poured into a glass bottle that is filled with cold water. Sulfate of iron and arabic gum are dissolved in another tank. The powdered nuts should be held in the water for several days until the water extracts from it all the tanning agent, while the sulfate of iron and arabic gum dissolve within several hours. Both solutions are stirred together, and after one or two days, the solution is drained to separate the liquid from sediments (http://www.scud.ru/pencil/ink_watery_powder/01.php).

Iron gall inks were **widely used** through the mid-20th century for a vast number of letters, manuscripts, pieces of art, music scores, drawings, maps, and public documents such as wills, marriage certificates, bookkeeping records, and real estate transactions. Iron gall inks are still being used by the U.S. Treasury for printing money (Irwin and Kampmeier 2003).

In other spheres, inks obtained by chemical means have replaced iron gall ink. Now, iron gall ink is primarily used by artists who are enthusiastic about reviving old methods—and seemingly by forgers of old documents (https://en.wikipedia.org/wiki/Andricus_kollari).

1.5.3 Production and use of shellac

Shellac is made of lacquer secreted by the females of lac scales belonging to the Kerriidae family; these organisms are parasitic on some tropical and subtropical trees. The lacquer is a source of three valuable natural products: resin, dye, and wax used in different economic branches.

1.5.3.1 Insects used in production of shellac and the history of its production

The **lac scales** are small (1–5 mm) insects that feed on the moisture of plants. The word *lac* in Hindi means "100,000" pointing to the vast number of insects needed for its production (Kondo and Gullan 2007). The insects, by means of special glands, generate the gumlike or resinous secretions. The resin forms a cocoon around the insect that is used to incubate the eggs it lays.

There are about 90 species of lac scales (Kondo and Gullanz 2007). Three species are of **industrial importance** (https://en.wikipedia.org/wiki/Kerriidae): *Laccifer lacca* (also called *Kerria lacca*), true lac scale; *Paratachardina decorella*, rosette lac scale; and *Paratachardina pseudolobata*, the lobate lac scale. However, *L. lacca*, the true lac scale, is the most significant. It is a little insect, 1–2 mm in size, that is widespread in South and Southeast Asia (Bangladesh, Myanmar, China, India, Sri Lanka, Malaysia, Nepal, Pakistan, and Taiwan). The global distribution of true lac scale *L. lacca* is shown in Map 1-3. The true lac scale (*L. lacca*) was also acclimatized to Transcaucasia (Azerbaijan, Georgia), North America (Mexico), and South America (Guyana).

It is parasitic on more than **400 host plants** belonging to many families: the cashew or sumac family (Anacardiaceae), custard apple family (Annonaceae), dogbane family (Apocynaceae), birch family (Betulaceae), gourd family (Cucurbitaceae), Ebenaceae family, coca family (Erythroxylaceae), spurge family (Euphorbiaceae), bean family (Fabaceae), walnut family (Juglandaceae), Lythraceae family, mallows family (Malvaceae), buckthorn family (Rhamnaceae), rose family (Rosaceae), citrus family (Rutaceae), willow family (Salicaceae), soapberry family (Sapindaceae), and Vitaceae family (https://ru.wikipedia.org/wiki/Kerria_lacca).

However, only some of these plants are used for the production of shellac. **In India**, for example, only 14 species of 113 plants inhabited by true lac scale, *Laccifer lacca*, are used for its production (Derry 2012); the **most common host trees** are (Yogi et al. 2015): dhak (*Butea monosperma*), ber (*Ziziphus mauritiana*), and kusum (*Schleichera oleosa*). In Thailand, they include the rain tree (*Samanea saman*) and pigeon pea (*Cajanus cajan*), while in China such plants include pigeon pea (*C. cajan*) and some hibiscus species, and in Mexico, the Barbados nut (*Jatropha curcas*) (https://en.wikipedia.org/wiki/Lac).

Shellac has been produced in India **for about 3,200 years** (Irwin and Kampmeier 2003). In ancient India, it was mainly applied for painting woolen and silk fabrics. Shellac is mentioned multiple times in different literature sources. For example, in the ancient Indian epos *Mahabharata*, a palace constructed entirely of dried shellac is described (https://en.wikipedia.org/wiki/Shellac). Shellac is described in the encyclopedia *Naturalis Historia* written in AD 77 by Pliny the Elder (Derry 2012).

In 1596, Garcia de Orte, physician of the Portuguese governor of India in Goa, described both lac resin and lac dye in his book *Conversations on the Simples, Drugs and Medicinal Substances of India* (Gibson 1942). Information on shellac was found in a report by the traveler John Huyghen van Linschoeten, sent in 1596 to the Portuguese king; it contained a complete list of the kinds of shellac used in those far-off days (Das and Singh 2014).

In 1220, shellac began to be used as an **artist's pigment** in Spain. In the thirteenth century, shellac was widely sold throughout the Mediterranean world. It was very popular in Italia, where it was widely applied for varnishing wedding trunks (cassone) (https://en.wikipedia.org/wiki/Shellac). From the beginning of the seventeenth century, shellac was commonly used in Europe as a dye because it was a cheaper alternative to cochineal. After that, in the second half of the 19th century, the production of synthetic aniline dyes began, and there was a loss of interest in the dyeing component of shellac. Then it came into use as a material for covering both new and antique furniture (Derry 2012).

1.5.3.2 Production of shellac

In the **process of digestion** of tree sap, lac scales (*Laccifer lacca*) secrete lacquer so they can attach themselves to the branches of the trees (Photo 1.5-6). Female lac scales provide the greatest amount of lacquer after conception, while the male scales essentially do not secrete it. Branches covered with lacquer are cut off for further processing, and some of them are transferred to other trees to be contaminated by lac scales.

The lac scale produces up to **four generations** a year, and lacquer is harvested the same number of times. The lacquer is removed by hand from the branches, and is washed in water, dried, charged to cloth bags, and placed into furnaces where the lacquer is heated to a temperature of 105°C. The melted lacquer is filtered through the cloth, and the waste is left in the bag (http://insectalib.ru/books/item/f00/s00/z0000013/st012.shtml).

The grade of lac (lacquer) is designated by **special terms**. *Sticklac* is the cocoons containing, besides the resin, the remainder of branches and insects; that is, it is raw material for production of shellac. Seedlac is the product obtained after the sticklac is milled, sieved, and washed many times to remove insect particles and different soluble substances. Shellac is the material in the form of thin, small chips that are used to produce industrial lacquer with alcohol (Marikovsky 2012). The proportions of components in different grades of lacquer are given in Table 1.5-3.

The weighting factor of conversion from sticklac to shellac is 0.58 (Sengupta et al. 2011). According to different data, in order to produce **one kilogram of shellac**, the following numbers of insects are needed: 37,400–198,000 (Irwin and Kampmeier 2003); 200,000 (https://en.wikipedia.org/wiki/Shellac); 220,000 (http://www.naturalhandyman.com/iip/infpai/shellac.html); or 300,000 (Singh 2006; Flinn 2011). The annual yield of sticklac per tree is as follows: 1–4 kg for dhak (*Butea monosperma*); 1.5–6 kg for ber (*Ziziphus mauritiana*); and 6–10 kg for kusum (*Schleichera oleosa*) (Sengupta et al. 2011).

Photo 1.5-6. In the process of digestion of tree sap, lac scales (*Laccifer lacca*) secrete lacquer so they can attach themselves to the branches of the trees. Branches covered with lacquer are cut off for further processing, and some of them are transferred to other trees to be contaminated by lac scales. Resin of the scale insect (*Laccifer lacca*) that is, shellac on the branch is shown.
Photo credit: By Jeffrey W. Lotz [CC BY 3.0 (http://creativecommons.org/licenses/by/3.0)], via Wikimedia Commons, 23 February 2012

India is the **main global producer** of shellac, Thailand is the **second** in order of importance, and Indonesia and China rank **third** and **fourth**, respectively. Other countries producing shellac include Bangladesh, Myanmar, Vietnam, Sri Lanka, Laos, and Mexico; however, their share is minor (https://en.wikipedia.org/wiki/Lac).

While shellac production in India has centuries-old traditions, this kind of activity is **relatively new** in other countries. Thailand began to extract shellac only during the early 1940s (Sengupta et al. 2011), while China began in the mid-1950s (http://www.fao.org/3/a-v8879e/v8879e08.htm). Elsewhere extraction began even later.

The **prosperity** of this branch was observed in the mid-1950s. At that time, India produced about 29,000 t of shellac annually. By the late 1980s, production had dropped to 7,000 t. During 1992–93, the world output of shellac was lowest; during this period, India produced only 4,500 t of shellac (Łagowska and Golan 2009). Later, production volumes began to increase in India as well as in Thailand and China. In the mid-1990s, Thailand produced about 7,000 t while China reached a level of about 3,000 t of shellac (http://www.fao.org/3/a-v8879e/v8879e08.htm).

The **volumes** of sticklac production in 2006–07 are given in Table 1.5-4.

Currently, India remains the leading producer of shellac. After a long decline in production—reaching 52% in the 1970s, 19% in the 1980s, and 4% in the 1990s—a boom occurred in the industry. For example, in 2013–14, 21,008 t were produced. Most production was in the Jharkhand state (58%), and substantial contributions were made by Chattisgarh (16.1%), Madhya Pradesh (11.9%), Maharashtra (5.6%), and Odisha (3.2%).

The remainder of the Indian territory (West Bengal, Meghalaya, Assam, etc.) was responsible for only 5.2% of production. Of the 21,008 t produced, 8,158.1 t were exported. The major importers in 2014 were the United States (25.23%), Switzerland (11.37%), Germany (9.97%), Pakistan (9.90%), and Bangladesh (7.33%) (Yogi et al. 2015).

Current data on shellac production volumes in **other countries** were not found. Most likely, the rates of shellac production growth in Thailand are higher than those in India, resulting in a gradual decline of India's share in the world market (https://en.wikipedia.org/wiki/Lac). The present situation in Indonesia is not clear. In China, production is concentrated mainly in the Yunnan Province, with some contribution by the Fujian province. The largest supplier of shellac is the Chuxiong DES Shellac Co., Ltd. (Yunnan

Table 1.5-3. The physical components of lac.[1]

	Sticklac (%)	Seedlac (%)	Shellac (%)
Resin	68	88.5	90.5
Dyestuff	10	2.5	0.5
Wax	6	4.5	4.0
Volatiles	4	2.5	1.8

[1] Derry (2012).

Table 1.5-4. Average production of sticklac across countries in 2006–07.[1]

Country	Production of sticklac (MT)
India	15,200
Thailand	8,000
Indonesia	8,000
China	1,000
Vietnam	400
Myanmar	50
Total	32,650

[1] (Sengupta et al. (2011).

Province, Mouding County), which produces about 1,000 metric tons a year (http://des-shellac.com/en/about.aspx).

1.5.3.3 *Use of shellac*

Shellac consists of resin, dye, and wax, so its use can be based on one or more than one component. The percentages of shellac use in the recent past and today are very different. For example, in the 1930s, one-half of shellac was used in the production of **gramophone records**. From 1921 to 1928, 18,000 t of shellac were used to produce 260 million gramophone records (https://en.wikipedia.org/wiki/Shellac). It was mainly used for this purpose from 1895 until the invention of vinyl in 1949; however, it was still used in some countries until the 1970s (https://en.wikipedia.org/wiki/Gramophone_record).

In the first half of the 20th century, shellac was widely used for **finishing wood** siding of walls and ceilings. Such decorator panel boards were comparable with today's gypsum board as far as frequency of use. Shellac was also frequently used for the manufacture of kitchen cupboards and parquet floors. It was widely applied for stiffening of shoe toes (pointes of ballet shoes) to keep dancers en pointe. Shellac was historically used as a protective covering for paintings (Irwin and Kampmeier 2003).

Shellac was also used to cover the pages of **books written in Braille** in order to protect them from wear when they were read. Before the large-scale implementation of plastic materials, shellac was applied as a molding compound in making picture frames, boxes, toilet articles, jewelry, ink wells, and even dentures. In museums, it was used to the mid-1960s to glue and stabilize bones of prehistoric animals. Early in the 20th century, shellac was applied to protect some military rifle stocks (Flinn 2011).

Alcohol solutions of shellac were once used to impregnate the **inductance coils** of electric motors and generators and transformer windings. The surfaces of single-layer coils were painted with a shellac solution, while multilayer coils were impregnated by dipping them into a solution. After drying and alcohol evaporation, the shellac glued the leads, providing supplementary insulation isolation and preventing the movement and vibration of leads in multilayer windings. As a result, running engines became less noisy and their service life increased.

All the present-day consumers of shellac can be subdivided into **three groups** (Irwin and Kampmeier 2003; Singh 2006; Farag 2010; Flinn 2011; Derry 2012; https://en.wikipedia.org/wiki/Shellac): (1) primary; (2) secondary; and (3) nonessential. **The primary consumers** of shellac are the following branches of industry: (1) pharmaceutical; (2) food; and (3) light industry. In the *pharmaceutical industry*, it is applied as glaze to cover tablets and pellets. Shellac can withstand gastric acids, so the medicinal component is released in the intestinal tract rather than in the stomach.

In the *food industry*, it is used under code E 904 as a material for confectionery coatings (envelopes for sweetmeats; in particular, chocolate-coated dragee: noisettes, chocolate-covered raisins, and others). Lac dye is applied to dye juice drinks, carbonated drinks, wines, candies, jams, and sauces. Shellac is used as a "wax" coating for fruits such as apples and citrus fruits to prevent moisture loss and prolong shelf life. Shellac is also used as binder for foodstuff stamp inks, such as for cheese and eggs.

In *light industry*, shellac has three intended purposes: (1) manufacture of headgear; (2) leather dressing; and (3) dyeing of fabrics. In the *first instance*, it is used to stiffen and waterproof felt hats, and to manufacture silk top hats and riding hats. Shellac is mixed with aqueous ammonia, and the mixture permeates the gauze fabric to provide the frame stiffness. In the *second case*, it is used to produce fluid creams that are used to impart shine to leather shoes and haberdashery. The *third purpose* is the manufacture of dyes, mainly for cotton and silk cloth; shellac gives a number of warm colors ranging from pale yellow through to dark orange-reds.

The secondary consumers of shellac are less common than the first group. They can include (1) production of protective coverings for wood items; (2) fabrication of decorative paints; (3) fabrication of abrasive wheels; and (4) fertilizer manufacturing. As a *protective covering,* shellac based on alcohol is applied in **two fields**: (1) furniture trade (because shellac is compatible with most other finishing materials, it is also used as a primer to prevent the extraction of tar from the wood after final finishing and to delay penetration of atmospheric moisture to the wood); and (2) topcoat for acoustic wooden instruments: violas, acoustic guitars, pianos, and others (French polish for finishing instruments—violas, guitars, and pianos—is made of shellac) and for improvement of their acoustic properties.

Shellac is used as a binding material for manufacture of *decorative paints* of the restricted lac color palette. Different tints of the yellow-brown palette are obtained when pigmented shellac and titanium dioxide are used. For fabrication of *abrasive wheels* used in fine work (grinding of soft glasses and other brittle materials with manual feed), a mixture of dry plaster powder, abrasive, and high-grade orange polishing shellac is heated and molded. In *fertilizer manufacturing*, granules of urea are coated with shellac to prevent their coalescence and to slow the release of fertilizer.

Nonessential consumers include the following: (1) manufacture of cosmetics (hair spray, lash mascara, nail color, perfume extract, and lipstick); (2) production of incandescent electric lamps; (3) use in pyrotechnics as a combustible substance, and for warning lights (for example, green light is composed of 85% barium chlorate and 15% shellac) and tracer ammunition (tracer composition is 55% barium nitrate, 35% magnesium, and 10% shellac); (4) dental technology (fabrication of individual custom impression trays for production of artificial crowns and dental prostheses); (5) cycling (fabrication of decorative handlebar tape and as a glue for tubular tires, in particular, for track racing); (6) fabrication of artificial flies for catching salmon (shellac is used to seal the cut threads in binding of flies); (7) watchmaking (owing to the low melting point of shellac—about 80–100°C—it is applied for fixing pallet fork stones in the mechanism and for attachment of small parts to the chuck); (8) as a binder in India ink; (9) displaying entomological collections (very small insects are not pinned but are stuck by using a gel adhesive mixture composed of 75% ethyl alcohol and shellac on special cartons; and they, in turn, are stuck on a pin); (10) restoring vintage fountain pens (silk tubes are soaked in shellac, replacing the damaged ink sac); (11) prevention of spontaneous damage to surface layers of string-plates of acoustic instruments made, in most cases, of softwood of resonant spruce; (12) impregnation of damaged pieces of canvases in the restoration of paintings; and (13) as a sealing wax.

1.5.4 Production and use of wax

Wax is a fatlike solid material of natural or synthetic origin. Natural waxes can be contemporary (animal and vegetative) and mineral. In this book, animal waxes produced by insects are of interest.

1.5.4.1 Insects producing wax

Producers of wax include many insects: bees, bumblebees, aphids, scale insects, and others. However, only different species of honey bees and two species of scale insects—Indian white wax scale (*Ceroplastes ceriferus*) and Chinese white wax scale (*Ericerus pela*)—are of **economic importance**. A certain role is also played by lac scales of the Kerriidae family that produce shellac. Wax accounts for 4–6% of its composition.

From the viewpoint of wax production, **honey bees** belonging to the genus *Apis* are the most well known and significant. This genus numbers seven species: western honey bee or European honey bee (*Apis mellifera*), black dwarf honey bee (*Apis andreniformis*), dwarf honey bee (*Apis florea*), giant honey bee *(Apis dorsata)*, Asiatic honey bee (*Apis cerana*), red bee of Sabah (*Apis koschevnikovi*), and Philippine honey bee (*Apis nigrocincta*). The range of honey bees is extremely wide, and they are distributed on all continents except for Antarctica.

The western honey bee, or **European honey bee** (*Apis mellifera*), having many subspecies, is the most significant and most common species of honey bee. These honey bees produce wax in eight wax-producing mirror glands on the bee's abdomen. After the wax is released from the glands, it stiffens in the form of small semitransparent wax scales about 3 mm long and 0.1 mm thick. In order to obtain one gram of wax, approximately 1,100 of these scales are needed (https://en.wikipedia.org/wiki/Beeswax). The bee has small brushes on its back legs that are used to remove the wax and carry it to the forelegs; then the wax is brought into the masticatory apparatus. The bees use the wax to construct combs where the larvae are laid and honey is stored (http://mednadom.com/statiy/pcheliniy-vosk/).

Most of the wax is produced by young worker bees aged 12–18 days. The size of the wax-producing glands depends on the bee's age, and after a great number of flights, they begin to gradually atrophy (https://en.wikipedia.org/wiki/Beeswax). Aside from the honey bees, other bees produce wax, but they are of no economic importance. For example, the solitary bees, which do not produce honey (eastern carpenter bee, *Xylocopa virginica*; alfalfa leafcutter bee, *Megachile rotundata*; hornfaced bee, *Osmia cornifrons*; and others) produce wax too, but in small quantities (http://roypchel.ru/pchely/vidy-pchel.html).

Only female bees produce wax, but both females and males of scale insects produce it. The **scale insects** are small insects that feed on the moisture of plants. Many of them are serious agricultural pests. The scale insects also have special wax-producing glands. They use the wax to protect egg capsules and the body from their fluid excrements (Qin 1997). Only two species are of **economic value**. The Chinese white wax scale *(Ericerus pela)* is more well known. It is one of the oldest insects used by humans (after silkworms and honey bees). The earliest data on the use of wax produced by the Chinese white wax scale, *E. pela*, are dated to the third century AD, while a detailed description of the wax manufacture technology is contained in a book published in the thirteenth century (Lung 2004).

This insect initially inhabited China (territories of the present-day provinces of Hunan, Sichuan, Yunnan, Zhejiang, and Xizang, and the Tibet autonomous region), Japan, South Korea, Vietnam, Russia (Primorsky Krai), and Brazil (Park et al. 1998; Ben-Dov 1993). The Chinese white wax scale has often been introduced to **other regions**. It was imported to Europe for the first time as early as 1615 (Lung Tsuen-Ni 2004). This insect feeds on 200 different species of host plants belonging to 98 genera and 36 families (Qi Jia and Ji-Fu Zhao 2004); however, the Chinese privet (*Ligustrum lucidum*) and Chinese ash (*Fraxinus chinesis*) are considered to be the best hosts (Zhang et al. 1993; Chen et al. 2015). The Chinese white wax scale can live under the following climatic conditions: annual average temperatures of 11°–16°C; annual rainfall of 800–1,200 mm; annual relative humidity between 65 and 75%; and annual sunlight of 1,900–2,500 h/year (Chen et al. 2007; Chen et al. 2015).

The range of the **Indian white wax scale** (*Ceroplastes ceriferus*) is much wider. It is distributed on all continents except Antarctica (https://www.eppo.int/QUARANTINE/Alert_List/deletedfiles/insects/Ceroplastes_ceriferus.doc). In Asia, the Indian white wax scale inhabits Cambodia, China, India, Indonesia, Japan, Malaysia, Myanmar, Philippines, Sri Lanka, Taiwan, Thailand, and Vietnam. In Africa, the range of these insects includes Malawi, Tanzania, and Uganda; in Central and South America, and the Caribbean, it includes Brazil, Chile, Jamaica, Panama, Puerto Rico, and the U.S. Virgin Islands. In Oceania, it includes Australia, Cook Islands, Fiji, Guam, New Caledonia, Papua New Guinea, New Zealand, Tonga, and Vanuatu, while in North America it includes many states of the United States and Mexico.

Over the last several years, it has been recorded in **some European countries**: Netherlands in 1999, Italy in 2001, and the United Kingdom (Lee et al. 2012). In the overwhelming majority of countries, its presence is undesirable because, apart from wax production, the Indian white wax scale is an invader of many fruit and ornamental plants; there are 122 such plant species in 46 families (https://www.eppo.int/QUARANTINE/Alert_List/deleted files/insects/Ceroplastes_ceriferus.doc).

1.5.4.2 Production of wax

The natural waxes produced by insects vary in physical and chemical properties. Information on some of these properties is given in Table 1.5-5.

It should be pointed out that the types of beeswax are very different. For example, the wax produced by the Asian species of bees (known as ghedda wax) is less acidic. The wax of stingless bees, Meliponinae, is distinguished by its deep-brown color, stickiness, resistance to breakage as compared with the wax of the European honey bee (*Apis mellifera*) (Bradbear 2009).

Pure beeswax is manufactured by remelting the wax-containing raw material. It consists of comb wax, cut tops of cells (comb capping), and growths on the walls of bee houses and frames. The **purest wax** is obtained by melting the comb capping (bees use it to seal the combs with honey). The wax obtained as a result of processing the old frames with combs is darker in color. Because several generations of larval bees were hatched in them, these frames contain residues of larval cocoons and bee bread (pollen), and such wax includes many admixtures.

Table 1.5-5. Comparative analysis of different types of wax produced by insects.[1]

Properties	Beeswax	Chinese wax	Shellac wax
Color	White to dark-yellow or gray	White to yellowish-white	Light yellow
Flavor	Honey-like	Practically odorless	–
Density	0.958–0.970 (at 15°C)	0.93	0.97–0.98
Melting point	62°–68°C	92°C	70°–80°C
Chemical composition	72% of compound ethers, 12–15% of hydrocarbons, 15% of carboxylic acids	95–97% of compound ethers, 1% of resin, 1% of alcohols, 1% of hydrocarbons	60–62% of compound ethers, 33–35% of alcohols, 2–6% of hydrocarbons

[1] https://en.wikipedia.org/wiki/Beeswax; http://data.cas-msds.com/Chinese%20Wax.html; http://www.xumuk.ru/encyklopedia/821.html

Table 1.5-6. Leading countries in the production of beeswax in 2012.[1]

Rank	Country	Quantity (MT)	Cost (in $1,000)
1	India	23,000	215,367
2	Ethiopia	5,000	46,818
3	Argentina	4,700	44,009
4	Turkey	4,235	39,655
5	Republic of Korea	3,063	28,683
6	Kenya	2,500	23,409
7	Angola	2,300	21,536
8	Mexico	1,990	18,633
9	Brazil	1,850	17,323
10	Tanzania	1,830	17,135

[1] http://faostat.fao.org/site/339/default.aspx.

Data on the production of beeswax by the 10 leading countries are given in Table 1.5-6.

The absence of China in the list comes as a **surprise**—all the more so because, in a catalogue of countries exporting wax for 2003 given in other publications of the FAO (Food and Agricultural Organization) (Bradbear 2009), it comes first with a quantity of 4,814 t (of 10,336 t of the world export) while the figures for 2014 indicate that China exported 11,000 tons in 2014 (Bogdanov 2016). Based upon the proportion that the quantity of wax ranges from 1.5 to 2.5% of that of produced honey (http://www.fao.org/docrep/x5326e/x5326e0c.htm#1.3.production), if the honey harvest in China reached 436,000 tons in 2012 (http://faostat.fao.org/site/339/default.aspx), production of wax should have been 6,540–10,900 tons. It should be understood that wax is frequently discarded, but this eventuality is unlikely to be characteristic of China.

In statistics on **international trade**, beeswax is usually grouped with other waxes produced by insects (http://www.fao.org/docrep/x5326e/x5326e0c.htm#1.3.production). It is evident that the share of beeswax predominates; however, quantitative indicators concerning the other kinds of wax are of interest. Information concerning shellac wax could not be found. As to the wax obtained from lac scales, Łagowska and Golan (2009) indicate a figure of 300 t as the annual average of its production in China.

There are data from several authors for different years with respect to the Szechuan Province—a **major wax-producing region** of the country—which are presented in an article by T.-K. Qin (1997). According to them, the harvest of wax reached 300 t early in the 20th century. In a publication dated 1913, it is said that the harvest of wax in off years was 3,000 t, while in good years it was twice as much. For the late 1930s, a figure of 2,800 t is given, while in a 1989 publication, it is said that, beginning in 1949, the maximum harvest was 590 t (Qin 1997).

There are **four basic methods** of wax processing (http://www.salkova.ru/Product_bee/Beewax/description.php): (1) process wax residue with the use of solar wax melters (raw wax material is melted by using heat obtained from sunbeams; in the melters, one can remelt only light combs); (2) steam melting (hot steam is fed into a chamber with primary raw wax material, and melted wax is drained in to a receiving tank); (3) water melting (raw material is boiled soft, and the wax floats to its surface); and (4) extraction (processing of the raw wax material using hot organic solvents: gasoline, carbon tetrachloride, or different petroleum solvents).

Production of wax using soft scale insects is quite complex and consists of **many stages** (Qin 1997). In April and May, the hibernating females of Chinese white wax scales (*Ericerus pela*) with their eggs (called egg capsules) are collected from the fields. In this case, they should meet **three criteria**: (1) color (red-brown); (2) flexibility (when the dorsum of the body is pressed, the affected part should go to the initial position); and (3) low moisture content (dryness). The egg capsules collected should be stored in a dry and cool place until they solidify. Then the large egg capsules containing large numbers of eggs are taken for creation of new cultures.

When the majority of nymphs are hatched out and begin to accumulate on the surfaces of **egg capsules**, they are ready for placement in small bags (three to six capsules in one bag). This process is called "wrapping insects". After that, some females crawl out to the surfaces of bags, and they are hanged on host plants. This process is called "hanging bags". When the insects disperse through the trees, control of natural enemies and handling of host plants are necessary until the next generation of egg capsules is ready for collection. The collected egg capsules are used for two purposes: (1) as a base for production of new egg capsules (i.e., females) and (2) as a source for wax production (Qin 1997).

The process of wrapping insects for obtaining the wax takes place later than that for production of egg capsules. This timing is related to the fact that the males always crawl out from the bags later than the females. The great threat to the insects is posed by strong winds, which may carry the larvae of males far from feeding trees. Therefore, in case of weather deterioration, the bags are removed from the trees and carried indoors. Nursing of insects includes a mass of other operations (moving the bags to other trees or branches in case of high crowding of larvae, collecting fallen bags and returning them to trees, controlling natural enemies, applying fertilizers into the circles around the host-tree trunks, etc.).

When the nymphs turn into pupae, wax secretion by the insects stops. Right after that, the **wax harvest** starts. Usually, this happens in the period of late August to early September. Wax is most readily collected during a light rain, after a rain, or in clear mornings before the dew has dried. If the branches were used for the first time, then wax is scraped from them. In case of recolonization of branches, the branches are cut away. The wax should be processed on the day of collection, and where that is not practicable it should be stored in a cool, ventilated location (Qin 1997).

Historically, the regions where egg capsules are obtained have been separated from regions where wax is produced; in many cases, they even have been situated in different provinces. An article by Herbert Beardsley in the November 1932 issue of *Nature* discusses special "wax caravans". Army of bearers transported bamboo baskets with egg capsules of the Chinese white wax scale (*Ericerus pela*) for a distance of 450 km from the Chien-Chang Valley up to the Sze-Chuan Mountains. To avoid the untimely development of nymphs, baskets were transported by night when it was cool (https://en.wikipedia.org/wiki/Chinese_wax). Now, depending on the distance, the egg capsules are conveyed by motor or air transport.

Information on the "productivity" of Chinese white wax scale (*Ericerus pela*) could not be found, but for the Indian white wax scale (*Ceroplastes ceriferus*), about 1,500 individuals are needed in order to obtain **1 g of wax** (http://www.cyberlipid.org/wax/wax0001.htm). As for shellac wax, it is produced by way of settling from alcoholic and water-alcoholic solutions of shellac. Then the precipitated sludge is filtered and dried (http://www.xumuk.ru/encyklopedia/821.html).

1.5.4.3 Use of wax in ancient times

The use of wax began in ancient times. Beeswax was used first. The lost-wax method of **casting metals** can be considered as its most important application. The wax was used to make the aftercast of an article (so-called duplicator stencil) that was covered with a heat-resistant material; after the duplicator stencil was melted, a mold (cavity) was obtained and this mold was filled with molten metal. Figures containing beeswax survived in royal Egyptian tombs dating from 3400 BC (Crane 1999). This method was known in other old high cultures, such as those in Sumer, India, and China. Many globally known ancient statues were made by using this method (Bogdanov 2016).

Wax produced by **scale insects** came into use much later. At least, Yang et al. (2011) state that the wax secreted by the Chinese white wax scale (*Ericerus pela*) was used for wax printing. Fabrics were coated with a thin layer of wax, and then a picture was applied to them by using a pointed stick. The fabric was dipped into a dye solution, and only those portions of fabric that had no protective layer of wax were painted. Afterwards, the process was repeated. Wax was used in print engraving, wax candle production, and Chinese medicine about one thousand years ago.

Candles of beeswax appeared earlier. Although Bogdanov (2016) states that they were used by the ancient Egyptians, it is evident that this idea is **mistaken**. The first reference to candles being made of beeswax goes back to the Han Dynasty (202 BC–AD 220) (https://en.wikipedia.org/wiki/History_of_candle_making). Prior to that, candles were made of different types of fat. The oldest intact candles of beeswax were found in the Alamannic (Germanic tribe) graveyard in Oberflacht, Germany, dating to the sixth to seventh century AD (https://en.wikipedia.org/wiki/Beeswax).

Information on when the use of **shellac wax** began could not be found. Because it was produced by way of precipitation from alcoholic and water-alcoholic solutions of shellac, and pure alcohol was obtained for the first time late in the eighth century—early in the ninth century by the Persian chemist Muhammad ibn Zakariya al-Razi—it can be assumed that its use began later.

The ancient Greeks and Romans created **wax masks** of deceased persons. Wax was also used in ancient Egypt for embalming dead bodies (https://en.wikipedia.org/wiki/Embalming). In ancient Greece and Rome, beeswax was used to stick together individual items and to manufacture writing implements and stamps for letters.

In those times, wax was applied in **sculpture** and **pictorial art**. So, encaustic painting was widely practiced; also known as hot-wax painting, it is a technique in which wax is the adhesive agent of paints. Pieces of wax mixed with paint on a metal palette were melted over a fire. The paints were laid on canvas, wood, and other materials with the use of not only brushes but also heated metal rods (https://en.wikipedia.org/wiki/Encaustic_painting).

From ancient times and through to the invention of paper, **wax tablets** were used for writing. These were plates made of a hard material (e.g., timber, bone, or baked clay) with a hollow that was filled with dark beeswax. On the plane, one could write using a pointed metal, woody, or bone stick. If necessary, the inscriptions could be erased and smoothed over, so the tablets could be used many times. The oldest wax tablet carved in ivory was found in 1986 in the course of excavations near Kaş in modern Turkey; it is dated to the fourteenth century BCE (https://en.wikipedia.org/wiki/Wax_tablet).

In Europe during the medieval period, sealing wax was made of beeswax, and it was largely used **to guarantee the authenticity** of the letter contained in the cover. The wax was melted and mixed with "Venice turpentine" obtained from the European larch tree (https://en.wikipedia.org/wiki/Sealing_wax). In the 18th and 19th centuries, the use of beeswax was essentially expanded in connection with industrial development. Apart from the manufacture of candles and artistic sculptures, wax was widely used for processing and care of wood, leather, and paper goods; in the pharmaceutical and cosmetics industries; in the manufacture of physical instruments and chemical apparatus; and for the packing of goods (Krivtsov and Lebedev 1993). In the second half of the 19th century, wax was used in the manufacture of some types of ammunition cartridges (the gasket separating gunpowder from the bullet was made of wax) (https://en.wikipedia.org/wiki/Cartridge(firearms)).

1.5.4.4 *Modern use of wax*

Today, most wax (about 80%) is used in beekeeping for manufacture of a wax foundation or honeycomb base (http://nazeb.ru/1454-vosk.html). These are thin (about 1 mm) wax sheets with pressed end caps and starter strips of comb cells on either side. The wax foundation is an important invention of modern beekeeping that makes it possible to hasten the process of comb production by bees. This in turn increases the harvest of honey at the expense of the bees' energy consumption dropping (https://en.wikipedia. org/wiki/Wax_foundation). The remaining wax goes to the following purposes (Crane 1990): pharmacy, 30%; cosmetics, 30%; making candles, 20%; and other purposes, 20%.

In **pharmaceutics**, beeswax is a necessary component in the manufacture of plasters, ointments, and creams; it also forms part of medical suppositories used in gynecology. In many countries, state standards provide for manufacture of pharmaceutical preparations only from beeswax (Krivtsov and Lebedev 1993). Among these preparations are adhesive strips, plaster of Spanish flies, mercurial plasters, soap plasters, camphor ointments, wax ointments, spermaceti ointments, lead ointments, and zinc ointments. Mastication of wax is used to eliminate inflammation in the mouth cavity. As a coating for pills, beeswax facilitates ingestion (Irwin and Kampmeier 2003). Beeswax is also used as a coating for dental flosses.

Beeswax is widely used in **cosmetology** as an ingredient in nutritional, cleansing, and whitening creams and face masks because it is close in composition to a number of components, such as sebum, and contributes to the formation of a waxlike film on the skin surface that prevents its desiccation.

The list of cosmetic products containing wax includes the **following** (http://www.korolevpharm.ru/ dokumentatsiya/syrevye-komponenty/264-beeswax-pchelinyj-vosk.html): (1) moistening preparations; (2) skin cleansers and preparations for skin care; (3) hair conditioners, lotions, and toners; (4) eye pencils and lash mascara; (5) dressings and plasters; (6) colognes, toilet waters, and deodorants; and (7) gels, creams, and suntan and after-sun lotions.

The candlemaking industry is the third (by volume) largest consumer of wax produced by insects. Candles made of these products include the following **types**: (1) votive; (2) household (utility); (3) taper; (4) tea; and (5) decorative. *Votive* candles are used in both Roman Catholic and Orthodox Churches. In Russia, the Holiest Synod (on May 4, 1882) determined that only candles made of pure beeswax should be used in churches. The use of Easter candles is widespread (these are also made of beeswax). Votive candles are used not only in churches. They are lighted in the houses of churchgoers in many situations, such as in cases of birth and chrismation of newborn infants, people dying, separation from a family member, in the event of a rainstorm, filling the first threshed grain, first putting cattle to pasture, and other events (http://izhchara.ru/gromnichnye-svechi/).

Household (utility) candles have the usual cylindrical form and are used for lighting during electricity blackouts. *Taper* candles also are cylindrical and have decorations and contain colorants and fragrances; these candles are used for faint lighting and as accessories. *Tea* candles are contained in small aluminum housings and are used as decorative lighting devices, for aroma lamps, and for heating teapots. *Decorative* candles are used exclusively for décor of rooms, and they can have different shapes as well as look like statuettes (http://ylik.ru/pcheliniy-vosk/izgotovlenie-svechej/#ixzz4500tbckK). Now, even where there is electric light, many people have candles in their houses. The preference for candles made of the beeswax lies in the facts that they burn longer and brighter than paraffin candles and, moreover, are odor free. They have been transformed from a basic source of light to a décor item, and they have taken on aesthetic value.

Apart from primary applications of wax produced by insects that have been listed, it is used in many other economic branches. Consumption for these needs is about 10 t a year (https://en.wikipedia.org/ wiki/Beeswax). **Other applications** are as follows (Krivtsov and Lebedev 1993; http://www.salkova. ru/Product_bee/Beewax/description.php; Irwin and Kampmeier 2003; Bradbear 2009; Bogdanov 2016; http://www.fao.org/docrep/x5326e/x5326e0c.htm#1.3.production):

1. **Food industry** (coatings for protection and conservation of fruits and cheeses against drying. Some chocolate sweets are coated with wax to prevent them from adhering to wrapping paper, and baking dishes for making sponge cake are greased with melted wax rather than with sunflower oil. Beeswax has been registered as nutrient additive E901).

2. **Neurosurgery** (production of bone wax to arrest hemorrhaging from bony tissue).

3. **Dentistry** (creation of typodonts).

4. Production of **lac varnishes** and mastic resins (component of natural polishing materials for furniture, wood items, parqueted floors, etc.).

5. **Perfumery industry** (wax is used to produce a fragrant ethereal oil applied for production of perfume extract; 1 t of wax provides 5 kg of this ether).

6. **Ophthalmology** (manufacture of preparations for cleaning contact lenses).

7. Production of **glues** (wax is used as a binding agent).

8. Production of **sealing compounds** (applied to protect locks of parquet planks and laminates against moisture).

9. Production of **waxed paper** (a waterproof paper with a thin wax coating on both sides, used for storage of foodstuffs and baked foods).

10. Production of **oil-based paints** (wax enters into their composition as a binding agent and increases the brightness and purity of color once it is dry).

11. Manufacture of **chalks and pencils** for children (they are nontoxic for children and do not cause allergic reactions).

12. Colored **hair chalks** (have a soft texture that does not damage hairs and are easily washed).

13. Manufacture of **shoe tar** applied to impregnate threads (waxed threads) in shoemaking.

14. Manufacture of **tree-wound paint** used to coat and heal graftage points and wounds that result when trees are trimmed or otherwise damaged.

15. **Electrical engineering**, where wax is used to manufacture electrets (electric analogs of permanent magnets); pastes for dielectric sensors; insulation; impregnation, jointing, and moisture-protective substances for galvanic batteries; accumulators; cables; inductance coils; and braiding of wire.

16. **Galvanoplastics** (manufacture of metallic duplicates of articles; wax is used to separate the original from an obtained metallic duplicate).

17. Production of **radios and telephones** (wax is necessary for the manufacture of low-power capacitors for low operating voltages).

18. **Polygraphy** (wax is a component of paints and is used to ensure that freshly printed newspaper does not dirty one's hands; it is also used to manufacture hard paper).

19. Manufacture of **rubber technical goods** (wax is used as to soften the rubber article).

20. **Textile industry** (wax is used for treating fabrics to impart certain properties).

21. **Aircraft industry** (beeswax is used to manufacture different impregnating and coating emulsions).

22. **Glass industry** (wax is used for hot casting of glasswork under pressure to smooth the surfaces of articles).

23. **Optical enterprises** (for protecting glasses while they are being engraved).

24. **Metallurgy industry**. Wax is used for the lost-wax method of casting metals. So-called duplicator stencils are coated from the outside with heat-resistant material; after the duplicator stencil is melted, the mold (cavity) is obtained, and this mold is filled with molten metal.

25. **Railway transport** (manufacture of compounds needed for greasing and impregnating leather articles: washers, collars applied in the brake systems of locomotives and motorized railcars; product for impregnating pads and cuffs made of leather used in braking equipment.

26. Manufacture of **bows and arbalests** (wax is used for greasing bow-strings made of synthetic threads, their windings, and guides for arrows).

27. Fabrication of **anatomical preparations** and moulages for training medical students.

28. Manufacture of **explosives** (wax incorporated into plastic explosives to lower their sensitivity to external effects such as shock, friction, and sparks).

29. Fabrication of **ski wax** (wax is one of the components).

30. **3D-modeling** (beeswax is used in the printing of three-dimensional laboratory models to check the working efficiency of developed units and installations).

31. **Grattage**, which is a method of making pictures by scratching inky paper or cardboard with a pen or other sharp instrument.

32. **Sewing** (waxing threads helps the sewing process—threads become smooth and pass easily through fabric).

33. Fabrication of **didjeridoos**, traditional musical wind instruments of aborigines of Australia (mouthpiece is made of beeswax to provide closer contact of the musician with it).

In many of the applications listed here, beeswax is not infrequently replaced by synthetic analogs.

In addition, **wax-figure museums** should be noted. These are collections of wax sculptures of known persons of the past and present. The first such institution was the wax-figure museum of the royal court of England, where 140 life-size figures were exhibited. The museum was opened in London in 1711 (https://en.wikipedia.org/wiki/Wax_museum). However, erstwhile, sculptures were created for funeral honors in European monarchies. In the medieval period, it was traditional to carry the body of a deceased member of the royal family in full raiment on the casket door, which was problematic in hot weather. As a substitute for the putrescent human body, they used wax figures in real clothes. In the museum of the Westminster Abbey in London, there is a wax figure of Edward III of England (died in 1377) (https://en.wikipedia.org/wiki/Westminster_Abbey).

By the end of the 19th century, most big cities had acquired their own wax-figure museums. **Madame Tussauds,** established in 1835 in London, now has branches in Amsterdam, Berlin, Hong Kong, Shanghai, and five cities in the United States: Las Vegas, New York City, Washington, DC, San Francisco, and Hollywood. Other known museums include the Canadian Royal London Wax Museum (Victoria, British Columbia) and the National Wax Museum in Dublin, Ireland. In 2014, the Mother's Wax Museum (named after Mother Theresa) was established in Kolkata, India (http://www.waxmuseumindia.com/index.html#header).

1.6

The Use of Insects as Food

Both the products of the vital activities of insects and the insects themselves can be used for food. To some extent, the use of the waste products of insects as food already has been considered. In Section 1.5.3.3, "Use of shellac", the use of shellac as a material for confectionery coatings and for coloring of juice drinks, carbonated drinks, wine, candy, jam, and sauces, as well as a "wax" coating for fruits and other uses is discussed. Shellac is the excrement of Indian lac insects.

In a like manner, information on the use of cochineal insects for food is contained in Section 1.5.2.3, "Production and use of carmine". This food coloring is incorporated in a variety of foodstuffs (cheeses, alcoholic drinks, canned fruits, confectionery, bakery and macaroni products, ice cream, desserts, meat and fish products, various appetizers, soups, sauces, dressings, mustards, and other foods). However, in these sections, the insects were considered as sources of specific ingredients that improve the marketable condition of products rather than as the staple food. In this chapter, we will be dealing with the use of insects for food—as the main meal.

1.6.1 Use of products of vital functions of insects

Bee honey is the main product of the vital functions of insects. It is flower nectar that has been partially digested in a bee's stomach—in essence, it is bee vomit. Honey contains 75–80% carbohydrates, 13–22% water, and a small quantity of different vitamins and microelements. According to different data, its energy value reaches 3,034 kcal/kg (https://www.fatsecret.com/calories-nutrition/usda/honey?portionid=45776&portionamount=1.000), 3,040 kcal/kg (Bradbear 2009), or 3,071–3,286 kcal/kg (Kappico et al. 2012).

Besides its direct use as food, honey is used as an ingredient in different dishes. Recipes including honey occur in the cuisines of almost all nations (except perhaps for some northern tribes and desert nomads). Most often, honey is used for baking (e.g., cakes, pies, and pastries) and preparation of desserts (made of fruit, berris, and nuts). In addition, it is widely used in the preparation of sauces, marinades for meat dishes, dressings for salads, aromatic additives to cocktails and teas, and sweeteners in some commercial beverages.

Honey is the critical raw material for alcoholic beverages known as mead (also, "honey wine" or "honey beer"). It is believed that production of these beverages began in about 2000 BC, in Greece (Ellis 2014). Mead was also known from many sources of ancient history throughout Europe, Africa, and Asia, and was especially popular in Eastern Europe and in the Baltic states (https://en.wikipedia.org/wiki/Mead). These beverages are created by fermenting honey with water, sometimes with various fruits, spices, grains, or hops. In many cultures, it was traditional that newlywedcouples drank mead not only during the wedding banquet but also for 30 days after it. Thus the term *honeymoon* (https://en.wikipedia.org/wiki/Honeymoon).

1.6.1.1 History of beekeeping

One can identify three stages in the history of bee keeping: (1) honey hunting; (2) wild-honey farming; and (3) beekeeping. Beekeeping refers to bee raising at specially equipped places (apiaries) in artificial bee houses: (a) box hive beekeeping (traditional beekeeping), bees are managed in fixed comb hives made of wood (bee trees, wild hives), clay, and other materials; and (b) movable comb hive beekeeping, in which honey bee colonies are raised in hives with movable frames.

Honey hunting is the search for nests of wild bees (combs) in their native habitats (hollows of tree trunks, cracks of cliffs, etc.). Honey and wax are extracted from those locations. Bees are repelled with smoke, and the trees or rocks are broken, frequently resulting in the physical destruction of the bee colony. Honey hunting was more often characteristic of nomadic tribes, which moved to other regions when all the available food and game resources in one location have been exhausted.

For tracking nests of wild bees, part of a comb with honey was used as bait. Two men stood some distance from one another with the bait. After filling their stomachs with honey, the bees flew to their homes. After observing the direction in which the bees moved, it was easy for hunters to converge on a common point where the nest was situated (http://apismelliferamellifera.0pk.ru/viewtopic.php?id=222).

Many **rock paintings** depicting honey hunting are known. Often, they are given the titles of the most ancient creations of this type. Ellis (2014) says that the first records of honey hunters are from rock paintings dating back to 15,000 BC. Gough (2008) argues that the pictures in the Cave of the Spider near Valencia, Spain, dating back to approximately 13,000 BC, are the oldest. They show a person climbing vines to harvest honey with bees flying around.

For the most part, honey hunting has become a thing of the past; however, it is **still practiced** among the aborigines in some regions of Africa, Asia, Australia, and South America. In particular, it is widely used in Nepal, where hereditary honey hunters gather it twice a year on rugged cliffs from the nests of the Himalayan honey bee (*Apis laboriosa*), the world's largest honey bee (Jung et al. 2015).

Later on, **wild-honey farming** emerged. Accumulating experience of contacts with bees, the hunters observed that if the nests of bees are not taken and if part of the procured food is left in the nest, then one can obtain honey from the same tree hollow for years with much less expenditure of time and labor. The men would gouge the holes in the thick trees (that is, to create artificial tree hollows) at heights of 4 to 15 m. Inside the hollows, skewers were mounted to fix the combs. All the basal branches of these trees were cut, which protected the wild hives from destruction by bears. The tops of the trees were also cut in order to increase their resistance to the wind (https://en.wikipedia.org/wiki/Honey_hunting).

Therefore, people switched from the incidental finding of nests of wild bees to organized honey hunting. This was not yet bee raising, but it defined a form of human economic activity that can be called the wild-honey-gathering system. The transition to classic wild-honey farming was possible with the appearance of iron tools in the early Iron Age (seventh to sixth centuries BC to the fourth century AD). It first occurred in the Middle East (about 7000 BC) (https://en.wikipedia.org/wiki/Beekeeping). The specific conditions of making wild hives resulted in the development of specialized instruments and accessories that are virtually the same among many nations and have been extant without changes since the ninth century (http://www.rnsp.su/istoriya-pchelovodstva/).

Wild-honey farming was replaced with **beekeeping**, that is, bee raising in specially equipped places (bee yards) with artificial hives. At first, so-called beekeeping in a hollow-log hive, or *traditional beekeeping*, occurred. The bees were kept in fixed-comb hives made of wood (hollowed-out logs, loggum hives), clay, and other materials. In hollowed-out logs, the bees develop just as in nature—that is, they seal honey into the natural combs they make; however, this method of honey production was not very efficient.

It is recognized that beekeeping originated 6,000 years ago in ancient Egypt (Nile River valley). The Egyptians usually used reed mats to make cylindrical mud hives that were stacked horizontally. These hives were used from at least 3100 BC (based on hieroglyphics) (Ellis 2014). The particular honey-making areas were in the upper reaches of the Nile, so the Egyptians transported hives made of reeds or burnt clay using large stranded rafts there. Then the Egyptians returned to collect the honey. The carriage of bees for honey production was also practiced in other countries. For example, the ancient Greeks transported bees to the Attica Peninsula, which was rich in honey plants, and islands of the Aegean

Sea, while the Romans transported them to islands in the Mediterranean Sea bays (http://www.rnsp.su/istoriya-pchelovodstva/).

Movable comb hive beekeeping became the final stage in beekeeping development. The beginning of this stage is considered to be 1814, when the Ukrainian beekeeper Petro Prokopovych invented the movable-frame hive with removable frames. This hive was reusable and afforded the opportunity to facilitate the work of taking care of bees and to make it possible to produce honey of superior quality (https://en.wikipedia.org/wiki/Petro_Prokopovych).

It must be noted that the 19th century is called the **"Golden Age" of beekeeping**, because, in this century, a variety of inventions were developed that sharply improved the efficiency of beekeeping. In 1851, Lorenzo Lorraine Langstroth of Philadelphia, referred to as the "Father of American Beekeeping", invented the first hive with movable frames where the frames were extracted from above by using a special device (https://en.wikipedia.org/wiki/L._L._Langstroth). In 1865, Major Franz Hruschka of Austria-Hungary (a man of Czech extraction) invented the honey extractor, used for honey extraction with centrifuges. Earlier on, honey was obtained by pressing the honeycombs (https://cs.wikipedia.org/wiki/Franz_von_Hruschka).

At the present time, beekeeping has developed economic and scientific branches. There are the International Federation of Beekeepers' Associations (Apimondia) with headquarters in Rome, Italy, and the International Bee Research Association (Cardiff, UK). In many countries, there are appropriate national organizations (http://www.honeyo.com/org-International.shtml), while local organizations exist in different regions. Sometimes, numerous organizations exist in a single region. For example, 18 local bee clubs exist in Florida alone (Ellis 2014).

1.6.1.2 Processing and production of honey

Honey bees sip at **two sources**: (1) nectar and (2) honeydew. **Nectar** is a sweet juice excreted by the nectarous glands of plants. Its main ingredient is a naturally occurring sugar that consists of approximately 55% sucrose, 24% glucose, and 21% fructose (https://en.wikipedia.org/wiki/Nectar). The remainder comprises amino acids, organic acids, minerals, vitamins, and aromatic compounds.

The compositions of the nectar of different plants vary. For example, the nectar of rape and dandelion consists largely of fructose and glucose, while plants of the bean family (Fabaceae) and the mint or deadnettle family Labiatae (acacia, clover, sage, and lavender) excrete nectar containing mainly sucrose. The concentration of sugars in the nectar depends on many factors (soil, temperature, humidity, time of year, etc.). The higher the humidity, the more abundant the nectar and the lower the concentration of sugar in it.

Bees prefer nectar with higher sugar contents—for example, about 50%—and will not gather it if the content is lower than 5%. Temperature is also of importance. The optimum temperature range is 10–30°C. Strong winds decrease the secretion of nectar; secretion also depends on the time of day. Maximum secretion of nectar is characteristic of midday hours and the beginning of the second half of the day (Bogdanov 2011).

Honeydew is a sticky sugar-rich liquid (in fact, urine) secreted by aphids, scale insects, and other insects that feed on sap. It appears on the leaves of trees and bushes, and it sometimes falls to the ground as very small drops. The sap of numerous species of plants is processed into honeydew. In particular, these plants include silver fir, maple, hazel, beech, ash, Eastern juniper, larch, apple, apricot, cherry, plum, peach, pear, lime, and others (https://en.wikipedia.org/wiki/List_of_honeydew_sources). The sap passes through the digestive system of insects, and excessive moisture is removed as drops of honeydew, which is harvested by bees.

Data vary concerning the quantity of honeydew secreted by the sucking insects. More often, quantities reach 300–400 l per hectare of forest, although some estimates indicate 1 ha of the forest can produced up to 10,000 l of honeydew in some cases (http://www.medoviy.ru/stat-mean-sort-1825-Pad_kak_syre_dlja_lesnogo_meda.html). The largest quantity of honeydew is produced on coniferous trees. Honeydew is a solution with variable concentration of sugars (5–60%), consisting mainly of sucrose. It also contains small amounts of amino acids, proteins, minerals, acids, and vitamins (Bogdanov 2011).

Nectar and honeydew are harvested from plants for the production of honey from the time of flowering of the first honey plants to the beginning of cold spells. Honey is the major food for bees, and without it the existence of honey bee colonies is impossible. The quantity of gathered honey depends on many factors: meteorological conditions, honey flow, and the power of the colony.

Generally, the period of **honey flow** can be divided into **two stages**: *supportive* (when bees gather nectar and honeydew for the life-sustaining activities of the bee colony itself, generally during the time of flowering of the first honey plants) and *productive* (procurement of honey for future use, more often in the period of flowering of the major honey plant within a region). In the nomadic method of beekeeping, honey flow is always higher, because when the flowering of one honey plant ends, the bee yard is transported to a place where other plants are flowering (Krivtsov and Lebedev 1993).

The quantities of nectar produced by different plants and, accordingly, **harvesting of honey by bees** differ greatly. For example, the mean volume of honey harvested from one hectare in the midland of European Russia can reach (http://www.kupi-uley.ru/rasteniya_medonosy.php) 20 kg (blackberry, apple), 50 kg (raspberry, rape), 100 kg (clover), 200 kg (maple, heather), 300–400 kg (acacia), 500–600 kg (coriander, willow herb), and even 1,000 kg (lime). The quantities of nectar gathered depend not only on the availability of honey plants in the vicinity of the bee yard but also on meteorological conditions. Rain, wind, and cold weather can reduce the quantities of nectar on plants; in addition, bees do not go out of their hives in bad weather. So poor weather conditions can reduce the honey flow three to four times (Ioyrish 1976).

If a bee fills its stomach with 50 mg of honey flow, then it should take 100,000 flights to gather 5 kg of nectar or honeydew for the production of about 250 g of honey. Each old bee of a medium-sized colony (with a total of about 10,000 worker bees) may take 10 flights to get food (Photo 1.6-1). The bees carry in their honey stomachs the nectar and honeydew gathered from plants to their colonies. In the honey stomach, nectar and honeydew mix with saliva and are enriched with its ferments. The bees in the hive take the honey flow and transfer it to each another and finally put it into combs. In the world as a whole, nectar is a source of higher monetary value. Only in some European countries, such as Greece, Switzerland, Turkey, Slovenia, and Austria, the value of honeydew as a honey source is comparable with that of nectar (Bogdanov 2011).

After the majority of the honeycombs are blocked, the activity of bees, **from the standpoint of humans**, ends and the harvesting stage starts. The ordinary **sequence of operations** is as follows (http://

Photo 1.6-1. Bee honey is the main product of the vital functions of insects. It is flower nectar that has been partially digested in a bee's stomach. European honey bee (*Apis mellifera*) workers clustered round their queen are shown.
Photo credit: Denis Anderson (CSIRO)

geolike.ru/page/gl_1921.htm; http://bortnic.ru/?p=464): (1) The beekeeper takes the frames with the sealed mature honey out of the hive and removes the bees with the use of a water pulverizer or fumigation of frames. (2) All the caps of the honeycombs are opened with a special tool—a fork—and the *unripe honey* is produced. (3) The unsealed honeycombs are placed into a honey centrifuge, and there the honey is extracted from the combs. (4) In honey settling, heavy particles gravitate to the bottom and the lightest admixtures, which commonly comprise 90% of the honey, are found at the surface; at temperatures of 18°–20°C, the honey settles over three days, and afterward, the upper layer is carefully removed and the honey is poured into other vessels in order to not disturb the particles gravitating to the bottom. This is the *refined honey.* (5) The honey is filtered, cleaning it of pollen grains, air bubbles, and other small impurities. *Filtered honey* is the final result. (6) Honey graining is performed in order to obtain a product of homogeneous fine crystalline and pasty consistency that can be spread on bread and other products. This product is *creamy honey.*

The specific features of honey production technology exist in different regions; however, these stages are on the whole characteristic of different countries. The single exception is related to the extraction of honey from combs. Currently, honey is mainly produced by centrifuging, except in the countries of Africa, where it most often is procured by pressing of combs (Bogdanov 2011).

1.6.1.3 *State of world beekeeping*

Beekeeping is practiced on a larger part of the Earth's surface than any other branch of agriculture; in fact, the success of some branches of plant cultivation depends on it. The global distribution of some bee species is shown on Map 1-4. At the present time, beekeeping is facing a lot of challenges. On the one hand, considerable scientific research has been aimed at, among other things, enhancing the efficiency of beekeeping. Also, from the beginning of 2013 to mid-2015, thousands of articles dealing with bees were

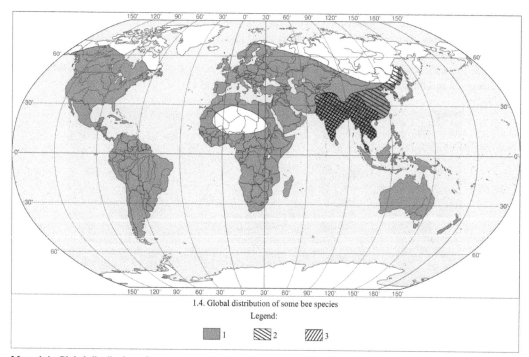

1.4. Global distribution of some bee species

Map 1.4. Global distribution of some bee species (http://commons.wikimedia.org/wiki/File:Apis_mellifera_distribution_map.svg?uselang=fr; http://commons.wikimedia.org/wiki/File:Apis_dorsata_distribution_map.svg?uselang=fr; http://commons.wikimedia.org/wiki/File:Apis_cerana_distribution_map.svg?uselang=fr) (*Apis cerana*).
Legend: (1) European honey bee *Apis mellifera*; (2) Asiatic honey bee *Apis cerana*; (3) Giant honey bee *Apis dorsata*

published in reviewed scientific journals, including 59 articles in two top journals (*Science* and *Nature*) (Crailsheim 2015). Sustained growth in honey production has been observed on a global scale (Fig. 1.6-1).

On the other hand, world beekeeping faces an array of problems. The **main problems** are as follows (Eroshenko 2013; Ellis 2014):

1. **Mass extinction of bees**. This phenomenon was first described in 2006. "Colony collapse disorder" is by the adult honey bee workers suddenly abandoning their hives, leaving behind food, brood, and queen. There is no single cause of bee extinction; however, a number of factors affect the mortality of insects, and only their proportions and interactions change (Cutler et al. 2014). These factors are (a) use of pesticides; (b) seed incrustation; (c) varroatosis; (d) herbicides; (e) stress on bees due to the use of mobile hives; and (f) virus diseases.

Pesticides constitute the greatest threat. From year to year, their toxicity increases. During the early 1990s, neonicotinoid insecticides appeared. Now they are used in more than 120 countries and on at least 140 agricultural plants (Lundin et al. 2015). These chemicals are 7,000 times more dangerous for bees—particularly for their nervous systems—than the notorious DDT. Neonicotinoids can be translocated into pollen and nectar. In addition, it is also worth noting that tractors were once used for air-spraying of pesticides, while now helicopters and other aircraft are used (Korzh 2010).

Seed incrustation is a process in which a mixture of components for creation of cover and, in this case, stimulant fertilizers, pesticides, colorants, and other chemicals are fixed on seeds by adhesive substances. In the 20th century, treatment of fields before and after flowering was used to minimize the deleterious effect of the preparations. However, now plant receive long-term protection due to incrustation, and the pollen proves to be "poisoned" for the entire period of flowering.

Varroatosis is a disease of honey bees caused by the varroa mite (*Varroa destructor*). It is second in order of importance as a cause of the extinction of bees (Lattorff et al. 2015). Varroatosis has been

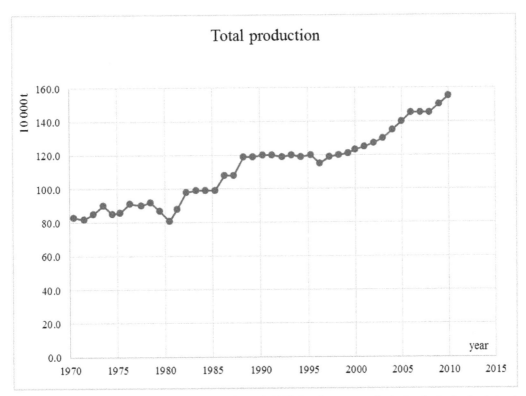

Fig. 1.6-1. Global honey production in 1970–2010 (Gu et al. 2002; http://ec.europa.eu/agriculture/evaluation/market-and-income-reports/2013/apiculture/chap3_en.pdf).

observed on all continents except for Australia (http://honeytechnology.ru/rukovoditel-mirovogo-soobshhestva-pchelovodov-zhil-ratiya-posetil-bashkortostan/). The situation was especially serious in the 1980s through 1990s. Some progress has been made since then; however, measures have failed as yet to eliminate this disease completely. The use of stronger and stronger preparations against the mites leads to the accumulation in the bees of the active substances of the acaricidal preparations. The acaricidal substances coumaphos and fluvalinate adversely affect the bees (Williamson et al. 2013).

Herbicides do not have a direct effect on the organism of the honey bee. Nevertheless, by destroying weeds, these preparations considerably reduce the diversity of plants on which the bees can feed, especially in regions with monocultural agriculture. This problem is very characteristic of the United States, where cornfields can stretch for hundreds of kilometers. To mitigate the effects of a monocultural diet, beekeepers use additional nutrition, but its effectiveness is low in many cases. This is because the standard sugar or honey syrup that was earlier used has been replaced with solutions of isoglucose (an artificial substitute for sugar made of the same corn).

Stress on bees is caused by increased production intensity. Hives are continually moved from place to place, and as a result of this, the stress on bees accumulates. In the United States, a colony of bees can be transported five to six times a year over a total distance of 20,000 km. It is considered to be absolutely normal to carry insects from California to Florida and vice versa. Such changes in the environment makes bees nervous and is an additional factor provoking the propagation of diseases and parasites (Eroshenko 2013).

Virus *diseases* not infrequently contribute to the destruction of bee colonies. At least twelve viruses affecting bees have been discovered in Europe (Simon-Delso et al. 2014). Often, viruses may remain latent during a particular time and populations of bees appear healthy. At times, bee pandemics arise that are related to the practice of using cheaper imported queen bees. When a European producer purchases and delivers queens from far away (for example, from China or Mexico), this provides a real opportunity for the transfer of diseases from continent to continent. In such cases, beekeepers use, antibiotics on a mass scale for prophylactic control of diseases. The antibiotics, on the one hand, do not always benefit the insects, and, on the other hand, they can contribute to the selection of microorganisms resistant to the veterinary preparations.

The above-mentioned factors result in decreases in the populations of bees. In Europe, this has occurred in the last 20–25 years, while in the United States this process began 60 years ago. Since 1961, the populations of bees have been cut in half. The number of bee colonies (beehives) was 6 million in 1947, 4 million in 1970, 3 million in 1990, and 2.5 million in 2013 (Fact Sheet 2014).

2. **Falsification of honey** is done by feeding sugar to bees and adding different products to honey: sugar, pollen, water, invert creams (flavored products similar in appearance, flavor, and taste to honey and made of starch-producing plants such as potatoes, rice, corn, wheat, topinambour, etc.; http://www.bees-products.com/?p=articles&j=27). Now, food chemistry has developed to such an extent that it is often problematic to distinguish natural honey from artificial honey just by taste. The highest volumes of honey falsification have been documented in China. At present, methods of identifying falsified honey are being developed (Herrero et al. 2013).

3. **Increasing legislative pressure.** Countries throughout the world have been legislating ever-increasing requirements as to the contents of foodstuffs. Meeting these requirements greatly complicates the lives of beekeepers, but the requirements do not always prevent the penetration of counterfeit products into the market. Twenty years ago, color, taste, and water content were the major criteria for choosing honey; now information on product taste or flowers on which it was gathered is not required when selling the honey.

However, the certificates on the contents of pesticides, antibiotics, and other components are necessary. The pollen of transgenic plants and the presence of alkaloids in honey are also considered. Certificates of conformance of honey to standards are issued by independent laboratories. Because issuing certificates is a profitable business, representatives of the businesses lobby in every way for the imposition of new requirements. Beekeepers often call this process "laboratory terror" (Eroshenko 2013).

4. **Increasing the average age of beekeepers**. An abundance of problems is resulting in a decline in the popularity of this business. Beekeeping attracts the attention of young people less and less. According to some estimates, the average age of modern beekeepers is 45–50 years. If this trend persists, there is a risk that beekeeping will lose its manpower in one to two generations (http://honeytechnology. ru/rukovoditel-mirovogo-soobshhestva-pchelovodov-zhil-ratiya-posetil-bashkortostan/).

At present, honey is likely produced in every country; however, the scales of production vary greatly. Information on honey production in the leading countries in 1961–2012 is given in Table 1.6.-1.

Even with shallow analysis of the data presented, the differences in the development of honey production are evident. Pronounced steady growth of honey production is observed in China, Turkey, Russia, Iran, and South Korea. Steady reduction of honey production is characteristic of the United States, which is caused, first of all, by the mass mortality of bees. Relative stability is characteristic of Spain and Angola. If the separate continents are considered, then their current shares in honey production are as follows: Asia, 43%; Europe, 23%; North America, 12%; Africa, 10%; South America, 10%; and Oceania and Australia, 2% (http://faostat.fao.org/site/339/default.aspx).

Table 1.6-1. Production of honey by major producing countries in 1961–2012, metric tons.[1]

Rank	Country	1961	1971	1981	1991	2001	2006	2012
1	China	53,000	85,000	115,000	208,000	251,600	332,600	**436,000**
2	Turkey	8,001	16,345	30,041	54,655	60,190	83,842	**88,162**
3	Argentina	20,000	18,000	38,000	54,000	80,000	105,000	**75,500**
4	Ukraine	–	–	–	–	60,043	75,600	**70,134**
5	United States of America	124,316	89,732	84,335	100,200	84,335	70,238	**66,720**
6	Russia	–	–	–	–	52,659	55,678	**64,898**
7	India	20,000	35,000	45,000	53,986	52,000	52,000	**61,000**
8	Mexico	24,000	35,024	70,557	69,495	59,069	55,970	**58,602**
9	Iran	–	–	–	12,000	26,600	36,039	**48,000**
10	Ethiopia	14,400	17,400	21,000	23,400	33,776	51,250	**45,905**
11	Brazil	7,749	–	–	18,668	22,220	36,194	**33,571**
12	Spain	9,068	8,197	14,501	25,300	31,617	30,661	**29,735**
13	Canada	15,902	23,594	34,769	31,620	35,388	48,353	**29,440**
14	Tanzania	–	8,000	–	20,000	26,500	28,000	**28,500**
15	South Korea	–	–	10,386	–	22,040	22,939	**25,000**
16	Romania	–	11,829	13,807	–	11,746 (2000)	18,195	**23,062**
17	Angola	20,000	15,367	16,000	20,000	24,000	23,000	**23,000**
18	Uruguay	–	–	–	–	–	–	**20,000**
19	Hungary	–	10,606	14,593		15,165 (2000)	19,714 (2005)	**17,000**
20	Central African Republic	–	–	–	–	–	–	**16,000**
	USSR	205,000	194,000	187,000	240,000	–	–	–
	Globe	–	**820,000**	**880,000**	**1,190,000**	**1,250,000**	**1,450,000**	–

[1] Shkenderov and Ivanov 1985; http://faostat.fao.org/site/339/default.aspx; Gu et al. 2002; Bradbear 2009; Bogdanov 2011.

There are also other significant differences between the continents (http://paseka.pp.ru/pchela-i-ulej/758-skhema-sovremennogo-pchelovodstva.html). For example, in comparing Europe and North America, one can identify many such differences. The New World has provided the most valuable inventions in practical beekeeping, whereas in the Old World, the most fundamental discoveries concerning the biology of bees have been made. Beekeeping in North America is, first of all, a business; beekeeping farms prevail there, and average harvests of honey are much higher. In Europe, the percent of nonprofessional beekeepers for which this work is no more than hobby is high, so the average number of hives per beekeeper is significantly lower.

There are significant differences in the equipment that is used. In European beekeeping, traditions often restrain the introduction of scientific developments. However, in North America, the technology of beekeeping develops without obstructions, and there is a tendency toward simplicity, conformance, and maximum possible mechanization. In the European countries, different hives are used and, even within the boundaries of one country, a large variety of "standard" hives can occur. These hives can contain frames and thus comb foundations of different shapes and sizes; many hives are very complex, which reduces the production of honey rather than increase it (http://paseka.pp.ru/pchela-i-ulej/758-skhema-sovremennogo-pchelovodstva.html).

1.6.2 Direct use of insects as food

1.6.2.1 History of entomophagy

It is evident that both ancestors of *Homo sapiens* and the primitive peoples themselves ate insects. Before humans acquired devices for hunting and fishery, insects were a substantial part of their diet. Coprolites (fossilized feces of ancient people) serve as evidence of this (https://en.wikipedia.org/wiki/Coprolite). Analysis of coprolites from many caves in the United States and Mexico showed the presence of many insects. For example, coprolites from caves on the Ozark Plateau, in the central United States, contain ants, beetle larvae, lice, ticks, and mites (https://en.wikipedia.org/wiki/Entomophagy).

According to data of Yi et al. (2010), insects were consumed in China 3,200 years ago. Locusts on sticks were served at banquets in the palace of the Assyrian king Ashurbanipal (668 BC–ca. 627 BC; van Huis et al. 2013). The use of wasp larvae and pupae since ancient times in Chinese culinary art is recorded in the book Tang Dynasty era (AD 618–907) (Feng et al. 2008). The earliest written evidence of entomophagy in the Near East and China dates to at least the second and first millennia CE (Schabel 2010).

The **first known mention** of entomophagy in Europe is present in a work by Aristotle (384–322 BCE), *Historia Animalium*, in which he noted that cicada females taste best after copulation because they are full of eggs. Insects were also consumed by the ancient Romans. For example, Pliny the Elder, in *Natural History*, composed in about 77 CE, wrote that the larvae of the great capricorn beetle (*Cerambyx cerdo*) were especially popular in the empire (https://en.wikipedia.org/wiki/Entomophagy).

In religious sacred books of the Christian, Jewish, and Islamic religions, there are also fragments related to entomophagy. The Bible tells about the use of locusts for food (most likely the desert locust, *Schistocerca gregaria*), as well as crickets and grasshoppers. The Islamic tradition of entomophagy extends to locusts, bees, ants, lice, and termites. In Judaism, only four species of locusts are recognized as kosher; that is, permitted for use as food (van Huis et al. 2013).

There are numerous references to the use of insects for food in literary works of the Middle Ages. For example, Leo Africanus of Morocco described in 1550 the use of locusts for food by nomads in Arabia and Libya (they eat cooked and dried locusts as well as pound them into flour for future consumption (https://en.wikipedia.org/wiki/Leo_Africanus). In his treatise of 1602, the Italian naturalist Ulisse Aldrovandi wrote about the use of fried silkworms by German soldiers in Italy. In 1737, the French naturalist René Antoine Ferchault de Réaumur described in his six-volume work *Memoirs of the History of Insects* (*Mémoires pour servir à l'histoire des insects*) the use of insects for food in different provinces of France (van Huis et al. 2013).

The practice of using insects as food in tropical areas of the earth has existed for many millennia. In modern Western society, this issue was considered by Gene R. DeFoliart (1925–2013), a professor in the entomology department at the University of Wisconsin, Madison. In 1988, he began to issue the *Food*

Insects Newsletter, which eventually reached audiences in at least 82 countries (http://labs.russell.wisc.edu/insectsasfood/about-dr-defoliart/).

1.6.2.2 Reasonability of using insects as food

The problem of shortages of food resources has always existed. The principal suppliers of foodstuffs for humanity are plant cultivation and cattle breeding. The population of the earth has been increasing and, accordingly, the needs of humanity for nutritive substances, especially for animal protein, have been growing. The complexity of this issue is due to the fact that the areas suited for cattle breeding are practically depleted. Therefore, increase in animal protein production will likely be accompanied with further deforestation, especially in tropical countries. For example, the establishment of cattle ranches resulted in deforestation of 38% of the total area of forests in Brazil in 1966–75 (Newman 1989).

At the present time, permanent pastures occupy 20% of the land area of earth; in addition, 24.5% of the land area of earth comprises deserts and semideserts used occasionally as pastures, and 6.7% is occupied by tundra deer pastures. In 2009, there were approximately 33 million km^2 of pastures (Food and Agriculture Organization of the United Nations 2012). Now, the possibilities for increasing pasture territory are practically exhausted. The insect farming is a mini-livestock enterprise. Similar enterprises are able to supplement the increasing demands of humanity for natural protein.

Advantages of enterprises for insect farming include the following (Wilson 2012; Mlcek et al. 2014): (1) they require minimum space; (2) insect farming requires little water, which is of the utmost importance for areas with watershortage in many regions of the world; (3) products not consumed directly by humans are used for feeding insects; (4) the demand exceeds the supply; (5) insects have short reproductive cycles; (6) investments show a faster payback; (7) in many cases, high financial returns are realized; (8) insects have considerable nutrition value for humans; (9) insects convert feed to protein efficiently; (10) the management of enterprises is quite simple; (11) insects are easily transported; and (12) insect farming does not require in-depth training.

On the whole, the extension of entomophagy practices can be recommended for **three reasons** (van Huis et al. 2013): (1) **health benefits**: (a) insects are a good alternative to the leading suppliers of animal protein, such as chicken, pork, beef, and even fish; (b) many species of insects have high contents of proteins and lean fats, as well as calcium, iron, and zinc; (c) insects are already part of the diet for many nations; (2) **environmental benefits**: (a) specific emissions of greenhouse gases by insects are far less than those for most other livestock—for example, methane is produced by only some groups of insects, such as termites and cockroaches; (b) insect rearing requires far less land and water than rearing other livestock; (c) ammonia emissions associated with insect rearing are also lower than those in other livestock enterprises; (d) because insects are cold-blooded animals, their efficiency in converting food into protein is very high; (3) **economic and social factors**: (a) harvesting and raising of insects are low-tech and low-capital investment options possible for even the poorest sections of society; (b) raising insects provides opportunities for subsistence for both urban and rural population; (c) raising of insects can be both low-tech and very difficult depending on the investment required.

According to data presented by Pimentel and Pimentel (2003), the production of 1 kg of animal protein in animal farming requires 6 kg of vegetable protein as food. The **feed-to-meat conversion rates** (quantity of eaten food for production of additional kilogram of weight) vary essentially depending on the species of animal and way it is managed. However, the common indices in the United States are as follows (Smil 2002): in order to produce an additional kilogram of live animal weight, 2.5 kg of feed is necessary for chicken, 5 kg for pork, and 10 kg for beef. The requirement for insects is **much less**. For example, for the production of 1 kg of live weight of crickets, only 1.7 kg of feed is required (Collavo et al. 2005). Furthermore, insects can feed on organic wastes.

Enormous funds are currently expended in the world every year to protect agricultural plants (first of all, cereal crops) against insect pests. For example, each year invasion of different species of locust to world agrosystems results in great economic losses. Different chemical and biological agents are used to control the insect populations; however, their efficiency is not high. The paradox of the situation lies in the fact that cereal crops contain no more than 14% protein while at the same time, the other food resources (insects), which contain up to 75% high-quality animal protein, are destroyed (Cerritos 2011).

The harvesting of pest insect species as food would be an ideal alternative. The **advantages** of harvesting pest insect species are as follows (Yen 2015): (1) increased plant food productivity; (2) provision of an additional food resource (insects); and (3) health and environmental benefits due to decreases in insecticide applications. Here, investigating the questions of economic efficiency and practicality is necessary, because pest outbreaks can be unpredictable and the mode of harvesting will depend on the target species (Mlcek et al. 2014).

A comparison of economic and environmental costs between farms for insect raising and traditional livestock enterprises was carried out by Dutch researchers from the Wageningen University (Oonincx and de Boer 2012). They analyzed the operation of a Dutch farm specializing in raising the larvae of the yellow mealworm beetle (*Tenebrio molitor*) and zophobas (*Zophobas morio*), which are used as food for poultry and fish, as well as for exotic domestic animals. They carried out a complex comparison of the larval and meat-and-dairy production using several parameters, including energy costs, feed efficiency, area of seated lands, and emission of greenhouse gases. The insects were given a combined feed containing oats, wheat middlings, soybeans, bread corn, maize, and brewers' yeast.

Studies showed that the **ecological costs** (contribution to the emission of greenhouse gases expressed in kilograms of emitted carbon dioxide per kilogram of produced protein) related to raising larvae are **essentially less** in comparison with expenses related to the production of meat and milk. In the case of milk, they proved to be 1.77–2.80 times less; in the case of chicken, 1.32–2.67 times less; and in the cases of pork and beef, 1.51–3.87 and 5.52–12.51 times less, respectively. **Energy consumption** for rearing (its major part goes toward the maintenance of temperature optimal for reproduction of insects) is also less, on average, by 40% in comparison with milk production and by 60%, 90%, and 130% in the production of chicken, pork, and beef, respectively. The overall land area used for rearing of larvae (it is composed of land used for the feed crop, feed storage area, farm area, etc.) is also less—1.81–3.23 times in comparison with land needed for the production of milk, 2.30–2.85 times in the case of production of chicken, 2.57–3.49 times in the case of pork, and 7.89–14.12 times in the case of beef production.

In addition, for the important index of feed efficiency (kilogram of consumed food per kilogram of growth), raising larvae also proved to be attractive. For insects, this index was 2.2, which is a little less than that for chicken (2.3) but it is much less than the indices for pork (4.0) and beef (up to 8.8).

In addition, high rates of **reproduction** (the female floor beetle lays up to 160 eggs over a period of 3 months, while female zophobas up to 1,500 eggs during the year) and very low reproductive ages (2.3 and 3.5 months, respectively) are characteristic of the insects. The results obtained allow one to unequivocally assert that, under conditions of essentially smaller ecological, energy, and economic costs, the farms for raising insect larvae can produce larger quantities of valuable food protein than traditional meat and dairy farms. In addition, successes of genetic engineering and further operational optimization (already reaching its limits in the meat and dairy industries) only decrease these costs and make raising insects more profitable (Oonincx and de Boer 2012).

Similar investigations (comparison of ecological costs related to farming the yellow mealworm beetle [*Tenebrio molitor*] and traditional types of livestock rearing) were carried out by Nadeau et al. (2014). Their results are presented in Table 1.6-2.

Many people take a dislike to insects, an attitude that is formed when they are children. Society has a prejudice against anything that cannot be considered as foodstuff. Often, people reject insects because they are "unclean". However, the majority of the edible insects, such as grasshoppers, locusts, and the larvae of butterflies and beetles, eat mostly fresh plant leaves or wood. For this reason, they are cleaner and more hygienic than crabs or lobsters, which eat carrion (Mitsuhashi 2010).

This prevalent attitude, thus, is no more than a prejudice. In time, many formerly unusual foodstuffs have become traditional. For example, eating frogs, originally a French dish, gradually became so popular throughout the world that a huge industry related to farming these amphibians has emerged. And in North Africa, fried or smoke-cured mopane moth caterpillars (*Gonimbrasia belina*), which costs 4 times more than conventional meat, are considered to be a delicacy. There the popularity of these larvae is so high that, during the season of their mass development, sales of beef and the meat of other animals

Table 1.6-2. Environmental impacts of insect rearing compared with traditional livestock raising.[1]

	Global warming potential (kg CO_2-eq/kg of edible protein)	Energy (MJ/kg edible protein)	Land (m²/kg edible protein)	Water (L/kg live weight)
Mealworms	14	173	18	No information
Beef	77–175	177–273	142–254	9, 700
Pork	21–54	95–237	46–63	2, 800
Chicken	19–37	80–152	41–51	1,500
Milk	25–39	36–144	33–58	800
Mealworms reared without energy input and on organic side streams	0.06	0.29	0.04	2.5

[1] Nadeau et al. (2014).

decreases markedly (https://en.wikipedia.org/wiki/Entomophagy). So one cannot exclude the possibility that traditions of using insects as food will gradually take hold in other regions.

It must be noted that people throughout the world unwittingly eat insects because it is impossible to avoid the pollution of foodstuffs with them or their parts. On average, throughout life, people consume unawares about 300 grams of insects together with other products (Europe preaches entomophagy 2010). Quite often, they occur in different greens (cabbage, lettuce, parsley, cauliflower, mange-tout, and radishes), mushrooms, many fruits (dates, figs, some breeds of cherries and plums, some breeds of apples and pears, different berries and walnuts), and different starchy products (for example, flour and cereals). In many countries, maximum admissible **levels of pollution** of foodstuffs with insects have been established; for example, in the United States, the *Defect Levels Handbook* (U.S. Food and Drug Administration 2014) was issued. Therefore, it is evident that people who dislike insects eat them in contaminated food (Mlcek et al. 2014).

1.6.2.3 Most commonly consumed insects

Different data are available on the quantity of insect species used as food (see Table 1.6-3).

One can see that, on the whole, the number of edible species tends to increase. It is evident that this information is not inclusive and, in time, their number will increase. Some evidence of the number of edible species of insects in some countries and regions were also found (see Table 1.6-4).

Table 1.6-3. Number of edible insect species in the world.

Number of species	Number of ethnic groups	Number of countries	Source	Notes
1,500	3,000	113	MacEvilly (2000)	
2,086	3,071	130	Ramos-Elorduy (2009)	
2,000			Dwi (2008)	
More than 1,900			van Huis (2013)	Two billion people
At least 1,681			Ramos Elorduy (2005)	
2,037			Jongema (2015)	
1,462			http://www.insectsarefood.com/resources.html	
More than 2,000			Malaisse (2005)	Representing at least 14 orders
More than 1,400			Nandasena et al. (2010)	
More than 2,000			Rumpold and Schlüter (2013)	

If the globe is considered as a whole, then the distribution of insects used for food **by taxa** is as follows (van Huis et al. 2013). The beetles (Coleoptera) (31%) occupy the first place; followed by butterflies (Lepidoptera) (18%); bees, wasps, and ants (Hymenoptera) (14%); grasshoppers, locusts, and crickets (Orthoptera) (13%); cicadas, leafhoppers, plant hoppers, scale insects, and true bugs (Hemiptera) (10%); termites (Isoptera) (3%); dragonflies (Odonata) (3%); and flies (Diptera) (2%). All remaining taxa account for 5%. As for the attribution of edible insects to the environment, 88% of species are terrestrial while 12% are aquatic insects (Yen 2015).

The use of insects relating to different taxa as food is characterized by their **own specificity**. Practically all butterflies and moths (Lepidoptera) are consumed as caterpillars; the hymenopterans (Hymenoptera) are largely used in their larval or pupal stages. As for the beetles (Coleoptera), both adult insects and larvae are used for food, while insects belonging to orthopteroid insects (Orthoptera), homopterans (Homoptera), termites (Isoptera), and true bugs (Hemiptera) are mainly consumed in the mature stage (Cerritos 2009). Spiders and scorpions are predominantly eaten in adulthood too.

Data are available on the number of edible species of insects within the **biogeographic regions**. Such regions are separated from others by significant barriers preventing the expansion of species (narrow passage, high mountains, deserts, oceans, or straits) and fauna within each biogeographic region (including insects) are characterized by a high degree of uniformity. Six major biogeographic regions exist (Humphries and Parenti 1999). Data on the number of the edible species of insects within different biogeographic regions are given in Table 1.6-5.

Table 1.6-4. Number of edible insect species in separate countries and regions.

Country	Number of species	Source
Botswana	27	Obopile and Seeletso (2013)
Laos, Myanmar, Thailand, and Vietnam	164	Yhoung-Aree and Viwatpanich (2005)
China	177	Feng et al. (2008)
China	178 (96 genera, 53 families, and 11 orders)	Chen et al. (2009)
Mexico	549	Ramos-Elorduy et al. (2008), taken from van Huis et al. (2013)
Central African Republic	96	Roulon-Doko (1998), taken from van Huis et al. (2013)
Thailand	194	Sirimungkararat et al. (2010)
Africa	470	Kelemu et al. (2015)

Table 1.6-5. Number of species of known edible insects in the biogeographic regions of the world.[1]

Region	Terrestrial	Aquatic
Africa	387	17
Australasia	90	5
Nearctic	75	15
Neotropical	608	82
Oriental	472	90
Palaearctic	289	52
Total	1,921	261

[1] Jongema (2014).

In general, insects are chiefly consumed in the subtropics and tropics. Only China, Japan, and Mexico can be considered as countries with expansion of entomophagy situated wholly or partially in the temperate zone. Even within the same countries, a tendency to increasing the use of insects for food is evident in the tropical zone. The **reasons** for this are as follows: (1) insects in the tropics are usually larger than those in temperate regions; (2) insects in the tropics often crowd together in great numbers, which eases their mass harvesting; (3) in the tropics, edible insects can be found throughout the year, while within the temperate zones, they go into hibernation to survive during cold winters; and (4) for many insect species in the tropics, harvests are predictable (van Huis et al. 2013).

1.6.2.4 *Nutritional value and chemical composition of edible insects*

The nutritional values of edible insects vary profoundly, which is caused, not least of all, by the wide diversity of species. However, even within one group of edible species of insects, the values can be vary greatly, depending on the metamorphic stage of the insects, their habitats, character of nutrition, and preparation and processing methods (drying, boiling, or frying). For example, adult insects (except for, maybe, butterflies) commonly have a higher content of protein than the younger instars. Grasshoppers in Nigeria that fed on bran siftings containing a large amount of valuable fatty acids were characterized by protein contents two times higher than those of insects feeding on corn (maize) (Ademolu et al. 2010).

The nutritional value of any foodstuff is determined, first of all, by the quantitative relation of proteins, fats, carbohydrates, microelements, and other substances it contains. The caloric content, digestibility, organoleptic characteristics, safety, absence of allergic responses, and other indices are also of importance. Data on the contents of proteins and fats in different species of insects are presented in Table 1.6-6.

Data on the energy content for different species of insects are presented in Table 1.6-7.

The key aspect of **protein metabolism** is the degree of digestibility of one or another protein by the human organism. Studies show that insect protein has good quality and high digestibility. Generally, its digestibility is 77–98%, which is only slightly lower than that of other proteins of animal origin (Mlcek et al. 2014). For example, the digestibility of protein of whole dried bees is 94.3%; moths (*Clanis bilineata*), 95.8% (Xia et al. 2012); fresh termites (*Macrotermes subhylanus*), 90.49%; and grasshoppers (*Ruspolia differens*), 85.67% (Kinyuru et al. 2010). The digestibility of insect proteins is higher than that of the majority of vegetable proteins.

Fats are necessary for the human organism, just as proteins are; first of all, they have a high caloric content. The human organism should always have fat stores. In most cases, the fat content of edible insects is between 10 and 50%. The fat content depends on many factors: species, reproductive stage, season, sex, life stage, habitat, and diet (Raksakantong et al. 2010). Usually, the fat content is higher at the larval and pupal stages, while the adult insects have a relatively low content. The females are fatter than the males. The content of essential fatty acids in insect fats is higher in comparison with animal fats (Chen et al. 2009).

Carbohydrates in insects are mainly present in chitin. The content of **carbohydrates** in the edible insects varies from 6.71% in stinkbugs to 15.98% in cicadas (Raksakantong et al. 2010). Studies have shown that a considerable amount of polysaccharides contained in them can improve the immune function of a human. As a form of low-calorie food, chitin is also of medical importance. In most cases, the chitinous coat of insects accounts for 5–20% of the dry weight (Chen et al. 2009).

Analyses have revealed that insects are rich in different **minerals**. For example, high contents of *potassium* and *sodium* are characteristic of cricket nymphs; *calcium,* phosphorus, and manganese are found in adult crickets; much *copper* is contained in the cavorting emperor (*Usta terpsichore*); and several species of aquatic true bugs and larvae of the yellow mealworm beetle are distinguished by their high content of iron (Sun 2008). The mineral composition depends significantly on the nutrition of the insects. For example, the content of calcium in wax worms (caterpillars of wax moths), larvae of the yellow mealworm beetle, pupae of the silkworm (*Bombyx mori*), and house crickets can be increased 5 to 20 times if the insects eat food with high concentrations of calcium (Mlcek et al. 2014).

Table 1.6-6. Contents of protein and fat in different insects.[1]

Insect	Species	Protein, %	Fat, %	Notes
Carpenter millers (Cossidae)	Several moths; e.g., cossid moth (*Endoxyla leucomochla*) witchetty (*witjuti*) grubs	38	40	Large, white wood-eating larvae of several moths
Mayflies (Ephemeroptera)	Larvae of mayfly (*Ephemerella jianghongensis*)	66.26	–	
True bugs (Hemipetra)	Larvae of giant water bug (*Sphaerodema rustica*)	73.52	–	
Orthoptera	Rice grasshopper (*Oxya chinensis*)	–	2.2	
Butterflies (Lepidoptera)	Larvae and pupae of cotton pink bollworm (*Pectinophora gossypiella*)	–	49.48	
Butterflies (Lepidoptera)	Larvae and pupae, Asian corn borer (*Ostrinia furnacalis*)	–	46.08	
Dragonflies (Odonata)	Adults and naiad	46–65	25.38	Range and average
Orthopterous insects (Orthoptera)	Adults and nymph	23–65	2.2	Range and average
Homopterans (Homoptera)	Adults, larvae, and eggs	45–57	27.73	Range and average
Bugs (Hemiptera)	Adults and larvae	42–74	30.43	Range and average
Beetles (Coleoptera)	Adults and larvae	23–66	27.57	Range and average
Butterflies (Lepidoptera)	Pupae and larvae	14–68	24.76	Range and average
Flies (Diptera)			12.61	Average
Hymenopterous insects (Hymenoptera)	Adults, pupae, larvae, and eggs	13–77	21.42	Range and average
Wasp	Larvae of black-bellied hornet (*Vespa basalis*)	53.18	–	
Wasp	Larvae of Asian giant hornet (*Vespa mandarinia mandarinia*)	54.59	–	
Wasp	Larva of banded paper wasp (*Polistes sagittarius*)	46.17	–	
Wasp	Larva of paper wasp (*Polistes sulcatus*)	57.88	–	
Orthopterous insects (Orthoptera)	Bombay locust (*Patanga succincta*)	24.4	1.5	
Orthopterous insects (Orthoptera)	Bombay locust (*Patanga succincta*)	27.6	4.7	
Cricket	Big-head cricket (*Brachytrupes portentosus*)	12.9	5.5	
Moths (Lepidoptera)	Larvae, pupae, and adults of silkworm (*Bombyx mori*)	50		

Table 1.6-6 contd. ...

...Table 1.6-6 contd.

Insect	Species	Protein, %	Fat, %	Notes
Butterflies (Lepidoptera)	Pupae of silkworm (*Bombyx mori*)	14.7	8.3	
Flies (Diptera)	Larvae or pupae of the housefly (*Musca domestica*)	50–60		
Termites (Isoptera)	–	32–38		Average
Scarab beetles	White grub, *Holotrichia* sp.	18.1	1.8	
Crickets	House cricket (*Acheta testacea*)	18.6	6.0	
Crickets	Taiwan giant cricket (*Brachytrupes portentosus*)	17.5	12.0	
Butterflies (Lepidoptera)	Caterpillar of bamboo borer (*Omphisa fuscidentalis*)	9.2	20.4	
Wasp	Larvae of hornet (*Vespa* sp.)	14.8	6.8	
True bugs (Hemiptera)	Edible stinkbug (*Encosternum delegorguei*)	30–36	51–53	Zimbabwe
Beetles (Coleoptera)	Larvae of yellow mealworm beetle (*Tenebrio molitor*)	18.7	13.4	
Beetles (Coleoptera)	Larvae of zophobas (*Zophobas morio*)	45		
Flies	Housefly (*Musca domestica*) pupae	62.5	19.2	
Flies	Black soldier fly (*Hermetia illucens*) larvae	56.1	12.8	
Flies	Black soldier fly (*Hermetia illucens*) pupae	52.1	19.7	
Orthopterous insects (Orthoptera)	House cricket (*Gryllus domesticus*)	70.6	17.7	
Beetles (Coleoptera)	Yellow mealworm (*Tenebrio molitor*)	52.0	33.9	
Beetles (Coleoptera)	Lesser mealworm (*Alphitobius diaperinus*)	64.8	22.2	
Beetles (Coleoptera)	Morio worm (*Zophobas morio*)	47.0	39.6	
Cockroaches (Blattoptera)	Six-spotted roach (*Eublaberus distanti*)	66.3	25.1	
Cockroaches (Blattoptera)	Death's-head cockroach (*Blaberus craniifer*)	65.0	22.0	
Cockroaches (Blattoptera)	Argentinean cockroach (*Blaptica dubia*)	64.4	24.5	Females

[1] Yen (2008); Chen et al. (2008); Feng et al. (2008); Dwi (2008); Xiaoming et al. (2010); Yhoung-aree (2010); Oonincx and de Boer (2012); van Huis et al. (2013); Nadeau et al. (2014); Bosch et al. (2014); Macadam and Stockan (2015).

Some insects are key sources of polyunsaturated fatty acids necessary for prophylaxis of many cardiovascular diseases. Studies have shown that the composition of unsaturated omega-3 and omega-6 fatty acids in mealworms compares with that for fish (and exceeds the values for cattle and hogs) (van Huis et al. 2013).

Apart from the evident food advantages, insects have also some **disadvantages,** as follows: (1) *risks* caused by consumption of insects in the inappropriate developmental stage or incorrect culinary preparation (for example, in consuming locusts and grasshoppers, the legs must be removed in order to avoid intestinal constipation, caused by the large spines on the tibia); (2) *toxicity* (for example, in southwestern Nigeria, epidemics caused by the use of African silkworm pupae [*Anaphe* spp.] emerge every year

Table 1.6-7. Examples of energy content of differently processed insect species, by region.[1]

Location	Common name	Scientific name	Energy content (kcal/100 g fresh weight)
–	Mopane moth (caterpillar)	*Imbrasia belina*	250
–	Taiwan giant cricket	*Brachytrupes portentosus*	121
–	Grasshopper nymphs and adults	Average for 20 species	416.8
Australia	Australian plague locust, raw	*Chortoicetes terminifera*	499
Australia	Green (weaver) ant, raw	*Oecophylla smaragdina*	1,272
Canada, Quebec	Red-legged grasshopper, whole, raw	*Melanoplus femurrubrum*	160
United States, Illinois	Larvae of yellow mealworm beetle (*Tenebrio molitor*), raw	*Tenebrio molitor*	206
United States, Illinois	Yellow mealworm beetle, adult, raw	*Tenebrio molitor*	138
Cote d'Ivoire (Ivory Coast)	Mendi termite, adult, dewinged, dried, flour	*Macrotermes subhyalinus*	535
Mexico, Veracruz State	Leaf-cutter ant, adult, raw	*Atta mexicana*	404
Mexico, Hidalgo State	Honey ant, adult, raw	*Myrmecocystus melliger*	116
Thailand	Field cricket, raw	*Gryllus bimaculatus*	120
Thailand	Giant water bug, raw	*Lethocerus indicus*	165
Thailand	Rice grasshopper, raw	*Oxya japonica*	149
Thailand	Grasshopper, raw	*Cyrtacanthacris tatarica*	89
Thailand	Pupae of silkworm (*Bombyx mori*), raw	*Bombyx mori*	94
The Netherlands	Migratory locust, adult, raw	*Locusta migratoria*	179
Thailand	Scarab beetles	*Holotrichia* sp.	98
Thailand	House cricket	*Acheta testacea*	133
Thailand	Taiwan giant cricket	*Brachytrupes portentosus*	188
Thailand	Bombay locust	*Patanga succincta*	157
Thailand	Caterpillar of bamboo borer	*Omphisa fuscidentalis*	231
Thailand	Pupae of silkworm	*Bombyx mori*	152
Thailand	Larvae of hornet	*Vespa* sp.	140
–	Larvae of yellow mealworm beetle	*Tenebrio molitor*	205.6
–	Larvae of zophobas	*Zophobas morio*	242.3
Thailand	Crickets		140.2

[1] Dwi (2008); Yhoung-aree (2010); van Huis et al. (2013); Nadeau et al. (2014).

because the heat-resistant enzyme thiaminase they contain results in disorders in the vascular system and abnormalities in locomotory functions; and (3) *allergies* (inhalation and/or contact with particles of insects at times leads to dermatitis, conjunctivitis, rhinitis, asthma, and contact urticaria) (van Huis et al. 2013).

1.6.2.5 Insect production methods

The edible insects can be obtained in **three ways**: (1) wild harvesting; (2) semi-domestication; and (3) farming of the insects. At present, wild harvesting predominates in the world, as 92% of known species of edible insects are obtained by this method. Six percent of species of edible insects are considered to be semi-domesticated, and only 2% of the species are reared (Yen 2015). It is evident that the second and third ways have huge potential to provide a more stable supply of foodstuffs.

Wild harvesting is the most ancient and most labor-consuming method. It is natural that, in the course of their life, people have accumulated a certain experience and awareness of the best season or time of day for harvesting, and they have also developed various appliances that improve harvesting efficiency. For example, the bugs *Nezara robusta* in southern Africa are harvested early in the morning when it is not hot and the insects are inactive; they are shaken down from the branches of trees using a long silambam with a bag attached at the end (Rousseau 2013). Under favorable circumstances, the harvesting of edible insects can be highly labor-efficient. According to research in Utah (USA), the harvesting of the locust *Melanoplus sanguinipes* provides an average return of 273,000 cal/hour per collector (Schabel 2010).

Semi-domestication, requires the selection of insects having the following characteristic *features* (van Huis et al. 2013): (1) short reproduction cycle; (2) large sizes; (3) gregarious behavior; (4) swarming; (5) high reproductivity; (6) good capacity for survival; (7) nutritional value; (8) potential for storage; (9) possibility to manipulate the habitats; (10) ease of cultivating host plants; (11) marketability; and (12) favorable cost-benefit ratio. The larger number of criteria met by one or another species of insects, the higher is its potential for semi-domestication. If crop plants can simultaneously host more than one species of edible insects, that is worth noticing.

As a rule, insects produced in semi-domestication are **not isolated** from wild populations. Sometimes, exceptions to this rule occur, and within any time of germination, the insects can be withdrawn from nature. In Venezuela, for example, palm weevil larvae (*Rhynchophorus palmarum*) are bred in plastic containers before they are released into the wild (Cerda et al. 2001). In using this way of producing insects, some actions related to habitat manipulation are usually pursued.

Different species of palm weevils are a **classic example** of semi-domesticated insects. Those are *Rhynchophorus palmarum* in Central and South America, *R. phoenicis* in Africa, and *R. ferrugineus* in Southeast Asia. In this case, methods of changing the environment are fairly simple: at a given time, the palms are hewed, and 1 to 3 months later, people return to the trunks to harvest the larvae (Choo et al. 2009).

In **farming other insects**, habitat changes consist, on the contrary, in host tree planting. In order to reduce natural mortality, the larvae can be protected from drought, heat, and predation by simple methods; for example, using protective shade cloths and leaves to cover branches and to shade houses. Sometimes, as for instance in the case of caterpillars of the emperor moth (*Gonimbrasia belina*), known as the mopane worm, they are protected against diseases. In the 1990s, their annual export from Botswana and South Africa reached hundreds of tons (https://en.wikipedia.org/wiki/Gonimbrasia_belina).

In Mexico, the eggs of aquatic true bugs (*Corisella*, *Corixa*, and *Notonecta* species), which lay them on aquatic vegetation in lakes, are highly valued. Here, changes in habitats consist of bundles of twigs, grasses, or reeds that are pressed down with stones to the bottoms of water bodies. After that, the females lay eggs on these bundles; the eggs can be easily harvested by extraction and shaking of the bundles (Parsons 2010). In some regions of Africa, a mix of humidified cellulose (e.g., paper, cardboard, and dried plant material) and soil is loaded in termite mounds for cultivation of termites. Termite mounds that are under construction are considered to be a perfect choice, but one can also obtain good results in this way with old termite mounds (van Huis et al. 2013).

Insect farming is a new and yet fairly rare insect production method. The insects are cultivated in captivity, in isolation from their natural populations, and their living conditions, diet, and food quality are controlled. This method has its benefits and drawbacks. On the one hand, it is the far more productive method for the production of consumable goods. On the other hand, the keeping of insects in confined spaces frequently results in various undesirable effects (van Huis et al. 2013).

One of the most adverse effects is **genetic deviations** due to inbreeding depression; genetic drift (changes in genetic resources of populations from generation to generation caused by some mechanisms other than natural selection); laboratory adaptation, so called founder effect (translation by the group of individuals of a large population of only part of the genetic diversity of this population; as a result, the initial and new populations can evolve in essentially distinguishing directions), etc. All of this often causes the insects raised on a commercial basis to no longer much resemble wild populations.

Insect farming for human consumption takes place in both tropical countries and states in the temperate zone. Cricket farming in Thailand, Laos, and Vietnam can serve as an example of rearing of insects in the *tropics*. In these countries, two species of crickets (native cricket [*Gryllus bimaculatus*] and house cricket [*Acheta domesticus*]) have been produced for many years. From the viewpoint of **production efficiency** (i.e., the output of final products), the native cricket is better; however, house cricket is tastier.

The **methods of farming** are very similar in these countries. Rings approximately 0.5 m in height and 0.8 m in diameter are placed in sheds in one's backyard and used as rearing units. A layer of rice husk is placed at the bottom of each ring, while chicken feed or other pet food, vegetables scraps from pumpkins and morning glory flowers, rice, and grass are used for nutrition.

In *temperate zones,* insects are reared predominantly for **domestic animals**. The main reared insects are mealworms (*Tenebrio molitor*). Because they are usually reared in confined spaces, climate control is frequently applied because high temperatures can result in the desiccation of soft-bodied larvae. The other reared insects include crickets and grasshoppers (Schneider 2009).

1.6.2.6 Insect processing prior to eating

After being wild-harvested or reared in a domesticated setting, the insects should be cooked. In some places, there is the practice of eating insects in the fresh form, but this now occurs quite rarely. One can identify the following variants of insect processing prior to eating: (1) eating of insects in the unaltered form; (2) processing into granular or paste forms; and (3) extraction from insects of particular food components.

When insects are consumed in the **unchanged form**, they are first subjected to freeze-drying, sun-drying, boiling, roasting, or frying (Photo 1.6-2). In general, in countries where people ate insects long

Photo 1.6-2. Common insect food stall in Bangkok, Thailand is shown. Counter-clockwise, from the back-left to the front— locusts, bamboo-worms, moth chrysalis, crickets, scorpions, diving beetles and giant water beetles. They are deep fried.
Photo credit: By User: Takoradee (Own work) [GFDL (http://www.gnu.org/copyleft/fdl.html) or CC-BY-SA-3.0 (http://creativecommons.org/licenses/by-sa/3.0/)], via Wikimedia Commons, 23 April 2006

ago, food habits have shifted towards Western diets (van Huis et al. 2013). Eating insects intact is more characteristic of tropical countries. The insects are usually eaten whole; however, for insects such as grasshoppers and locusts, some their body parts (for example, wings and legs) are preliminarily separated. Already, many edible-insect cookbooks have been published; for example, the books by Ramos-Elorduy (1998), Menzel and D'Aluisio (2004), Gordon (2013), and van Huis et al. (2015).

Processing into the **granulated and paste forms** is more often used in countries where consumers are not accustomed to eating whole insects. Grinding or milling are normal methods for processing of a great number of products, and they are often used for reworking insects. In these forms, they are better accepted by consumers. The resulting powders and pastes usually have been added to other products with low protein contents to increase their nutritional value. However, this method was also used in countries with long-standing traditions of entomophagy. For example, in Thailand and Laos, powder of the ground giant water bug (*Lethocerus indicus*) is a main ingredient of a very popular chili paste.

The extraction from insects of **certain food components** is preferable in societies where the use of insects for food has no long-standing traditions. Often, people are receptive to an idea of the nutritional value of insects, but they prefer to be blind to what they eat. This method of processing insects is acceptable to inhabitants of North America and Europe. However, at times it has created problems for people having allergic responses (Mlcek et al. 2014). In most cases, the proteins are extracted from the insects, but the extraction of fats, chitin, minerals, and vitamins is also possible. Up to now, processes for extraction of individual food components have been too expensive, and the further development of cost-efficient and practical methods of their commercial use is needed (van Huis et al. 2013).

Investigations of edible insects are being carried out in different institutions in many countries. Some of them are listed in Table 1.6-8 below.

At this time, the importance of edible insects in providing protein is essential primarily for some nationalities in tropical countries. Some data are given in Table 1.6-9.

Table 1.6-8. Organizations conducting edible-insect research.

Country	Organization	Founders and scientific leaders
The Netherlands	Wageningen University, Plant Sciences Group, Laboratory of Entomology	Professor Arnold van Huis
United States	Montana State University	Professor Gene DeFoliart, Dr. Florence Dunkel
Denmark	University of Copenhagen, Faculty of Sciences	Dr. Nanna Roos
Thailand	Khon Kaen University, Entomology Division	Professor Yupa Hanboonsong
China	Chinese Academy of Forestry, Research Institute of Resource Insects, Kunming, Yunnan province	Dr. Ying Feng
Kenya	Jaramogi Oginga Odinga University of Science and Technology	Professor Monica Ayieko
Mexico	National Autonomous University of Mexico, Institute of Biology, Faculty of Sciences	Professor Julieta Ramos-Elorduy
Laos	National University of Laos, Faculty of Agriculture	–

Table 1.6-9. Share of protein from insect consumption.

Region, community	Share of total protein intake, %	Source	Notes
Democratic Republic of the Congo, indigenous Gbaya	15	Kelemu et al. (2015)	
Northwest Amazon	5–7	van Huis et al. (2013)	12–26% during May to June
Southern Africa, Asia, Australia, and South America; various indigenous groups	5–10	MacEvilly (2000)	
Amazon region, local communities	8–70	Paoletti and Dufour (2005)	From insects and several other invertebrates such as spiders and earthworms
Certain African communities	5–10	Ayieko and Oriaro (2008)	

1.6.3 Insects as animal feed

As was noted in Section 1.1.4, insects are a food resource for many wild animals. Therefore, the question of their use as food for domestic animals arises naturally enough. Domestic animals include a vast number of different animals, which can be classified based on different criteria. For example, one can subdivide domestic animals according to their **place of residence**: (1) domestic animals kept in special structures (cattle yard, horse stable, doghouse); and (2) domestic animals living directly in people's houses.

Subdivision based on practical usefulness to humans also is logical. In this case, animals providing immediate material benefits (sources of food, materials, etc.) or performing work (transportation of goods, guarding property, etc.) will be assigned to one category. The other category includes animal companions that amuse, give pleasure, and may even communicate with people. In this section, we will mainly consider animals on this principle. It will be understood that the limits of categories are often conventional. For example, some may rear rabbits to produce meat and fur, and the others may keep a rabbit at home as a pet.

It must be said that the production of animal feed is a major segment of the economy. For example, in 2015, world feed production reached 965 million tons with a total cost of US$460 billion. Information of the volumes of fodder production for different categories of animals is given in Table 1.6-10.

We will try to gain insight as to what degree the insects are used in fodder production and the prospects of their further use.

Table 1.6-10. Volumes of fodder production for different categories of animals.

Species	Total feed (million metric tons)
Poultry	439
Pigs	256
Ruminants	196
Aquaculture animals	41
Pets	22
Equine animals	11
Total	**965**

1.6.3.1 Insects as feed for domestic animals and livestock

The borderlines between these categories are not always clear. For example, a horse can be both an agricultural and a companion animal. Nevertheless, such animals (i.e., having multiple uses) are to the fullest extent represented in livestock production and aquaculture; the species include mammals, birds, fish, and some other hydrobionts.

To a large extent, these animals feed on commercially produced foods using tailor-made components (for example, combination fodder for agricultural animals and fish). At the present time, the **major ingredients** of food for animals and fish are fish meal, fish oil, soybeans, and several other grains. Now, about 10% of the world fishery catch is expended for production of fish meal, which is predominantly used in aquaculture (Food and Agriculture Organization of the United Nations 2014).

The **nutritional properties** of insects are very high, and they can be successfully substituted for many ingredients used in the production of foods (fish, soybeans, etc.). Black soldier flies, common housefly larvae, silkworms, and yellow mealworms have been recognized as major species for the commercial production of foods. To a lesser degree, grasshoppers and termites are also prospective species (van Huis et al. 2013).

For example, for the production of freshwater prawns in Ohio (USA) submerged foods for catfish are used, which consist of soybean meal, cottonseed meal, menhaden fish meal (*Brevoortia tyrannus*), corn grain, etc. The production of foods consisting of **black soldier flies**, their fecal masses, and wheat middlings to raise prawns has begun there. Comparative studies did not reveal gustatory differences between prawns reared with the use of traditional and these new foods (Tiu 2012).

However, the use of insects as live food rather than processed food is more common. Traditionally, they are used in raising poultry. Prospective insects for use in this industry include acridoid grasshoppers, crickets, cockroaches, termites, stinkbugs, cicadas, aphids, scale insects, psyllids, beetles, caterpillars, flies, fleas, bees, wasps, and ants. For example, in villages of Guinea, Togo, Burkina Faso, and India, termites are used to feed fowl, including chickens, as well as to feed ostriches in farms across Africa.

In many countries, maggots—**larvae of the common housefly** (*Musca domestica*)—are widely used as food for poultry. They are successfully applied in Nigeria, Togo, Cameroon, Russia, and South Korea. The maggots are fed both in the fresh form and in the dried state; the latter form facilitates storage and transportation. Studies have shown that maggots can be substituted for fish meal in the production of broiler chickens. For example, in South Korea, the importance of maggots in feeding broiler chickens was analyzed, and it was found that the carcass quality and rate of live weight gain are optimal at 10–15% maggots in the ration (Hwangbo et al. 2009).

In India, conventional feed accounts for 60% of the total cost of raising poultry; furthermore, there is a shortage of such foods as maize and soybeans used for their preparation. The harvesting of insect agricultural pests and their application as an ingredient in the production of foods make it possible to reduce the consumption of harmful pesticides for their control, to improve yields, and to redirect the **saved grain** for feeding of humans.

In this connection, in India, the possibilities of using **acridoid grasshoppers** as food were investigated. In particular, the nutritional content of **four species of acridids** was studied: common short-horned grasshopper (*Oxya fuscovittata*), common Indian grasshopper (*Acrida exalata*), rice grasshopper (*Hieroglyphus banian*), and short-horned grasshopper (*Spathosternum prasiniferum prasiniferum*). The studies showed that all of acridids investigated are characterized by higher protein contents than conventional soybean and fish meal and may well be used as an ingredient in the production of foods. The common short-horned grasshopper (*O. fuscovittata*) was recognized to be promising, and it is reasonable not only to harvest it but also to rear it, in view of its elevated rates of fecundity and fertility (Anand et al. 2008).

Many species of fish feed on insects, so the use of insects as food in aquaculture is absolutely natural. For the time being, the share of insects in food for household fish is insignificant. The insect species considered **most suitable** for the production of combination fodder are black soldier fly larvae (*Hermetia illucens*), common housefly larvae (*Musca domestica*), silkworms (*Anaphe panda*), yellow mealworms, grasshoppers (Acrididae), and termites (*Kalotermes flavicollis*) (European Food Safety Authority Scientific Committee 2015).

Black soldier fly larvae are already used successfully for feeding red seabream (*Pagrus major*) (Iwai et al. 2015), rainbow trout (*Oncorhynchus mykiss*), channel catfish (*Ictalurus punctatus*), and blue tilapia (*Oreochromis aureus*). In the case of rainbow trout, the larvae may be substituted for 25% of fish meal and 38% of fish oil (St. Hilaire et al. 2007). In China, Mormon cricket meal (*Anabrus simplex*) is used as food in freshwater aquaculture as well as in pig breeding and fur farming (farm-grown mink) (Irwin and Kampmeier 2003).

1.6.3.2 Insects as pet food

A vast number of various animals feed on insects (see Sections 1.1.4.1–1.1.4.6). Some of them are used as pets, which do not procure food in the wild and which should be fed. In order to have an idea of the size of the problem, data for Europe are presented. Currently, the number of pet-food-producing companies in Europe is 650 plants. The estimated numbers of pets (millions) in Europe are as follows: cats, 99.2; dogs, 81.0; decorative birds, 54.7; decorative fish (number of aquaria), 12.65; small mammals, 29.3; and reptiles, 7.3 (FEDIAF 2015). It also should be mentioned that some insects, spiders, and scorpions are kept as pets (see Section 1.8.2) and must be fed.

Cats are the most numerous pets. These are carnivorous animals, so animal protein is important to them (the source in cat food is slaughterhouse byproducts, the internal organs of animals). Cat food also contains animal fats, vegetable components (different cereal ingredients such as corn [maize] and rice) with the addition of vitamins, minerals, and other useful elements. Insects generally are not used in producing cat food. However, around the world, the share of insects in the **diet of feral cats** is about 6% (Plantinga et al. 2011), so the need to use insects for feeding domestic cats is evident.

Dogs rank next to the cats in pet populations; the world population of pet dogs is roughly 300 million (http://4ento.com/2015/05/07/pet-food-challenges-why-need-change/). Some insects are usually present in the diet of wild relatives of dogs (wolves, coyotes, jackals, foxes, etc.); however, they generally are not used in the commercial production of dog foods. Studies on estimating the protein quality of insects as potential ingredients for dog foods have already been carried out (Bosch et al. 2014).

Decorative birds kept as the pets can be phytophagous and entomophagous. In practice, there are no birds that feed on only vegetable food, and even species considered to be phytophagous (canaries, parrots, crossbills, bullfinches, weavers, etc.) need animal protein during some period of life. To a greater or lesser degree, most decorative birds are entomophagous (reedbirds, orioles, kinglets, wrens, nightingales, thrushes, warblers, bluethroats, starlings, redstarts, larks, robins, tits, flycatchers, waxwings, etc.). Rearing of insects to feed them has been a long-felt need.

Food for **aquarium fish** is no different, for the most part, from that used in aquaculture. A vast number of the artificial dry and frozen natural feeds are on the market; however, the significance of rearing of insects in this case will grow in future. As for **mammals**, people frequently keep rodents as pets (chinchillas, mice, rats, hamsters, squirrels), monkeys, ferrets, pygmy pigs, guinea pigs, hedgehogs, etc.). **Reptiles** typically kept as pets include tortoises, iguanas, chameleons, agamas, geckos, and snakes. To a greater or lesser degree, the great majority of domestic mammals and reptiles are entomophagous.

In conclusion, insects are insufficiently used in the production of foods for pets. A lack of living insects is observed to a greater extent. Among the **most popular insects** for use as pet food are crickets (tropical house cricket, *Gryllodes sigillatus*; two-spotted cricket, *Gryllus bimaculatus*; and house cricket, *Acheta domesticus*); mealworms (superworm, *Zophobas morio*; lesser meal black beetle, *Alphitobius diaperinus*; and yellow mealworm beetle, *Tenebrio molitor*); locusts (migratory locust, *Locusta migratoria*); sun beetles (*Pachnoda marginata peregrina*); wax moths (greater wax moth, *Galleria mellonella*); cockroaches (orange-spotted cockroach, *Blaptica dubia*; speckled or lobster cockroach, *Nauphoeta cinerea*); and maggots of the housefly, *Musca domestica*.

Use of Insects in Agriculture and Forestry

In agriculture and forestry, insects are used as agents for the **biological control** of pests and weeds. Biological control takes advantage of antagonistic interspecific interrelations between the living organisms. It involves the use against pests of their natural enemies: **parasites** and **predators** (which cause death or damage to the pest individuals). Biological control often makes it possible to significantly decrease the density of populations of pests and weeds, thus lowering their injuriousness. Recall that pests are organisms that have detrimental effects on human welfare; this concept is closely related to human activity. In natural systems, however, there are no harmful or useful species.

Attempts at the biological control of the insect pests and weeds were analyzed by Johnson (2000b). These studies revealed the following: (1) more than 1,300 attempts resulted in the full or partial control of the target species; (2) in all, more than 135 species of insect pests and weeds were substantially to completely suppressed by introducing their natural enemies; (3) throughout the 20th century, there was an average of 1.4 successful attempts at biological control per year; (4) taxa of insects can be rated with respect to the successfulness of their application in biological control as **follows**: (a) homopterous insects (Homoptera); (b) butterflies and moths (Lepidoptera); (c) beetles (Coleoptera); (d) flies (Diptera); (e) hymenopterans (Hymenoptera); (f) long-horned grasshoppers (Orthoptera); (g) true bugs (Heteroptera); and (h) earwigs (Dermaptera).

The following **conclusions** were also drawn from this study: (1) there is a close correlation between the successfulness of attempts and the number of scientific studies performed; (2) throughout the world, success with the use of parasitoids was attained 4 times more frequently than with the application of predators; (3) most successes were related to the control of scale insects; (4) the maximum number of successful attempts at control was were accomplished on the prickly pears (*Opuntia* sp.; Johnson 2000b). A total of more than 6,000 species of insects have been tested and released as biological control agents of insect pests and weeds (http://www.cals.ncsu.edu/course/ent425/text01/impact1.html).

1.7.1 Control of pests

The essence of biological control of insect pests lies in the restriction of their numbers to the **economic threshold** of injuriousness. This value for each pestiferous species is individually determined. For example, in the case of the Colorado potato beetle (*Leptinotarsa decemlineata*), the threshold level of the number of pests in the period of the plants' flowering responsible for losses of approximately 5% of the potato yield is on average five imagoes and five larvae per plant (Kakharov 2008).

Natural enemies—including predators, parasitoids, and parasites—are used to decrease the populations of harmful insects. The difference between these types of natural enemies lies in the fact that the **predator** feeds on the prey and, as a rule, kills it at once. Over the course of its life, the predator eats up many prey individuals. The predatory species—entomophagous species—include 16 orders among the insects with incomplete metamorphosis (dragonflies, mantids, stoneflies, grasshoppers, earwigs, thrips, true bugs) and those with complete metamorphosis (beetles, lacewings, alderflies, scorpion flies, caddis flies, lepidopterans, hymenopterans, dipterans). The predatory bugs, thrips, beetles, lacewings, flies, and hymenopterans are the most important to biological control.

Parasitoids kill the host *gradually* by feeding on its tissues in the course of their development. The larvae of parasitoids hatch from eggs and develop inside as well as on the body of the host. A parasitoid causes the inevitable death of the host because, in the course of its development, the larvae of parasitoids entirely consume the host. When parasitoids reach adulthood, they become gregarious. About 25% of all insect species on earth are parasitoids (http://www.biologyguide.ru/gbids-420-3.html).

For the most part, **parasites** do not kill their hosts; otherwise, they would inevitably disappear. The parasitic forms of insects (parasites and parasitoids) are encountered in five orders: beetles, twisted-winged insects, butterflies, hymenopterans, and flies; that is, in the orders of insects with complete metamorphosis. Parasitic hymenopterans and flies are of the greatest practical importance (http://mylektsii.ru/9-1665.html).

In carrying out biological control, choosing a species that will optimally reduce the numbers of a particular pest is of great importance. There is no accurate algorithm for identifying effective enemies; success is attainable only as a result of enumeration of the possibilities. However, there are some criteria that allow us to narrow the search limits. Eleven such **criteria** have been identified (Maximova 2014):

1. *Ecological compatibility.* The ecological requirements of the target species and the entomophagous insect should be close.

2. *Synchronization in time.* The life cycles of the host and parasite should be synchronized; the reproductive stage of the parasite should be active simultaneously with the receptive stage of the host.

3. *Fast positive response* to the increase in host density. Of prime importance is a capacity for rapid increases in the number of parasites in response to increases in the population of the target species.

4. *High reproductive potential.* The entomophagous insect should have a high reproductive rate and/or a short time of reproduction.

5. *Capacity for search.* The useful organism should be capable of using the host population at low densities of the host.

6. *Capacity for resettlement.* In selecting an entomophagous species for introduction, it is desirable that one or several introductions are sufficient for expansion across the host range.

7. *Specificity toward the host and compatibility with it.* Most cases of successful biological suppression have been related to the use of entomophagous insects that are highly specific toward the hosts. Monophages or limited oligophages are preferable.

8. *Availability of additional food sources.* The imagoes of parasites frequently require the presence of appropriate additional hosts.

9. *Superparasitism.* In order to realize its possibilities in the fight, the entomophagous insect should have a minimum of natural enemies.

10. *Suitability for keeping in captivity.* Before it is released into the wild, the entomophage is usually reared artificially in order for it to populate the maximum territory and account for the deaths of some individuals.

11. *Safety for the ecosystem.* Introduced species must have harmful impact on target species only.

The **introduction of an entomophage** is a complex process that includes a number of stages. The introduction of parasitic insects is the best example of this process; for the most part, the introduction programs were related to just parasites. **These stages** are as follows (Maximova 2014): (1) *identification of the pest* (ascertainment of the species and estimation of the damage they inflict); (2) search of the literature *on the biology of the pest*, its distribution, ecological requirements, and other characteristics; (3) search of the literature *on natural enemies* of the pest; (4) field *investigations of the pest* and its enemies; (5) *prediction of the success* of a biological control program; (6) *collection of useful insects*; (7) their *transportation* taking into consideration a number of conditions (choice of the optimized route and points of transport changes, maintenance of required temperatures, containers that prevent damage and escape of the insects, choice for transportation of the most suitable stage of the insects, and other considerations); (8) *quarantine* (imported species mates in captivity during several generations to eliminate diseased insects and tramp individuals of other species); (9) *reproduction*; (10) *release and colonization* (abiotic conditions are necessary that are suited for the entomophage at the site of introduction, and additional hosts or sources of additional nutrition may be needed; (11) *control collections* (availability of the entomophage at the point of release for a period of 3 years suggests success); and (12) *appraisal of results* (to calculate the economic efficiency and to determine reasons for success or failure).

1.7.1.1 Control of agricultural pests

Both predatory and parasitic insects are used to control agricultural pests. Using the weaver ant (*Oecophylla smaragdina*) to protect Chinese and Southeast Asian citrus orchards can be considered (https://en.wikipedia.org/wiki/Weaver_ant) to be the first experience of using **predatory insects** for biological control of agricultural pests. As early as the third century AD, nests of this ant were sold near Canton (now, Guangzhou) to protect citrus cultures from pests such as the Lychee giant stink bug (*Tessaratoma papillosa*) (Johnson 2000a).

The next landmark is the twelfth century AD. At that time in Yemen (Middle East), ants of an unknown species were applied on a large scale to protect date palm stands (Maximova 2014). At about the same time, the benefit of lady beetles for controlling aphids and scales was acknowledged (Johnson 2000a).

The use of **parasitic insects** began much later. In the first known instance, in the 11th century in China, parasitic tachinid flies were used to control populations of harmful insects. The next known attempt of the biological control of agricultural pests with the use of parasitic insects occurred in the 17th century, when ichneumonoid parasites were correctly found to be as a means of fighting flies and beetles (Orr 2009).

Over time, the purely empiric arrangements for using insects for pest control become **scientifically justified**. In 1602 in Europe, the parasitic wasp *Cotesia glomerata* laying its eggs into the larvae of the cabbage butterfly (*Pieris brassicae*) and cardinally lowering the numbers of the pest was described. Now it has been established that infestations of populations of the larvae of the cabbage butterfly with this parasite not infrequently reaches 90% and more (Saulich and Mussolin 2013).

A paper by the Italian naturalist Ulisse Aldrovandi, entitled, "De Animalibus Insectis" (1602), became the first publication regarding biological control. This paper generalized all that had been written earlier about insects, and it also described, for the first time, parasitism among insects. Aldrovandi described the attack of the white butterfly parasite (*Cotesia glomerata*) on the small cabbage white butterfly (*Pieris rapae*) (Zamotajlov 2012). In 1634, the English naturalist Thomas Muffet published an illustrated guide to the classification and lives of insects, Theatre of Insects (*Insectorum sive Minimorum Animalium Theatrum*) (https://en.wikipedia.org/wiki/Thomas_Muffet).

In 1734, R.A.F. de Réaumur in his *Mémoires pour servir à l'histoire des insectes* explained three ways parasitic insects enter the body of their prey: (1) the parasite's eggs are introduced from plant leaves, (2) the parasite fixes its eggs on the body of the caterpillar, and (3) the parasite lays eggs in the body of the caterpillar. He illustrated the larvae of the ichneumon fly *Apanteles glomeratus* inside a pierid (whites family) caterpillar. In addition, he proposed the introduction of the eggs of hoverflies (Syrphidae) for the control of aphid populations in greenhouses (Egerton 2006).

Before the end of the 19th century, insects were rarely used for pest control. The impetus to the **large-scale implementation** of biological control methods (although the term *biological control* was apparently used for the first time by the late Harry Smith of the University of California in 1919) was given by the successful experience in control of **cottony cushion scale** (*Icerya purchasi*). This scale was accidentally introduced to California with planting material in 1869, and by 1886 it had become a serious danger to citrus cultures in the region (Orr 2009). The insect settles along leaf veins and sucks the phloem sap, which results in defoliation and dieback of twigs and small branches (http://www.biocontrol. entomology.cornell.edu/success.php). The effects of activity of the cottony cushion scale was comparable with that of the *Phylloxera* attack in France (see item 2.1.1.3).

An officer of the U.S. Department of Agriculture, C.V. Riley, proposed to resolve this problem by introducing natural enemies of the scale to the infested area. In 1888, he sent his worker Albert Koebele to Australia to collect natural insect enemies of *Icerya purchasi*. A. Koebele found that the **vedalia ladybird beetle** (*Rodolia cardinalis*) is a natural enemy of the cottony cushion scale. From November 1888 to March 1889, he sent five shipments of the beetles to California.

In all, 514 vedalia ladybird beetles were received and released into citrus groves in California (http://www.biocontrol.entomology.cornell.edu/success.php). It was not necessary to construct farms for the "mass production" of vedalia ladybird beetles, because they had reproduced themselves very well (Chauvin 1970). By 1890, the problem of cottony cushion scale in California was resolved. Expenditures for this project, aside from the workers' wages, were only $1,500, while the benefits reached millions of dollars annually (Orr 2009).

The introduction of entomophages for pest control is seldom limited to one species. As a rule, **several species** of insects are selected, and their efficiency is examined under new conditions. For example, in the case of the cottony cushion scale, the parasitic fly *Cryptochaetum iceryae* was introduced in addition to the vedalia ladybird beetle. This fly became a major factor in controlling pests in coastal areas of California. Later on, it was successfully acclimatized for controlling the cottony cushion scale in Chile (Nartshuk 2003).

During the early 1970s, major damage to wheat in Chile was caused by two aphid species: English grain aphid (*Sitobion avenae*) and rose grain aphid (*Metopolophium dirhodum*). The Chilean government initiated a program introducing their natural enemies (parasitic wasps and predatory insects). In 1975–80—from South Africa, Canada, and Israel—five species of **predatory** insects were delivered, and in 1976–81, nine species of **parasitic** wasps of the families Braconidae and Aphelinidae were introduced from Europe, California (USA), Israel, and Iran (Sampaio et al. 2009). The total number of the parasitic insects released in Chile reached four billion (Peshin et al. 2009).

An analogous situation was also observed in Brazil. There, the parasitoids *Aphidius ervi, A. rhopalosiphi, A. uzbekistanicus,* and *Praon volucre* delivered from Europe, Israel, and Chile in 1978–92 were used for biological control of the English grain aphid (*Sitobion avenae*) and rose grain aphid (*Metopolophium dirhodum*). As a result of a sharp decrease of the pests' numbers, $40 million a year was saved due to reductions in the quantities of insecticides that were used (Sampaio et al. 2009).

Decisions to introduce entomophages for the purpose of biological control are preceded, as a rule, by long-term studies. For example, in Cuba, the broad mite (*Polyphagotarsonemus latus*) causes major damage to pepper production in greenhouses. Five species of predators of the broad mite (*Neoseiulus barkeri, N. californicus, N. cucumeris, Amblyseius swirskii,* and *A. largoensis*) were examined. The predatory mite *A. largoensis* was recommended as the best candidate for introduction, due to its short life cycle, high fecundity rate, and adequate efficiency in searching for *P. latus* as prey (Rodríguez et al. 2015).

However, **errors** in choosing agents for biological control can lead to serious consequences. For example, the multicolored Asian lady beetle (*Harmonia axyridis*) was introduced in 1916 from Asia to North America and northwest Europe as a predator of aphid pests. However, it gradually changed into a notorious global pest. The multicolored Asian lady beetle became a threat to native biodiversity because it began to eat, in addition to aphids and scales, ten other species of insects (Koch 2003). In addition, the multicolored Asian lady beetle became a noxious household pest and a minor agricultural pest (to fruit production and processing) (Orr 2009).

In its native habitats in Asia, this lady beetle hibernates in rocky cracks. Because the numbers of such places are insufficient in many new areas it inhabits, the beetles often to winter in residential houses in large numbers. The multicolored Asian lady beetle has a foul smell, it stings, and it often falls from ceilings into dishes containing food and beverages (Huelsman et al. 2002). In addition, contact with sensitized individuals not infrequently results in different allergic responses, including allergic rhinoconjunctivitis, asthma, pruritus, urticaria, angioedema, and anaphylaxis (Goetz 2008).

However, the use of the insect entomophages for control of agricultural pests exerts great **positive effects**. Now, the biological control of pests is applied over approximately 3.5 million km^2 (350 million ha), which corresponds to about 8% of the area of agricultural lands, and it has very high benefit-cost ratios of 20–500:1 (van Lenteren 2012). By 2006, more than 5,000 introductions of about 2,000 species of natural enemies of pests were carried out in 600 areas of 196 countries (van Lenteren et al. 2006; Orr 2009). For example, for control of more than 90 species of harmful insects, about 500 species of parasites and predators had been delivered to the United States from other countries by 1956, and 100 species of them have become naturalized. In Canada, 220 species of entomophages have been resettled and 50 of them have acclimatized (Maximova 2014). The contribution of biological control to yield gain is estimated at about 20% (Bengtsson 2015).

1.7.1.2 Control of forest pests

Both predatory and parasitic insects are also used in the control of forest pests. Because agricultural plants are of far more value to humans than those in forests, the latter's pests (and, accordingly, their control) have received little attention. The **scientific work** in this area began in the 18th century. In 1701, the great Dutch naturalist Anton van Leeuwenhoek sketched and described a parasite of the willow sawfly. The Swedish naturalist Carl Linnaeus proposed using the forest caterpillar hunter (*Calosoma sycophanta*) for control of some forest pests (Zamotajlov 2012).

The mid-19th century marks **the beginning of the practical application** of insects for the suppression of forest pests. In 1840, the French naturalist F. Boisgiraud used the forest caterpillar hunter (*Calosoma sycophanta*) to eradicate gypsy moth larvae which had settled in all the poplars in his village. For this purpose, he collected in nature a large number of ground beetles and released them to trees. This release had a positive effect (Pushkin 2015). In 1844, the famous German entomologist Julius Theodor Christian Ratzeburg published the book *Die Ichneumonen der Forstinsekten in entomologischer und forstlicher Beziehung* where he drew attention to the importance of parasitoids in suppressing forest insect populations.

In 1868, the gypsy moth (*Lymantria dispar*) was introduced in the United States (see Section 2.6, Invasions of insects). The first large-scale outbreak of this pest in the United States took place in 1889. The Bureau of Entomology, U.S. Department of Agriculture—including researchers L.O. Howard, W.F. Fiske, and C.G. Hewitt—carried out large-scale programs introducing natural enemies of the gypsy moth (1905–14, 1922–23) and brown-tail (*Euproctis chrysorrhoea*) (1909–1911) in the United States and Canada. Damage by the gypsy moth was reduced, and brown-tails were entirely suppressed (Coppel and Mertins 1977).

Unintended introductions of various pests are not infrequent; therefore, resettlement of their natural enemies from their former habitats is needed periodically. The choice and introduction of agents of biological control are preceded by **long-term studies**, which were mentioned earlier. The emerald ash borer (*Agrilus planipennis*) was observed in North America for the first time in 2002. This phloem-feeding beetle has already killed tens of millions of ash trees and threatens to kill most of the 8.7 billion ash trees throughout North America (https://en.wikipedia.org/wiki/Emerald_ash_borer).

At first, the search for the borers' natural enemies was conducted in its natural range (China), and in 2007 three species of **parasitic nonstinging wasps** (*Spathius agrili, Tetrastichus planipennisi,* and *Oobius agrili*) were selected and delivered. For 7 years (2008–14), their interrelations with the emerald ash borer were studied in six stands of eastern deciduous forest in southern Michigan (Duan et al. 2015). In 2015, the parasitic wasp *Spathius galinae* was approved for release (Bauer et al. 2015).

However, the release of the natural enemies of pests does not always guarantee success in pest control. **Three resulting variants** are possible: (1) complete success; (2) partial success; (3) failure.

The literature includes many descriptions of the **successful** introduction of natural enemies for control of pests (Yang et al. 2014; Matosevic et al. 2016). One example is the case of the red turpentine beetle

(*Dendroctonus valens*). It is believed that this beetle was transported to China from the West Coast of the United States in 1983 in rough logs designed for the manufacture of pit props (excavation supports). The first large outbreak was observed in Shanxi Province in 1999. Subsequently, this beetle spread quickly to the adjacent provinces of Hebei, Henan, and Shaanxi. By 2003, over 500,000 ha of Chinese pine stands had been infected with this pest, and about 4 million trees had been destroyed (Sun et al. 2004). The Chinese responded with the mass rearing and release of the predatory beetle *Rhizophagus grandis*, which greatly reduced the pest population (Vega and Hofstetter 2015).

An example of **partial success** is the introduction of the parasitoid wasp *Lathrolestes thomsoni* for control of the amber-marked birch leaf miner (*Profenusa thomsoni*). In 1991, this pest was first found in Alaska, where it became a widespread and damaging pest to birches. In 2003, the parasitoid wasp *L. thomsoni* was chosen as a biological control agent. From 2004 through 2008, 3,636 imagoes of *L. thomsoni* were released to birch forests in Alaska. At the sites of release, affected leaves were reduced by 30–40%; however, a cardinal decrease in the number of pests was not attained (Soper et al. 2015).

Publications devoted to **failures** of biological control are quite scarce. The lack of such papers by no means reflects the general picture: successes are discussed readily, while problems are marginally described. A typical example of a failure is attempts to control the balsam woolly adelgid (*Adelges piceae*). These small wingless insects were transported from Europe to the United States in about 1900. They inflict damage to the balsam fir (*Abies balsamea*) and the Fraser fir (*Abies fraseri*).

For example, in Great Smoky Mountains National Park (USA), the balsam woolly adelgid has destroyed about 95% of the Fraser firs, creating "**ghost forests**" (https://en.wikipedia.org/wiki/Balsam_woolly_adelgid). In efforts to control this pest, six predatory insects (three beetles and three flies) were introduced to the United States from Japan, China, India, Pakistan, and Europe. They were released in the New England region and in the states of North Carolina, Washington, and Oregon. However, the results of this introduction proved to be unfavorable (Vail et al. 2001; Livingston and Pederson 2010).

Various species of **ants** are of great importance as agents of biological control. Their important feature is their capability to be switched to more abundant species of prey. Thanks largely to this quality, ants efficiently protect forests against many needle- and leaf-eating insects. However, a large number of ants is required for this purpose. Therefore, species with high colony densities are most effective (for example, the red wood ant [*Formica rufa*] or ant [*Formica imitans*] form colonies that can include millions and tens of millions of individuals).

Artificial resettlement of ants to the focuses of outbreaks of pests is needed. There are **two ways** to accomplish this resettlement: (1) moving the entire anthill (i.e., liquidation of the nest in the previous place and its transfer to the new section of forest); and (2) resettlement of part of the anthill. In the second case, withdrawal of not more than one-fourth of its dome volume is permissible. The important point is selection of a whole anthill or its part for the resettlement.

A new place should meet the following **conditions** (Zakharov 2015): (1) illumination (the anthill should be lighted by the sun for several hours each day); (2) water regime (one cannot place the nests within areas where flooding is likely); (3) food reserve (it is better to confine the settlement to groups of trees that include many different species; sections of the forest where the trees vary in age are also favorable); (4) competition (the presence of competitive species of ants is undesirable); (5) placement (it is desirable to place the nest to the south of a trunk of a mature tree, the butt of which is lighted by the sun; or placement on an old dry stub is possible, but the anthill should take in the whole stub in such a way that the stub is covered with a layer of construction material not thinner than 10 cm; in case of high numbers of woodpeckers and forest gallinaceous birds, it makes sense to cover the anthill with small boughs of coniferous trees); and (6) time (it is better to perform the resettlement in the morning hours when temperatures are equal to less than 20°C; one cannot settle ants in a new place just before or during rain).

1.7.2 Weed control

Broadly speaking, weeds are plants that are undesirable on territories and water bodies that are used by humans for the economic activity. For example, about 2,200 species of economically significant weeds have been counted in the United States. More than half of these species, rather than being aboriginal species, were delivered to the country accidentally (as seeds or forages, or with ship ballast) or were deliberately introduced as ornamental plants.

Weeds can be categorized according to **different criteria,** such as groups of plants (grasses, perennial shrubs, and trees), inflicted damage (agricultural and environmental weeds, etc.), and environmental conditions (terrestrial and aquatic weeds). A common feature of all weeds is that they complicate life for humans in one way or another. It is natural that humans attempt to control weeds in every possible way. One of these is the use of phytophagous insects, which by feeding on the weeds lower their number. This method is fairly widespread and not infrequently results in success. By 2000, 41 species of weeds had been successfully controlled by way of the introduction of insects or pathogens in different regions of the earth (McFadyen 2000). Some of them are listed in Table 1.7-1.

The frequent and successful use of insects for the biological control of weeds is caused by the following **factors** (Reimer 2000): (1) there are numerous species of phytophagous insects; (2) these insects frequently show selectivity to host plants when they can feed exclusively and essentially on a particular weed; and (3) under any ecological conditions, weeds have a broad range of natural enemies.

The following sections consider the use of insects for control of terrestrial and aquatic weeds.

1.7.2.1 Control of terrestrial weeds

Terrestrial weeds affect agriculture (yield depression, deterioration of crop quality, difficulties in harvesting of crops, increases in fuel consumption and depreciation of agricultural machines, creating conditions for reproduction of agents of diseases, destructive insects, and rodents), cattle breeding (danger to life and health of livestock, deterioration of the quality of cattle breeding products—milk, meat, wool), the food industry (difficulties in processing products and deterioration of the quality of the obtained products), and the forestry industry and forest restoration (obstruction of restoration and further development of cultivated forest plants) (Govorushko 2012).

Generally, terrestrial weeds do great **economic damage**. The direct crop losses from weeds are estimated at US$10 billion annually. The cost of weed control is $6.2 billion/year, including $3.6 billion for herbicides (Reimer 2000).

Table 1.7-1. Major recent successes in biological control of weeds.[1]

Common name	Latin name	Region
Golden wreath wattle	*Acacia saligna*	South Africa
Hamakua pamakani	*Ageratina riparia*	Hawaii
Alligator weed	*Alternanthera philoxeroides*	Subtropics
Nodding thistle	*Carduus nutans*	Canada
Jack in the bush	*Chromolaena odorata*	Africa, Asia
Black sage	*Cordia curassavica*	Malaysia
Water hyacinth	*Eichhornia crassipes*	Tropics, subtropics
Harrisia cactus	*Harrisia martinii*	Australia
Klamath weed	*Hypericum perforatum*	Hawaii, continental United States
Giant sensitive plant	*Mimosa invisa*	New Guinea
Water lettuce	*Pistia stratiotes*	Tropics, subtropics
Salvinia	*Salvinia molesta*	Tropics, subtropics
Tansy ragwort	*Senecio jacobaea*	Australia, United States
Rattlebox	*Sesbania punicea*	South Africa
Noogoora burr	*Xanthium occidentale*	Australia
Skeletonweed	*Chondrilla juncea*	Australia

[1] van Lenteren (2012).

In contrast to the control of pests, the use of insects to control weeds has no centuries-long history. The **first attempt** to use insects to control weeds is thought to have been the introduction in 1863 of mealybugs (*Dactylopius ceylonicus*) from northern India to southern India to control the prickly pear cactus (*Opuntia vulgaris*) (Omkar 2016).

The **first published paper** devoted to the biological control of weeds (Lantana Weed Project) came out in 1902 and was devoted to control of wild sage (*Lantana camara*), specially imported to Hawaii in 1885 as an ornamental plant. However, it very quickly became a noxious weed (Davis et al. 1992).

As in the case of the biological control of pests, the introduction of any helpful phytophagous insect is seldom limited to one species. So, 23 species of phytophagous insects were delivered from Mexico to Hawaii for the biological control of **wild sage** (*Lantana camara*) at the beginning of the 20th century, and 8 of these species have become naturalized (Thomas and Ellison 2000). However, even the careful selection of biological control agents does not always guarantee success. For example, for the biological control of wild sage, a total of 36 biocontrol agents have been used in 33 regions of the world. However, none of these attempts has ended in success (https://en.wikipedia.org/wiki/Lantana_camara).

Another ailed attempt at weed control was the fight against **ragweed** (*Ambrosia artemisiifolia*) in southern former USSR; this weed was transported to the USSR from North America. In the period from 1969 to 1979, more than 30 species of natural enemies of ragweed were selected in Canada and the United States. After detailed study, four species of insects (olive-shaded bird-dropping moth [*Tarachidia candefacta*], tephritid fruit fly [*Euaresta bella*], ragweed leaf beetle [*Brachytarsus tomentosus*], and another ragweed leaf beetle [*Zygogramma suturalis*]) were selected and introduced in this region. However, despite all efforts, the program of ragweed control failed (Esipenko 2012).

Examples of the **successful** biological control of weeds are well known. One of them is the control of **St. John's wort** (*Hypericum perforatum*). It was accidentally delivered from Europe to North America and Australia, where it displaced indigenous pasture grasses over thousands of square kilometers and became a key problem for cattle farmers. Leaf beetles (*Chrysolina hyperici* and *C. geminata*) delivered from Europe have helped to substantially reduce the numbers of this weed (Izhevsky et al. 2014). Successful cases of biological control of the musk thistle (*Carduus thoermeri*) and puncturevine (*Tribulus terrestris*) were described in a paper by Vail et al. (2001). The successful use of the cactus moth (*Cactoblastis cactorum*) to control prickly pears (mostly *Opuntia stricta*) in Australia illustrate Photos 1.7-1, 1.7-2. This case will be described in the Section 1.11.7. Monument to *Cactoblastis cactorum* in Dalby, Queensland, commemorating the eradication of the prickly pear in the region, is shown on Photo 1.7-3.

Sometimes, biological control agents that are effective in one region prove to be **useless** in others. For example, seeds of the diffuse knapweed (*Centaurea diffusa*) entered North America from Europe, presumably as an admixture to alfalfa seeds. Now this plant is widespread in 19 states of the United States, and it is a harmful weed having detrimental effects on cattle breeding (its thistles can damage the mouths and digestive tracts of animals that attempt to feed on it) and on agriculture (greatly reducing crop yields and purity) (https://en.wikipedia.org/wiki/Centaurea_diffusa). The introduction of the gallflies (*Urophora quadrifasciata* and *U. affinis*), which has been effective in other areas, in many cases did not reduce weed density (Myers 2000).

Another example is control of the **tansy ragwort** (*Senecio jacobaea*). The seeds of this plant, which is widely distributed in Eurasia, entered North America, Australia, and New Zealand in the 18th century and early in the 19th century, mainly with soil used as ballast (Brown 1988). The tansy ragwort became widespread on pastures, native meadows, and open forests.

Because the plant is poisonous and is a serious hazard to horses, cattle, and poultry (Markin and Littlefield 2008), repeated attempts to control it were undertaken using different phytophagous insects. Most often, the noctuid moths (cinnabar moth [*Tyria jacobaeae*]) and two species of leaf beetles (*Longitarsus jacobaeae* and *L. ganglbauer*) were used for this purpose. However, they had little impact on the weed density (Myers 2000).

Therefore, when insects are used for weed control, **three results** are possible (van Lenteren 2012): (1) complete control, meaning that no other methods of control are required on this territory (it does not mean that the weed is eradicated); (2) substantial control, when other methods of control are also necessary but less effort is required (for example, less herbicide is needed); and (3) negligible control,

Photo 1.7-1. Prickly pears (mostly *Opuntia stricta*) were imported into Australia in the 19th century for use as a natural agricultural fence and in an attempt to establish a cochineal dye industry. They quickly became a widespread invasive species, rendering 40,000 km² (15,000 square miles) of farmland unproductive. Picture shows a typical view of the site, overgrown with prickly pear.
Photo credit: CSIRO

Photo 1.7-2. In 1926, larvae of the cactus moth (*Cactoblastis cactorum*) were delivered from Argentina and dispersed over the cactus thickets from an aircraft. In seven years, the cactus moth had almost completely destroyed the cactus thickets in the region. Prickly pear site after treatment with *Cactoblastis cactorum* is shown.
Photo credit: CSIRO

when the use of phytophagous insects results in no significant lowering in the numbers of weeds, so other control measures are necessary.

The overall **success rates** of weed biological control programs using insects have been increasing. At the present time, they have reached 50–80% in South Africa, Australia, and Hawaii (McFadyen 2003). This is largely because the numbers of released insects were insufficient in the early control programs, but now they are bred in great numbers before release.

From an **economic point of view**, the control of weed numbers by means of insects is completely justified. Commonly, only one-third of the introduced species become naturalized under new conditions, and not more than one-third of successfully acclimatizing ones exercise significant influences over the growth of the target plant. Therefore, no more than 10% of attempts at this type of biocontrol are successful (McFadyen 1998).

However, the financial effects from one successful introduction can be so great (annual incomes exceeding total expenditures by 1 to 2 orders of magnitude) that they compensate for the costs of many

Photo 1.7-3. There is a monument to *Cactoblastis cactorum* in Dalby, Queensland, commemorating the eradication of the prickly pear in the region.

Photo credit: By Saintrain (Own work) [GFDL (http://www.gnu.org/copyleft/fdl.html) or CC-BY-SA-3.0 (http://creativecommons.org/licenses/by-sa/3.0/)], via Wikimedia Commons, 13 February 2008

failed projects (McFadyen 1998). For example, control of St. John's wort in California using beetles (*Chrysolina hyperici* and *C. quadrigemina*) lowered the weeds' numbers by 99%. As a result, an estimated $3.5 million per year was saved from 1953 through 1959 in this state (http://www.biocontrol.entomology.cornell.edu/success.php). In addition, less use of herbicides reduces environmental pollution; this "ecological efficiency" is as important as the economic advantages (Zaitsev and Reznik 2004).

1.7.2.2 Control of aquatic weeds

Aquatic weeds develop in lakes, storage reservoirs, rivers, irrigation and drainage canals, and other water bodies. They affect water transport (mechanical resistance of plants to ships' movements) (Photo 1.7-4), crop husbandry (invasion of rice fields), fishery (difficulties in using fishing nets, seines, and other fishing gear), hydroenergetics (reductions in the head and hydropower generation in hydropower plants), water supply (reductions of water flow into water supply intakes), water transfer (reductions in volumes of transferred waters), and recreational activities (limiting swimming and sailing) (Govorushko 2012).

In contrast to terrestrial weeds, which greatly affect crop husbandry and cattle breeding, the impacts of aquatic weeds are wider and dispersed in many kinds of human activity. The other feature of aquatic weeds is their predominant distribution in tropical and subtropical regions. Of all kinds of biological control (control of agricultural pests, forest pests, and terrestrial weeds), the control of aquatic pests (weeds) is the newest, with a history of only slightly over half a century.

The first attempt of using insects for biocontrol of aquatic weeds is the successful fight against the alligator weed (*Alternanthera philoxeroides*) in the United States in 1964 (Cuda 2009a). This weed, originally from Argentina, was delivered through ballast tanks to a number of countries: Australia, China, New Zealand, Sri Lanka, Thailand, and the United States (https://en.wikipedia.org/wiki/Alternanthera_philoxeroides).

In 1959, the U.S. Agricultural Research Service began searching for insects that are natural enemies of the two most pernicious aquatic weeds in waterways in the southeastern United States: alligator weed (*Alternanthera philoxeroides*) and water hyacinth (*Eichhornia crassipes*). In the period 1960–62, these studies were carried out in South America, where the insects most effective for the control of these weeds were identified and studied. For the alligator weed, three species of insects (alligator weed flea beetle [*Agasicles hygrophila*], alligator weed thrips [*Amynothrips andersoni*], and alligator weed stem borer [*Vogtia malloi*, now *Arcola malloi*]) were selected and introduced into the United States during 1964–71. Where these insects were released, they caused considerable reductions in the numbers of the weeds.

Photo 1.7-4. A ship blocked by water hyacinth *Eichhornia crassipes* in Winam Bay (Kisumu port, Kenya) is shown. In the countries adjacent to Lake Victoria, abundant masses of water hyacinth are called 'green icebergs' since, driven by seasonal trade winds, they alternately migrate to the shores of Kenya, Tanzania, and Uganda. The port of Kisumu was blocked by water hyacinth from late December till early May every year.
Photo credit: N.R. Robertson, Aquarius Systems Inc., January 1999

More recent studies showed that less than 400 ha of a total area of 39,255 ha occupied by alligator weed in 1963 in 10 southern states in the United States remained by 1981 (Vail et al. 2001). The alligator weed flea beetle (*Agasicles hygrophila*), which later on was successfully used for biological control in Australia, China, Thailand, and New Zealand, proved to be most successful in the fight against alligator weed (https://en.wikipedia.org/wiki/Alternanthera_philoxeroides). Nevertheless, it is not as tolerant to cold as other species; therefore the populations of the alligator weed flea beetle die off during severe winters in regions with temperate climates (Cuda 2009b).

Yet another example of successful biological weed control is related to using the salvinia weevil (*Cyrtobagous salviniae*) for control of kariba weed (*Salvinia molesta*). It was first applied for cleaning Lake Moondarra (near Mount Isa, Queensland, Australia); the surface of the lake was covered with a solid layer of kariba weed 30 or 40 to 60 cm thick. In 1980, the salvinia weevil was released into the lake. Its larvae fed on the roots, rhizomes, and buds of the plants, leading to their die-off, sinking to the bottom, and decomposition.

The weevil has destroyed tens of thousands of tonnes of the weed (Hangay and Zborowski 2010). As early as May 1981, the lake was entirely clear of salvinia (Gurr and Wratten 2000). Later on, the salvinia weevil was successfully used for control of kariba weed in 13 tropical countries (https://en.wikipedia.org/wiki/Salvinia_molesta). In many cases, its application reduced the numbers of kariba weed by 90% or more in less than one year (https://en.wikipedia.org/wiki/Cyrtobagous_salviniae).

Insect control of the water hyacinth (*Eichhornia crassipes*) (Photo 1.7-5), water lettuce (*Pistia stratiotes*), and red water fern (*Azolla filiculoides*) has been fairly successful in a majority of countries (Cuda et al. 2008). However, in a number of cases, attempts at biological control of aquatic weeds end with no success. For example, hydrilla (*Hydrilla verticillata*) is one of the most pernicious aquatic weeds in many tropical and subtropical countries (Haller 2009). The Asian hydrilla leaf-mining fly (*Hydrellia pakistanae*) is often used to control this weed. This control method can be fairly effective in most hot regions, but its efficiency decreases sharply in regions with cooler climates (https://en.wikipedia.org/wiki/Hydrellia_pakistanae). Therefore, the search for the most effective insect for its control has continued unsuccessfully for more than 30 years (Bownes 2014).

In many cases, one cannot talk of the complete and ultimate control of an aquatic weed, and weeds as a whole, in a country. Generally, the issue is only about reducing their numbers below the level of injuriousness. In addition, new, never-before-seen weeds frequently arise. For example, control of three species of aquatic weeds is most urgent now in New Zealand (Paynter 2013): lagarosiphon (*Lagarosiphon major*), hornwort (*Ceratophyllum demersum*), and egeria (*Egeria densa*). Insects capable of disrupting their numbers are being sought.

The South African insects leaf-mining fly (*Hydrellia lagarosiphon*) and shoot-tip mining midge (*Polypedilum* sp.) were considered the most effective agents of biocontrol for lagarosiphon. To control egeria, the *Hydrellia* fly, which inhabited South America and caused heavy defoliation was chosen (Walsh et al. 2013). As for hornwort (*Ceratophyllum demersum*), no insect has been identified that is able to effectively control its numbers. Use of the pathogenic fungus *Fusarium* sp., which damages the plant, and the grass carp (*Ctenopharyngodon idella*), which can feed on the plant, are possible control methods (Curt et al. 2010).

Photo 1.7-5. Introduction of natural insect predator mottled water hyacinth weevil, *Neochetina bruchi*, became an effective measure in eliminating the water hyacinth in Lake Victoria.
Photo credit: CSIRO

1.8

Insects as a Form of Entertainment

Up to now, we have considered the positive aspects of human interaction with insects when the insects provided some practical use. However, insects do not always provide material benefit; on the contrary, they often require nonrefundable financial expenses. Nonetheless, such a form of interaction refers to the positive aspects, as when people derive, to a greater or lesser degree, moral satisfaction from interacting with insects. Examples of these types of interactions include hobbies, entertainment, and other leisure activities in which insects play a key role.

1.8.1 Insect-based tourism

In recent years, tourism for the main objective of observing wild animals has grown in popularity. Tourists are attracted first of all to **big game** animals. However, even those tourists coming to look at big game often take interest in insects, too. For example, Hluhluwe-Imfolozi Park in South Africa features Africa's "big five" game: elephant, rhinoceros (black/hook-lipped and white/square-lipped), Cape buffalo, lion, and leopard (https://en.wikipedia.org/wiki/Hluhluwe–Imfolozi_Park).

However, 95% of tourists visiting the park also expressed an interest in information on invertebrates, in particular, monarch butterflies, morpho butterflies, dragonflies, and jewel beetles (Schowalter 2013). A certain segment of the market related to the observation of animals accounts for the tourists who come purposefully to look at insects. This tourism is widespread, both in developed countries (Taiwan, Japan, Great Britain, Australia, New Zealand) and developing ones (Indonesia, Mexico, Malaysia, Thailand, India, South Africa).

Abundance of entomofauna varies greatly from country to country. For example, about 40,000 species live in India, but only 450 species in Greenland (http://www.apus.ru/site.xp/04905205605512 4054050050056124.html). Clearly, the countries with rich entomofauna located, mainly, in the tropics and subtropics have advantages in development of insect-based tourism. An indirect measure of the abundance of entomofauna in different countries is the number of pages in Wikipedia dedicated to insects; for example, upon the request "Insects of Indonesia" and "Insects of China", references to 111 and 128 pages, respectively, appear.

A potential for development of the entomo-tourism can be determined by some taxa of insects. For example, in Australia, these are jewel beetles (more than 2,000 species), fireflies, stick insects; in New Zealand fireflies; in Tanzania, South Africa, and Ghana butterflies (Tisdell and Wilson 2012); and in Malaysia butterflies, ants, and termites (Hamdin et al. 2015). Often, butterflies become the focus of the tourist destination. For example, there is widespread tourism based on observation of butterfly migration. Observation of this migratory behavior among butterflies is rare: it is known for approximately 250 species, only about 20 of which conduct regular and distant flights.

The **monarch** (*Danais plexippus*), the best-known migrant butterfly, migrates long distances every year (Photo 1.8-1). It is widely distributed in South America, Central America, North America, Australia, and Oceania (https://en.wikipedia.org/wiki/Monarch_butterfly). Tourism related to this species of butterfly is very popular. For example, in the Monarch Butterfly Biosphere Reserve, situated in the Mexican state of Michoacán, up to 100,000 butterflies accumulate on the trees, and the total number of butterflies can reach 50 million (Photo 1.8-2) (https://en.wikipedia.org/wiki/Monarch_Butterfly_Biosphere_Reserve). The guided tours to the reserve last from November to March, and up to 250,000 Mexican and foreign tourists, predominantly from the United States, Canada, Spain, France, Germany, and Japan, participate in these tours (Lemelin 2009, 2012a).

In Australia, guided tours aimed at observations of butterflies and fireflies are popular. Tourism related to **butterflies** is concentrated in the Australian Butterfly Sanctuary, which is the largest butterfly

Photo 1.8-1. The monarch (*Danais plexippus*), the best-known migrant butterfly, migrates long distances every year. It is widely distributed in South America, Central America, North America, Australia, and Oceania. Butterflies in flight are shown. Photo credit: By Raina Kumra, nyc, USA (flickr, copyright Raina Kumra, nyc, USA) [CC BY 2.0 (http://creativecommons.org/licenses/by/2.0)], via Wikimedia Commons, 15 September 2008

Photo 1.8-2. Tourism related to this species of butterfly is very popular. The Monarch Butterfly Biosphere Reserve located on the border of Michoacan and State of Mexico, up to 100,000 butterflies accumulate on the trees, and the total number of butterflies can reach 50 million. Tree completely covered in butterflies is shown.
Photo credit: Bfpage at English Wikipedia [CC BY 3.0 (http://creativecommons.org/licenses/by/3.0)], via Wikimedia Commons, 26 February 2000

light aviary in the Southern Hemisphere, in Kuranda (far north Queensland). The sanctuary was opened in 1987 and, since then, more than a million people have visited it (http://www.kuranda.org/listing/australian-butterfly-sanctuary/).

Tourism related to **fireflies**, which was developed in four eastern states—Queensland, New South Wales, Victoria, and Tasmania—is confined to humid months of year (between October and March). In this country, many species of fireflies and other species are observed (Wildlife Tourism Challenges 2009). For example, at Springbrook National Park (National Bridge Section), in Queensland, the endemic species of glow worm (*Arachnocampa flava*), giving off a blue-green bioluminescence, is the species attracting the tourists (Tisdell and Wilson 2012).

In **New Zealand**, tourism related to fireflies is even more widespread than in Australia. There, the object of the tourist interest is the New Zealand glow worm (*A. luminosa*), being also the endemic species. More than 20 habitats of this insect are known in this country; however, the Waitomo Caves in the North Island and the Te Ana-au Caves in the South Island are most frequented by tourists (https://en.wikipedia.org/wiki/Arachnocampa_luminosa). For example, the Waitomo Caves attract more than 400,000 tourist visitors annually and, during the summer months, the visitation reaches 2,000 people per day (Saikim et al. 2015).

In Endau-Rompin National Park (state of Pahang, Malaysia), tourism related to insects is organized. Early on, in three sections of the park, field observations were carried out on 16 groups of insects, including ants, termites, wasps and bees, butterflies, moths, flies, mosquitoes, beetles, bugs, leafhoppers and cicadas, cockroaches, mantises, stick insects, crickets, grasshoppers, and dragonflies.

According to their results, ants and termites turned out to be the most **advanced** species in development of tourism in all three sections of the park. Further, park visitors were asked which aspects of the insects were of principal interest to them. Seventy-three of the 117 (62%) respondents wanted to know about ants' communication system, 60 (51.1%) about their defense system, 53 (45.3%) about their morphology, and 49 (42%) about their foraging behavior (Hamdin et al. 2015).

1.8.2 Insect-based leisure

Various types of insect-based leisure activities are also common. The **difference** between tourism and leisure with respect to insects lies in the fact that, in the case of tourism, insects are observed in their natural environment (in most cases, in conservancies, reserves, national parks, and other protected areas); whereas leisure activities occur with vacationers, in parks, insectariums, pavilions, and so forth, where the insect habitat was artificially created. At times (e.g., in museums, exhibitions), visitors observe insects only in their prepared form; nevertheless, even this kind of observation of insects is of interest for people. Consequently, we consider insect-based leisure separately from the analogous insect-based tourism.

For example, people go to **insectariums** to observe certain insects. Insectariums are special rooms designed for the keeping and raising of insects, and they receive frequent visitors. Modern insectariums present large buildings with several rooms and complex accessories for control of lighting, temperature, humidity, air exchange, and the like. For example, in the Hiroshima City Forest Park Insectarium, there is a "Papillon Dome", in which about 500 butterflies from 10 different species flutter about. Expositions of about 50 different species of live insects are held in western Japan as well as exhibitions devoted to seasonal themes, insect interaction experiences, and outdoor insect observation activities (http://www.hiroshima-navi.or.jp/en/sightseeing/shizen_koen/dobutsu_shokubutsu/21754.php). Every year, millions of people worldwide visit the hundreds of insectariums and pavilions of butterflies (Lemelin 2012b).

Butterflies enjoy the greatest popularity among people; therefore, there are more establishments exhibiting butterflies than those exhibiting other insects. The **Butterfly Park** (Photo 1.8-3) and **Insect Kingdom in Singapore** is widely known. More than 2,500 free-flying butterflies from more than 50 species inhabit this park. Besides butterflies, the park also houses more than 3,000 species, including the largest Hercules beetle (*Dynastes hercules*) (http://www.focussingapore.com/singapore-tourism/butterfly-park-insect-kingdom.html).

Photo 1.8-3. The Butterfly Park and Insect Kingdom in Sentosa island (Singapore) is widely known. More than 15,000 butterflies, representing over fifty species living here. These butterflies range from the 25 millimeters (1 in) *Eurema sari* to the 150 mm (6 in) *Papilio iswara*. The Insect Kingdom house has 3,000 species of rare insects from around the world, including a 160 mm *Dynastes Hercules* beetle. Picture shows entrance to Butterfly Park.
Photo credit: Yu.S. Govorushko, 28 February 2013

Another example of this type of establishment is the 21-meter-long **butterfly house** in Coff Harbour (New South Wales, Australia), which opened in 1995 (http://www.butterflyhouse.com.au/). Thamesis, New Zealand, features a tropical butterfly and orchid house (http://www.butterfly.co.nz/index.html).

The richness of entomofauna of the tropical rain forests in the Sabah State (Malaysia) reflects many objects related to insects. Among them are the butterfly farm at Poring Hot Springs, the exhibition of insects in the Gunung Alab Resort, and the Insect Museum of the Sepilok Forest Research Centre (http://www.etawau.com/Life/Gallery/Insects.htm). Because of the many species of butterflies inhabiting Taiwan, it is frequently called the "Butterfly Kingdom". In the town of Puli, Nantou province, the popular **MuhSheng Museum of Entomology** displays more than 16,000 insect species, including the broad-tailed swallowtail (*Agehana maraho*), swiftmoth (Hepialidae), and *Pazala timur chungianus* (http://eng.taiwan.net.tw/m1.aspx?sNo=0002114&id=5263).

In 1981, in Hampshire county, near Southampton in southern England, the **New Forest Butterflies Farm** was established. During the first 2 months, more than 52,000 people visited the farm. At that time, the farm consisted only of a section of tropical butterflies. Now, it has a garden of tropical butterflies, a true jungle occupied by butterflies and their larvae from around the world; a garden of British butterflies, where the threatened and endangered species of butterflies of Great Britain are raised; and ponds of dragonflies and Insectarium exhibiting many species of bird-eating spiders, scorpions, stick insects, tropical cockroaches, and grasshoppers (http://nacekomie.ru/forum/viewtopic.php?t=12).

Many **insect festivals** worldwide attract lots of people are every year held. One of the world's oldest known insect festivals is the Festadelgrillo (the Festival of Crickets) in Florence, Italy. Other examples include the Dragonfly Citizen Summit in Japan, the annual Texas Butterfly Festival, butterfly festivals in Tamaulipas state, Mexico (Hvenegaard et al. 2012), and in Kolhapur, Maharashtra state, India (http://ladybirdconsulting.co.in/past-events.html#BFFR), and the annual Festival of a Million Fireflies held from late May through early July at Purushwadi, Maharashtra, India (http://www.grassroutes.co.in/#fireflies-festival).

In Irun City, northern Spain, there is a **permanent exhibition of butterflies**, which contains about 7,000 butterfly species from the Basque Country and many regions of the world. Many of them are accompanied with detail information (distribution map, detailed description). The tour rounds off with a screening of a documentary on the butterflies' biological cycle (http://tourism.euskadi.eus/en/museums/exhibition-on-worldwide-butterflies/aa30-12375/en/).

1.8.3 Use of insects as pets

For most people, the very thought of the presence of insects, spiders, scorpions, and so on, in their houses seems to be unpleasant. However, many people consider the idea of keeping insects and other invertebrates as pets fascinating. It seems that **crickets and bush crickets** were among the first insects that were kept indoors. The first-known mention of this is in an epigram dating to 600 B.C.E. found in Ancient Greece, which speaks of a girl and her dying pet cricket (van Huis et al. 2013).

In Asian cultures, keeping insects indoors is more accepted and has a **centuries-old tradition**. A large body of data suggests that thousands of years ago, singing crickets and katydids were pets in China, Japan, and other Asian countries. In addition, keeping insects indoors in this region is as habitual as keeping dogs or cats indoors in the West. Night-singing crickets are sometimes used in the East to signal the presence of intruders because crickets stop singing when disturbed by other noise (https://www.si.edu/Encyclopedia_SI/nmnh/buginfo/pets.htm).

It is necessary to provide pet insects with **habitat conditions** close to their environmental ones (proper humidity, temperature) and appropriate keeping and care (feeding, cleanliness). The goal is facilitated by the fact that the development cycle of captive animals is not closed (i.e., reproduction is not needed) (Löser 2001). Generally, invertebrates require **much less care** than cats or dogs or caged animals (e.g., rabbits) because they take up minimum space, require less food, do not need daily walks, and do not lose hair or give off a disagreeable odor.

Millions of species of invertebrates exist; however, by no means can all of them be kept as pets. Among the invertebrates that are most often kept indoors are (Clark 2012; https://www.si.edu/Encyclopedia_SI/nmnh/buginfo/pets.htm): (1) walking-sticks, (2) praying mantises, (3) cockroaches, (4) beetles, (5) ants, (6) centipedes, (7) millipedes, (8) scorpions, (9) spiders, (10) caterpillars, and (11) mealworms.

To keep each species of insect, specified **requirements** must be met; not meeting these requirements can have deplorable consequences. Let's consider the examples of mantises and hissing roaches (*Gromphadorhina portentosa*). The **feeding** regimen is important. For example, an adult mantis should feed once every 1 to 2 weeks and, most significantly, should not eat to excess (http://www.hintfox.com/article/bogomolovie-osobennosti-ih-zhizni-i-povedenija.html). For cockroaches, there should always be food in the terrarium; otherwise, the insects will eat each other (http://www.syl.ru/article/187887/new_madagaskarskie-tarakanyi-shipyaschie-soderjanie-foto). The terrariums should also be properly equipped. For example, mantis terrariums should contain small branches and submerged stumps because they like to hang upside down watching for prey. Hissing roaches need hiding places made of paper or cardboard in the form of little houses, boxes, and other structures.

Maintenance of the appropriate **temperature and humidity conditions** is also needed. For example, comfortable conditions for cockroaches to live and reproduce include air temperature in the range of 25°C to 32°C and 65% humidity. At temperatures of 18°C to 20°C, the insects will live but will not reproduce. For adequate development of mantises, average room temperature must be 23°C to 25°C.

The presence of **water** in the terrarium can be essential. Mantises need to have a drinking cup continuously filled with fresh water; however, it is strictly forbidden to place a water container in a terrarium with Madagascar hissing cockroaches because they will inevitably drown. For cockroaches, water is supplied in the form of water-soaked sponge or carded web.

The correct **number of insects** in the terrarium is also of considerable importance. Cockroaches are usually kept as colonies including several dozens of individuals. Mantises, however, must be kept as individuals because of their aggressiveness to each other; even with the sufficient amount of food, they may eat their competitors (http://www.hintfox.com/article/bogomolovie-osobennosti-ih-zhizni-i-povedenija.html).

1.8.4 Insects as a spectacle

Many people are interested in observing insects, and some leisure activities are related to this. The best-known **kinds of entertainment** include (1) flea circuses, (2) cockroach races, and (3) insect fighting.

There were periods in history when these leisure activities were highly popular; however, their popularity has strongly, and most likely, irrevocably dropped. Nevertheless, these kinds of entertainment have been preserved in some places.

However, these **specific kinds of observations** occur largely under conditions that are unnatural for the insects, and the insects are sometimes forced to do things that are absolutely uncharacteristic of them. It should be remembered that this form of entertainment is based on human nature but that it may be considered barbarian treatment of insects. From a legal viewpoint, these events **violate the rights of animals** and are outlawed in accordance with criminal codes in a number of countries.

1.8.4.1 *Flea circuses*

It is unknown when flea circuses first appeared. One assumption is that they were invented by medieval prisoners. **Prisons** had an abundance of fleas, and the prisoners suffered from boredom and insect bites. From the prisons, the blood-sucking acrobats came to trade fairs and, later, to the palaces (https://d3.ru/comments/329344/). According to other sources, the trained fleas existed as far back as Ancient Egypt (Abrams 2009). The first references to flea performances were related to **watchmakers**, who used them to display their metalworking skills. In 1578, Mark Scaliot produced a lock and chain that were attached to a flea (http://www.fleacircus.co.uk/History.htm).

Nevertheless, flea circuses acquired distinction in the 17th century. Late in the 18th and early in the 19th centuries, flea circuses were hugely popular in Europe. Many European monarchs were not indifferent to this kind of entertainment and, in France, a **court flea circus** was even established where the insects fought with swords, juggled with balls, rode on a merry-go-round, and walked on a wire, showing the marvels of training. It miraculously survived during the French Revolution and in 1905 court flea circus was closed.

The first **registered** flea circus was owned by the Italian impresario **Louis Bertolotto**. It took place during the early 1820s on Regent Street, London. The circus, advertised under the name of "extraordinary exhibition of industrious fleas" (https://en.wikipedia.org/wiki/Flea_circus), featured fleas dragging chariots and even showed the mockeries of current events of the time.

Flea circuses were integral to numerous trade fairs up to the 1960s. The flea circus at Belle Vue Zoological Gardens, Manchester, England, was still operating in 1970. However, the interest in them has sharply **dropped** since then, for reasons that are not evident. It may well be that similar shows became unattractive for present-day spectators or that the fleas or their trainers were harder to come by. In our time, trained fleas are infrequently remembered, and only a limited number of people have seen them (http://www.zooclub.ru/chlen/bloshinyi_cirk.shtml). Nevertheless, the contemporary state of this art is not as depressed as it might seem.

Modern flea circuses include the following (https://en.wikipedia.org/wiki/Flea_circus): (1) Professor A.G. Gertsacov's Acme Miniature Flea Circus, playing in Canada and the United States from 1996 (http://www.trainedfleas.com/); (2) Swami Bill's Flea Circus, featured at the Denver County Fair (US) (http://www.denvercountyfair.org/#!special-events/c457); (3) Professor B's Flea Circus, playing in northern California (US) in recent years (http://www.playland-not-at-the-beach.org/entertainer_bio_prof_b_01.html); and (4) the Flea Circus Birk, performing annually at Munich Oktoberfest (Germany).

The **Cardoso Flea Circus** existed from 1994 to 2000. The first public performance of the circus took place in October 1996 at San Francisco's Exploratorium. The demonstration was accompanied by audio and video for a larger projection for the wide audience (https://en.wikipedia.org/wiki/Mar%C3%ADa_Fernanda_Cardoso). The artists of the María Fernanda Cardoso company were "able" to walk a tightrope, be living cannonballs, juggle, pull in harness play trains, and play golf. In spring of 2013, a tragedy took place in **the Floh Circus Birk**: its entire company, consisting of 300 artists, died in their transportation box because of cold weather. However, a German entomologist has donated 50 fleas to Cardoso (http://ru.focus.lv/life/zhivotnye/v-germanii-pogibla-truppa-bloshinogo-cirka).

In the first flea circuses, **human fleas** (*Pulex irritans*) were solely used as artists because they respond to training more easily than other insects. However, it became more difficult in time to find human fleas and, more recently, **cat fleas** (*Ctenocephalides felis*) came to be used. The Cardoso Flea Circus has employed cat fleas exclusively (https://en.wikipedia.org/wiki/Mar%C3%ADa_Fernanda_Cardoso). The average lifespan of fleas is about 1 year. Subtracting the duration of growth (6 months) and training (usually 3 months), only 3 months remain for the performance (Abrams 2009).

Fleas are not trained in the way mammals or birds are trained. For example, fleas are not observed to display their proclivity for leaping or walking. The flea circus imposes high requirements on the keepers of fleas rather than on the fleas themselves. Many things depend on the **circus's requisite details** and on the mastership of the trainer in their application. The difficulty is to affix these requisite details to the insect (e.g., to put a yoke twisted from the very finest silver or copper wire on the chest of the tiny leaper, with the intent that the insect cannot break away from it) (http://www.zooclub.ru/chlen/bloshinyi_cirk.shtml).

Over the course of its life, a flea may learn **just one trick**, with the whole performance based on the trainer's movements and circus's requisite details. For example, fleas **are "trained" not to leap** by placing them into a container with a cover. After numerous failed attempts, a flea stops leaping. Around its neck is wrapped a thin golden, silver, or copper wire, which forces the insect to the ground (a flea lives with it the rest of its life) in order to force it to move only horizontally.

In order for fleas **to fight with swords**, it is necessary to attach a sword to the tarsus of the "swordsman". The insect senses the presence of a living being and attempts to jump on it, moving its legs and appearing to the onlooker to be brandishing a sword (https://d3.ru/comments/329344/). The effect of **music-making** is achieved as follows: miniature musical instruments are glued to the flea performers and their enclosure is then heated. The fleas fight to escape, giving the appearance of playing the instruments (https://en.wikipedia.org/wiki/Flea_circus).

1.8.4.2 Cockroach races

Cockroach races exemplify a typical **game of chance** with all its attributes: uncertainty of outcome, possibility of stakes, real fighting, spectators, and visual appeal. In our time, it is an exotic form of entertainment: not many people have seen a cockroach race or taken part in one, and the number of people who have placed bets on a cockroach race is far less. However, **in the past**, these kinds of entertainment were highly popular. Evidence of cockroach races can be found in the historical records of Ancient Egypt, Byzantium, and the Roman Empire, where a personal cockroach lived in each noble house and regularly participated in these competitions (http://gambledor.com/ru/articles/cockroach-racing).

There is also an opinion that this kind of entertainment originated with sailors arriving from the Caribbean (https://otvet.mail.ru/question/14358814). They had a lot of free time, an enormous number of cockroaches in the freight from tropical countries, and the desire to entertain themselves one way or another during their long voyages (http://klop911.ru/tarakany/o-tarakanax/tarakani-bega.html). By the way, these points of view do not contradict one another. This kind of gambling entertainment arrived in Russia in 1812 and, several decades later, became a commercial event in its own right. This is evidenced by descriptions of cockroach races that can be found in the novel *Adventures of Nevzorov, or IBIKUS* (A. Tolstoy; 1924), *Beg* [*Race*] (M. Bulgakov; 1927), and some stories by Arkady Averchenko (http://gambledor.com/ru/articles/cockroach-racing).

In **Turkey**, this kind of game of chance achieved maximum prosperity in the first half of the 20th century. Literature describes cockroach races in this country; here, a business related to breeding and supply of race cockroaches grew rapidly. The Kurd Shaheen Fesah, whose business of prospered in the mid-20th century, was one of the best-known cockroach breeders. On a special farm he kept up to half a million cockroaches, and most known cockroach sports stars in Turkey were purchased from him (http://klop911.ru/tarakany/o-tarakanax/tarakani-bega.html).

Australia is now considered the leader in the field of cockroach races. The foundation of this tradition was laid by students in 1982 from Brisbane (Queensland) who held cockroach competitions in a hostel, which attracted a great many onlookers. The interest in these competitions was such that the cockroach races became the real show, and, up to now, the festival called the "Australia Day Cockroach Races" is held every year on January 26 at the Story Bridge Hotel, an event that coincides with special fancy balls called "Miss Cocky Races" (http://www.cockroachraces.com.au/). Today, cockroach races are officially run in all major Australian cities.

This kind of entertainment also occurs in other countries. For example, its popularity has grown recently in the **United States** (Photo 1.8-4), where, not infrequently, cockroach races coincide with entomological exhibitions. At Loyola University, in Baltimore, Maryland, the competition is called "Madagascar Madness: The Running of the Roaches," and attracts a great number of onlookers (https://en.wikipedia.org/wiki/ Cockroach_racing). At Purdue University, in Lafayette, Indiana, the annual cockroach races are held at the university's Bug Bowl (https://extension.entm.purdue.edu/bugbowl/events.php).

In the United States, cockroach races are also popular events held prior to the presidential election; they are held every 4 years by the Pest Control Association of New Jersey. The race is meant to predict the outcome of the presidential election; for example, in June 2012, in New Jersey, the namesake cockroaches of Barack Obama and John McCain were raced to determine the winner (https://en.wikipedia.org/wiki/ Cockroach_racing).

In **Russia**, several agencies rent accessories for the provision of cockroach races. All that is needed is to call such a company and make an order. Then, at the appropriate time, the cockroaches themselves, race tracks, and necessary attributes are delivered to the indicated address, as well as a specialist capable of conducting the competition (http://gambledor.com/ru/articles/cockroach-racing). Similar services are also rendered in Ukraine (http://www.bugdesign.com.ua/roach_race/).

The **Madagascar hissing cockroach** (*Gromphadorhina portentosa*) is the most preferred species used in the races. In the Caribbean, the **Central American giant cave cockroach** (*Blaberus giganteus*), which has a smaller body but comparable friskiness when compared with the Madagascar cockroach, is in good demand. Traditional Russian red cockroaches or German cockroaches (*Blattella germanica*) and Oriental cockroaches (*Blatta orientalis*) are unwanted for similar amusements because they are too small and too unpredictable with respect to race direction and speed (http://klop911.ru/tarakany/o-tarakanax/ tarakani-bega.html).

Generally, cockroaches are very fast-running insects. Regarding speed for short distances, Madagascar hissing cockroaches are second only to tiger beetles (Cicindelinae), being able to run the distance equal to 50 lengths of their body per second. If humans ran at this rate of speed, the champion sprinters would be able to run 100 meters per second (https://en.wikipedia.org/wiki/Cockroach_racing).

As for the **rules for holding** of cockroach races, they are clearly prescribed. There are three kinds of competitions (https://en.wikipedia.org/wiki/Cockroach_racing): (1) sprint; (2) steeplechase; (3) Gold Cup event. The competitions in **sprint** are most popular and held as follows. The cockroach-drome

Photo 1.8-4. Cockroach races in our time is an exotic form of entertainment: not many people have seen a cockroach race or taken part in one, and the number of people who have placed bets on a cockroach race is far less. However, in the past, these kinds of entertainment were highly popular. Cockroach racing at Las Vegas is shown.
Photo credit: By Sean Savage (Flickr: Cockroach Racing at Lost Vegas) [CC BY-SA 2.0 (http://creativecommons.org/ licenses/by-sa/2.0)], via Wikimedia Commons

(structure made of transparent channels about 1.5 meters in length) is installed on a horizontal surface. Containers with the runners are placed alongside the race track; names of the "athletes," their ages, and some unusual features are written on each container.

In general, five to eight cockroaches take part in a heat. After that all bets are off; each cockroach is placed into the individual channel. At the same time, a bright light goes on at the starting point, frightening the insect. The cockroach tries to find a more obscure place and runs to the opposite end of the track. According to the rules, the keepers may pinch the insects using small rods and originating noise for maximal frightening. It is prohibited to touch the participants with rods (http://gambledor.com/ru/articles/cockroach-racing).

The **steeplechase** uses a modernized race track with small obstacles that confound the insects and prevent them from continuously running forward. The **Gold Cup event** is an Australian invention (in particular, such competitions are held at the cockroach races in Brisbane). In turn, it has a fundamentally different format and requires from the participants not only speed. At the center of a 6-meter ring, a container with cockroaches is situated. At the command of "Start," the container is lifted and the "athletes" scatter in every direction. The participant that reaches the edge of the ring first wins (https://en.wikipedia.org/wiki/Cockroach_racing).

As with all athletes, cockroaches **should be trained for the competitions**. In essence, the training activity comes down to **three items** (http://gambledor.com/ru/articles/cockroach-racing): (1) training of speed performances (to teach the insect to run as quickly as possible, it is placed on a round bottle that rotates continuously on its axis); (2) compliance with nutrition prescription (many "athletes" keep to diets including apricot halves, beer, white wine, and coffee and are sprayed with water once a week); (3) starvation prior to competitions (there is a belief that if a cockroach is not fed for 10 days before the heat, it will more quickly reach a finish).

The organizers of competitions and keepers of the race cockroaches try to be original in **choice of the nicknames** for "sportsmen". "Soft Cocky", winning in 1982 in the heat in Australia, is considered the best-known champion. There is even a special Hall of Fame for stars of the cockroach sport with the most unusual names, such as Cocky Balboa, Cocky Dundee, Drain Lover, and Priscila-Queen of the Drains.

1.8.4.3 Insect fighting

Insect fighting is another **game of chance (gambling),** in which insects wrestle each other, similar to other kinds of blood sports such as dog fighting, cock fighting, and so on. In competitions of this kind, not only insects but also other arthropods (e.g., Arachnida, Myriapoda) can take part.

Many years ago when people lived in **caves**, they were surrounded by a variety of insects (lice, fleas, bedbugs, ants, mosquitoes, and flies) and spiders. For the most part, they were nuisances for people: they bit and sucked blood. In turn, people ate insects in times of famine. However, when their bellies were full and food was abundant, people sought out bugs to observe their actions. It is assumed that the first fights observed by ancient people were the **fights between red ants and black ants** (Zillen 2008).

Now, the crickets, spiders, and beetles take part most often in such fights, with ants and wasps more rarely used for this purpose. Different forms of fights between insects are most popular in China, Japan, Vietnam, and Thailand (https://en.wikipedia.org/wiki/Insect_fighting). Little by little, such leisure activity is becoming popular for youth in western countries, where it is known colloquially as **bug fighting**.

Cricket fighting is most popular in China, where its history dates back centuries. It is assumed that it became the traditional Chinese pastime during the Tang Dynasty (618–907 A.D.) (https://en.wikipedia.org/wiki/Insect_fighting); however, historic references and documentary evidence refer to the beginning of the Song dynasty (960–1278 A.D.). One piece of evidence is three pieces of cricket carriers (special containers for moving fighting crickets), discovered in the course of excavation of the tomb of South Songin Zhen Jiang, Jiangsu province, in 1964 (Jin 2011).

Two Chinese **historical figures** who were passionate about cricket fighting are the famous Cricket Minister, Jia Shi-Dao (1213–1275), who was under an accusation of failure to execute his duties because of his all-consuming passion for cricket fighting, and Cricket Emperor, Ming Xuan-Zhong (ca. 1427–1464), during whose reign crickets became the primary tribute of the palace (Jin 2011). In the beginning, cricket fighting was an indulgence of emperors; however, it later became popular among commoners (https://en.wikipedia.org/wiki/Insect_fighting).

In the history of China, two periods have existed when cricket fighting was **forbidden**: during the Qing Dynasty (1644–1911) (Huis van et al. 2013) and during the Cultural Revolution (1966–1976) (Cricket Fighting Contests in China 2011). During these periods, cricket fights were held illegally. Now, they are again widely distributed, mainly in large cities like Shanghai, Beijing, Tianjin, and Guangzhou, where numerous cricket fighting clubs and societies prosper. It should be noted that although cricket fighting in China is permitted, betting on crickets is illegal. For example, of the 300,000 to 400,000 cricket enthusiasts in Shanghai alone, about 90% are interested in betting on the fights (Huis van et al. 2013). In Hong Kong, cricket fighting is banned (Areddy 2009).

Generally, the **cricket collection and breeding** industry was developed in the country. Crickets are freely sold at street markets; in Shanghai alone there are a dozen cricket markets. In 2010, 400 million yuan (US$63 million) were spent on crickets in China (https://en.wikipedia.org/wiki/Cricket_fighting).

Fighting crickets are always male. They normally live a little more than 100 days, ending in autumn. The Chinese term for cricket fighting is approximately translated as the "autumn enjoyment" (Areddy 2009). In the country, there is a system of regional competitions held by the Association for Cricket Fighting in summer at 25 locations around China. They are crowned by the national championship held in Beijing every year by the Association for Cricket Fighting within 2 days in late September.

Every approved competitor has a right to run up to 35 insects and all of them are weighed and labeled before the beginning of competitions. The cricket fights are organized according to **weight class**. Within each weight class, the pairs of fighters are determined by lot. The competitions are held according to the Olympic principle: the loser is eliminated while the winner passes to the next stage (https://en.wikipedia.org/wiki/Cricket_fighting).

In the fighting container, there is a partition dividing it into two parts. The keepers irritate the whiskers of their crickets with a straw stick, making them more aggressive. When both crickets are quite irritable, the partition is lifted and a match starts. The cricket that first begins **avoiding contact**, runs away from battle, stops chirping, or is thrown from the fighting container is considered the loser. The competitions result rarely in the death of insects; however, the losses of extremities are a routine practice. Prized crickets become famous and, after their death, are buried with all due honors, while the cricket-loser is simply thrown out on the street (Areddy 2009).

Crickets from a few counties of the northeastern Shandong Province can be considered the best fighters. The crickets have **genealogical tables** and are carefully bred by skilled specialists. Each cricket usually has its own clay pot and diet that includes ground shrimp, red beans, goat liver, and maggots. At night before the fight, for raising the fighting spirit, female crickets are dropped into the pot (https://en.wikipedia.org/wiki/Cricket_fighting).

The Japanese burrowing cricket (*Velarifictorus mikado*) is considered to be **best suited** to the fights. However, due to the migration of the Chinese to other parts of the world and development there of this kind of spectacle, other species are used. For example, in cricket fighting in Philadelphia, the fall field cricket (*Gryllus pennsylvanicus*) is usually used (Jin 2011). The Jerusalem cricket, genus *Stenopelmatus*, inhabiting the western United States and parts of Mexico, is also highly valued for their jumping ability and aggression (https://en.wikipedia.org/wiki/Insect_fighting).

In some regions of the world, **spider fights** are held. They are most popular in Japan, Singapore, and Philippines. The fights in **Japan** and **Philippines** are held between females of various species of orb-weaving spiders. In Japan, the black and yellow garden spider (*Argiope aurantia*) is commonly used, while in Philippines, the species *Neoscona punctigera* takes part most often in the battles. The female spiders kill their rivals unless they run away or receive assistance from the keeper (human handler). In the fights of spiders in **Singapore**, male jumping spiders (Salticidae; most often, *Thiania bhamoensis*), take part. Male spiders battle only for dominance and, as a rule, the loser runs away, although it can sometimes lose one or more legs in the fight (https://en.wikipedia.org/wiki/Spider_fighting).

Rules for running the competitions are quite simple. The spiders are placed on the different ends of a stick and pushed toward one other (Photo 1.8-5). They meet in the middle (midway) and kick off a battle (Overton 1998). The fact of the fight termination is specified in advance. The lethal fights end with bite of

Photo 1.8-5. Spider fighting occurs in different forms in several areas of the world. Among them are the Philippines, Japan, and Singapore. The fights that occur in the Philippines and in Japan are staged between females of various species of web weavers. Female spiders will kill a rival if the loser does not quickly flee or receive the aid of a human handler. The contests that are staged in Singapore are fights between male jumping spiders. The males fight only for dominance, and ordinarily the loser will flee, though sometimes they will lose a leg in the fight. Two spiders engaging in a fight to the death are shown. Photo credit: By Kguirnela (Own work) [CC BY-SA 4.0 (http://creativecommons.org/licenses/by-sa/4.0)], via Wikimedia Commons, 24 August 2007

one of the spiders resulting in paralysis. The nonlethal fights end when one spider falls off the stick (one or several times, depending on agreement). Any contest is a tournament in which the last two victorious spiders meet in the finale (Sekine 2002).

In Aira (Kajiki, central Kagoshima prefecture, Kyusyu Island, Japan), spider fights called Kumo Gassen have been held every year for **more than 420 years** (https://en.wikipedia.org/wiki/Kajiki, Kagoshima). Training to participate in the competitions is carried out in advance. The strongest spiders are found and thoroughly cared for. The keeper feeds them with insects such as grasshoppers and beetles. Sometimes the keeper sprays the spiders with shochu (Japanese traditional hard liquor made from grains and vegetables). In addition, a vitamin solution is injected into the insects that will become the prey of the spider. These competitions are events on a national scale with many participants; in the competitions held on January 16, 2002, about 400 spiders took part. They lasted 7 hours and attracted about 3,000 onlookers. A spider beauty contest was also held (Sekine 2002).

In **Philippines**, spider fighting is most popular among the rural Filipino children, especially in the Bisaya region. They collect spiders from bushes or trees in the early morning or at dusk, when the spiders spin their webs. Between fights, spiders usually live in separate cells of boxes made from discarded cardboard. Spider fights are prohibited there now because they have a negative effect on achievement of school children: the children spend long time hunting and training spiders and have no time for homework and lessons. Adults also practice this kind of entertainment and their bets may go as high as 50,000 pesos (about US$1,000) (Overton 1998).

In Southeast Asia and Japan, **beetle fighting** is popular. Most often, the male stag beetle (Lucanidae), rhinoceros beetle (Dynastinae), or Goliath beetle (*Goliathus*), for their sheer size, take part in them. Of the separate species, the brown rhinoceros (or fighting) beetle (*Xylotrupes gideon*) and Japanese horned beetle, or kabutomushi (*Allomyrina dichotoma*) are most popular. The beetle-fighters are highly valued by fans of this sport. For example, a brown rhinoceros beetle (*X. gideon*) may be sold in northern Thailand for 2 to 4 euros, depending on size and aggressiveness. For well-tried champions, people may even pay up to 200 euros (Micke 1996).

Beetle fighting can be held on a log, an offshoot of a plant, a stump, or a circle drawn in the dirt. Forcing beetles to fight requires patience and differs greatly from the fighting of other species of insects. During beetle fights, small noisemakers are often used that imitate the female's mating call (fighting beetles are male). The fight is considered **terminated** if one insect pushes the other out of the ring, if one insect throws the other from the plant offshoot, if one insect runs out of the ring (which happens rarely), or if one insect falls on its back. From time to time, insects fight to the death (https://en.wikipedia.org/wiki/Insect_fighting).

Sometimes, **ant fighting** is practiced. Fire ants (*Solenopsis*) and army ants are used for this purpose. This kind of fight requires the patience of the organizers, who must carefully handle and instigate insect fighting by causing them to bite each other (https://en.wikipedia.org/wiki/Insect_fighting).

1.8.5 Photo hunting for insects

One of the most diverse **hobbies** people have is the photography of insects. Naturally, some people do this as part of their professional duties, but in an overwhelming majority, nonprofessionals photograph insects for the sole reason that this is of interest to them. Sometimes these enthusiasts are not involved in entomology or biology and therefore have often no fundamental knowledge either of their subject matter (e.g., life cycle of insects, features of their structure, character of nutrition) or of photographic technique (e.g., choice of camera angle, composition of picture, using different cameras).

Although there has been no shortage of available information about biological features of insects because popular entomological literature was always in sufficient quantity, **information on photography** of insects, specifically, was rare for a long time. However, with the development of the Internet, this problem was solved; now one may find a great many reference books and manuals about insect photography (Afanasyev 2012; Hussey 2010; Phillips 2004; Plonsky 2002; Cole 2011). In addition, books devoted to this topic are now common (West and Ridl 1994; Thompson 2003, 2007; Busch and Sheppard 2012).

Photographers first took pictures of insects early in the 20th century; back then, their number was insignificant. However, with the development of technologies such as new models of cameras, macro lenses, and color film, the number of insect photographers has **gradually increased**. Nowadays, with the large-scale implementation of digital technologies, the photographing of insects has become much simpler, and this has caused a rapid increase in the number of people interested in this hobby.

It is a challenging task to photograph insects. Because of their small size, the photographer must get **as close as possible** and patiently wait for the insect to turn to provide the desired angle. In some cases (e.g., photographing insects hunting prey), the period during which the interesting photo can be snapped is very short. Here, it is essential to be at the right place, at the right time.

It is essential for photographers to be aware of the **way of life of the insect** which they intend to shoot. Knowing when and where the subject can be and what it eats allows the photographer to define more exactly the shooting location and time. For example, to shoot a butterfly, it is necessary to ascertain in advance which flower nectar is preferred by that butterfly, where it lays eggs, and so forth, and the same can be said for all insects. Once the mode of life, behavior, and sometimes the structure of the insect is ascertained, it will be clearer to know when and where to go to capture them (Plonsky 2002).

The first hours after sunrise and the last hours before sunset are **preferable** for shooting photographs of insects. In the morning, insects are slow-moving and, therefore, one can mount a tripod and shoot with long exposures. Animals and plants are often covered with dew, which creates interesting effects. In the evening, activity of insects decreases again, and the setting sun provides a chance to get good shots. Also, in the evening, there is higher probability of finding interesting subjects of hunting and interaction between insects. The daylight hours are little suited for photographing insects, as the sun provides too much glare (http://www.rosphoto.com/portfolio/foto_nasekomyh_praktika-325).

A couple of tips about how **to get access to** an insect are given by Plonsky (2002): (1) walk slowly, be patient, and take the time to simply observe the insects to understand their behavior; (2) some insects are more tolerant than others (this difference concerns both individuals of different species and ones of the same species); (3) go at a crawl and avoid casting a shadow on the insect (if it flies away, remain motionless within 1 or 2 minutes because insects will often return to the same place); (4) be alert, and make all settings on the camera in advance, because, most often, you will only have one chance for the shot.

There are also **ethical norms** to follow when photographing insects. For example, one cannot say that a photograph was shot in nature if it was done in a studio or if dead insects were posed as if they were alive. Sometimes, to perform a shooting, an insect is caught and chilled in the refrigerator or placed for some time in an entomological killing jar, with the intent that not to kill but only to quiet the insect. Then, the sleeping insect is placed in the shooting location and the photographer waits for it to awaken. Often the insect does not awaken from the narcosis but instead dies, making the shot look unnatural. A set of questions concerning ethical norms in the photography of insects is considered by Wild (2013).

1.8.6 Collecting of insects

Most people, at some time in their life, have the desire to build a **collection** of something: it may be a collection of coins, postage stamps or matchbox labels, postcards, shells, stuffed animals, birds' eggs, autographs, jewels. Such a desire can arise in childhood, at a mature age, or even late in life. Children in rural localities often begin to collect insects. In most cases, this interest is forgotten as the person ages, leaving only masses of gaudy wings, fragments of insect legs and bodies, and a few rusty pins (Haseman 2007).

However, not all people give up on their childhood passion completely and, sometimes, it becomes a lifetime hobby, even if the person's profession is in no way related to entomology. Take, for example, the famous banker and financier **Lionel Walter Rothschild**. Throughout his childhood, he caught and collected insects, giving special precedence to butterflies. Over the course of his life, he has financed many entomological and zoological expeditions, discovering a considerable number of new species.

The number of explorers hired by him to seek out and collect butterflies has exceeded 400 (Irwin and Kampmeier 2003). Rothschild willed his core collection (about 2,250,000 butterflies) to the British Museum in London, thanks to which their collection of Lepidoptera became the world's largest (https:// en.wikipedia.org/wiki/Walter_Rothschild,_2nd_Baron_Rothschild). Another example is the famous Russian-American author Vladimir Nabokov, who, from the age of 7 until his dying day, collected butterflies (Marsh 2016).

It is unclear when the **collecting of insects began** (Photo 1.8-6). It is known that it was widely popular in the Victorian age (1837–1901). Inside houses of those days as a rule were cabinets that displayed exhibits of seashells, petrified remains, minerals, and dried insects. The collecting of insects left an imprint on European culture (e.g., in the song by Georges Brassens "La chasse aux papillons" ["The Hunt for Butterflies"]). This hobby is one of the most commonly encountered kinds of collecting (Photo 1.8-7). Now, it is especially popular among Japanese youths (https://en.wikipedia.org/wiki/Insect_collecting).

The **entomological collection** is an assemblage of insects chosen on the established principles, dried, pinned, and stored in special boxes. The collection of insects can be divided into two groups: private collections and collections for research. The principles of insect selection in each category are commonly different. The objects into the **private collections** are selected in accordance with the color range, sizes, personal preferences (i.e., the basis is formed by the showiness, aesthetic qualities and diversity of collected insects). **Research collections** are stored in the funds of the research institutions, universities, and museums. They are collected by researchers, and the taxonomic belonging is commonly their main criterion (Golub et al. 2012).

Entomological collection can be also divided into the following **types** (http://dic.academic.ru/ dic.nsf/ruwiki/1525388): (1) general collection (collection of insects from around the world, group of countries or particular country; the most commonly encountered type of entomological collections); (2) exploratory collection (collection of insects in which some special questions are considered); (3) specialized collection (includes entomological material constructed on specific principles and covering the relatively small part of any taxon or geographical area but presented with maximal completeness); (4) thematic collection (assemblage of insects elaborating on a specific topic, such as butterflies with scientific [Latin] names invoking ancient Greek gods and heroes of myths); (5) educational collection

Photo 1.8-6. Hobby collecting of insects is one of the most commonly encountered kinds of collecting and it began many years ago. Canvas oil painting "The Butterfly Hunt" (1874) by Berthe Morisot is shown.
Photo credit: https://www.wikiart.org/en/berthe-morisot/the-butterfly-hunt-1874

Photo 1.8-7. Nowadays as well, there many people gather insects. The photo shows a person which is sweep-netting for insects in the sward on the meadow just north of Starvelarks Wood (Essex, England).
Photo credit: John Rostron [CC BY-SA 2.0 (http://creativecommons.org/licenses/by-sa/2.0)], via Wikimedia Commons, 30 April 2005

(explains some entomological concepts, such as differences in structures of the representatives of basic insect orders).

A vast number of entomological collections exist throughout the world. The **largest** of these are listed on websites of insect and spider collections (Evenhuis 2016). For example, the U.S. National Entomological Collection stored, mainly, in the National Museum of Natural History in Washington DC, contains 35 million specimens. Although most of these specimens are kept dry, some groups (e.g., spiders and all soft insects are stored in alcohol). In most cases, the collections are arranged by taxa; lower categories (genus, species) are arranged in alphabetical order and, for selection of taxa, organized by the countries of origin within every species (http://entomology.si.edu/Collections.html).

The world's largest **collection of butterflies** (about 10 million specimens from all over the world) is stored in the Entomological Museum Witt, located in Munich, Germany. It was established in 1980 by businessman Thomas J. Witt for studying the moths (Heterocera). Beginning in 2000, the museum has been part of the Zoologische Staatssammlung München (Bavarian State Collection of Zoology) (https://en.wikipedia.org/wiki/Museum_Witt).

1.8.7 Insects as a fishing bait

As insects are the main kind of food for many species of fish, they are widely used as bait in fishing (Photo 1.8-8). The most effective are those which are preferred, at some time, by fish (while the fish's tastes can change for many reasons). The **optimal bait** is the insect that is most often eaten by fish. However, if there is overabundance of any particular kind of food, the fish begins to feed on it with less

Photo 1.8-8. As insects are the main kind of food for many species of fish, they are widely used as bait in fishing. For instance, Lesser Wax Moth caterpillars are used as bait for trout fishing. *Achroia grisella* caterpillars, length 13–16 mm are shown.
Photo credit: By Rasbak (Own work) [GFDL (http://www.gnu.org/copyleft/fdl.html) or CC BY-SA 3.0 (http://creativecommons.org/licenses/by-sa/3.0)], via Wikimedia Commons, 12 December 2011

desire. To better appreciate the feeding preferences of fish, one needs to find out what sort of insects are found at the water surface and in the air nearby or to study of the stomach of caught fish.

Sometimes imitations of various insects are used for fishing. In case of **artificial bait**, the primary goal of the fisherman is to not attract the fish by smell or taste but to deceive it and to force it to ingest the fish hook preliminarily because the bait looks and behaves like fish prey. The first reference to fishing with the use of artificial bait (artificial flies) is contained in the poem by the Roman poet Martial (born 38–41; died ca. 103) (Macadam and Stockan 2015).

Information on using different species of insects as bait, methods of their procuring, keeping, putting on hook, and so on, can be found on **many websites** (http://sputnik-rybolova.org.ua/zhivotnie-nazhivki/nasekomie-i-ich-lichinki-v-kachestve-nazhivki; http://naperekate.narod.ru/alesfishi224.html).

The table below presents insects used as a bait for fishing and most characteristic species of fishes drawn by them. It is important to bear in mind that it is difficult to make a complete and comprehensive list of such insects in view of the **vast variety** of their species and the **great number** of species of fishes feeding on them. In addition, the regional tasteful predilections cannot be excluded when the same species of fish prefers different insects under different natural conditions. Therefore, the information contained in the Table 1.8-1 can be considered as an example.

In fishing practice, combined baits are widely used because a "sandwich" of two or three different baits is very attractive for fish. Among the popular combinations of bait from insects are (http://www. all-fishing.ru/index.php?name=pages&op=view&id=2308): bloodworm and maggot; bloodworm and caddisfly; caddisfly and grasshopper; house fly and maggot; house fly and grasshopper; butterfly and grasshopper.

Table 1.8-1. Insects used as bait. (http://wiki.fishingplanet.com/index.php?title=Worms_%26_Insects_Baits, www. myfishinghome.com; http://kombat.com.ua/pitanie/pitanie2.html; http://belkamfish.com/master/600/11.htm; http://nik-fish. com/different/podruchnaya-nazhivka-nasadka.html; http://fishingwiki.ru/Наживки; http://fishermenfrompinsk.ru/rybalka/ poleznye-sovety-rybakam-4/assortiment_primanok_rybolova-32.php)

English name	Latin name	Fish species	Note
larva of May beetle	*Melolontha*	chub, ide, asp	Late spring – Early autumn
bloodworm	Chironomidae	roach, perch, bream	all year round
maggot of blowflies brown color	Calliphoridae	bream, whitebream, whitefish, ide, roach, crucian carp	all year round
maggot of blowflies red color	Calliphoridae	perch, bleak, bream, white bream, whitefish, ide, roach	
larva of common greenbottle (pinkie)	*Lucilia caesar*	roach, ide, bleak	
lesser wax moth caterpillar	*Achroia grisella*	trout	
mole cricket	Gryllotalpidae	catfish, chub, ide, barbel, cyprinidfish	catching of large fish with leger rig
mayfly	Ephemeroptera	roach, perch, bream, chub, gudgeon	
larva of caddisfly	Trichoptera	trout, bream, barbel, perch, bleak, sneep, roach, chub	
larva of bark beetle	Ipidae, Scolytidae	large roach, bream, perch, ide, lenok, grayling, chub, ide, bleak, ruffe	
ant	Formicidae	grayling, roach, rudd, ide	Summer
pupae of ants (ant eggs)	Formicidae	carp, crucian carp, bream, rudd, ide, roach	
fly	Muscidae	perch, roach, chub, rudd	
lubber grasshopper	*Brachystola magna*	ide, chub, asp, catfish, rudd, dace, roach, grayling	
dragonfly	Odonata	chub, ide, large roach, rudd, perch, trout, salmon	Angling from the water surface
larva of dragonfly	Odonata	large roach, perch, tench, rudd, ide	Best of times for fishing – early morning, sunrise
gadfly	Oestridae	chub, roach, ide, rudd, lenok	
larva of gadfly	Oestridae	cyprinid fish, grayling	Fishing with leger rigs
larva of yellow mealworm beetle (mealworm)	*Tenebrio molitor*	roach, white bream, dace, crucian carp	
wasp larva	Vespidae	perch, bleak, bream, white bream, white fish, ide, roach dace, chub, bleak, sneep	

Table 1.8-1 contd. ...

...Table 1.8-1 cont.

English name	Latin name	Fish species	Note
larva of rhinoceros beetle	Dynastinae	ide, chub, catfish, pike	
ground beetle	*Carabus*	trout, lenok, grayling	
larva of rat-tailed maggot	*Eristalis*	bream, perch	
European field cricket	*Gryllus campestris*	perch, trout, bluegill	
desert cricket	*Gryllus desertus*	barbel, chub, ide, rudd	
bee larva	Apidae	trout	
caterpillar of small tortoise-shell butterfly	*Vanessa urticae*	chub, ide	
caterpillar of cabbage butterfly	*Pieris* spp.	chub, ide	
larva of long hornbeetle	Cerambycidae	bream, roach, large ruffe, grayling, lenok	
larva of bark beetles	Ipidae, Scolytidae	chub, ide, roach, perch, bream, bleak, ruffe	
horsefly	Tabanidae	bleak, sneep, roach, chub	
larva of horsefly	Tabanidae	perch, pike, ruffe, bream	
larva of water beetle	Dytiscidae	perch, roach, skimmer bream	
sawtoothed grain beetle	*Oryzaephilus surinamesis*	chub	
colorado potato beetle	*Leptinotarsa decemlineata*	chub	
dobsonfly	Megaloptera	trout, grayling, ide, roach	

1.9

Other Uses of Insects

From the distant past until the recent times, people have used insects in a variety of other diverse ways. For example, fire beetles (family Elateridae), which inhabit tropical and subtropical America, including parts of Texas and Florida, have two luminescent spots on their bodies. The best-known species, the headlight elater *Pyrophorus noctilucus* (https://en.wikipedia.org/wiki/Pyrophorus_(beetle), has the highest **light intensity** among these insects (up to 45 mLb). Such luminosity is enough to read a book next to the beetle (http://www.zin.ru/animalia/Coleoptera/rus/incoel.htm). The first Europeans settling in South America used these beetles to light their huts and to fill their lamps. Traveling at night through the jungle, residents of these localities tied fire beetles to their toes to light the path (Akimushkin 1993).

Diamphidia, or Bushman arrow-poison beetle, is an African genus of leaf beetle that is related to the Colorado potato beetle. The larvae and pupae of these beetles release a strong poison **diamphotoxin** (https://en.wikipedia.org/wiki/Diamphotoxin), which the San people (Bushmen) of southern Africa apply to the heads of hunting arrows. Even a low dose of this poison results in the paralysis of muscles and **death** of large mammals (Chaboo 2011).

In Africa, furnaces for copper smelting were made of **termite mounds**. For this purpose, the middle part of the termite mound was hollowed out and, from below, a hole for the molten metal tapping and ash pit were made. The ore and coal were loaded from above. The termite mounds were durable and served for many years (http://insectalib.ru/books/item/f00/s00/z0000013/st012.shtml).

Dermestid beetles and their larvae are used in **natural history museums** to clean animal skeletons. The following conditions are needed: (1) appropriate temperature regimen (25°C to 30°C); (2) darkness in the room (normal feeding and reproduction of these beetles happen only in darkness); (3) skeletons to be cleaned should be dry and not mouldy (Zaslavsky 1966).

It is difficult to designate Sections 1.9.1 and 1.9.2 as "positive" aspects of insects' effect on human activity. However, even these kinds of insect use should be elucidated, in order to cover all aspects thoroughly.

1.9.1 Insects as weapons of war and terror

Insects have played a role in military operations for centuries. There are **three variants** of their use to do damage to the enemy (https://en.wikipedia.org/wiki/Entomological_warfare): (1) using biting insects (bees, wasps, hornets) for direct descent upon the enemy; (2) distributing agricultural insect pests within enemy territory to destroy their crops; (3) releasing insects infected with a pathogenic agent into a specific area where they can infect people and animals.

The **first variant** of the use of insects as weapons is the oldest. In the opinion of entomologist Jeffrey Lockwood (2008), people first used insects in conflicts as early as the Stone Age, when wasp nests were thrown into the caves of the enemy. Documentary evidence of using insects in war, referring to the 4th century B.C.E., is found in the tract "Siege Defense", where the first military theorist of antiquity Aeneas Tacticus reported: "There were situations when a number of people have created a trouble for people being in the undermining by drawing wasps and bees into the burrowed tunnel" (http://xlegio.ru/sources/aeneas-tacticus/de-obsidione-toleranda.html).

Many more **historical examples** of the use of insects in military operations are known. For example, in describing the unsuccessful siege of the Parthian fortress Hatra (Parthian Empire) by Roman troops in 198 C.E., the historian Herodian said that defenders of the fortress threw pottery vessels at the Romans that contained insects (exact type unknown) that caused deadly wounds by stinging exposed body parts and faces of the intruders. According to some, the insects were bees or wasps, but Jeffrey Lockwood believes that they could have been scorpions rather than insects (Rousseau 2014).

In the siege of the fortress of Acre in 1191, during the **Third Crusade**, the King of England Richard the Lionheart successfully used clouds of bees housed in large clay pots against the enemy. Using launching devices, his troops threw hundreds of these clay beehives at the Muslim enemy, causing them to run from the bees (Ioyrish 1976). In 1289, the Austrian Duke Albert I of Habsburg unsuccessfully attempted to carry out an assault on the Hungarian fortress of Nemetujvar (now Güssing, Austria). Beehives were thrown from the fortress walls, and this became the decisive weapon (Rousseau 2014).

Many other cases of the use of stinging insects (predominantly bees) in warfare are presented by Lockwood (2008).

There is no evidence proving **the second variant**—the purposeful **distribution of agricultural pest insects**. During the American Civil War (1861–1865), the Confederacy (southern states) accused the Union (northern states) of intentionally introducing the harlequin cabbage bug (*Murgantia histrionica*), but these accusations were not proved. In World War II, Germany and Great Britain examined the possibility of an entomological attack against each other by way of the mass distribution of the Colorado potato beetle (*Lepinotarsa decemlineata*) in enemy territory; however, this intention was not practically realized (https://en.wikipedia.org/wiki/Entomological_warfare).

The **third variant** of using insects as weapons is well confirmed. One of the first successful attempts involved the plague epidemic in Asia Minor in the 14th century, known as the "Black Death", which caused the deaths of 30–60% of Europeans. Plague first appeared in Europe when fleas—transmitters of plague—were introduced as a kind of biological warfare from the Crimean city of Kaffa (now Feodosia).

During World War II, **Japan** widely used insects as weapons on the territory of China. **Unit 731**, the Japanese unit of biological warfare (Photo 1.9-1), used plague-infected fleas and flies and cholera-coated flies to infect the Chinese population (Lockwood 2012). The Japanese Unit 731 program planned to produce 5 billion plague-infected fleas per year. Japanese military distributed these insects, dropping bombs filled with the infected fleas from low-flying aircraft. Depending on various sources, the epidemics resulting from these actions caused the deaths of 440,000 (Endicott and Hagerman 1999) to 580,000 Chinese (https://en.wikipedia.org/wiki/Unit_731 0).

During the Cold War, entomological weapons were used both by the Soviet Union and by the United States. In the Soviet Union, methods of distribution of diseases of livestock animals included the use of ticks to carry the foot-and-mouth disease virus and the bacterium *Chlamydophila psittaci* resulting in avian chlamydiosis in chickens. The United States Army worked out a plan for establishment of an entomological warfare facility designed for the monthly production of 100 million yellow fever–infected mosquitoes (https://en.wikipedia.org/wiki/Entomological_warfare).

The use of insects for **entomological bioterrorism** is, in principle, no different from their military use. Insects are efficient delivery vehicles of the precursors of diseases because insects can be easily collected and their eggs transported, without detection. Allegedly, a similar act of bioterrorism was the release in California of the Mediterranean fruit fly (*Ceratitis capitata*), causing extensive damage to a wide range of fruit crops (https://en.wikipedia.org/wiki/Ceratitis_capitata). The responsibility for it was taken by the "ecoterrorist organization" calling itself "The Breeders". The 1989 attack was estimated to have cost $60 million in eradication efforts (https://en.wikipedia.org/wiki/1989_California_medfly_attack).

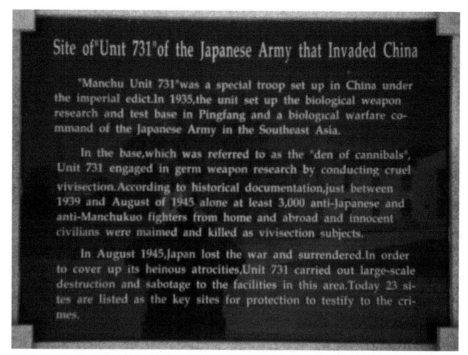

Photo 1.9-1. Unit 731 was officially known as the Epidemic Prevention and Water Purification Department of the Kwantung Army. It was a covert biological and chemical warfare research and development unit of the Imperial Japanese Army that undertook lethal human experimentation during the Second Sino-Japanese War (1937–1945) of World War II. Depending on various sources, the epidemics resulting from these actions caused the deaths of 440,000 to 580,000 Chinese. Memory plate for the atrocities of Unit 731 at the site next to Pingfan,Heilongjiang province, China is shown.
Photo credit: By Markus Källander (Unknown) [Copyrighted free use], via Wikimedia Commons

1.9.2 Use of insects in torture and execution

Many varieties of torture by insects are known, distinguished by cruelty and refinement. In most cases, the naked victim was tied to a tree or post; covered with honey, or something else to attract insects; and left to be tortured by numerous insects, most often, flies. In Siberia (Russia), this practice took place in the summer; within 24 hours, numerous gnats (e.g., mosquitoes, midges, gadflies, gadbees) turned the victim into a tumid mannequin.

Similar torture with the use of **ants** was another popular method. Usually, a naked person to be tortured was tied up to a post or on the ground between four stakes near an ant hill; honey or other sweet substance was then poured over his or her body, and other people waited while the swarm of ants surrounded the body. Attracted by the sweet snack, the insects first ate the honey and then took on the body of the captive, biting him mercilessly. This torture method was the preferred trick of the "Bloody" Countess Elizabeth Báthory de Ecsed (1560–1614), who punished her maidservants for the tiniest mistakes and faults in such a manner (https://en.wikipedia.org/wiki/Elizabeth_B%C3%A1thory).

There was also a variety of torture by tickling by means of a **beetle**, which was most commonly encountered in India. A small beetle was placed on a man's penis or a woman's nipple and covered with half a nut shell. After a while, tickling caused by the motion of the insect's legs on the victim's body became so unsupportable that the interrogated confessed to anything at all (http://gothic.com.ua/ukrrus/forum/index.php?showtopic=6460).

In many regions of Africa and South America, species of **army ants** that rip through everything in their path are widespread. Their colonies can number 15 million individuals (https://en.wikipedia.org/wiki/Army_ant). People on the spot have used (and possibly continue to use) them for torture by tying the victim in the path of these insects. In a relatively short time, the victim's body was eaten down to the bone (http://ucrazy.ru/interesting/1327197690-pytki-nasekomymi.html).

In Asia, **tarantulas** and **scorpions** were used for torture, instead of ants. By the end of the 1970s (period of the Pol Pot communism collapse), in prisons of Cambodia, in addition to modern torture devices (e.g., electroshock), containers of scorpions were placed using special tongs on the nude women and men being tortured (http://ucrazy.ru/interesting/1327197690-pytki-nasekomymi.html).

Tortures using insects were with pleasure applied to women because women are the **most fearful of arthropods**. Tortures with insects in the form of tickling occurred when a naked captive was placed next to a container of insects, the floundering of which resulted in aversion, panic, fear, and more painful feelings of the victim. Frequently, the captive was at the ready to admit to crimes of whatever sort as a result of such travails.

The ancient Persian method of execution called a **scaphism** (from ancient Greek σκάφη [boat]) is well known. The person to be executed was stripped naked and pinioned inside a narrow boat or hollowed-out tree trunk. From above, he or she was covered with a matching boat or tree trunk, leaving hands, feet, and head exposed. The prisoner was force-fed milk and honey to cause diarrhea. In addition, to attract insects, the person's body was coated with honey.

The unlucky individual was then set adrift in the stagnant pond or left in a sunny spot. The victim's feces accumulated inside the boat container, attracting a growing number of insects, which slowly devoured the flesh and laid eggs, resulting in gangrene. In order to prolong the torment, the victim would be **fed every day** (http://mentalfloss.com/article/23038/9-insane-torture-techniques).

According to Plutarch, the young soldier named **Mithridates**, who accidentally killed Cyrus the Younger, son of Persian King Darius II, was executed in such a manner in 401 B.C.E. The luckless youth died only after suffering from this torture for 17 days (https://en.wikipedia.org/wiki/Scaphism).

1.9.3 Use of insects in rituals

There is little or no difference between the use of insects in rituals and their use in torture. The difference consists only in motivation: the individual being tested in the ritual is subjected to torture (sometimes willingly), in the furtherance of a particular goal. Most often, insects are used in **initiation rites** or rites of passage, marking entrance or acceptance into a group or society (https://en.wikipedia.org/wiki/Initiation).

The **initiation rite to become a warrior** for boys of the Sateré-Mawé tribe in the Brazilian Amazon serves as such an example (https://en.wikipedia.org/wiki/Maw%C3%A9_people). When the boys turn 13 years old, they must go through a rite of passage, the basis of which is being tested by venomous ants. Each boy must place both hands into gloves or mitts made of woven leaves into which bullet ants (*Paraponera clavata*) have been placed (Photos 1.9-2, 1.9-3). The ants possess a very strong venom, and the force of their sting exceeds that of any wasp or bee. On the scale of the Schmidt sting pain index, it corresponds to the highest level (4+) and is described as causing "waves of burning, throbbing, all-consuming pain that continues unabated for up to 24 hours" (https://en.wikipedia.org/wiki/Paraponera_clavata).

Each boy has to withstand 10 minutes of **bullet ant stings** to his hands inside the woven mitts. The boys are prohibited from crying because crying shows a lack of determination. To complete the rite, the boys must wear the mitts on their hands 20 times within several months. After each of these events, the hands are paralyzed for some time, and the boy feels ill and writhes uncontrollably with pain. The only "protection" provided is a coating of charcoal on the hand, supposedly to confuse the ants and inhibit their stinging (Galvan 2014).

In the Brazilian Amazon, there are **female initiation rites** for girls of the Indian tribes in the Tupí-Guaraní family. These rites were timed to the girl's menarche (Balee 2000) and related to stings of the nesting in soil-raiding ants (*Neoponera commutata*). In the southern California and northern Mexico, the indigenous tribes for centuries performed the **ritual of appropriation of dream helpers** through hallucination. For this purpose, the young men, under the supervision of an elderly member of the tribe, consumed large quantities of live, unmasticated red harvester ants (*Pogonomyrmex californicus*). This resulted in the prolonged state of unconsciousness during which the dream helpers supposedly visited the young men (https://en.wikipedia.org/wiki/Insects_in_culture).

Some Indian tribes still follow certain rituals, called ordeals. Young men may be **married** if they will withstand the so-called "ant court". In scant attire, these men go to hammocks and baskets with

Photo 1.9-2. *Paraponera clavata* is a species of ant, commonly known as the bullet ant, named for its potent sting. It inhabits humid lowland rainforests from Nicaragua and the extreme east of Honduras and south to Paraguay. This ant species is used for the initiation rite to become a warrior for boys of the Satere-Mawe tribe in the Brazilian Amazon.
Photo credit: By Hans Hillewaert/, via Wikimedia Commons

Photo 1.9-3. When the boys of the Satere-Mawe tribe turn 13 years old, they must go through a rite of passage, the basis of which is being tested by venomous ants. Each boy must place both hands into gloves or mitts made of woven leaves into which bullet ants (*Paraponera clavata*) have been placed. The picture shows British naturalist Steve Backshall participating in an initiation rite.
Photo credit: https://en.wikipedia.org/wiki/Paraponera_clavata, 01 January 2008

ants are put on their bodies. The ants may move freely across the body and bite the martyrs ruthlessly. If young men do not say a word or move a muscle, then they have a right to get married (http://ucrazy.ru/interesting/1327197690-pytki-nasekomymi.html). Yet another example of ritual in some nations is related to the application of butterfly to the forehead in the hope of increasing level of intelligence (Sviridov 2011).

The Mursi people of southwestern Ethiopia, close to the border with south Sudan, use insects in a unique way. They apply **ritual tattooing** to the abdominal skin in the form of symmetric lines consisting of separate cicatrical tissue, or swellings. These bosses are characterized by both different sizes and diverse forms: spherical, oval, linear. Such "tattooing" is applied as follows. Larvae of diverse land and aquatic insects are pushed through an incision in the skin.

They begin to develop there but the organism, coming to grips with newcomers, encapsulates the "outsiders" with its connective tissue and they finally perish, leaving behind under the **skin swellings** in different sizes and shapes. Knowing this, the Mursi alternate especially the points of their introduction against each other, depending on the final "pattern" that the Mursi wanted to achieve (http://planeta.by/article/752).

1.9.4 Use of insects as decorations

Insects have always been popular motifs found in jewelry—necklaces, hairpins, earrings, and so on. In more recent times, real insects came to be used in such adornments. The existing practice of their application has three variants: (1) whole dried insect; (2) separate pieces of insects; and (3) living insects. Most often, beetles are used for these purposes.

For the **first variant** of using insects, jewel beetles or metallic wood-boring beetles (Buprestidae) are the most relevant. They are named for the presence of metallic luster commonly appearing in the integument of body, which may be gold, bronze, green, yellow, or red. Often, there are varicolored spots (https://www.zin.ru/animalia/coleoptera/rus/incobu_.htm). The colors of the jewel beetles are persistent; therefore, the beetles keep quite well in the dry state. Big, gorgeous specimens of jewel beetles are frequently used in the manufacture of brooches, bracelets, and other jewelry (http://kungrad.com/nature/fauna/zlatki/). Many tribes of Africa, South America, and Oceania use the dried jewel beetles in the making of beads (http://dic.academic.ru/dic.nsf/ruwiki/88741).

For the best safekeeping, the beetles are often given a coat of lacquer or encapsulated in a synthetic resin (i.e., every effort is exerted for preservation to make the ornate devices). Insects preserved **in amber** are highly valued (Photo 1.9-4). In early times, Phoenician merchants paid 120 swords and 60 hand jars for flies embedded in amber. Early in the 19th century, amber containing insects was very fashionable in France and Russia (Srebrodolsky 1984). Now, there is a wide market for imitations of insects embedded in amber; methods are considered in detail by Manukyan (2013).

The bright, gorgeous wing cases of beetles are used most often as **separate fragments of insects** in jewelry designs. Their use, called "beetle wing art", has the long-standing traditions in Thailand, Myanmar, India, China, and Japan. Depending on the region, different species of jewel beetles are used for these purposes, but the beetles belonging to genus *Sternocera* are traditionally the most highly prized (https://en.wikipedia.org/wiki/Cultural_entomology). Insect galls and morphobutterfly and dragonfly wings have also been incorporated into jewelry designs (Irwin and Kampmeier 2003).

Iridescent beetle wings are used not only for the creation of jewelry designs but also as ornaments in **paintings** and **textiles**. Nancy Armstrong's book, *Victorian Jewelry*, mentions 37 yards of fabric spangled with beetles and butterflies (Armstrong 1976). Such fabrics were manufactured in many towns of Europe, but they were to a large extent made in India, initially in Punjab and Delhi. In the 1851 Great Exhibition of the Industry of All Nations in London, an Irish lace dress decorated with whole beetle wings was presented (Rivers et al. 2011).

Some people prefer to wear **living insects** as ornaments, and there are countries where such ornaments are traditional. For example, in India and Sri Lanka, women may decorate themselves with jewel beetles (*Chrysochroa ocellata*). On solemn occasions, such living jewelry (2.5–3.7 cm [1.5 inch] long) may be worn; sometimes a small chain is attached to one leg anchored to the clothing to prevent escape

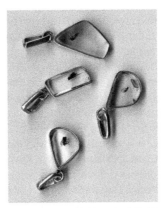

Photo 1.9-4. As decorations, insects preserved in amber are highly valued. Baltic Amber silver pendants with fossil insects (Diptera) are shown.
Photo credit: A.R. Manukyan (Kaliningrad Amber Museum, Russia)

(https://en.wikipedia.org/wiki/Cultural_entomology). In the late 19th century, click beetles (*Pyrophorus noctilucus*) were in fashion in Paris; they were used as evening ornaments and called "living diamonds" (Akimushkin 1993).

However, this tradition was most of all developed in **Mexico**, where the practice of using beetles as ornaments has existed over a period of many centuries. It is especially strong on the Yukatan Peninsula, where buffalo clasp beetles (*Dynastes fibula*) are widely distributed. These beetles, because of their size, superior elasticity of wing cases, substantial lifetime, and total endurance, are ideally suited for manufacturing brooches from them.

The beetles are adorned with artificial or true precious stones, and thin chains are attached to the wing cases to make them creep and not get lost. **Attached to the dress** of its owner, the beetles spend the rest of their short lives simply walking and sparkling as they move. If given the correct care, the beetle may live a year and then some (Kukharenko 2013).

Beetles are not the only insects used as living jewelry. For example, in 2006 the fashion designer of avant-garde American Gothic fashion, Jared Gold, created a collection of brooches using the **Madagascar hissing cockroaches** (*Gromphadorhina portentosa*) decorated with imitation gems (https://en.wikipedia.org/wiki/Jared_Gold). Care instructions and a special mixture for feeding the cockroaches were attached to the jewelry. In localities where balm crickets (cicadas) live, children often fasten the insects to their skin or their clothes. As for beetles, the species belonging to the family Zopheridae, are most suitable to serve as living brooches. These beetles mostly inhabit Mexico and southern United States, where they are called iron-clad beetles (Kukharenko 2013).

1.9.5 Use of insects to search for deposits

Insects can be used for searching for **ore deposits**. Ore is a natural formation containing compounds of useful components (minerals, metals); an ore deposit is an accumulation of ore with concentrations making extraction of the valuable components economically attractive. Ores occur rarely on the earth's surface, being found more often at some depth within the interior. Provided that the depth is not great (tens of meters), the search for deposits involves taking surface samples and if the concentration of the useful component is found to be elevated, drilling is performed with the aim of finding the ore.

Such insects as **termites** and **ants** construct their nests using different locally available materials, including soil and ground particles delivered from the underground horizons. So, some termites inhabiting in deserts, burrow in the soil in vertical tunnels the length of which may reach 40 meters (http://animalreader.ru/termitnik-shedevr-inzhenernogo-iskusstva.html). If the termite mounds are located not far from the ore body, then the material of termite mounds contains the elevated concentrations of the useful component (Photo 1.9-5). Therefore, based on the analysis of samples taken from the termite mound or anthill, one may determine whether there is an ore deposit nearby.

For example, in the vicinity of the **Garden Well gold deposit**, in Western Australia, the concentration of gold in the surface soils is, on the average, 2 mg/t (2 parts per billion). However, in 22 mounds formed by the termite *Tumulitermes tumuli*, the average concentration of gold was 7.4 mg/t (7.4 parts per billion); in the mounds of the subterranean termite *Schedorhinotermes actuosus*, 8.4 mg/t (8.4 parts per billion); and in the nest of ant species *Rhytidoponera mayri*, 24.4 mg/t (24.4 parts per billion).

Analysis of the material of termite nests makes searches for gold significantly **less expensive** and much safer for the environment. Today, geologists use expensive drilling for these purposes (Stewart and Anand 2014). Termites do not especially choose to bring the gold to their nests; however, soil particles in the underground layers (delivered by the termites from depths of 3 to 13 feet) have a higher gold content and the insects thereby serve involuntarily as artisans (Stewart and Anand 2012).

Termites are also able to serve as indicators of deposits of **other** mineral resources. The Vila **Manica copper deposit** in Mozambique was discovered by using termites. Later, the largest **kimberlite (diamond) mine** in the world, in Jwaneng, Botswana, and several **gold prospects** in southern Africa were found by termite mound sampling (http://www.afrol.com/articles/10447).

In Bastar (Madhya Pradesh, India), an analysis of material of the termite mound was acknowledged to be very promising for searching for **tin** deposits. It is also assumed that this method can be used for

Photo 1.9-5. Termites construct their nests using different locally available materials, including soil and ground particles delivered from the underground horizons. If the termite mounds are located not far from the ore body, then the material of termite mounds contains the elevated concentrations of the useful component. Therefore, based on the analysis of samples taken from the termite mound or anthill, one may determine whether there is an ore deposit nearby. Termite Mounds in the Bungle Range in Western Australia are shown.
Photo credit: By Ouderkraal (Own work) [CC BY-SA 3.0 (http://creativecommons.org/licenses/by-sa/3.0)], via Wikimedia Commons, 5 January 2012

discovery of the deposits of heavy metals like niobium, tantalum, tungsten, and so on, because the termite mounds there have a good spatial distribution (Surya Prakash Rao and Raju 1984).

At present, the method of searching for deposits of commercial minerals by means of termites is increasingly used worldwide; a new science, termed *geozoology*, has even been created for its further development.

However, it should not be supposed that this method is new. As early as precolonial times, between 1502 and 1880, many generations of gold diggers in West Africa used the **work of termites** to study the deeper layers of soils for the discovery of gold and other precious metals and minerals. Today's geologists only rediscovered this ancient and inexpensive method in the 1960s. For example, the Canadian company North Atlantic Nickel, using this method in searching for gold in Mali, discovered many previously unknown areas of ancient artisan workings between 1997 and 2003. Since 2004 it called North Atlantic Resources Ltd. (http://www.afrol.com/articles/10447).

1.9.6 Use of insects for recycling of organic waste

Organic waste is waste containing **organic substances** released by animals and plants. Organic waste is formed during operation of many industrial, agricultural, housing, utility, and service enterprises. Each kind of enterprise and facility is characterized by its own **structure of organic waste**. For example, organic waste from wood in wood-processing and woodworking enterprises includes tree branches, trimmings, bark, chips, cuttings, and wood dust.

Food scraps as a kind of organic waste of the human population consist predominantly of fruit and vegetable peelings, meat and fish scraps, bones, bread, dairy products, eggshells, and so on. Disposal of organic waste is also an issue for enterprises involving animal husbandry, crop raising, cellulose and paper, tanning, and food (e.g., sugar, beer, dairy and meat, canning, and confectionary) industries.

In many cases, these wastes are **stocked** at sites specially intended for this purpose, occupying considerable areas and contaminating the environment. Every year, **1.3 billion tons** of organic waste are produced (Veldkamp et al. 2012). Often, the use of insects allows us to decide effectively the question of their disposal. Processing organic waste with the use of **flies** is most efficient because fly larvae raised with the use of organic waste grow very quickly; larvae size increases 300 to 500 times within a week. At the end of 1 year, the biomass from a pair of flies and their offspring reaches about 87 t in the case of full realization of their genetic potential (Ernst et al. 2007).

At present, across the globe, the problem of **animal husbandry waste** disposal is particularly acute. The **black soldier fly larvae** (*Hermetia illucens*) are able to transform the proteins and other nutritive substances of manure into more valuable biomass (e.g., animal feedstuff). In storage of cattle manure, these larvae reduce the content of phosphorus by 61–70% and nitrogen by 30–50%.

During field tests performed in **pig-breeding complexes** in Georgia, United States, it was found that the black soldier fly larvae, in the act of digesting pig manure, reduced nitrogen by 71%; phosphorus by 52%; potassium by 52%; and aluminum, boron, cadmium, calcium, chromium, copper, iron, lead, magnesium, manganese, molybdenum, nickel, sodium, sulphur, and zinc by 38–93%. Therefore, larvae are able to reduce the polluting potential by 50–60%, or more (Liu et al. 2008).

For the disposal of animal husbandry wastes, the **common house fly** (*Musca domestica*) is also used; this insect is especially effective for processing pig and chicken manure (Nartshuk 2003). When settling 1 t of native pig manure (fresh manure without additives of urine or water from drinking-bowls) with common house fly eggs, 60 to 100 kg of fly larvae biomass and 640 to 700 kg of biohumus are produced in 5 to 6 days. The biomass of common house fly larvae is complete protein feed for pigs, calves, poultry, fur animals, and fish, which is discussed in Section 1.6.3. Insects as Animal Feed (Ernst et al. 2007).

The common house fly larvae can be used to process not only feces but also **other organic wastes**, including wastes of cellulose-and-paper and wood-processing production, straw and stubble remains (roots and lower parts of stems), screenings, weeds, and so on. In the course of processing, liquid excrement or a solution containing cellulose-decomposing microorganisms is poured over the milled waste, and vegetable stock subjected to preliminary microbiological treatment is also introduced.

Afterward, the mixture is allowed to ferment, and fly larvae are introduced (http://www.findpatent. ru/patent/216/2163586.html). There is also a somewhat different method of phytomass processing that uses **not only** fly larvae but also larvae of locust, grasshopper, plant lice, or other insect pests (http://ru-patent.info/21/60-64/2160004.html).

The larvae of **blow flies** (family Calliphoridae) are used in the disposal of meat and fish industry waste, while black soldier fly larvae are used to process organic waste such as coffee bean pulp, vegetables, distillers' waste, and fish-processing byproducts (van Huis et al. 2013).

1.10

Businesses Related to Insects

According to Dale Carnegie, the formula of enrichment is expressed by the following brief phrase: "Find a demand and meet it". It is evident that insects play an important role in many kinds of human activity. Correspondingly, business related to insects is also oriented to satisfy demands of people for insects or for services provided by them. The distribution and scale of different kinds of such businesses vary widely: some are on the upswing, while others are on the decline. Nevertheless, monetary flow in business amounts to billions of dollars, so it makes sense to identify the specific types.

Just as in the other spheres of the economy, there are both **limited specialized organizations,** oriented solely to one market segment (e.g., provision of different species of insects to collectors, tourism and leisure related to insects, production of living jewelry), and **universal firms,** rendering a wide range of different services of an entomological nature. Worldwide Butterflies, for example, established in 1960, delivers butterflies and moths to organizations for research, school and private hobby collections, medical investigations, and photography, as well as providing professional and nonprofessional entomological equipment (http://www.wwb.co.uk/).

Most developed kinds of insect-related business involve the rearing of insects for purposes of biological control, plant pollination, scientific studies, human and animal nutrition, collectors, or tourism and leisure; and the manufacturing of entomological equipment.

1.10.1 Rearing of insects for biological control

The mass rearing of insects for purposes of **biological control** started in the late 1960s. By 1996, there were 64 companies worldwide producing natural enemies for biological control of pests. By region, they were distributed as follows: Western Europe 26, North America 10, Central Europe 8, Russia 5, Asia 5, South America 5, Australia/New Zealand 5 (van Lenteren et al. 1997).

By 2004, there were already 85 **commercial producers** of natural enemies for augmentative forms of biological control (van Lenteren 2012). This market volume reached about $50 million in 2000, while the year-on-year growth was 15–20% (van Lenteren and Bueno 2003). It is challenging to estimate the present-day scale of this business. This is related to the fact that insects are usually included in the biopesticide category (along with other arthropods and microorganisms), where they play an imperceptible role in the general structure of biopesticides. For example, in 1995, 92% of the bioinsecticide market was occupied by gram-positive, soil-dwelling bacterium (*Bacillus thuringiensis*) (Jijakli 2010).

In the 1990s, this kind of business in **Europe** involved the following figures: there are 750 workers in the sphere of rearing of insects for biological control of pests. Only in three companies did the staff exceed 50 employees, while the personnel of most organization numbered fewer than 10. The three largest companies serviced more than 75% of the biological control market. Total sales were $15 million in 1987 and $60 million in 1991. About 90% of the total sales was accounted for by 29 of the more than 80 species of natural enemies (van Lenteren et al. 1997).

The majority of insects acting as natural enemies of pests were used in Europe for biological control in **greenhouses**, with the exception of ladybirds of *Harmonia* spp. and egg parasitoids (*Trichogramma* spp.), which were applied in the open fields. The **most used** agents of biological control in greenhouses included a parasitic wasp *Encarsia formosa* (25% of the total market), predatory mite *Phytoseiulus persimilis* (12%), and *Amblyseius cucumeris* (12%).

The other indicator characterizing this kind of business is the various groups of pests which the natural enemies were reared to inhibit. 84% of all expenses were related to four groups of pests: white flies (*Aleurodidae*), thrips (*Thripidae*), spider mites (*Tetrarhynchus*), and aphids (*Aphididae*) (Peshin and Dhawan 2009). In comparison with the United States, at that moment, many more species of insects acting as natural enemies of pests were purchasable in Europe, which is explained by a far larger greenhouse industry, and, in addition, large differences in prices for biological control agents (van Lenteren 2012).

At present, business related to the rearing of insects for biological control is **growing rapidly**. The expenses just for research studies in this field are worth $80 to $90 million per year in North America alone (Cohen 2001). The earlier situation of the European commercial biological control suppliers being larger than their American counterparts remains the same (van Lenteren 2012). Now, the companies Koppert Biological Systems in the Netherlands (www.koppert.nl) and Biobest in Belgium (www.biobest. be) are the **most powerful producers** in the world market.

There are also many medium-sized and smaller companies. For example, the database of the Bio-Integral Resource Center in Berkeley, California lists many firms rearing entomophagous insects along with producers of various pesticides (2015 Directory of Least Toxic Pest Control Products 2015). It contains an alphabetized list of the scientific names of beneficial insects and other arthropods, which lists the suppliers of each species, and also includes a separate list of suppliers, which provides complete contact information. The mission of **Association of Natural Biocontrol Producers (ANBP)** is to address key issues of the augmentative biological control industry through advocacy, education, and quality assurance. In particular this site contains information about members of association.

1.10.2 Production of insects for human and pet food

Despite the fact that entomophagy is part of long-standing traditions in many regions of the world, the business related to production of foodstuffs and fodders is a relatively new phenomenon. For the time being, the production of **pet food** predominates. The use of insects for food is often considered exotic and these companies survive mainly through the sale of insects as pet food (van Huis et al. 2013). Nevertheless, production of insects is quite profitable in the regions where insects are habitual food.

Such an example is provided by **Thailand**, one of the few countries of the world having developed a sector of rearing of insects to use for food. Currently, more than 20,000 insect farming enterprises are incorporated in the country. The essential part of them comprises small family enterprises providing the farmers and members of their families with means of subsistence (Hanboonsong et al. 2013). All told, 194 species of insects are used for food in this country (Photo 1.10-1) (Sirimungkararat et al. 2010). The most frequently used insects and prices for them are given in Table 1.10-1.

Rearing of crickets (in the north) and palm weevil larvae (in the south) is commonly practiced. For example, **cricket farming** in Thailand was started in 1998. Despite the fact that, in the country, its own species of crickets exist, a tastier house cricket (*Gryllus domesticus*) imported from temperate regions in Europe and the United States is mainly reared. For period of 2007 to 2011, production was on average around 7,500 tons per year. Rearing of **palm weevil larvae** for home consumption by local people started in 1996. After that, this insect became a popular food item for people; farms for its rearing appeared in 2005. In 2011, in the south 120 farmers produced 43 tons of palm weevil larvae (Hanboonsong et al. 2013).

Photo 1.10-1. Thailand is one of the few countries of the world having developed a sector of rearing of insects to use for food. Now, more than 20,000 insect farming enterprises are incorporated in the country. 194 species of insects are used for food in this country. Photo shows most popular consumed insects in Hua Hin town.
Photo credit: I. Arzamastcev (Pacific Geographical Institute, Vladivostok, Russia), 20 April 2017

Table 1.10-1. Insects most commonly marketed and consumed in Thailand (Hanboonsong et al. 2013).

Common name	Scientific name	Seasonal occurrence	Wholesale price/kg (THB), fresh
Bombay locust	*Patanga succincta*	August–October	220–250
Oriental migratory locust	*Locusta migratoria manilensis*	June–July	220–250
Domestic house cricket	*Acheta domesticus*	All year (from farmed sources)	80–100
Common/field cricket	*Gryllus bimaculatus*	All year (from farmed and harvested sources)	100–120
Common/field cricket	*Teloegryllus testaceus*	All year (from farmed and harvested sources)	100–120
Mole cricket	*Gryllotalpa africana*	May–July	150
Short-tailed cricket	*Brachytrupes portentosus*	October–November	120
Giant water bug	*Lethocerus indicus*	July–October	10 (male); 8 (female)
Predaceous diving beetle	*Cybister limbatus*	July–October	120–140
Water scavenger beetle	*Hydrous cavistanum*	July–October	120–140
Bamboo borer	*Omphisa fuscidenttalis*	Aug–Nov	300
Silkworm pupae	*Bombyx mori*	All year (from farmed sources)	120
Scarab beetle	*Holotrichia* spp.	May–August	150
Red ant/weaver ant	*Oecophylla smaragdina*	March–May	300
Palm weevil larvae	*Rhynchophorus ferrugineus*	All year (from farmed sources)	250–300

US$1.00 = THB30.00 approximately (March 2013).

A key example of this kind of business is provided by Haocheng Mealworms, Inc., from another Asian country, **China**. While raising insects in Thailand is exclusively oriented to the domestic market and designed to feed people, this company produces insects as fodder for animals and its products are designed, to a great extent, for foreign consumers. It was established in 2002, is based in Xiangtan, Hunan province, and specializes in rearing of larvae of mealworms (*Tenebrio molitor*), superworms (*Zophobas morio*), and fly maggots (prepupae) (Veldkamp et al. 2012). Accordingly, its products are presented by these larvae in the most diverse forms (living, dried, frozen, conserved, pulverized).

The company exports 200 tons of dried mealworm per year to the United States, United Kingdom, Germany, France, Denmark, Japan, Australia, Korea, South Africa. Mealworms and superworms can be used as additives to fodder of pets, including birds, dogs, cats, frogs, turtles, scorpions, and goldfish. According to company data, mealworms can also be used as human food: incorporated into bread, flour, instant noodles, pastry, biscuits, candy, and condiments and used directly in dishes on the dining table (http://www.hcmealworm.com/).

Within Europe, the **Netherlands** has the most businesses of this kind, with 18 companies producing insects. The volume of production varies from hundreds of kilograms per month to 2,000 kg per month. The **main species** included among the reared insects are the yellow mealworm (*Tenebrio molitor*), lesser mealworm (*Alphitobius diaperinus*), superworm (*Zophobas morio*), house cricket (*Acheta domesticus*), banded cricket (*Gryllus sigillatus*), field cricket (*Gryllus campestris*), migratory locusts (*Locusta mirgatoria*), fruit flies *Drosophila* spp., greater waxmoths (*Galleria mellonella, Pachnoda marginata*), Turkestan cockroach (*Shelfordella tartara*), and the Dubia cockroach (*Blaptica dubia*). The insects are produced mainly for **pet stores** and **zoos**; however, some companies grow insects as feed for livestock and fish, and three companies—Kreca, Meertens, and vande Ven—produce insects for **human consumption** (Hanboonsong et al. 2013).

Kreca was established in 1978. It breeds and rears 12+ species of insects. Among them are crickets (*Gryllus sigillatus* and *Acheta domesticus*); mealworms and superworms (*Zophobas morio, Alphitobius diaperinus*, and *Tenebrio molitor*); locusts (*Locusta migratoria*), Turkestan cockroach (*Shelfordella tartara*), fruit flies, and waxmoths.

At first, the company has reared the insects only for the animal feed industry, but since 2007 the subdivision Kreca Ento-Food BV was organized to produce insects for **human consumption**. The company offers insects either as blanched and frozen, as blanched and freeze-dried, or as a freeze-dried flour product (http://www.kreca.com).

In 2014, the Dutch feed company Coppens and Dutch **insect producer Protix Biosystems** signed an agreement to include insect ingredients in the forage fodder. It provided for the inclusion of 200 tons of fat and 300 tons of protein of the larvae of the black soldier fly (*Hermetia illucens*) into 15,000 tons of compound feed. Protix produces 2.5 to 3 tons of insects per week (Why are insects not allowed in animal feed? 2014).

An example of an **African company** developing industrial-scale insect farming is AgriProtein (Republic of South Africa), a company that aims to satisfy the growing demand for animal feed. The species used by the company is the **common house fly** (*Musca domestica*). The production process starts with rearing stock flies in sterile cages, each holding more than 750,000 flies. For rearing, different kinds of waste, including human waste (feces), abattoir blood, and food waste are used. The single female fly can lay up to 1,000 eggs over a 7-day period and, from them, the larvae emerge. The larvae are collected before they turn into pupae, and then dried, milled, and packed. The end target of company is 100 tons of larvae per day (van Huis et al. 2013).

A number of companies producing fodders and foodstuffs from insects exist also in **North America**. For example, **EnviroFlight** (Yellow Springs, Ohio State, USA) uses the co-product from breweries, ethanol production, and preconsumer food waste as a feedstock for black soldier fly larvae (*Hermetia illucens*). The frass—waste created by insects after digesting the feedstock—becomes a high-protein, low-fat feed for omnivorous species, such as tilapia, freshwater prawns, and catfish. The material is also beneficial as a protein source for poultry, swine, and cattle (http://www.enviroflight.net/). The larvae themselves are cooked or dried and added to the fodders for predatory fish such as rainbow trout, perch, bass, and bluegill (van Huis et al. 2013).

The other American companies include Big Cricket Farms (Youngstown, Ohio) (http://bigcricketfarms.com/), All Things Bugs, LLC (Athens, Georgia) (http://allthingsbugs.com/), Chapul (Salt Lake City, Utah) (www.chapul.com/); Cricket Flours LLC (Portland, Oregon) (http://www.cricketflours.com/about/), raising crickets exclusively for human consumption; and Hotlix (Pismo Beach, California), producing confectionery products from insects (http://hotlix.com/candy/index.php?route=common/home), etc.

Examples of **Canadian companies** include Next Millennium Farms (Campbellford, Ontario) (http://entomofarms.com/), which produces 60 tons of crickets annually (http://www.fastcompany.com/3037716/inside-the-edible-insect-industrial-complex), and Enterra Feed (Langley, British Columbia) (http://www.enterrafeed.com/), which produces the animal feedstuffs using black soldier fly larvae (Zanolli 2014).

Some **restaurants** have insect dishes on their menus. In Vancouver, British Columbia, there are two restaurants (Vij's Restaurant and Rangoli Restaurant) that offer cricket dishes (https://en.wikipedia.org/wiki/Entomophagy).

1.10.3 Manufacturing of entomological equipment

Entomological equipment includes specialized instruments, appliances, and devices for the collection, treatment, preparation, and storage of entomological material (i.e., insects). It can be subdivided into the following **categories** (https://ru.wikipedia.org/wiki/Энтомологическое_оборудование): (1) equipment for collection and temporary storage of caught insects; (2) equipment for treatment and preparing of the entomological material; (3) equipment for storage of insects.

The **first category** includes different insect nets (e.g., aerial insect net, sweep net, for catching insects sitting on the grassland vegetation by "mowing"; aquatic net, tree bark net, for collection of insects from tree bark); beating sheets (applied for catching insects, knocking them from trees and bushes); aspirator (device for catching small insects) (Photo 1.10-2); different traps and other devices for catching insects (funnels, pitfall traps, bottle traps, malaise traps, flight interception traps, and other passive types of insect traps); killing jars (special devices for killing of insects); killing agents such as ether, chloroform, and ethyl acetate; butterfly bags; and glassine envelopes (https://en.wikipedia.org/wiki/Insect_collecting).

Photo 1.10-2. Entomological equipment includes specialized instruments, appliances, and devices for the collection, treatment, preparation, and storage of entomological material (i.e., insects). It can be subdivided into the following categories: (1) equipment for collection and temporary storage of caught insects; (2) equipment for treatment and preparing of the entomological material; (3) equipment for storage of insects. Use a powered aspirator to collect mosquitoes in a northern Thailand jungle is shown.
Photo credit: By U.S. Air Force photo/Capt. Tim Davis (USAF Photographic Archives (Image permalink) [Public domain], via Wikimedia Commons

Photo 1.10-3. Equipment for treatment and preparing of the entomological material includes spreading boards, tools for preparation, tweezers, scalpels, entomological pins, mounting boards, pergamine setting strip, curator's blocks, pinning blocks, mounting glue, and magnifiers. Insect mounting equipment is shown. Curved insect forceps, fine forceps, entomological pins, minuten, precut cards, plastazote, glue brush you can see.
Photo credit: By Notafly (Own work) [CC BY-SA 3.0 (http://creativecommons.org/licenses/by-sa/3.0)], via Wikimedia Commons, 28 April 2008

The **second category** includes spreading boards (devices for spreading winged insects to be mounted for display in entomological collections), tools for preparation, tweezers, scalpels, entomological pins (special needles, predominantly, steel, varnished, with brassy rounded head, applied to mount the insects), mounting boards (fragments of heavy paper, cardboard, or polyvinyl chloride for pasting on of beetles and other insects), pergamine setting strip (designed for spreading of wings of insects), curator's blocks (for convenient organization of the entomological instrument on the table), pinning blocks (used for aligning of the height of labels' arrangement or specimen on the entomological pin), mounting glue, and magnifiers (Photo 1.10-3).

Entomological collecting boxes, display cases, insect storage systems (cabinets, drawers, and pinning trays), and drawers exemplify the **third category**. Descriptions of all types of entomological equipment can be found on these websites: www.entosphinx.cz/, www.insectnet.eu/, http://www.watdon. co.uk/acatalog/entomological-equipment.html. Comprehensive insight into the diversity of kinds of entomological equipment can be obtained from the catalogue of products of one of the companies-producers (BioQuipBooksCatalog 2012).

Some of this equipment is used only by professional entomologists; the rest of the equipment is used by nonprofessional entomologists; the number of nonprofessionals greatly exceeds the number of professionals. It is difficult to determine the approximate **number of people** in the world who are potential purchasers of this equipment. Entomologists in the world number in the thousands, and more than 3,500 entomologists act only in the field of the insect systematics (Zamotajlov et al. 2015). The Entomological Society of America has nearly 7,000 members (http://www.entsoc.org/resources/faq).

The number of members of the Russian Entomological Society in 2012 reached about 2,000; the Entomological Societies of France and Ukraine had 600 members (2007) and 400 members, respectively. On the one hand, not all entomologists need entomological equipment, but on the other hand, not many users of this equipment are members of any formal entomological society.

Nevertheless, a wide range of people need entomological equipment; consequently, a **number of firms** are striving to satisfy this demand. A list of several of them is given in Table 1.10-2.

Naturally, companies producing entomological equipment differ in terms of their assortment of output products, scope of specialization, character of mutual relations with purchasers, and so on. BioQuip Products, Inc., is an example of a company with a wide **assortment** of entomological equipment. Ward's

Table 1.10-2. Some producers of entomological accessories.

Name	Location	Website
BioQuip Products, Inc.	Rancho Dominguez, CA, USA	http://www.bioquip.com/
Ward's Natural Science Establishment	Rochester, NY, USA	http://www.wardsci.com/
Ento Sphinx s.r.o.	Pardubice, Czech Republic	www.entosphinx.cz/
Rose Entomology	Benson, AZ, USA	http://www.roseentomology.com/
Paradox company	Cracow, Poland	http://www.insectnet.eu/contact.php
Australian Entomological Supplies Pty. Ltd.	Bangalow, N.S.W., Australia	http://www.entosupplies.com.au/equipment/field
B&S Entomological Services	Armagh, N. Ireland, UK	http://www.entomology.org.uk/
Watkins and Doncaster	Leominster, UK	http://www.watdon.co.uk/acatalog/entomological-equipment.html
TWC of America, LLC	Adell, WI, USA	http://www.entoproducts.com
Entomopraxis	Barcelona, Spain	www.entomopraxis.com
BiolinxLabsystems Pvt. Ltd.	Mumbai, Maharashtra, India	http://www.indiamart.com/biolinx-labsystems/search.html?ss=entomological%20equipment
H.L. Scientific Industries	Ambala, Haryana, India	http://www.hlscientific.in/entomologicals-equipments.html
Industrial ND Commercial Services	Nacharam, Hyderabad, India	http://www.indiamart.com/industrial-commercial-services/entamology-equipments.html
Labcare Scientific	Singanallur, Coimbatore, India	http://www.labcarescientific.in/entomology-product.html

Natural Science Establishment, along with all of the Indian manufacturers of the entomological segment, occupies only a small part of the equipment line. Most manufacturers distribute their products with the assistance of a dealer; however, Rose Entomology, recently closed, was a firm that dealt directly with purchasers and used no distributors or dealers.

1.10.4 Providing hobbyists and collectors with insects

There is entomological equipment that is designed mainly for hobbyists and collectors; however, we will be dealing with businesses related to providing collectors with insects, often species that are missing from their own collections. For example, a new collector can begin his or her collection with the acquisition of a full set of insects. Services of this kind are in demand; therefore, companies aiming to meet this demand exist.

The **first** of these companies, **Staudinger**, was established by the famous German entomologist Otto Staudinger (1830–1900). It was engaged in the sale of butterflies and beetles all over the world through the beginning of World War II. Its services were used by the **greatest collectors** of the past: Grand Duke Nikolay Romanov, establishing the national collection of the Russian insects, and Baron Rothschild, possessing the world's largest private collection. Earlier entomologists and simple collectors got material for their collections either independently, going on expeditions, or exchanging with other collectors. By the late 19th century, these early collectors could purchase the required insects, according to the Staudinger catalogue containing more than 1,000 items (https://en.wikipedia.org/wiki/Otto_Staudinger).

In 2001, the annual **turnover of trade** in collectors' insects was estimated at $200 to $300 million (Khodorych 2002). There has been a **reduction** in volumes of insect sales and number of insect collectors (Huis van et al. 2013); however, 3 to 4 million collectors all over the world provide a steady demand (main buyers are collectors in Czech Republic, Germany, France, United States, and Japan). In contrast to professional entomologists and insect collectors, most of the emerging collectors' money is spent on big, brightly colored tropical insects to be souvenirs or interior decorating items (Khodorych 2002). Some **firms** currently selling insects are provided in Table 1.10-3.

Insects of outstanding commercial interest belong, for the most part, to a few butterfly (Photo 1.10-4) and beetle families. Most insects are still collected in the **wild**; however, recently, the **rearing** of insects is on the rise. Farm-raised insect varieties include common butterfly species pupae and beetles. Japan and Taiwan have strong beetle-farming communities (mainly, stag beetles Lucanidae, flower chafers Cetoniidae, and rhinoceros beetles Dynastidae). These countries also features industrial-scale production of insect-rearing materials, several insect shops in larger cities, and a number of magazines devoted to the topic of beetle-growing.

More than 100 companies are now involved in **raising insects** in tropical countries (e.g., Malaysia, Indonesia, Madagascar, Costa Rica, Philippines, Papua New Guinea). For scientists and professional collectors representing only a small part of 3 to 4 million, the artificially reared insects are not of great value.

Table 1.10-3. Some providers of insects for collections.

Name	Location	Web-site
Butterflies and Things	North Olmsted, OH, USA	https://www.butterfliesandthings.com/
Insect-Sale	Chia Yi, Taiwan	http://www.insect-sale.com/
ALD Entomologie	Paris, France	http://boutique.ald-entomologie.fr/
Insects4Sale	Saint James, MO, USA	http://www.insects4sale.com/
The Bugmaniac	Makassar South Sulawesi, Indonesia	http://www.thebugmaniac.com/
God of Insects	Hastings-on-Hudson, NY, USA	http://www.godofinsects.com/
Insect Frame UK	London, UK	https://www.insectframe.co.uk/
Amazonian Butterflies E.I.R.L.	Satipo, Junin, Peru	http://www.amazonianbutterflies.com/
Asiahouse24	Neumünster, Germany	http://www.asiahouse24.de
Australian Insect Sales	Queensland, Australia	http://www.ozinsects.com/
Butterfly Workx	Dunnellon, FL, USA	http://www.butterflyworkx.com
Collector's Secret	Paris, France	http://www.collector-secret.com/index.php/top-auctions/
Exotic-Insects	Bali, Indonesia	http://www.exotic-insects.com
Insect-Collections.ru	Moscow oblast, Russia	http://insect-collections.ru/
Insectopia.ru	Moscow, Russia	http://www.insectopia.ru/
Hundred Ways	Sergiyev Posad, Moscow Oblast, Russia	http://hundredways.ru/babochki-i-nasekomye-v-ramkah.html
Bugdesign	Ukraine, Kiev	http://www.bugdesign.com.ua/
Entomodesign	Voronezh, Russia	http://www.entomolog.ru/
Giradys Indonesia	Malang, East Java, Indonesia	http://www.giradis-insect.com
The Insect Collector	Manching, Bavaria, Germany	http://www.theinsectcollector.com
Insect-Supply.com	Voronezh, Russia	http://www.insect-supply.com
Insect-Supply.eu	Jaworzno, Poland	http://www.insect-supply.eu
Insect-Trade.eu	Prague, Czech Republic	http://www.insect-trade.eu
Interinsects	Guetersloh, Germany	http://www.interinsects.com
Kingdom of Beetle Taiwan	Taipei, Taiwan	http://screw-wholesale.myweb.hinet.net/
Nature et Passion	Paris, France	http://www.nature-et-passion.com
Thorne's Insect Shoppe LTD	London, Ontario, Canada	http://www.thornesinsects.com

Photo 1.10-4. Insects of outstanding commercial interest belong, for the most part, to a few butterfly and beetle families. Dried butterflies selling in shop in Jakarta, Indonesia.
Photo credit: S.M. Govorushko, 10 of December 2015

As a rule, scientists and professional collectors are interested in the little-studied species or species needed for the pursuance of a specific scientific study. Many collectors want to have in their collections those butterflies that may be inappropriate for mass rearing because of their low-key visual appearance or because creating conditions for the species to thrive caused difficulties. For example, the Central Asia high-altitude Apollo butterfly *Parnassius autocrator* lives only in the mountains of Afghanistan and Tajikistan at heights of up to 3,000 m (https://en.wikipedia.org/wiki/Parnassius_autocrator).

Small and unattractive, but interesting in the context of scientific interests of the entomologist, this butterfly is of immeasurably higher value for the scientist than the bright and readily available one. The simplest way to add these insects to a collection is to catch them in the wild (Khodorych 2002).

The range of **prices** for insects is very great. If the prices per pupa are low—from a few cents to a few dollars—the prices of the adult insects can be very high. The price for insect depends mainly on the rarity of the species and the **size** of the individual. For example, the specimen of one of the world's largest beetles, long-horned beetle *Macrodontia cervicornis*, 162 mm long, is sold for €1,600, while an individual 170 mm long is estimated at €4,000. Similarly, the relict longhorn beetle *Callipogon relictus* from North Korea, 83 mm long, costs €430, while an individual longer by only 1 mm is sold for €460 (http://www.insects.de/anzeigen.php?art=CCE).

A message in the mass media described a living male stag beetle from the Dorcus family (Lucanidae), 80 mm long, distinguished by a large set of fearsome pincer-like jaws and a gleaming shell, that sold for 10 million yen ($90,000) (http://www.thaibugs.com/?page_id=690). However, this beetle was designed for use as a breeding stud for future rearing of this species (Izhevsky et al. 2014).

During the early 1990s, after the fall of the Soviet Union, there grew a **buoyant demand** for butterflies and beetles living in the former Soviet territory. European collectors reaped the benefits of a branching network of catching and reselling to the Western world of these formerly inaccessible insects. At that time, the proceeds of catchers were estimated at $5,000 to $12,000 per season (April through October), while the total volume of oversea sales reached between $1 million and $1.25 million a year (Weisman et al. 1999).

However, recently the wave of former Soviet Union insect availability has receded, and several sharp drops in prices and volume of sales has occurred because of market oversupply. In some cases, the sales and catching of insects were illegal. Some species in Europe are forbidden for sale and catching by the international agreements (Convention on International Trade in Endangered Species of Wild Fauna and Flora, the Bern Convention (Convention on the Conservation of European Wildlife and Natural Habitats), or by European Union directives. Total export volume of insects from the former Soviet Union territory is not more than $300,000 to $400,000 a year; several most lucky dealers can earn up to $30,000 a year (Khodorych 2002).

Additional problems in the trade of insects involve the difficulties of moving the living specimens **across borders**; there is a risk whenever a new species is unintentionally introduced to a new region. The practical way to solve this issue is through the support of new government agencies such as the Insect Farming and Trading Agency, in Papua New Guinea (van Huis et al. 2013).

The development of collections, predominantly private, occurs also during entomological **exhibitions** and **trade fairs**, the number of which reaches approximately 100 European countries annually. For example, every fall since 1939, the Amateur Entomologists' Society has held a fair in Sunbury-on-Thames, Surrey, England (https://www.amentsoc.org/events/exhibitions.html). Twice a year (in March and October), 2-day entomological fairs are held in Prague.

Other fairs include the Juvisy Insect Fair in Paris, which has existed for more than 100 years, and the International Insect Exchange Fair, in Frankfurt am Main, where thousands of interested people meet (http://entomolog.narod.ru/news_ent_yarmarki.html). Insect exhibitions and trade fairs are also held in Berlin, Dessau, Braunschweig, Ingolstadt, Vienna, Munich, Modena, and Lyon.

For entomologists wishing to build their own collections, travel agents organize trips focused on the gathering of specimens for collections. For example, Amazon Insects, in Peru, offers a wide variety of trips in the Amazon Basin (to different places, with various traveling times, for specific insect species, ranging from one person to large groups). The company provides the expertise, experience, and logistics, at the local, national, and international level, to be the collector's in-country agent available to provide professional assistance in any situation (http://www.amazoninsect.com/collecting_trips.html). Similar services are also offered in Ukraine (http://entomology.com.ua/travels/?route=list-region) and Taiwan (http://classifieds.insectnet.com/hsx/classifieds.hsx?session_key=&search_and_display_db_button= on&db_id=32262&query=retrieval).

1.10.5 Butterfly houses

Insects were first raised for scientific purposes or for biological control of agricultural and forest pests. In the late 20th century, with the increase in number of collectors, **farms for rearing of insects**, predominantly butterflies, came to be established in a number of countries. The best known of them are located in the Penang Island, Malaysia; in Bulolo, Papua New Guinea; and in Satipo, Peru (Kaabak and Sochivko 2012). In the 21st century, rearing of insects is increasingly realized for aesthetic and commercial purposes. In this paragraph, businesses related mainly to the demonstration of insects will be discussed.

Enterprises established for the extensive demonstration of insects have different names, the **most common** of which is "butterfly houses"; these are insectaria with special rooms designed for keeping, rearing, and breeding of insects. Insectaria allow scientists to observe the peculiarities of insect life, while raising insects (e.g., entomophages) to be used in biological control of pests or for laboratory experiments. Over time, showrooms began to appear in the insectaria, and a certain part of their budgets was given to demonstration of insects. Later on, specialized enterprises, predominantly targeting the commercial demonstration of insects and called butterfly houses, were organized. However, many butterfly houses have also pursued the combination of scientific, educational, environment-oriented, and other purposes together with the commercial constituent.

The exposition of the tropical butterflies on the Guernsey Island (English Channel) can be considered the **first** such enterprise. In the mid-1970s, the commercial tomato industry went into decline and vast areas of plastic houses came to be empty. Mr. David Lowe, who lived on the Island, came up with the idea of using the plastic houses to raise tropical plants and butterflies imported from Southeast Asia, to improve the tourist business. In 1977, the first butterfly house was opened on the island; it was a big success, quickly compensating for the financial expenses (Irwin and Kampmeier 2003).

The **high profitability** of this business resulted in mass construction of butterfly houses in the United Kingdom; several of them were established by Mr. Clive Farrell. By the late 1980s, there were more than 50 butterfly houses in the United Kingdom (http://www.butterfly-insect.com/whoweare.php).

Nowadays, butterfly houses are a well-developed segment of the tourist industry (Photo 1.10-5). Butterfly houses and butterfly gardens all over the globe are visited by 40 million people each year. It is

Photo 1.10-5. The photo shows the entrance in first Indian Butterfly park in Bennargatta national park. It was opened 25 November 2006, occupies 7.5 acres (30,000 m²) and houses a butterfly conservatory, a museum, and an audiovisual room. The butterfly conservatory is 10,000 sq ft (1,000 m²). Within the conservatory, the environment has been designed to support over twenty species of butterfly. It is a humid tropical climate, with an artificial waterfall and appropriate flora to attract butterflies. The conservatory leads to a second and third dome, which house a museum containing dioramas and exhibits of carefully preserved butterflies.
Photo credit: Rameshng at Malayalam Wikipedia [CC BY-SA 3.0 (http://creativecommons.org/licenses/by-sa/3.0)], via Wikimedia Commons

estimated that the **global turnover** of the butterfly house industry is on the order of $100 million (Boppre and Vane-Wright 2012). A list of some of these butterfly houses is presented in Table 1.10-4.

The butterfly gardens can exist on their own, or they can be parts of zoological gardens. For example, the **largest butterfly garden** in Europe is in the Emmen Zoo (the Netherlands); it opened in 1985 and now receives visitors throughout the year. To maintain the abundance of living butterflies, at least 70 butterflies should come into being from pupae. The pupae are delivered to Emmen Zoo **by mail** from farms breeding the butterflies in Costa Rico, Malaysia, and the Philippines.

In the garden, the constant temperature of 25°C is maintained. At the lower temperature, the butterflies are unable to fly and fall into transfixion. They feel comfortable at a higher temperature (32°C), but that temperature is too hot for visitors. Therefore, if the air temperature increases to 28°C, the cooling system automatically switches on. At night, the temperature is lowered to 20°C. The humidity in the garden is always high, as in the tropics (http://www.danaida.ru/obsh/sad.htm).

Such gardens are oriented to the **general public**; therefore, the mainly large and eye-catching butterfly species are presented in them (van Huis et al. 2013). In 1997, the first International Conference of Butterfly Exhibitions was held; these conferences have been held regularly ever since. In 2001, the International Association of Butterfly Exhibitions was established. In 2008, the IABE officially welcomed butterfly farmers and suppliers to the organization and it was renamed the International Association of Butterfly Exhibitors and Suppliers (http://iabes.org/).

1.10.6 Other kinds of business

A variety of other kinds of business exist related to insects. Because insects are frequently used in **scientific research** (see Sections 1.2.1 to 1.2.3), there is a constant demand for their production. This demand is mainly met by the insectaria (rooms for rearing and keeping of insects). The **Montreal Insectarium** (Quebec, Canada) established in 1990 by the Canadian entomologist Georges Brossard is the greatest in the world (https://en.wikipedia.org/wiki/Montreal_Insectarium). Modern insectaria present the large buildings with several rooms and complex accessories for regulation of lighting and temperature.

Another kind of business is the rearing of insects for **pollination** of plants. It is known that the decline of populations of pollinator insects has been observed recently; consequently, a number of insect

Table 1.10-4. Some butterfly houses (Prepared by author, based on numerous Internet sources).

Name	Location	Web-site
	Asia	
The Kindom of Butterfly	South Sulawesi, Indonesia	http://travee.co/1098
Kuala Lumpur Butterfly Park	Kuala Lumpur, Malaysia	http://klbutterflypark.com/
Butterfly Conservatory of Goa	Ponda, Goa, India	http://www.bcogoa.org/
Butterfly garden	Wetland Park, Hong Kong	http://www.wetlandpark.gov.hk/en/exhibition/reserve_butterfly.asp
Butterfly Park	Bangalore, India	bannerghattabiologicalpark.org/butterfly_park.html
Simply Butterflies Conservation Center	Bilar, Bohol, Philippines	simplybutterfliesproject.com/conservation_center.html
Phuket Butterfly Garden & Insect World	Phuket, Thailand	www.phuket.com/attractions/phuket-butterfly-garden.htm
Malacca Butterfly and Reptile Sanctuary	Ayer Keroh, Malacca, Malaysia	www.butterflyreptile.com/
Penang Butterfly Farm	Penang, Malaysia	www.penang.ws/penang-attractions/butterfly-farm.htm
Butterfly Park Bangladesh	Chittagong, Bangladesh	www.bangladeshbutterflypark.com.bd/
Singapore Zoological Gardens	Singapore	www.zoo.com.sg/exhibits-zones/fragile-forest.html
Butterfly safari Park	Thenmala, Kerala, India	http://www.kerala-traveller.com/2008/02/kerala-butterfly-safari-park-thenmala.html
Butterfly garden, Kadoorie Farm and Botanic Garden	Hong Kong	www.kfbg.org/
Butterfly house, Ocean Park	Hong Kong	http://www.oceanpark.com.hk/en/park-experience/attractions
	Europe	
AlarisSchmetterlingspark	Sassnitz, Rügen, Germany	http://www.alaris-schmetterlingspark.de/
The North Somerset Butterfly House	Weston-super-Mare, United Kingdom	www.nsbutterflyhouse.com/
Schmetterlinghaus in the Imperial Garden	Vienna, Austria	http://schmetterlinghaus.at/en/
Bordano Butterfly House	Friuli Venezia Giulia, Italy	www.turismofvg.it/Museums/House-of-butterflies-of-Bordano
Bornholm Butterfly Park	Bornholm, Denmark	bornholm.info/en/bornholm-butterfly-park
Collodi Butterfly House	Tuscany, Italy	www.holidaytuscany.it/eng/visit-tuscany/butterfly-house.php
Butterfly Arc	Veneto, Italy	http://www.micromegamondo.com/casadellefarfalle/index.php?centro=2
Butterfly Botania	Joensuu, Finland	www.botanicgardens.fi/
Butterfly Garden	Grevenmacher, Luxembourg	www.papillons.lu/en/
Mariposario de Benalmádena	Andalusia, Spain	www.mariposariodebenalmadena.com/
Magic of Life Butterfly House	Aberystwyth, Ceredigion, Wales, UK	www.magicoflife.org/
London Butterfly House	London, UK	https://en.wikipedia.org/wiki/London_Butterfly_House

Table 1.10-4 contd. ...

...Table 1.10-4 contd.

Name	Location	Web-site
	Europe	
Golders Hill Park	London, UK	https://www.tripadvisor.ru/ShowUserReviews-g186338-d211599-r318858741-Golders_Hill_Park-London_England.html
Oasidelle Farfalle di Milano	Lombardia, Italy	http://www.milanoperibambini.it/in-citta/musei/364-loasi-delle-farfalle.html
Passiflorahoeve	Vlindertuin Harskamp, Netherlands	http://www.passiflorahoeve.nl/home-en/
Seaforde Gardens and Butterfly House	Seaforde, County Down, Northern Ireland	www.seafordegardens.com/
Stratford Butterfly Farm	Stratford-upon-Avon, Warwickshire, UK	www.butterflyfarm.co.uk/
Straffan Butterfly Farm	Straffan, Republic of Ireland	www.straffanbutterflyfarm.com/
VlinderkasDiergaardeBlijdorp	Rotterdam, Netherlands	https://www.diergaardeblijdorp.nl/en/
VlindertuinDierenparkEmmen	Emmen, Netherlands	http://www.passiflorahoeve.nl/home-en/
Wrocławskie Zoo	Wrocław, Poland	http://www.zoo.wroclaw.pl/index.php?lang=2
Butterfly House	Praid, Romania	www.lepkehaz.ro/en/the-butterfly-house-in-praid
	North America	
The Butterfly Farm	Aruba,Sint Maarten, Saint Martin	www.thebutterflyfarm.com/
La Marquesa Forest Park Butterfly House	Guaynabo, Puerto Rico	www.puertoricodaytrips.com/la-marquesa-forest-park/
Mariposario Chapultepec	Mexico City, Mexico	http://www.mariposario.org.mx/
Montreal Insectarium	Montreal, Quebec, Canada	http://espacepourlavie.ca/en/insectarium
Cambridge Butterfly Conservatory	Cambridge, Ontario, Canada	www.cambridgebutterfly.com/
Niagara Parks Butterfly Conservatory	Niagara Falls, Ontario, Canada	http://www.niagaraparks.com/niagara-falls-attractions/butterfly-conservatory.html
Victoria Butterfly Gardens	Brentwood Bay, British Columbia, Canada	butterflygardens.com/
Newfoundland Insectarium	Deer Lake, Newfoundland and Labrador, Canada	http://nlinsectarium.com/
Butterflies & Blooms	Baldwin, Georgina, Ontario, Canada	www.bluewillowgarden.com/conservatory.html
Academy of Natural Sciences of Drexel University	Philadelphia, USA	www.ansp.org/
Ashland Nature Center Butterfly House	Hockessin, USA	www.delawarenaturesociety.org/AshlandDirections
Aveda Butterfly Garden	Minnesota Zoo, Apple Valley, USA	http://mnzoo.org/
Bear Mountain Butterfly Sanctuary	Jim Thorpe, USA	bearmountainbutterflies.com/
BernieceGrewcock Butterfly and Insect Pavilion	Omaha, USA	www.omahazoo.com/exhibits/butterfly-insect-pavilion/
Bioworks Butterfly Garden	Tampa, USA	http://www.mosi.org/
Blooming Butterfly Garden	St. Paul, USA	http://www.comozooconservatory.org/
Brookside Gardens	Wheaton, USA	http://www.montgomeryparks.org/brookside/

Table 1.10-4 contd. ...

...Table 1.10-4 contd.

Name	Location	Web-site
	North America	
Butterflies and Plants: Partners in Evolution	Washington, USA	butterflies.si.edu/
Butterfly Conservatory and Insect Zoo	Manhattan, USA	https://www.k-state.edu/butterfly/
The Butterfly Farm	Birmingham, Alabama, USA	https://www.birminghamzoo.com/
Butterfly Garden	The Bronx, USA	https://bronxzoo.com/exhibits/butterfly-garden
Butterfly Garden	Boston, USA	www.mos.org/exhibits/butterfly-garden
Butterfly House	Detroit, Michigan, USA	detroitzoo.org/animal-habitat/butterfly-garden/
Butterfly House	Dayton, Ohio, USA	www.metroparks.org/butterfly-house/
Butterfly House	Mackinac Island, USA	www.originalbutterflyhouse.com/
Butterfly House	East Lansing, USA	http://www.michigan.org/property/michigan-state-university-horticultural-gardens-and-butterfly-house/
Butterfly House	Chesterfield, USA	http://www.missouribotanicalgarden.org/visit/family-of-attractions/butterfly-house.aspx
Butterfly House San Antonio Zoo	San Antonio, USA	www.sazoo-aq.org/attractions/butterflies/
Butterfly House	Whitehouse, USA	http://www.wheelerfarms.com/
Butterfly Landing	Boston, USA	http://www.zoonewengland.org/franklin-park-zoo/exhibits/butterfly-landing
Butterfly Magic	Tucson, USA	https://www.tucsonbotanical.org/
The Butterfly Palace and Rainforest Adventure	Branson, Missouri, USA	www.thebutterflypalace.com/
Butterfly Pavilion	Los Angeles, USA	http://www.nhm.org/site/explore-exhibits/special-exhibits/butterfly-pavilion
Butterfly Pavilion	Westminster, Colorado, USA	https://www.butterflies.org/
The Butterfly Place	Westford, USA	https://butterflyplace-ma.com/
Butterfly Wing	Reiman Gardens, Ames, USA	http://www.reimangardens.com/
Butterfly Wonderland	Scottsdale, USA	butterflywonderland.com/
Butterfly World	Vallejo, USA	https://www.sixflags.com/discoverykingdom/attractions/butterfly-habitat
California Academy of Sciences	San Francisco, USA	http://www.calacademy.org/
Cecil B. Day Butterfly Center	Pine Mountain, USA	http://www.callawaygardens.com/things-to-do/attractions/day-butterfly-center
Cockrell Butterfly Center & Insect Zoo	Houston, USA	www.hmns.org/cockrell-butterfly-center/
Florida Museum of Natural History Butterfly Rainforest	Gainesville, USA	http://www.flmnh.ufl.edu/
Frederik Meijer Gardens	Grand Rapids Township, USA	www.meijergardens.org/
Key West Butterfly and Nature Conservatory	Key West, USA	www.keywestbutterfly.com/
Living Conservatory exhibit	Raleigh, North Carolina, USA	naturalsciences.org/living-collections/living-conservatory
Magic Wings Butterfly Conservatory	South Deerfield, USA	www.magicwings.com/

Table 1.10-4 contd. ...

...Table 1.10-4 contd.

Name	Location	Web-site
	North America	
Magic Wings Butterfly House	Durham, USA	http://www.carrboro.com/butterflyhouse.html
Marshall Butterfly Pavilion	Phoenix, USA	https://www.dbg.org/butterfly-exhibit
Monsanto Insectarium	St. Louis, USA	https://www.stlzoo.org/visit/thingstoseeanddo/discoverycorner/insectarium/
The Montgomery Zoo	Montgomery, AL, USA	www.montgomeryzoo.com/
Orange County Native Butterfly House	Newport Beach, USA	encenter.org/visit-us/butterfly-house/
Panhandle Butterfly House	Navarre, USA	panhandlebutterflyhouse.org/
Puelicher Butterfly Wing	Milwaukee, USA	https://www.mpm.edu/plan-visit/exhibitions/permanent-exhibits/first-floor-exhibits/puelicher-butterfly-wing
Sertoma Butterfly House	Sioux Falls, USA	https://butterflyhousemarinecove.org/
Tradewinds Park Butterfly World	Coconut Creek, USA	https://butterflyworld.com/
Tropical Butterfly House	Seattle, USA	https://www.pacificsciencecenter.org/exhibits/tropical-butterfly-house/
Frederick Meijer Gardens	Grand Rapids, USA	www.meijergardens.org/
	Oceania	
Otago Museum Discovery World Tropical Forest	Dunedin, New Zealand	http://otagomuseum.nz
Australian Butterfly Sanctuary	Kuranda, Australia	australianbutterflies.com/
Coffs Harbour Butterfly House	Coffs Harbour, Australia	www.butterflyhouse.com.au/
Melbourne Zoo butterfly enclosure	Melbourne, Australia	www.zoo.org.au/melbourne/highlights/butterfly-house
	South America	
Águias da Serra Borboletário	São Paulo, Brazil	http://borboletario.aguiasdaserra.com.br/
El Mariposario	Quindío, Colombia	www.jardinbotanicoquindio.org/
San Jose Butterfly Farm	San José, Costa Rica	http://www.butterflyfarm.co.cr/en/the-butterfly-farm/butterfly-farm-costa-rica-tour-information/index.html
MariposarioMindo	Mindo, Ecuador	www.mariposasdemindo.com/
Cali Zoo	Cali, Colombia	www.zoologicodecali.com.co/
Pilpintuwasi Butterfly Farm	Iquitos, Peru	http://www.amazonanimalorphanage.org/

species are now cultured primarily to ensure sufficient pollinators in the field, orchard, or greenhouse at bloom time.

For example, bumblebees are bred, reared, and packaged for sale to growers for pollinating vegetable crops (particularly tomatoes) grown under greenhouse and plastic tunnel conditions (Irwin and Kampmeier 2003). Bumblebees for pollination accounted for $10 million in 1991 in Europe (van Lenteren et al. 1997).

In some countries, keeping living insects as **pets** is now a popular hobby, especially in Japan and Taiwan, where many stores sell living insects—for the most part, beetles. For example, in Osaka there is a big store called Insect Shop Global, with numerous of species for sale as pets (http://www.globalosaka.co.jp). It is not surprising for Japan that the greatest focus is placed on beetles of the genus *Dorcus*, stag beetles family. A large variety of rather rare *Dynastes* species (for example, *D. myashitai*, *D. maya*, *D. hyllusmoroni*, and others) occurs. The beetles are sold individually or in pairs, or as larvae (http://www.beetlebreeding.ch/beetle-shops-of-japan-insect-shop-global-in-osaka/).

Because the insects are frequently used as **bait in fishing**, some companies focus on that demand. A great supply of artificial bait is available on the market (a list of the UK stores selling the artificial bait can be found here: http://www.fish-uk.com/online_fishing_tackle_and_bait_s.htm). Many stores also sell real insects (Photo 1.10-6), but in this case the choice is much narrower, predominantly maggots and casters of blue bottle fly *Calliphora vomitoria* that are alive (http://www.ebay.co.uk/sch/Live-Fishing-Bait/179963/bn_10030959/i.html), dried (http://stores.ebay.co.uk/Norfolk-Pet-Supplies/Fishing-Bait-/_i.html?_fsub=1276540016), or frozen (http://www.wormsdirectuk.co.uk/acatalog/maggots.html).

Insects—specifically, butterflies—are frequently used in **wedding** ceremonies and other celebrations. Worldwide, newly married couples and their relatives find the release of butterflies to be a romantic and memorable tradition. The release of the beautiful, multicolored butterflies has a variety of names ("bunch of butterflies", "show of butterflies", "salute of butterflies") and it can be carried out at different times during the wedding (http://babochkindom.ru/articles/286623; http://vsebabochki.ru/salut/).

Many companies render **services of this kind**: Butterfly Release Company, Orlando, Florida (http://www.butterflyreleasecompany.com/); Swallowtail Farms, Inc., El Dorado Hills, California (http://www.swallowtailfarms.com/); Amazing Butterflies Tamarac, Florida (www.amazingbutterflies.com/wedding.htm). Painted lady butterflies (*Vanessa cardui*) and monarchs (*Danaus plexippus*) are the **most commonly used** for such releases (Irwin and Kampmeier 2003). In North America alone, 11 million individual butterflies have been released per year during weddings and other ceremonies (Boppre and Vane-Wright 2012).

Photo 1.10-6. The insects are in great demand for fishermen. For the satisfaction of this demand, many companies sell insects for their use in fishing. In places of mass recreation, there are even special vending machines that sell live bait installed. A live bait vending machine, Brighton Recreation Area, Michigan is shown.
Photo credit: By Dwight Burdette (Own work) [CC BY 3.0 (http://creativecommons.org/licenses/by/3.0)], via Wikimedia Commons

1.11

Insects in Culture

1.11.1 Insects in fiction

Many books about insects have been written for children of different ages. These **include** (1) factual books about the lives of insects (biology, habits, etc.) showing varying degrees of detail and (2) fiction (novels, stories, tales). In literature, insects usually are given human qualities: abilities to think and to talk, and different other human characteristics.

Fewer books related to insects have been written for **adults**. Several of them are listed in Table 1.11-1.

Table 1.11-1. Some literary works related to insects and arachnids.[1]

Kind of insect	Name of novel, story, tale, play	Author	Year of publication	Note
Ant	Ants	Manuel Komroff	1932	Novel
Ant	Three Thousand Red Ants	Lawrence Ferlinghetti	1970	Play
Ant	Leiningen Versus the Ants	Carl Stephenson	1937	Short story
Bee	The Swarm	Arthur Herzog	1974	Novel
Beetle	The Metamorphosis	Franz Kafka	1915	Novel
Beetle	The Gold-Bug	Edgar Allan Poe	1843	Short story
Bug	The Golden Bug	Orson Scott Card	2007	Short story
Bug	Bugs	Theodore Roszak	1981	Novel
Bug	The Bug Wars	Robert Asprin	1979	Novel
Bumble-Bee	Bumble-Bees	Walt Whitman	1882	?
Butterfly	Morpho Eugenia	A.S. Byatt (Dame Antonia Susan Duffy)	1992	Novel
Butterfly	Papillon	Henri Charrière	1969	Novel
Butterfly	The Collector	John Fowles	1963	Novel
Butterfly	The Aurelian	Vladimir Nabokov	1930	Short story
Butterfly	Father's Butterflies	Vladimir Nabokov	2000	Addendum to the novel The Gift
Butterfly	Doctor Faustus	Thomas Mann	1947	Novel

Table 1.11-1 contd. ...

...Table 1.11-1 contd.

Kind of insect	Name of novel, story, tale, play	Author	Year of publication	Note
Butterfly (Death's-head Hawkmoth)	The Sphinx	Edgar Allan Poe	1846	Short story
Caterpillar	Alice in Wonderland	Lewis Carroll	1865	Fairytale
Caterpillar	Caterpillars	Edward Benson	?	Short story
Cricket	The Cricket Boy	P'u Sung-ling	1952	Short story
Flea, beetle, moth, spider, wasp	Harry Potter and the Chamber of Secrets	Joanne Rowling	1998	Novel
Fly	The Fly	Katherine Mansfield	1922	Short story
Fly	Fly	Richard Chopping	?	Tale
Fly	The Flies	Jean-Paul Sartre	1944	Play
Hornet	An Egyptian Hornet	Algernon Blackwood	1915	Short story
Louse, fly, bee, ant, beetle, spider, moth, butterfly, wasp	Harry Potter and the Order of the Phoenix	Joanne Rowling	2003	Novel
Sphinx moth	The Silence of the Lambs	Thomas Harris	1988	Novel
Spider	Kiss of the Spider Woman	Manuel Puig	1976	Novel
Spider, beetle, fly, bug, flea	Harry Potter and the Goblet of Fire	Joanne Rowling	2000	Novel
Spider, beetle, grub, maggot, moth, wasp, bedbug, fly, mosquito, cockroach	Harry Potter and the Half-Blood Prince	Joanne Rowling	2005	Novel
Spider, beetle, midge, moth, dragonfly, fly, cockroach	Harry Potter and the Deathly Hallows	Joanne Rowling	2007	Novel
Spider, moth, beetle, fly	Harry Potter and the Philosopher's Stone	Joanne Rowling	1997	Novel
Spider, moth, blowfly, beetle, caterpillar, dragonfly	Harry Potter and the Prisoner of Azkaban	Joanne Rowling	1999	Novel
Wasp	Wasps	Aristophanes	422 BC	Play
Wasp	The Wasp and the Canary	Artie Knapp	2006	Short story

[1] http://www.waitingman.net/BugLit.html; https://en.wikipedia.org/wiki/Cultural_entomology; http://bijlmakers.com/proverbs-and-quotes/harry-potter/, with additions by the author.

Different species of insects are represented in this list quite evenly; however, butterflies, spiders, and flies are more **prevalent**.

1.11.2 Insects in poetry

The number of pieces of poetry related to insects is **as numerous** than that of prose works. To a greater degree, they are also found in children's poetry, although they occur quite often in adult poetry as well. In Table 1.11-2 some poetic works related to insects are shown.

It may be noted that butterflies, bees, and flies are mentioned **most frequently**. They are followed by spiders.

Table 1.11-2. Some pieces of poetry related to insects and arachnids.[1]

Kind of insect	Name of poem or rhyme	Author	Note
Ant	The Metaphysics of Ants	Tripp Howell	Poem
Ant	Ant	Kenneth Rexroth	Poem
Ant	1 ant, 2 ants, 3 ants	Douglas Florian	Poem
Ant, fly, bee	Every Insect	Dorothy Aldis	Poem
Bee	Five Busy Honey Bees		Rhyme
Bee	Limerick: There Was an Old Man in a Tree	Edward Lear	Poem
Bee	What Does the Bee Do?	Christina Rossetti	Poem
Bee	Lying in Grass	Hermann Hesse	Poem
Bee	The Honey Bee	Ted Hughes	Poem
Bee	Ode to Bees	Pablo Neruda	Poem
Bee	The First Days	James Wright	Poem
Bee	One Little Honey bee by My Window Flew	Elsa Gorham Baker	Poem
Bee	Humble Bee	Ralph Waldo Emerson	Poem
Bee, gnat	To Autumn	John Keats	Poem
Beetle	Forgiven	Alan Milne	Poem
Beetle	The Beetle (Chrząszcz) (Photo 1.11-1)	Jan Brzechwa	Poem
Beetle	Beetle	Sue Chehrenegar	Poem
Beetle	The Beetle Crawls on Leaf and Twig	Sue Chehrenegar	Poem
Beetles, millipedes	Naturalist	Victor Fet	Long poem, 2007
Bug	Bugs	Meish Goldish	Poem
Bug	Little Bugs	Rhoda Bacmeister	Poem
Bug	Bugs Come Out in Spring	Sue Chehrenegar	Poem
Bug	Bugs	Rhoda Bacmeister	Poem
Butterfly	From Cocoon Forth a Butterfly	Emily Dickinson	Poem
Butterfly	Two Butterflies Went Out at Noon	Emily Dickinson	Poem
Butterfly	Two Tortoiseshell Butterflies	Ted Hughes	Poem
Butterfly	Life and Death of a Butterfly	Pablo Neruda	Poem
Butterfly	Obsidian Butterfly	Octavio Paz	Poem
Butterfly	To a Butterfly	William Wordsworth	Poem
Butterfly	Another Song of a Fool	William Butler Yeats	Poem
Butterfly	Poem about Butterfly	Shashikant Nishant Sharma	Poem
Butterfly	Fly Fly Butterfly	Aileen Fisher	Poem
Butterfly	Fuzzy, Wuzzy, Creepy Crawly Caterpillar Funny	Lillian Vabada	Poem
Butterfly	Creepy Crawly Caterpillar	Mary Dawson	Poem
Butterfly, beetle, and bee	Calico Pie	Edward Lear	Poem
Butterfly, grasshopper	The Butterfly's Ball, and the Grasshopper's Feast	William Roscoe	Poem
Caterpillar	Caterpillar	Christina Rossetti	Poem
Caterpillar	The Caterpillar	Guillaume Apollinaire	Poem
Cranefly	A Cranefly in September	Ted Hughes	Poem

Table 1.11-2 contd. ...

...Table 1.11-2 contd.

Kind of insect	Name of poem or rhyme	Author	Note
Cockroach	Tarakanishche	Korney Chukovsky	Poem
Cricket	The Little Mute Boy	Federico Garcia Lorca	Poem
Cricket	Song for September	Robert Fitzgerald	Poem
Cricket	Stars and Cricket	Octavio Paz	Poem
Cricket	What Makes the Crickets 'Crick' All Night	Helen Wing	Poem
Dragonfly	The Dragon-fly	Alfred Lord Tennyson	Poem
Dragonfly, scorpion, ant	Epic of Gilgamesh	Unknown author	Epic poem
Firefly	Firefly	Meish Goldish	Poem
Firefly	I Like to Chase the Fireflies	Grace Wilson Coplen	Poem
Flea	The Flea	Guillaume Apollinaire	Poem
Flea	Fleas Interest Me So Much	Pablo Neruda	Poem
Flea	The Flea	John Donne	Poem
Fly	The Daddy Long-Legs and the Fly	Edward Lear	Poem
Fly	The Fly	Guillaume Apollinaire	Poem
Fly	The Fly	William Blake	Poem, 1794
Fly	Las Moscas (The Flies)	Antonio Machado	Poem
Fly	40,000 Flies	Charles Bukowski	Poem
Fly	A Considerable Speck	Robert Frost	Poem
Fly	The Fly	Karl Shapiro	Poem
Fly	A Fly Flies By, Quicker Than the Eye!		Poem
Fly, mosquito, spider, beetles, other insects	Mukha-Tsokotukha	Korney Chukovsky	Poem
Fly, spider	Caught	Elise Partridge	Poem, 2002
Gnat	A Swarm of Gnats	Hermann Hesse	Poem
Gnat	Gnat-Psalm	Ted Hughes	Poem, 1967
Grasshopper	The Grasshopper	Guillaume Apollinaire	Poem
Grasshopper	In the Likeness of a Grasshopper	Ted Hughes	Poem
Grasshopper, cricket	On the Grasshopper and the Cricket	John Keats	Poem
Grasshopper, ant	Strekoza i Muravei (A Grasshopper and the Ant)	Ivan Krylov	Fable
Insect	Death Is Like the Insect	Emily Dickinson	Poem
Insect	Iliad	Homer	Epic poem
Insect	To-Day, This Insect	Dylan Thomas	Poem
Ladybird	Ladybird Beetle	Sue Chehrenegar	Poem
Ladybug	Ladybug, Ladybug Fly Away Home		Rhyme, 1744
Ladybug	Lady Bug Wears	Dee Lillegard	Poem
Ladybug	Ladybugs All Dressed in Red	Maria Fleming	Poem
Ladybug	Little Red Bug, Oh So Cute	Susan M. Paprocki	Poem
Locust	Himeros	Robert Fitzgerald	Poem
Locust	The Locust Swarm	Hsu Chao	Poem

Table 1.11-2 contd. ...

...Table 1.11-2 contd.

Kind of insect	Name of poem or rhyme	Author	Note
Louse	To a Louse, On Seeing One on a Lady's Bonnet at Church	Robert Burns	Poem, 1786
Mosquito	Pesky Mosquito	Meish Goldish	Poem
Scorpion	The Scorpion	Hilaire Belloc	Poem
Scorpion	Na skorpiona (On a Scorpion)	Victor Fet	Poem, 2002
Snake fly	Rhaphidioptera	Bruce Noll	Poem
Spider	The Tragedy at Mini-Beastie Hall	Josie Whitehead	Poem
Spider	Little Miss Muffet		Rhyme
Spider	The Spider As an Artist	Emily Dickinson	Poem
Spider	Arachne	Thom Gunn	Poem
Spider	The Spider	Cesar Vallejo	Poem
Spider	I Have a Little Spider	Janet Bruno	Poem
Spider	The Spider Weaves a Sticky Web	Amy Goldman Koss	Poem
Spider, fly	The Spider and the Fly	Mary Howitt	Poem
Wasp	Wasps	Anne Ruddick	Poem

[1] http://static1.squarespace.com/static/54dae2b6e4b011d9d8f95d9e/t/54f90909e4b0d13f30b89ea7/1425606921096/BugPoems.pdf; http://www.dltk-kids.com/crafts/insects/songs.htm; https://www.flickr.com/photos/squatbetty/6244800448/; http://www.cals.ncsu.edu/course/ent425/text01/impact2.html; with additions by the author.

Photo 1.11-1. Chrząszcz (beetle, chafer) by Jan Brzechwa is a poem famous for being one of the hardest-to-pronounce texts in Polish literature, and may cause problems even for adult, native Polish speakers. "In the town of Szczebrzeszyn a beetle buzzes in the reed...". Monument to the poem's hero is shown.
Photo credit: By MaKa~commonswiki (Own work) [GFDL (http://www.gnu.org/copyleft/fdl.html), CC-BY-SA-3.0 (http://creativecommons.org/licenses/by-sa/3.0/) or CC BY 2.5 (http://creativecommons.org/licenses/by/2.5)], via Wikimedia Commons, July 2006

1.11.3 Insects in films

The insects are widely present both in feature and animated films. Below, some films in which insects are not only mentioned but play the key or relatively **significant roles** are presented (Table 1.11-3).

Table 1.11-3. Some animation and theatrical films related to insects.[1]

Kind of insect	Name of movie	Creator	Year
Ant	Antz	Eric Darnell, Tim Johnson	1998
Ant	The Naked Jungle	Byron Haskin	1954
Ant	Them!	David Weisbart	1954
Ant	Empire of the Ants	Bert I. Gordon	1977
Ant	Glass Trap	Ed Raymond (Fred Olen Ray)	2005
Ant	The Bone Snatcher	Jason Wulfsohn	2003
Ant	The Hive	David Yarovesky	2014
Ant	Legion of Fire: Killer Ants!	George Manasse	1998
Ant, moth	Bug Bites: An Ant's Life	Michael Schelp	1998
Bee	Bee Movie	Simon J. Smith, Steve Hickner	2007
Bee	Deadly Invasion: The Killer Bee Nightmare	Rockne S. O'Bannon	1995
Bee	The Swarm	Irwin Allen	1978
Beetle	Caved In: Prehistoric Terror	Richard Pepin	2006
Beetle	Metamorphosis	Chris Swanton	2012
Bug	Bug	William Friedkin	2006
"Bug" (fictional alien insects)	Starship Troopers	Paul Verhoeven	1997
Butterfly	The Blue Butterfly	Léa Pool	2004
Butterfly	Papillon	Franklin J. Schaffner	1973
Butterfly	The Collector	William Wyler	1965
Caterpillar	Alice in Wonderland	Walt Disney	1950
Cockroach (fictional alien)	Men in Black	Barry Sonnenfeld	1997
Cockroach	War of the Coprophages (twelfth episode of the third season of the science fiction television series The X-Files)	Kim Manners	1996
Cockroach	Joe's Apartment	John Payson	1996
Cockroach	Mimic	Guillermo del Toro	1997
Cockroach, cricket	Pinocchio	Ben Sharpsteen, Hamilton Luske	1940
Cockroach	The Hellstrom Chronicle	Ed Spiegel, Walon Green	1971
Fly	The Fly	David Cronenberg	1986
Fly	Lord of the Flies	Harry Hook	1990
Housefly	The Amityville Horror	Andrew Douglas	2005
Lady beetle	Ladybug Ladybug	Frank Perry	1963
Locust	The Good Earth	Sidney Franklin, Victor Fleming, Gustav Machatý	1937

Table 1.11-3 contd. ...

...Table 1.11-3 contd.

Kind of insect	Name of movie	Creator	Year
Mosquito	Skeeter	Clark Brandon	1993
Moth	Doctor Doolittle	Betty Thomas	1998
Numerous kinds of insects	Maya the Bee	Alexs Stadermann	2014
Numerous kinds of insects	A Bug's Life	John Lasseter	1998
Numerous kinds of insects	Minuscule	Hélène Giraud, Thomas Szabo	2006
Numerous kinds of insects	Angels & Insects	Philip Haas	1995
Sphinx moth	The Silence of the Lambs	Jonathan Demme	1991
Stag beetle	Lucanus Cervus	Ladislas Starevich	1910
Tsetse fly	Congo Crossing	Joseph Pevney	1956

[1] Compiled by the author with the use of numerous Internet sites.

It may be noted that the insects in the animated films are depicted mainly as **anthropomorphic characters**; that is, they are endowed with human qualities (abilities to think and to talk; have their own personality). As for feature films, insects appear in monster movies as **horrid creatures**. Among individual species of insects, note that ants, cockroaches, and butterflies are most commonly shown.

1.11.4 Insects in music

Insects are widely represented in music also, with varying degrees of prominence. In some musical compositions, the insects are only mentioned (for example, the musical *Kiss of the Spider Woman* by John Kander, or the song "Bullet with Butterfly Wings" by Billy Corgan), while in others they are minor characters (for example, the song "Butterfly Kisses" by Bob Carlisle and Randy Thomas). However, in many of them, the insects occupy a **central position** (for example, "Flight of the Bumblebee" by Nikolai Rimsky-Korsakov or "Song of the Flea" by Modest Mussorgsky). A number of musical compositions related to insects are presented in Table 1.11-4.

Even with a superficial glance at this table, one can see that the **butterflies** comprise about 30% of the list of insects. Grasshoppers, spiders, and bees follow far behind.

1.11.5 Insects in painting

From the time of appearance of human beings, they were always surrounded by insects. The interactions between people and insects was also embodied in pictorial art. Initially, these were the **cave paintings** of primitive people, such as the cave paintings in Cuevas de la Araña en Bicorp in Valencia, Spain, which date back approximately 10,000 years. They show a person ascending vines to harvest honey with bees flying around (https://en.wikipedia.org/wiki/Cultural_entomology). Egyptian frescoes dating back 3,500 years featuring butterflies are also known. In Table 1.11-5, more recent examples of pictorial art related to insects are presented.

Not all of the artists listed in Table 1.11-5 were just painters. For example, Johann Jacob Scheuchzer was also a researcher (medicine, mathematics, paleontology) and depicted the insects from a scientific point of view. Similarly, numerous pictures of insects made by Leonardo da Vinci are related mainly to examination of their flight.

Numerous **publications** have been devoted to insects in pictorial art; for example, articles by Dicke (2004), Kritsky and Mader (2010), Kritsky et al. (2013). Dicke (2004) visited 180 art museums in 13 countries (Western Europe and the United States) and found 3,045 works of art related to insects. For the most part, these were paintings, although a few were sculptures.

The paintings dated from the 13th century to modern times. Dicke found that the images of insects were more often found on **Dutch still lifes** of the 17th and 18th centuries, as well as in surrealistic pictures and works performed in the modernist style (Jugendstil works). Naturally, the "commitments" of different

Table 1.11-4. Some musical compositions related to insects and arachnids.[1]

Kind of insect	Composition	Composer	Year	Note
Ant	The Ants Go Marching	Robert D. Singleton	1990	Song
Bedbug	The Bedbug	Dmitri Shostakovich	1929	Music to the comedy of same name by Mayakovsky
Bee	A Sleepin' Bee	Harold Arlen	1954	Song
Bee	The Bee and the Rose (L'abeille et la rose)	Georges Bizet	1854	Song
Blue-tail fly	Jimmy Crack Corn (Blue Tail Fly)	Folk song	1846	Song
Bug	Insects, Bugs and Other Species	Dora Cojocaru	2013	Arr. for three clarinets
Bumblebee	Flight of the Bumblebee	Nikolai Rimsky-Korsakov	1899–1900	Part of opera The Tale of Tsar Saltan
Bumblebee	Baby Bumblebee	?	?	Song
Butterfly	Someone Saved My Life Tonight	Elton John/Bernie Taupin	1975	Song
Butterfly	The Butterfly Song (If I Were a Butterfly)	Brian M. Howard		Song
Butterfly	Butterfly Kisses	Bob Carlisle/Randy Thomas	1997	Song
Butterfly	Butterfly	Edvard Hagerup Grieg	1868	Lyric piece of Piano concerto in A minor
Butterfly	The Butterfly Lovers	He Zhanhao/Chen Gang	1959	Violin Concerto
Butterfly	Papillons	R. Schumann	1829–31	A musical portrayal of events in Jean Paul's novel Die Flegeljahre
Butterfly	Butterfly Wings	Fryderyk Chopin	1836	Etude for Piano No. 21, Op. 25,9 (Butterfly Etude)
Butterfly	Butterfly	Edvard Grieg	1903	Op. 43, No. 1
Butterfly	Les papillons	François Couperin	1713	Instrumental compositions for harpsichord
Butterfly	The Butterfly	Sergei Prokofiev	1921	Song
Butterfly	Madama Butterfly	Giacomo Puccini	1904	An opera in three acts
Butterfly	Seven Butterflies (Sept papillons)	Kaija Saariaho	2000	Piece
Butterfly	The Butterfly Lovers	Chen Gang, He Zhan-Hao	1959	Violin concerto
Butterfly	Bullet with Butterfly Wings	Billy Corgan	1995	Song
Cicada, ant	The Cicada and the Ant	Henri Sauguet (Pierre-Henri Poupard)	1941	Song
Cockroach	The Cockroach (La Cucaracha)	Spanish folk corrido	ca. 1910	Folk song

Table 1.11-4 contd. ...

...Table 1.11-4 contd.

Kind of insect	Composition	Composer	Year	Note
Cricket	Terrapin Station (Grateful Dead)	Jerry Garcia/Robert Hunter	1977	Song
Cricket	The Cricket (Le grillon)	Maurice Ravel	1906	Song from cycle Natural Histories (Histoires naturelles)
Flea	Song of the Flea	Modest Mussorgsky	1879	Mephistopheles' Song in Auerbach's Cellar
Flea	Song of the Flea	Ludwig van Beethoven	1809	No. 3, Aus Goethes Faust (Mephistos Flohlied)
Fly	From the Diary of a Fly	Béla Bartók	1926	For piano (Mikrokosmos Vol. 6/142)
Fly	The Fly	William Blake/Benjamin Britten	1965	Song
Dragonfly	Die Libelle (ja) (The Dragonfly)	Josef Strauss	1866	Polka Mazurka, Op. 204
Grasshopper	El Grillo	Josquin des Prez	1505	Song
Grasshopper	Parade of the Grasshoppers	Sergei Prokofiev	1935	Twelve easy pieces; extracts arranged for orchestra in the suite Summer Day, Op. 65 bis
Grasshopper, ant	The Grasshopper and the Ant	Jacques Offenbach		
Grasshopper, dragonfly	Grasshoppers and Dragonflies	Op. 87; composed by Sergei Prokofiev	1945	Ballet Cinderella (Act I, No. 14)
Honey bee	Tupelo Honey	Van Morrison	1971	Song
Insect	The Night's Music	Béla Bartók	1915	Piano suite
Insect	Insect Symphony	Kalevi Aho	1988	Symphony
Insect	Chorus of Flowers and Insects	Pyotr Tchaikovsky	1869–70	Choral composition
Insect	Piano Sonata No. 10	Alexander Scriabin	1913	Piano sonata
Locust	Moses in Egypt (Mosè in Egitto)	Gioachino Rossini/Andrea Leone Tottola	1818	Three-act opera
Maggots	Shattered (Rolling Stones)	Mick Jagger/Keith Richards	1978	Song
Spider	The Spider's Feast (Le festin de l'araignée)	Albert Roussel	1912	Music for ballet-pantomime
Spider	Kiss of the Spider Woman	John Kander	1990	Musical based on Manuel Puig novel
Spider	Spider's Feast (Le festin de l'araignée)	Albert Roussel	1912	Ballet-pantomime
Termite	Lay My Love	Brian Eno/JohnCale	1990	Song
Wasp	The Wasps	Ralph Vaughan Williams	1909	Incidental music

[1] Compiled by the author with the use of numerous internet sites.

Table 1.11-5. Some paintings related to insects and arachnids.[1]

Insects portrayed	Artist	Years of life	Examples
Ant	Maurits Cornelis Escher	1898–1972	*Ant*
Bee, butterfly, dragonfly	Clive Uptton	1911–2006	*"i" for Insects*
Beetle	Jan Fabre	1958–	*Heaven of Delight*[2]
Beetle	Jacob van Hulsdonck	1582–1647	*Nectarines and Grapes in a Basket on a Table, with Plums, Oranges, a Butterfly and a Beetle*
Beetle, bee, butterfly, grasshopper, locust	Nick Taggart	?	*Grasshopper*, 2004; *Bee*, 2003; *Beetle*, 2002; *Butterfly*, 2004
Beetle, bug, butterfly	Andy Warhol	1928–1987	*Happy Bug Day*
Beetle, butterfly	Jan Brueghel, the Younger	1601–78	*A Crown Imperial, a Peony and Other Flowers in a Wooden Tub with Butterflies and Beetle*
Beetle, butterfly, dragonfly, grasshopper	Ambrosius Bosschaert, the Elder	1573–1621	*Still-Life of Flowers*
Beetle, butterfly, fly	Jan van Huysum	1682–1749	*Flowers, Fruits and Insects*
Beetle, butterfly, fly, cockchafer, dragonfly, spider	Jan van Kessel, I	1626–79	*Insects and Fruit; A Dragon-fly, Two Moths, a Spider and Some Beetles, with Wild Strawberries* (Photo 1.11-2)
Beetle, butterfly, ladybug	Abraham Mignon	1640–79	*Still Life with Fruits, Foliage and Insects*
Beetle, butterfly, spider	Jan Mortel	1652–1719	*Still Life with Fruit and Butterflies; Still Life with Fruit and Insects*
Beetle, ladybug, butterfly, ant	Charley Harper	1943–	*Insect Diversity*
Butterfly	Henri Matisse	1869–1954	*Boy with a Butterfly Net*
Butterfly	Carl Spitzweg	1808–85	*The Butterfly Hunter*
Butterfly	Wilhelm von Kaulbach	1805–74	*Butterflies*
Butterfly	Vasily Dmitrievich Polenov	1844–1927	*The Dragonfly*
Butterfly	Leopold Carl Muller	1834–92	*Egyptian Girl with a Butterfly*
Butterfly	Ambrosius Bosschaert II	1609–45	*Peaches, Grapes, a Pear and White Currants in a Wan-li Kraak Porcelain Dish, with Shells, a Lizard and a Butterfly on a Ledge*
Butterfly	Adriaen Coorte	1665–1707	*Three Peaches on a Stone Ledge, with a Red Admiral Butterfly*
Butterfly, dragonfly	Otto Marseus van Schrieck	1614/1620–78	*Flowers, Insects and Reptiles*, 1673

Table 1.11-5 contd. ...

...Table 1.11-5 contd.

Insects portrayed	Artist	Years of life	Examples
Butterfly, dragonfly, beetle, bee	Rachel Ruysch	1664–1750	*Rose Branch with Beetle and Bee, 1741*
Butterfly, beetle, ant	Johann Jacob Scheuchzer	1672–1733	*Physica Sacra*
Butterfly, beetle, cockchafer bug, bee	Jan Brueghel the Elder	1568–1625	*A Crown Imperial, a Peony and Other Flowers in a Wooden Tub with Butterflies and Beetles*
Butterfly	Van Gogh	1853–90	*Two White Butterflies,1889*
Butterfly, fly, dragonfly, grasshopper	Balthasar van der Ast	1593–1657	*Flowers in a Vase with Shells and Insects, ca. 1628*
Butterfly, moth, beetle, wasp, dragonfly, ant, mite, etc.	Jan Davidsz de Heem	1606–84	*Still Life with Fruit and Butterflies*
Butterfly, stag beetle	Jacques de Gheyn II	1565–1629	*Three Butterflies and Stag-Beetle*
Dragonfly, moth, beetle, stag beetle	Joris (Georg) Hoefnagel	1542–1601	*Dragonfly, Two Moths and Two Beetles*
Fly	Ilya Kabakov	1933–	*Queen Fly*
Fly, ant, bee, butterfly, grasshopper	Salvador Dali	1904–89	*Ant Face*
Grasshopper (katydid), locust, cricket	Charles Ephraim Burchfield	1893–1967	*Insects at Twilight; The Insect Chorus*
Grasshopper, ant, beetle, wasp	Maurits Cornelis Escher	1898–1972	*Grasshopper*
Long-horned beetle, cicada, ant, dragonfly, butterfly, fly, bee	Leonardo da Vinci	1452–1519	*The Long-Horned Beetle and Dragonfly Study*
Stag beetle	Hans Hofmann	1880–1966	*Stag Beetle*
Stag beetle	Giovannino de' Grassi	1350–98	*Creation of the World*
Stag beetle	Albrecht Dürer	1471–1528	*Stag Beetle*
Stag beetle	Stefan Lochner	1410–51	*Altar of the City Patron*
Stag beetle	Giovanna Garzoni	1600–70	*Vase with Flowers, a Peach and a Butterfly*
Stag beetle, butterfly	Peter Binoit	1590–1632	*Apricots in a Woven Basket, Carnations in a Glass Vase, Raspberries in a Wan-Li Bowl, with Red Currants, Other Fruit, and Insects on a Stone Ledge*
Stag beetle, beetle	Georg Flegel	1566–1638	*Food with Beetles and Titmouse; Still Life with Stag Beetle*

[1] Compiled by the author with the use of numerous internet sites.
[2] Was created with 1.6 million Buprestidae beetles.

Photo 1.11-2. The interaction between people and insects was also embodied in pictorial art. The painting entitled "A Dragon-fly, Two Moths, a Spider and Some Beetles, With Wild Strawberries" by Jan van Kessel the Elder (1626–1679) is shown. The insects include Wasp beetle, top left; clouded border moth, top right; migrant hawker dragonfly and cardinal beetle, center left; magpie moth, center right; cockchafer, lower left.
Photo credit: http://www.flickr.com/photos/47071837@N02/8266423381/ by curry15 (Jan van Kessel the Elder [Public domain], via Wikimedia Commons)

painters to insects are very different. Some artists depicted only a single insect, while others depicted over 100 insects in a single work of art. In the picture by Jan van Kessel (Europe—the continents) (1664), for example, Dicke counted **115 insects**.

The situation with the number of pictures devoted to insects is similar. Some painters created only one picture with images of insects, but the artistic legacies of others include more than a hundred such pictures. **Salvador Dali** in particular paid a great deal of attention to insects. Kritsky et al. (2013) listed **59 of his pictures** with images of insects. In fairness it must be said that he did portray biodiversity. In most of his paintings, he preferred to show flies, ants, and grasshoppers. In some, works, bees, butterflies, harvestmen, and damselflies are also present.

Sometimes, the degree of detail of the insect images in pictures by painters is so great that entomologists can easily determine **their species**. For example, in the painting by Jan Davidz de Heem titled *A Glass Vase of Flowers and Cornstalks*, S. Segal identified the following **arthropods:** (1) an orange tip, *Anthocharis cardamines*; (2) a brimstone, *Gonepteryx rhamni*; (3) the yellow underwing, *Noctua pronuba*; (4) a larva of a carabid beetle; (5) a brown ant, *Formica rufa*; (6) a pierid caterpillar; (7) a velvet mite, *Trombidium* sp.; (8) a vapourer caterpillar, *Orgyia antiqua*; (9) a crane fly, *Pachyrhina crocata*; (10) a damselfly, *Coenagrion puella*; (11) a banded longhorn beetle, *Strangalia maculata*; (12) a *Paravespula vulgaris* wasp; (13) an earwig, *Forficula auricularia*; and (14) a digger wasp, *Ammophila sabulosa* (Segal 1990).

Dicke (2004) evaluated the representation of different **insect and arachnid orders in** paintings. By far, Lepidoptera (butterflies, moths, tineids) (1,800) rank first, followed by Diptera (flies, mosquitoes, black flies, etc.) (700), Hymenoptera (bumblebees, wasps, bees, ants) (650), Coleoptera (beetles) (600), Odonata (dragonflies, damselflies) (400), Orthoptera (grasshoppers, crickets, locusts) (250), Arachnida (spiders, scorpions, ticks, mites) (250), Dictyoptera (cockroaches, mantids, termites) (70), and Dermaptera or earwigs (30).

The paintings with diverse insects and arachnids (beetles, grasshoppers, spiders, mosquitos, butterflies, bees, bumblebees, scorpions, and many more) are very commonly encountered. As a rule, an insect shown in a painting is considerably magnified, and many people like that. For the most part, paintings related to insects are colorful and positive.

1.11.6 Insects in sculptures

A sculpture is a three-dimensional work of art. In principle, monuments and statues meet this definition; the boundaries between these concepts are very conditional. Nevertheless, I thought it good to consider these categories in separate paragraphs. In this section, we consider mainly works of art that are kept in museums and private collections.

Materials used to create sculptures are most diverse. In addition to traditional materials (loam, wax, stone, metal, wood, bone, etc.), ceramics, different polymers, glass, papier-mâché, celluloid, Galalith, and other materials are used. Even such nonconventional types of sculpture as textile sculptures (three-dimensional images of fabric) and works of sculpture created of garbage and different wastes have been produced. Their **sizes** are also markedly different (from miniature sizes of a few centimeters to quite large ones measuring several meters).

As a rule, monuments and statues are larger and are displayed out of doors. Table 1.11-6 gives examples of sculptures of insects created by different authors in the 18th through 21st centuries.

Table 1.11-6. Some sculptures related to insects and arachnids.[1]

Describable insects	Name of sculptor	Years of life	Examples
Ant, flea, a house fly, mosquito, Colorado potato beetle, leafhopper	Alfred Keller	1902–55	*Mosquito in Flight* (1937, 60:1 scale); *Myrmica rubra with Aphid* (1944, 100:1 scale)[2]
Ant, grasshopper	Germaine Richier	1904–59	*Mantis; The Spider*
Beetle	Diether Roth	1901–98	*Spice Windows*, 1971
Beetle (screech beetle)	Unknown sculptor; installed in the British Museum, Great Britain	Assumed creation time, period of Ptolemies (305–30 BC)	Sculpture made of greenstone, with length of 152.5 cm and height of 91.5 cm; brought from Constantinople (today's Istanbul), where it arrived in the times of the late Roman Empire when Constantinople was its capital (since AD 330)
Beetle, spider, butterfly, bee, wasp, locust, dragonfly, mantid	Mike Libby	1976–	Sculptures of insects from mechanical or electronic components
Bug, bee	Christy Rupp	1949–	*Bee with Toxic Pollen*, 1999
Butterfly	Antoine-Denis Chaudet	1763–1810	*Amor and Psyche*
Butterfly, beetle, ant, dragonfly, grasshopper	Annemieke Mein	1944–	*Christmas Beetles*, 1981; *Grasshopper Flight*, 1988; *Dance of Mayflies*, 1988[3]
Butterfly, beetle, ant, dragonfly, grasshopper	Anna Collette Hunt	?	*Stirring the Swarm* (ceramic installation), 2011
Cricket	Amy Youngs	1968–	*Cricket Call*, 1998; *Holodeck for House Crickets*
Fly, bee, dragonfly, spider	Tom Friedman	1965–	*Monster Fly*, 2008; *Bee*, 2007
Fly, beetle, bug	Mark Oliver	1966–	*Lamp Fly, Gullet Beetle, Muscle Bug* (series of insects called *Litterbugs*)
Fly, mosquito	Julio Gonzalez	1876–1942	*Still Life II*
Spider	Louise Bourgeois	1911–2010	*Spider* (1994). Steel, glass, water, and ink. 44.76 × 77.95 × 64.76 inches
Spider, dragonfly, grasshopper	Anneli Arms	1935–	*Sphinx Moth*. Wire, foil, polyurethane, acrylics; 45 × 40 × 16 inches

[1] Compiled by the author with the use of numerous internet sites.

[2] Created models of insects for Museum of Natural History in Berlin, Germany, used such materials as papier-mâché, celluloid, and galalith.

[3] Textile sculpture.

1.11.7 Insects in monuments and statues

For our purposes, a monument is a structure that is devoted to any historical event related to insects. Statues of insects appear in various places; they are not placed in these locations incidentally. However, for the most part, their erection was **not related to a real event**; the insects can be symbols of some populated locality, characters of literary works, etc. Examples of monuments and statues related to insects are given in Table 1.11-7.

For all intents and purposes, only **three historical events** have been related to monuments of insects. The Boll Weevil Monument in Enterprise, Alabama, USA, was the **first**. It is considered to be the first monument in the world raised in honor of the boll weevil (*Anthonomus grandis*). The monument consists of a statue of a woman holding a pedestal with a boll weevil perched on top. The boll weevil was indigenous to Mexico, but it appeared in Alabama in 1915. By 1918 farmers were losing whole cotton crops to the beetle. On the brink of ruin, the farmers had to cross over to the other crops; they chose peanuts. The farmers were much surprised that the peanut generated unprecedented profits. After a year, the farmers had recovered all their losses. And three years after, in 1919, they raised a monument to their benefactor—the boll weevil. At first, the beetle itself was absent from the monument; it appeared only in 1949. The monument was created by Luther Baker.

The **next historical event** related to an insect occurred in Australia. Prickly pears (mostly *Opuntia stricta*) were imported into Australia in the 19th century for use as a natural agricultural fence and in an attempt to establish a cochineal dye industry. They quickly became a widespread invasive species, rendering 40,000 km^2 (15,000 square miles) of farmland unproductive.

For a long time, attempts to exterminate the prickly pear were unsuccessful. It was rooted out, burned, and exterminated with chemical pesticides; however, these measures were unsuccessful. In 1925, larvae of the cactus moth (*Cactoblastis cactorum*) were delivered from Argentina and dispersed over the cactus thickets from aircraft. In seven years, the cactus moth had almost completely destroyed the cactus thickets in the region. This case is often cited as an example of successful biological pest control.

A monument in Humboldt County (California, USA) was erected in honor of the **victory over St. John's wort** (*Hypericum perforatum*). It was imported from Europe to America as an ornamental and medicinal plant. In the absence of natural predators, St. John's wort became a pernicious weed in the western United States. In order to eradicate it, the leaf beetle *Chrysolina geminata* was introduced from Europe to California in 1945.

One more event memorialized by the installation of a monument, in Hungary, is the 50th anniversary of the Colorado potato beetle penetration into Europe. This beetle was first discovered in Europe in Leipzig, Germany, in 1876–77. Afterwards, the Colorado potato beetle was carried to the European continent two more times, but it was possible to eliminate it. However, in 1918 during World War I, the beetle was able to establish itself near Bordeaux (France), and then it started to invade other European countries.

The **fifth monument** to an insect—the firefly—was erected in 2004 in the heart of London; admittedly, the beetle is depicted in the company of many animals, including horses, elephants, dogs, mules, pigeons, bears, dolphins, and monkeys. The monument was erected in memory of animals that took part in the wars of the 20th century. The inclusion of the firefly in this group is explained by the fact that British soldiers, in the days of World War I, read maps in its light (Quinn 2011).

Other cases are not related to real events. Some statues are devoted to **fictional characters of poems** (for example, for Tsokotukha Fly from Korney Chukovsky's poem or to a cricket from "The Beetle" poem by the Polish poet Jan Brzechwa. Creation of other statues is related to the fact that the insect is a **symbol** of a populated area (for example, a bee for the town of Medyn in Kaluga Oblast, Russia, or a mosquito for Komarno, Manitoba, Canada). Sometimes, a statue of an insect is installed for **promotional purposes** (for example, an ant with a radio set near the gatehouse of the radio manufacturer Ryazan in Russia, or a statue of a Colorado beetle for advertising a chemical pesticide in Dnipropetrovsk, Ukraine).

These examples do not exhaust the whole list of monuments and statues devoted to insects. The **leading city** with monuments to insects is Seoul, which has five such monuments.

Table 1.11-7. Some monuments and statues related to insects and arachnids.[1]

Describable insect	Location	Notes
Ant	Göteborg, Sweden	1976, Jan Åke Jonsson (1921–92)
Ant	Seoul, South Korea	
Ant	Bologna, Italy	Installed in a city park
Ant	Dubai, UAE	
Ant	Belokurikha, Altai Krai, Russia	
Ant	St. Petersburg, Russia	Sculpture *Ants—Workers*
Ant	Ryazan, Russia	2013, installed beside the Ryazan radio manufacturer; a radio set hangs behind the ant's back
Ant	Phoenix, Arizona, USA	
Ant, dragonfly	Abakan, Republic of Khakassia, Russia	Sculpture *Dragonfly and Ant*
Bee	Medyn, Kaluga Oblast, Russia	2008, A. Kalachinsky
Bee	Moscow, Russia	2005, Sergei Soshnikov
Bee	Ternopil, Ukraine	2010, Dmitry Pilipyak
Bee	Minsk, Belarus	2005, Pavel Voinitsky
Bee	Gifu, Japan	1960
Bee	Hidalgo, Texas, USA	1992, Jerome Vettrus (The World's Largest Killer Bee statue). The First Colony of Africanized honey bees or Killer Bees entered the United States in Hidalgo on Oct. 15, 1990. The first attacks in May of 1991 were in Brownsville about 50 miles east. The first fatality in 1993 was in Harlingen about 35 miles east
Bee	Tisdale, Saskatchewan, Canada	1993
Bee	Swarzędz, Poland	Near Bee-Keeping Museum
Bee	Ufa, Russia	2012, floral sculpture *Burzyan Bee*
Bee	Ust-Kamenogorsk, Kazakhstan	Project designer Tatyana Agafonova
Bee	Vuchkovo village, Zakarpattia Oblast, Ukraine	
Bee	Gomel, Belarus	
Beetle (boll weevil)	Enterprise, Alabama, USA	1919; the monument stands more than 13 feet (4.0 m) tall; creator was Luther Baker (however, the boll weevil itself appeared in it only in 1949) (Photo 1.11-3)
Beetle (Colorado potato beetle)	Berdyansk, Zaporizhia Oblast, Ukraine	
Beetle (Colorado potato beetle)	Dnipropetrovsk, Ukraine	2003; statue's height is 2 m
Beetle (Colorado potato beetle)	Hungary	Erected in 1998 in honor of 50-year jubilee of the "arrival" of the beetle in Europe from America
Beetle (Hercules beetle)	Kuala Lumpur, Malaysia	
Beetle (Hercules beetle)	San Diego, California, USA	
Beetle (Hercules beetle)	Grand Canyon, USA	Northwest of Phoenix, Arizona
Beetle (Hercules beetle)	St. Louis, Missouri, USA	

Table 1.11-7 contd. ...

...Table 1.11-7 contd.

Describable insect	Location	Notes
Beetle (Hercules beetle)	Singapore	
Beetle (Hercules beetle)	Bristol, United Kingdom	
Beetle (mantis)	Seoul, South Korea	
Beetle (rhinoceros beetle)	Currumbin, Australia	
Beetle (rhinoceros beetle)	Tokyo, Japan	
Beetle (rhinoceros beetle)	Kiev, Ukraine	
Beetle (screech beetle)	Donetsk, Ukraine	
Beetle (screech beetle)	Luxor, Egypt	Statue *Sacred Scarab* created in the times of Pharaoh Amenhotep III government (1388—1353/1351 BC) (Photo 1.11-4)
Beetle (screech beetle)	Perm, Russia	
Beetle (screech beetle)	Tanzania	
Beetle (screech beetle)	Ashborough, North Carolina, USA	
Beetle (screech beetle)	Anyang, South Korea	
Beetle (screech beetle)	London, United Kingdom	
Beetle (screech beetle)	Brussels, Belgium	
Beetle (stag beetle)	Hong Kong, China	
Beetle (stag beetle)	Sheffield, United Kingdom	
Beetle (stag beetle)	Sumatra Island, Indonesia	
Beetle (stag beetle)	Bristol, United Kingdom	
Beetle (stag beetle)	Yokohama, Japan	
Beetle (stag beetle)	Japan	
Beetle	Penang Island, Malaysia	
Beetle	Australia	
Beetle	Calgary, Canada	*Beetle of Millennia*
Beetle	Tenerife Island, Canary Islands, Spain	
Beetle	Washington, USA	
Beetle	Novgorod, Russia	Composition *A Beetle at Grass* is situated near the Novgorod drama theater beside the Volkhov River
Beetle	Seoul, South Korea	
Butterfly (South American cactus moth)	Dalby, Queensland, Australia	In a park by the Myall Creek, which runs through the town of Dalby
Butterfly	Gomel, Belarus	

Table 1.11-7 contd. ...

...Table 1.11-7 contd.

Describable insect	Location	Notes
Butterfly	Berlin, Germany	Composition *Large Divided Oval: Butterfly*, sculptor Henry Moore, made of bronze by 750th Anniversary of Berlin, weight is 8 t.
Butterfly	Onaway, Michigan	2002, a steel statue of a monarch butterfly hovering over some flowers
Butterfly	St. Louis, Missouri, USA	25,000-pound monarch butterfly fiberglass statue
Butterfly	Kiev, Ukraine	Forged butterfly in shoes and with manicure file near a local beauty salon
Butterfly	Salt Lake City, Utah, USA	*Butterfly Angel*, custom bronze statue
Butterfly	Perm, Russia	2011, Karlis Ile
Cricket	Zamość, Lublin Voivodeship, southeastern Poland	2002, devoted to the piece of poetry character, Jan Brzechwa; The Beetle
Cricket	Te Kuiti, New Zealand	
Dragonfly	Taganrog, Russia	
Dragonfly	Donetsk, Ukraine	
Dragonfly	Salekhard, Yamalo-Nenets Autonomous Okrug, Russia	Rustam Ismagilov
Fly	Querétaro, Mexico	*The dead fly* 700 x 550 x 600 cm; creator, Florentijn Hofman
Fly	Moscow, Russia	
Fly	St. Petersburg, Russia	The fly takes a photo of its natural food—a heap of manure
Fly	Sochi, Russia	A monument to *Buzzy-Wuzzy Busy Fly* (a fly sits in a pumpkin and holds in its tarsus a coin that it found)
Fly	Astana, Kazakhstan	A monument to *Buzzy-Wuzzy Busy Fly* (a great fly walks across a field in search of its coin)
Fly	Donetsk, Ukraine	A monument to *Buzzy-Wuzzy Busy Fly*, forged statue (Buzzy-Wuzzy Busy Fly comes home with its purchase—samovar [fire pot])
Glow-worm	London, United Kingdom	A monument to animals that perished in the war; includes horses, dogs, dolphins, elephants, pigeons, and even glow-worms
Grasshopper	Rethymno, Crete, Greece	
Grasshopper	Greetsiel, Lower Saxony, Germany	
Grasshopper	Vilnius, Lithuania	On a wall of building
Grasshopper	Mexico City, Mexico	
Grasshopper	Salem, Oregon, USA	1988; creator, Wayne Chabre; grasshopper mounted in a head-down position on the exterior brick wall of a building; the brazed copper sculpture measures 5 feet (1.5 m) x 2.5 feet (0.76 m) x 2 feet (0.61 m)
Hercules beetle	Daejeon, South Korea	
Lady beetle	Suburb of Milwaukee, Wisconsin, USA	On a wall of building
Lady beetle	Tokyo, Japan	
Lady beetle	Seoul, South Korea	
Lady beetle	Warszawa, Poland	

Table 1.11-7 contd. ...

...Table 1.11-7 contd.

Describable insect	Location	Notes
Lady beetle	Volgograd, Russia	
Lady beetle	Millau, France	
Locust	Sarreguemines, northeastern France	
Mosquito	Dobrino settlement, Khanty-Mansi Autonomous Okrug	2006
Mosquito	Usinsk, Komi Republic, Russia	2012, metallic statue, 2 m high
Mosquito	Kronstadt, town near St. Petersburg, Russia	
Mosquito	Suwon, South Korea	
Mosquito	Shetland Islands, United Kingdom	
Mosquito	Alaska, USA	
Mosquito	Berdyansk, Zaporizhia Oblast, Ukraine	For the fact that nonbiting midge, more correctly, its bloodworm, restores the health properties of silt in the Sea of Azov
Mosquito	Novosibirsk, Russia	
Mosquito	Komarno, Slovakia	Made of stainless steel and rotates as a weathercock around its axis; wingspread is 460 cm; situated on a stoned postament
Mosquito	Komarno, Manitoba, Canada	Sculpted in 1984, it is made of steel and has a wingspan of 15 feet
Mosquito	Ventspils, Latvia	
Mosquito	Noyabrsk, Yamalo-Nenets Autonomous Okrug, Russia	2006, Valery Chaly; composition made of the written-off metallic parts
Spider	St. Petersburg, Russia	Forged figure on a wall of building
Spider	Bilbao, Spain	
Spider	Ottawa, Canada	1999; bronze-steel female spider up to 9 m high bearing a bag with 26 marble eggs; Louis Bourgeois (1911–2010)
Spider	Havana, Cuba	
Spider	Ann Arbor, Michigan, USA	
Spider	New York, USA	
Spider	Seoul, South Korea	
Spider	Tokyo, Japan	
Tick	Würchwitz village, Saxony-Anhalt, Germany	This monument can be called a monument to *Mite-Earner*. The sauce to a cheese is added just by secretions of the special mites to a box where the cheese is placed until it acquires its peculiar taste and flavor
Water strider	Central-eastern Nevada, USA	1985; Installation of Michael Heizer (1944–); *Water Strider*; compacted earth, concrete, steel (7.2 x 42.7 x 33.5 m)

[1] Compiled by the author with the use of numerous internet sites.

Photo 1.11-3. A monument in Enterprise, Alabama is shown. "After the boll weevil destroyed (1910–15) the area's cotton, locals began diversified farming. In gratitude for the resulting prosperity, the city erected a monument to the boll weevil in 1919."
Photo credit: Saintrain at English Wikipedia [CC BY-SA 3.0 (http://creativecommons.org/licenses/by-sa/3.0) or GFDL (http://www.gnu.org/copyleft/fdl.html)], via Wikimedia Commons

Photo 1.11-4. Several species of the dung beetle, most notably the species *Scarabaeus sacer* (often referred to as the sacred scarab), enjoyed a sacred status among the ancient Egyptians. Granite scarab beetle statue in Karnak temple, Luxor, Egypt is shown.
Photo credit: By Wouter Hagens (Own work) [GFDL (http://www.gnu.org/copyleft/fdl.html) or CC BY-SA 3.0 (http:// creativecommons.org/licenses/by-sa/3.0)], via Wikimedia Commons, 2 February 2010

1.11.8 Insects in heraldry

Heraldry is a special historical discipline studying coats of arms; that is, **distinctive signs of entities** that are transferred by succession. The owners of coats of arms can be most diverse (individuals, social stratum, kin, city, country, etc.). Images of insects are not common in heraldry. Where they are used, their heraldic meaning reflects the qualities and signs of particular insects. Examples of coats of arms related to insects are listed in Table 1.11-8.

It is evident that **bees** predominate on coats of-arms. They are included on more than 300 emblems. The presence of bees represents industry, skill, delicate work, exactness, and elocution. Images of ants suggest diligence. The **butterfly** on emblems is identified with carelessness and fickleness, the **dragonfly**

Table 1.11-8. Some emblems related to insects.[1]

Portrayed insect	Owner of the coat of arms	Notes
Ant	Ahja rural municipality, Estonia	Põlva County
Ant	Bjärträ parish, Sweden	Now part of Kramfors municipality
Ant	Formiga municipality and town, Brazil	Central-west Minas Gerais state
Ant	Brekendorf municipality, Germany	Schleswig-Holstein
Ant	Fulleren commune, France	Haut-Rhin department, Alsace, northeastern France
Ant	Multia municipality, Finland	Province western Finland
Ant	Village Saint-Maurice-sur-Moselle, France	Lorraine region
Ant	Sewen city, France	Alsace region
Ant	Marwitz city, Germany	Federated state Brandenburg
Ant	Zeschdorf municipality, Germany	District Märkisch-Oderland, federated state Brandenburg
Bee	115th Engineer Battalion	Utah Army National Guard, US Army, American Fork, Utah, USA
Bee	d'Abeille family	Provence, France
Bee	Town Abejorral, Colombia	Antioquia department
Bee	Village Abella de la Conca, Spain	Province of Lleida, autonomous community of Catalonia
Bee	Mehedinţi county, Romania	Near border with Serbia and Bulgaria
Bee	Alçay-Alçabéhéty-Sunharette commune, France	Pyrenées-Atlantiques department, Aquitaine region, southwestern France
Bee	Alozaina town and municipality, Spain	Málaga province, autonomous community of Andalusia, southern Spain
Bee	Alvesta municipality, Sweden	Kronoberg County, southern Sweden
Bee	Arcachon commune, France	Gironde department, southwestern France
Bee	Levallois-Perret commune, France	Northwestern suburbs of Paris
Bee	Princess Beatrice of York, United Kingdom	First daughter of Prince Andrew, Duke of York and Sarah Ferguson, Duchess of York (born 1988)
Bee	Salamanca province, Spain	Autonomous community of Castilla y León, western Spain
Bee	British writer Sarah, Duchess of York	Former wife of Prince Andrew, Duke of York, born October 15, 1959
Bee	Arroba de los Montes municipality, Spain	CiudadReal, Castile-LaMancha
Bee	City of Aschères-le-Marché, France	Loiret department in north-central France
Bee	Attichy commune, France	Oise department, northern France
Bee	Former Bacup Municipal Borough Council, England	The former local authority of Bacup, Lancashire
Bee	Bains-les-Bains commune, France	Vosges department, Lorraine, eastern France
Bee	Elba island, Italy	Largest island of the Tuscan Archipelago
Bee	Bar-sur-Aube commune, France	Sub-prefecture of the Aube department
Bee	Barrow-in-Furness Borough Council, United Kingdom	Local government district with borough status in Cumbria, England

Table 1.11-8 contd. ...

...Table 1.11-8 contd.

Portrayed insect	Owner of the coat of arms	Notes
Bee	Bartninkai town, Lithuania	Marijampole County, southwestern Lithuania
Bee	Battigny commune, France	Meurthe-et-Moselle department, northeastern France
Bee	Bazancourt commune, France	Marne department, northeastern France
Bee	Sir David Stuart Beattie (1924–2001)	14th Governor-General of New Zealand (1980–85)
Bee	Béjar town and municipality, Spain	Province of Salamanca, autonomous community of Castile and León, western Spain
Bee	Municipality Belvis de la Jara, Spain	Province of Toledo, Castile-La Mancha
Bee	Bénamenil village, France	Meurthe-et-Moselle department, Lorraine region
Bee	Besseges city	Region of Languedoc Roussillon, department Gard
Bee	Blackburn town, United Kingdom	Lancashire, England
Bee	Bagneaux-sur-Loing commune, France	Seine-et-Marne department, Île-de-France region, north-central France
Bee	Lyon city, France	Rhône-Alpes region, east-central France
Bee	Marcellus commune, France	Lot-et-Garonne department, southwestern France
Bee	Ménil-la-Horgne commune, France	Meuse department, Lorraine, northeastern France
Bee	Rive-de-Gier commune, France	Loire department, central France
Bee	Rochegude commune, France	Drôme department, southeastern France
Bee	Saint-Médard-d'Eyrans commune, France	Gironde department, Aquitaine, southwestern France
Bee	La Chaux-de-Fonds city, Switzerland	La Chaux-de-Fonds, canton of Neuchâtel
Bee	Boissy-Saint-Léger commune, France	Val-de-Marne department, southeastern suburbs of Paris
Bee	Borne municipality and a town, Netherlands	Eastern Netherlands
Bee	Boulieu-lès-Annonay commune, France	Ardèche department, southern France
Bee	Commune Bourg-de-Péage, France	Drôme department, Rhône-Alpes region, southeastern France
Bee	Cruz Machado municipality, Brazil	State of Paraná, Southern Region of Brazil
Bee	Igrejinha city, Brazil	State Rio Grande do Sul
Bee	Canela town, Brazil	Serra Gaúcha, Rio Grande do Sul
Bee	Dois Irmãos municipality, Brazil	State Rio Grande do Sul
Bee	Breteuil-sur-Noye commune, France	Oise department, northern France
Bee	Brioude commune, France	Haute-Loire department, Auvergne region, south-central France
Bee	Bugeat commune, France	Corrèze department, central France
Bee	Bègles commune, France	Gironde department, southwestern France
Bee	Sir James Campbell, 2nd Baronet of Ardkinglass	c.1666–5 July 1752, was a British member of Parliament
Bee	Cap-d'Ail commune, France	Alpes-Maritimes department, southeastern France
Bee	Caudrot commune, France	Gironde department, southwestern France
Bee	Cenotillo Municipality, Mexico	State of Yucatán

Table 1.11-8 contd. ...

...Table 1.11-8 contd.

Portrayed insect	Owner of the coat of arms	Notes
Bee	Charmont commune, France	Val-d'Oise department, Île-de-France region
Bee	Chaudrey commune, France	Aube department, north-central France
Bee	Clan Maxton	Scottish clan
Bee	Les Clouzeaux commune, France	Vendée department, Pays de la Loire region, western France
Bee	Risaralda department, Colombia	Paisa region
Bee	Bockenheim, Germany	City district of Frankfurt am Main
Bee	Jasov Abbot Ambros Martin Strbal, Slovakia	Born 1972
Bee	Princess Beatrice of York, United Kingdom	Born 1988, first daughter of Prince Andrew, Duke of York, and Sarah Ferguson
Bee	Burnley Borough Council, United Kingdom	The local government authority for the Borough of Burnley, Lancashire, England
Bee	David Douglas Crosby, Canada	Born June 28, 1949, Canadian prelate of the Roman Catholic Church
Bee	Grzegorz Kaszak, Poland	Born February 24, 1964, is the bishop of Sosnowiec
Bee	Settlement Gvozdevka, Russia	Voronezh oblast
Bee	Settlement Kadom, Russia	Ryazan oblast
Bee	Kirsanov town, Russia	Tambov oblast
Bee	Klimavichy, Belarus	Mahilyow oblast
Bee	LaUnión, Spain	Region of Murcia in the southeast of Spain
Bee	Lebedyan town, Russia	Lipetsk oblast
Bee	Lipetsk town, Russia	Lipetsk oblast
Bee	Manchester City Council, United Kingdom	Greater Manchester, England
Bee	Bishop Maurizio Gervasoni, Italy	Born December 20, 1953, bishop of Vigevano, Italy
Bee	Medyn town, Russia	Kaluga oblast
Bee	Settlement Novoe Mesto, Russia	Bryansk oblast
Bee	Novokhopersk town, Russia	Voronezh oblast
Bee	Pestravsky rayon, Russia	Samara oblast
Bee	Piergiorgio Bertoldi, Italy	Born July 26, 1963, Italian archbishop apostolic nuncio of the Holy See
Bee	Province of Salamanca, Spain	Autonomous community of Castile and León, western Spain
Bee	Rossendale Borough Council, United Kingdom	Local authority for the Rossendale district of Lancashire, northwestern England
Bee	Sandovo rayon, Russia	Tver oblast
Bee	Sarah Ferguson	Prior to her marriage to Prince Andrew, Duke of York
Bee	Shatsk town, Russia	Ryazan oblast
Bee	Tambov town, Russia	Tambov oblast
Bee	Tambov oblast, Russia	
Bee	Royal and National Academy of Pharmacy of Spain	

Table 1.11-8 contd. ...

...Table 1.11-8 contd.

Portrayed insect	Owner of the coat of arms	Notes
Bee	Bishop Vincenzo Apicella, Italy	Bishop of Velletri-Segni
Bee	Cocumont commune, France	Lot-et-Garonne department, southwestern France
Bee	Colroy-la-Grande commune, France	Vosges department, Lorraine, northeastern France
Bee	Condé-sur-Iton commune, France	Eure department, northern France
Bee	Cormenon commune, France	Loir-et-Cher department, central France
Bee	Coucy-lès-Eppes commune, France	Aisne department, Picardy, northern France
Bee	Courtisols commune	Marne department, northeastern France
Bee	D'Huison-Longueville commune, France	Département de l'Essonne
Bee	Daubeuf-près-Vatteville commune, France	Eure department, northern France
Bee	Kirchenpingarten municipality, Germany	District of Bayreuth, Bavaria
Bee	Dzhankoy town, Russia	North of Crimea
Bee	Elbeuf commune, France	Seine-Maritime department, Haute-Normandie region, northern France
Bee	Ennery commune, France	Val-d'Oise department, Île-de-France, northern France
Bee	Aranzazu town and municipality in Colombia	Department of Caldas
Bee	City and municipality of Fusagasugá; Colombia	Department of Cundinamarca, central Colombia
Bee	Guijuelo municipality, Spain	Province of Salamanca, Castile, and León
Bee	Herce town and municipality, Spain	La Rioja province, northern Spain
Bee	Iglesiar rubia, Spain	Castile and Leon
Bee	La Granja commune, Chile	Santiago Province, Santiago Metropolitan Region
Bee	Masa, Spain	Castile and León
Bee	Melipilla commune and capital city, Chile	Santiago Metropolitan Region
Bee	Moratilla de los Meleros municipality, Spain	Province of Guadalajara, Castile-La Mancha
Bee	Paymogo town and municipality, Spain	Andévalo comarca, province of Huelva
Bee	Department of Risaralda, Colombia	West-central region of the country
Bee	San Ramón of Chile	Santiago Province, Santiago Metropolitan Region
Bee	Terradilos de Sedano, Spain	Province of Burgos, Castile and León
Bee	Villadoz municipality, Spain	Province of Zaragoza, Aragon
Bee	Colonia department, Uruguay	Southwestern Uruguay
Bee	Estissac commune, France	Aube department, north-central France
Bee	Eugénie-les-Bains commune, France	Landes department, Aquitaine, southwestern France
Bee	Christophe de Thou, France	1508–82, the first president of the Parliament of Paris
Bee	Fixin city, France	Department Côte d'Or, region of Bourgogne
Bee	Fourchambault commune, France	Nièvre department, central France
Bee	Fuente Obejuna city	Province of Córdoba, Andalucía, Spain
Bee	Grand'Combe-Châteleu commune, France	Doubs department, Franche-Comté region, eastern France

Table 1.11-8 contd. ...

...Table 1.11-8 contd.

Portrayed insect	Owner of the coat of arms	Notes
Bee	Heillecourt commune, France	Meurthe-et-Moselledepartment, northeastern France
Bee	Hemelveerdegem village, Belgium	Municipality of Lierde, Denderstreek, province East Flanders
Bee	Hilary Mary Weston, Canada	26th lieutenant governor of Ontario (1997–2002), born 1942
Bee	Högersdorf municipality, Germany	Schleswig-Holstein
Bee	Honigsee municipality, Germany	Schleswig-Holstein
Bee	Hoogeveen municipality and town, Netherlands	Northeastern Netherlands
Bee	Hummuli borough, Estonia	Valga County, southern Estonia
Bee	Jörg Immendorff painter and sculptor, Germany	1945–2007
Bee	55th Infantry Division, France	French army formation during World War I and World War II
Bee	Jednorożec Gmina (administrative district), Poland	Przasnysz County, Masovian Voivodeship, east-central Poland
Bee	Kārķi parish administrative unit, Latvia	Valka district
Bee	Ladbergen municipality, Germany	District of Steinfurt, North Rhine-Westphalia
Bee	Ladrillar municipality, Spain	Las Hurdes, province of Caceres, Extremadura
Bee	Lamelouze city, France	Department Gard, region of Languedoc Roussillon
Bee	Le Lamentin town, French West Indies	Outre-Mer department, Martinique
Bee	Lamotte-Beuvron commune, France	Loir-et-Cher department, central France
Bee	Lappion commune, France	Aisne department, Picardy, northern France
Bee	Laval-Atger commune, France	Lozère department, southern France
Bee	Lemud commune, France	Moselle department, Lorraine, northeastern France
Bee	Levallois-Perret commune, France	Northwestern suburbs of Paris
Bee	Viļāni town, Latvia	Rēzekne district, eastern Latvia
Bee	Madriat commune, France	Puy-de-Dôme department, Auvergne
Bee	La Meilleraye-de-Bretagne commune, France	Loire-Atlantique department, western France
Bee	Settlement Sladki	Labinsk raion, Krasnodar krai
Bee	Vidzy, Belarus	Vitebsk oblast
Bee	Gordeeskiy district, Russia	Bryansk oblast
Bee	Floirac commune, France	Gironde department in Aquitaine in southwestern France
Bee	Kārķi parish administrative unit, Latvia	Valka district
Bee	Milanówek town, Poland	Grodzisk Mazowiecki County, was replaced on June 10, 1997
Bee	Peney-le-Jorat municipality, Switzerland	District Gros-de-Vaud, canton Vaud
Bee	Hârtop commune, Romania	Suceava County
Bee	Varėna city, Lithuania	Alytus County
Bee	Settlement Privol'e, Ukraine	Donetsk Oblast
Bee	Vilshanka district, Ukraine	Kirovograd oblast, central Ukraine

Table 1.11-8 contd. ...

...Table 1.11-8 contd.

Portrayed insect	Owner of the coat of arms	Notes
Beetle	Belmont-sur-Lausanne municipality, Switzerland	District of Lavaux-Oron, canton of Vaud
Beetle	Eyragues city, France	Department of Bouches-du-Rhône, southeast of France
Beetle	Rombly commune, France	Pas-de-Calais department, Nord-Pas-de-Calais region
Beetle	Rostoklaty village and municipality, Czech Republic	Kolín district, Central Bohemian Region
Beetle	Barneveld town and a municipality, Netherlands	In the province of Gelderland
Beetle	Romairon municipality, Switzerland	District of Jura-Nord Vaudois in the canton of Vaud
Bumblebee	Hummeltal municipality, Germany	District of Bayreuth, Bavaria
Butterfly	Kapelle municipality and town, Netherlands	Zuid-Beveland, southwestern Netherlands
Butterfly	Bady Bassitt city, Brazil	São Paulo state
Butterfly	Daigny commune, France	Ardennes department, northern France
Butterfly	Monnières commune, France	Loire-Atlantique department, western France
Butterfly	Kirchseeon town, Germany	Upper Bavarian district of Ebersberg
Butterfly	Peguerinos municipality, Spain	Province of Ávila, Castile, and León
Butterfly	Perho municipality, Finland	Province of Western Finland, central Ostrobothnia region
Butterfly	Skrudaliena parish, Latvia	Daugavpils municipality
Butterfly	Slatina village, Czech Republic	Svitavy district
Butterfly	Chříč village and municipality, Czech Republic	Plzeň-North district, Plzeň region
Butterfly	Svilajnac town and municipality, Serbia	Resava region, central Serbia
Butterfly (mulberry butterfly)	Fergana province, Russian Empire	1890
Dragonfly	Nykvarn town and municipality, Sweden	Stockholm County
Dragonfly	Foucrainville commune, France	Eure department, Haute-Normandie region, northern France
Dragonfly	Mognéville commune, France	Meuse department, Lorraine, northeastern France
Dragonfly	Bonnefamille commune, France	Isère department, southeastern France
Dragonfly	Brides-les-Bains commune, France	Savoie department, Rhône-Alpes region, southeastern France
Dragonfly	Kiili Parish municipality, Estonia	Harju County, northwestern Estonia
Dragonfly	Persan commune, France	Val-d'Oise department, Île-de-France, northern France
Dragonfly	Unterspreewald district, Germany	Dahme-Spreewald, Brandenburg
Dragonfly	Gierałtowiczki village, Poland	Administrative district of Gmina Wieprz, Lesser Poland Voivodeship
Dragonfly (Photo 1.11-5)	Mănăstirea Humorului municipality, Romania	Suceava County
Dragonfly	Verești commune, Romania	Suceava County

Table 1.11-8 contd. ...

...Table 1.11-8 contd.

Portrayed insect	Owner of the coat of arms	Notes
Fly	Haute-Isle commune, France	Val-d'Oise department, Île-de-France, northern France
Fly	Muchow municipality, Germany	Ludwigslust-Parchim district, Mecklenburg-Vorpommern
Fly	Pressy commune, France	Pas-de-Calais department, Nord-Pas-de-Calais region
Grasshopper	Bure commune, France	Meuse department, Lorraine, northeastern France
Grasshopper	Buenavista del Norte municipality and town, Spain	North coast of Tenerife island
Grasshopper	La Llagosta municipality, Spain	Province of Barcelona, autonomous community of Catalonia
Grasshopper	Ustka town, Poland	Middle Pomerania region, northwestern Poland
Horsefly	Aspin-en-Lavedan commune, France	Hautes-Pyrénées department, southwestern France
Horsefly	Villebon-sur-Yvette commune, France	Essonne department, Île-de-France, northern France
Horsefly	Thaon-les-Vosges commune, France	Vosges department, Lorraine, northeastern France
Ladybird	Gujan-Mestras commune, France	Gironde department, southwestern France
Mayfly	Tlustice village, Czech Republic	Central Bohemian Region
Mosquito	Komárov, Czech Republic	Part of town of Opava, Moravian-Silesian region
Mosquito	Muckental municipality, Germany	Baden-Württemberg
Wasp	Wespen village, Germany	District of Salzlandkreis, Saxony-Anhalt (now it is part of the town of Barby)
Wasp	Osa town, Russia	Perm Krai
Wasp	Paranga urban-type settlement, Russia	Republic Mariy El

[1] Compiled by the author with the use of numerous internet sites.

Photo 1.11-5. Insects are present most often in the coats-of-arms of territorial entities (cities, settlements, territories, districts). Coat of arms of Mănăstirea Humorului, Suceava County, Romania is shown.
Photo credit: By Primăria comunei Mănăstirea Humorului [Public domain], via Wikimedia Commons

with flippancy, while the **fly** in heraldry is a sign of war. **Ladybugs** are considered to be a lucky symbol; they personalize the arrival of spring and are valued by the farmers because they eat crop pests.

Insects are present most often in the coats of arms of **territorial entities** (cities, settlements, territories, districts). However, they are encountered also as emblems of **armed forces** (for example, the 115th Engineer Battalion), **persons of distinction** (for example, Princess Beatrice of York, United Kingdom), and **public agencies** (for example, the Rossendale Borough Council, United Kingdom).

As for the representation of particular countries in this list, the most common is **France**, which is mentioned in the table 74 times. Spain (20), Russia (18), and Germany (15) follow.

1.11.9 Insects and arachnids in numismatics

Numismatics is the study of coins, bank notes, medals, and tokens. The use of insects' images in numismatics started long ago. The **first coins** with images of beetles, bees, and scorpions were issued at the end of the seventh century BC. They were made with small nubbles of electrum, a natural mixture of gold and silver (http://insects.org/ced1/numis.html).

Images of insects on coins were characteristic of the **ancient Greeks** to the greatest extent. Bees, beetles, butterflies, cicadas, ants, grasshoppers, and praying mantises were included. For the most part, the images were small and were symbols of families or local governors responsible for stamping the coins. However, insects were essential parts of the design of some coins.

For example, a cult of adoration of the fertility goddess Artemis existed in ancient time. In her honor, a temple—one of the "seven wonders of the world"—was erected in the ancient Greek town of Ephesus, on the western coast of Asia Minor, today İzmir Province, Turkey. The bee was a sacred symbol of Artemis, and priestesses of Artemis were called "bees of the goddess". For almost six centuries, a bee was the heart of the design of Ephesian coins (https://en.wikipedia.org/wiki/Ephesus). In total, more than **300 kinds** of ancient Greek coins featuring insects and arachnids are known (http://insects.org/ced1/numis.html).

The decay of Greek rule and the rise of the Roman Republic and Empire resulted in a **drastic reduction** in the number of insects' images on coins. None of the Roman coins had an insect as the key element of its design. However, until 44 BC, they appeared often on coins of the Roman Republic in the form of **small symbols**. About 200 known coins of the Roman Republic included images of insects. The coming of the Roman Empire after Julius Caesar led to the almost total disappearance of entomological subjects on Roman coins (http://www.insects.org/ced1/numismatic-entomology.html).

After Julius Caesar, the insects practically disappeared from coins. Images of **scorpions** appeared occasionally on coins only in some Roman colonies (former Greek territories).

The Renaissance era, beginning in the fifteenth century, greatly influenced the **stamping of coins**, which again became a mode of creative expression. The production of medals (coin-like metal disks with no monetary value) and tokens (unofficial substitutes for coins, usually issued by merchants) also began.

Nevertheless, insects never again reached their prominence in the classic Greek coining. Until today, their appearance on coins has been **quite rare**. Within the past five centuries, fewer than 100 different types of coins with images of insects were issued. Only over the last several years, an increase in the number of issued coins with images of insects became apparent. However, for the most part, these were collectors' coins rather than real ones; these coins were related to **propaganda** for measures aimed at wildlife conservation. The issue of paper money with images of insects was also related to environmental campaigns (http://insects.org/ced1/numis.html).

As for medals and tokens, a great diversity of insects and arachnids have been pictured on them. At this time, more than 2,000 different medals and tokens can be assigned to this category. The quick growth of their number also can be explained by keen interest in the environment (http://www.insects.org/ced1/numismatic-entomology.html).

Table 1.11-9 presents descriptions of some 20th- and 21st-century coins related to insects.

Table 1.11-9. Some coins related to insects.[1]

Portrayed insects	Species (common name, Latin name)	Country, year of issue	Face value	Note
Ant	Giant bull ant, *Myrmecia nigriceps*	Australia, 2010		Aluminum bronze
Ant	Venomous bull ant, *Myrmecia gulosa*	Australia, 2014	1 dollar	Aluminum bronze
Ant	Leaf-cutter ant, *Atta* sp.	Chad, 2015		Copper/nickel + brass ring
Ant	Red wood ant, *Formica rufa*	Latvia, 2003	1 lats	Copper-nickel
Ant	Ant	Sint Eustatius of Caribbean Netherlands, 2013	5 cents	Copper
Ant	Red wood ant, *Formica rufa*	San Marino, 1974		Aluminum
Ant	Red wood ant, *Formica rufa*	San Marino, 1986		Gold
Ant	Australia's bull ant, *Myrmecia pyriformis*	Tuvalu, 2015		Silver
Ant	Jack jumper ant, *Myrmecia pilosula*	Zambia, 2010		Copper-nickel
Ant	Bullet ant, *Paraponera clavata*	Zambia, 2010		Copper-nickel
Ant	Red harvester ant, *Pogonomyrmex barbatus*	Zambia, 2010		Copper-nickel
Ant lion	Ant lion (larva), *Hagenomyia micans* (order Neuroptera)	Palau, 2006		Gold
Bee	Western honey bee, *Apis mellifera*	Canada, 2015		Silver
Bee	Western honey bee, *Apis mellifera*	Cook Islands, 2014		Gold-plated silver
Bee	Western honey bee, *Apis mellifera*	Italy, 1919		Copper
Bee	Western honey bee, *Apis mellifera*	Italy, 1953		Aluminum
Bee	Western honey bee, *Apis mellifera*	Luxembourg, 2013	5 euros	Aluminum/bronze + silver ring
Bee	Western honey bee, *Apis mellifera*	Malta, 1972		Aluminum
Bee	Western honey bee, *Apis mellifera*	Morocco, 2011	10 santimat	Brass
Bee	Western honey bee, *Apis mellifera*	Sint Eustatius of Caribbean Netherlands, 2013	5 dollars	Brass + copper-nickel ring
Bee	Western honey bee, *Apis mellifera*	Norway, 1960		Copper-nickel
Bee	Western honey bee, *Apis mellifera*	Poland, 2010	10 zlotych	Brass
Bee	Western honey bee, *Apis mellifera*	San Marino, 1974		Aluminum
Bee	Western honey bee, *Apis mellifera*	San Marino, 1986		Gold
Bee	Western honey bee, *Apis mellifera*	San Marino, 2001		Gold
Bee	Western honey bee, *Apis mellifera*	Slovenia, 1993		Aluminum
Bee	African honey bee, *Apis mellifera scutellata*	South Africa, 2011		Gold
Bee	Western honey bee, *Apis mellifera*	Togo, 2005		Silver
Bee	Western honey bee, *Apis mellifera*	Ukraine, 2010		Gold
Bee	Western honey bee, *Apis mellifera*	Vatican City, 1992		Aluminum
Bee	African honey bee, *Apis mellifera scutellata*	Zambia, 2010		Copper-nickel

Table 1.11-9 contd. ...

...Table 1.11-9 contd.

Portrayed insects	Species (common name, Latin name)	Country, year of issue	Face value	Note
Beetle	Sacred scarab, *Scarabaeus sacer*	The Isle of Man, 2008	1/25 crown	Diameter: 13.92 mm Weight: 1.24 g Alloy: 999/1000 Au
Beetle	Sacred scarab, *Scarabaeus sacer*	The Isle of Man, 2008	1 crown	Diameter: 38.60 mm Weight: 28.28 g Alloy: 925/1000 Ag
Beetle	Sacred scarab, *Scarabaeus sacer*	Republic of the Fiji Islands	1 dollar	Diameter: 40.0 mm Weight: 20.00 g Alloy: 999/1000 Ag Coin is partly gold-plated using 24-karat gold
Beetle	Sacred scarab, *Scarabaeus sacer*	Egypt, 2010	1 pound	Diameter: 23.0 mm Weight: 8.5 g bimetal
Beetle	Sacred scarab, *Scarabaeus sacer*	Egypt, 1977	5 piasters	Diameter: 25.0 mm Weight: 4.5 g Copper-nickel
Beetle	Sacred scarab, *Scarabaeus sacer*	Republic of Equatorial Guinea (Annobón Island)	50 ecuele	Weight: 15.55 g Alloy: 999/1000 Ag
Beetle	Seven-spot ladybird, *Coccinella septempunctata*	San Marino, 1974	2 lire	Weight: 0.8 g Diameter: 18.0 mm Aluminum
Beetle	Seven-spot ladybird, *Coccinella septempunctata*	Island country Niue (New Zealand), 2013	2 dollars	Weight: 28.28 g Diameter: 41.0 mm Alloy: 925/1000 Ag Covered with gold
Beetle	Seven-spot ladybird, *Coccinella septempunctata*	Canada, 2011	25 cents	Weight: 12.61 g Diameter: 35.0 mm Nickel-plated steel
Beetle	Seven-spot ladybird, *Coccinella septempunctata*	Canada, 2011	20 dollars	Diameter: 38.0 mm Weight: 31.39 g Alloy: 999/1000 Ag
Beetle	Seven-spot ladybird, *Coccinella septempunctata*	Australia, 2010	50 cents	Diameter: 36.6 mm Weight: 15.591 g Alloy: 999/1000 Ag
Beetle	Seven-spot ladybird, *Coccinella septempunctata*	Australia, 2010	1 dollar	Diameter: 30.6 mm Weight: 13.8 g Aluminum bronze
Beetle	Seven-spot ladybird, *Coccinella septempunctata*	Republic of Palau, 2012	1 dollar	Diameter: 11.0 mm Weight: 0.5 g Alloy: 999/1000 Au
Beetle	Seven-spot ladybird, *Coccinella septempunctata*	The Cook Islands	5 dollars	Diameter: 38.61 mm Weight: 31.1 g Alloy: 999/1000 Au
Beetle	Seven-spot ladybird, *Coccinella septempunctata*	Poland	7 ducats	Diameter: 27.0 mm
Beetle	Seven-spot ladybird, *Coccinella septempunctata*	Poland, 2012	1 ducat	Diameter: 27.0 mm Weight: 8.5 g

Table 1.11-9 contd. ...

...Table 1.11-9 contd.

Portrayed insects	Species (common name, Latin name)	Country, year of issue	Face value	Note
Beetle	Seven-spot ladybird, *Coccinella septempunctata*	Australia, 2014	1 dollar	Diameter: 25.0 mm Weight: 9.0 g Aluminum bronze
Beetle	Seven-spot ladybird, *Coccinella septempunctata*	Australia, 2014	1 dollar	Diameter: 25.0 mm Weight: 11.6 g Alloy: 999/1000 Ag
Beetle	Stag beetle, *Lucanus cervus*	Poland, 1997	2 zloty	Diameter: 27.0 mm Weight: 8.15 g bronze
Beetle	Stag beetle, *Lucanus cervus*	Poland, 1997	20 zloty	Diameter: 38.61 mm Weight: 28.28 g Alloy: 925/1000 Ag
Beetle	Stag beetle, *Lucanus cervus*	Pridnestrovian Moldavian Republic	100 rubles	Diameter: 32.0 mm Weight: 14.14 g Alloy: 925/1000 Ag
Beetle	Stag beetle, *Lucanus cervus*	Poland, 2010	10 pennies	Diameter: 32.0 mm Weight: 13.0 g
Beetle	Stag beetle, *Lucanus cervus*	Russia, 2013	50 rubles	Diameter: 27.0 mm Weight: 8.0 g bimetal
Beetle	Stag beetle, *Lucanus cervus*	Russia, 2013	50 rubles	Diameter of silver token: 27.0 mm
Beetle	Stag beetle, *Lucanus cervus*	Russia, Udmurtia, 2013	10 kopecks	
Beetle	Stag beetle, *Lucanus cervus*	Republic of the Fiji Islands, 2014	10 dollars	Diameter: 50.0 mm Weight: 62.2 g Alloy: 999/1000 Ag
Beetle	Rhinoceros beetle, *Oryctes nasicornis*	Palau, 2006	1 dollar	Diameter: 11.0 mm Weight: 0.5 g Alloy: 999/1000 Au
Beetle	Rhinoceros beetle, *Oryctes nasicornis*	Australia, 2012	1 dollar	Diameter: 30.6 mm Weight: 13.8 g Aluminum bronze
Beetle	Beetle, *Themognatha chevrolati*	Cook Islands, 2000	1 dollar	Diameter: 40.0 mm Weight: 31.1 g Alloy: 999/1000 Ag
Beetle	Rosalia longicorn, *Rosalia alpina*	Slovakia, 2001	500 crowns	Diameter: 40.0 mm Weight: 33.63 g Alloy: 925/1000 Ag
Beetle	Beetle, *Rosalia coelestis*	Russia, 2012	2 rubles	Diameter: 33.0 mm Weight: 17.00 g Alloy: 925/1000 Ag
Beetle	Common cockchafer, *Melolontha melolontha*	Poland, 2009	Ducat	Diameter: 27.0 mm Weight: 8.4 g
Beetle	Common cockchafer, *Melolontha melolontha*	Poland, 2012	Ducat	Diameter: 23.0 mm
Beetle	Horned dung beetle, *Copris lunaris*	Russia, 2013	5 kopecks	
Beetle	Hercules beetle, *Dynastes hercules*	Bonaire Island, 2013	5 cents	

Table 1.11-9 contd. ...

...Table 1.11-9 contd.

Portrayed insects	Species (common name, Latin name)	Country, year of issue	Face value	Note
Beetle	Colorado potato beetle, *Leptinotarsa decemlineata*	Bonaire Island, 2013	10 cents	
Beetle	Rainbow stag beetle, *Phalacrognathus muelleri*	Australia, 2014	1 dollar	Diameter: 25.0 mm Weight: 9.0 g Aluminum bronze
Beetle	Convergent ladybug, *Hippodamia californica*	Australia, 2010		Aluminum bronze
Beetle	Rhinoceros beetle, *Oryctes rhinoceros*	Australia, 2012	1 dollar	Aluminum bronze
Beetle	Firefly	Canada, 2015		Silver
Beetle	French flower chafer, *Trichius gallicus*	San Marino, 1974		Aluminum
Bug	Assassin bug, *Reduvius* sp.	Sint Eustatius of Caribbean Netherlands, 2013		Copper-nickel
Bug	Pond skater, family Gerridae	Sint Eustatius of Caribbean Netherlands, 2013		Copper-nickel
Bumblebee	*Bombus* sp.	Republic of Palau, 2011	2 dollars	Diameter: 35 mm Weight: 15.5 g (1/2 oz.) Alloy: 925/1000 Ag
Bumblebee	Northern bumblebee, *Bombus polaris*	Canada, 2012		Copper-nickel
Bumblebee	Big-eyed bumblebee, *Bombus confusus*	Lithuania, 2008		Silver
Bumblebee	Bumblebee, *Bombus* sp.	Palau, 2011		Silver
Bumblebee	Bumblebee, *Bombus fragrans*	Russia, Tatarstan, 2013	50 kopecks	Copper
Butterfly		Republic of Palau, 2010	1 dollar	Diameter: 38.61 mm Weight: 28.28 g Alloy: 925/1000 Ag
Butterfly	Cairns birdwing butterfly, *Ornithoptera priamus*	Australia, 2010		Aluminum bronze
Butterfly	Wanderer butterfly, *Danaus plexippus*	Australia, 2012	1 dollar	Aluminum bronze
Butterfly	Ulysses butterfly, *Papilio ulysses*	Australia, 2014	1 dollar	Aluminum bronze
Butterfly	Short-tailed swallowtail, *Papilio brevicauda*	Canada, 2006		Silver
Butterfly	Eastern tailed blue butterfly, *Cupido comyntas*	Canada, 2013		Copper-nickel
Butterfly	Arcius bluetail butterfly, *Phetus arcius*	Congo Democratic Republic, 2002		Silver
Butterfly	Dark greenish butterfly, *Euphaedra* sp.	Congo Democratic Republic, 2002		Silver
Butterfly	Red and black butterfly, *Cymothoe sangaris*	Congo Democratic Republic, 2002		Silver
Butterfly	Ulysses butterfly, *Papilio ulysses*	Cook Islands, 2000		Silver

Table 1.11-9 contd. ...

...Table 1.11-9 contd.

Portrayed insects	Species (common name, Latin name)	Country, year of issue	Face value	Note
Butterfly	Hawk moth	Annobon Islands of Equatorial Guinea, 2013		Brass (copper-nickel ring)
Butterfly	Swallowtail	Annobon Islands of Equatorial Guinea, 2013		Bronze + brass ring
Butterfly	Emperor moth, *Saturnia pavonia*	British Guernsey, 1997		Silver
Butterfly	Brimstone butterfly, *Gonepteryx aspasia*	British Guernsey, 1998		Silver
Butterfly	Blue swallowtail, *Papilio manlius*	Mauritius, 1975	25 rupees	Silver
Butterfly	Wanderer butterfly, *Danaus plexippus*	Mexico, 1987	100 dollars	Silver
Butterfly	Stoplight catone, *Catonephele numilia*	Bonaire Island of Netherlands Antilles, 2012	1 cent	Copper
Butterfly	Zebra longwing, *Heliconius charithonia*	Bonaire Island of Netherlands Antilles, 2012	10 cents	Copper-nickel
Butterfly	Pink-spotted cattleheart, *Parides photinus*	Bonaire Island of Netherlands Antilles, 2012	5 cents	Copper-nickel
Butterfly	Angled sulphur, *Anteos maerula*	Bonaire Island of Netherlands Antilles, 2012	5 cents	Copper-nickel
Butterfly	Arcas cattleheart, *Parides arcas*	Bonaire Island of Netherlands Antilles, 2012	50 cents	Copper-nickel
Butterfly	Red postman, *Heliconius erato*	Bonaire Island of Netherlands Antilles, 2012	1 dollar	Bronze
Butterfly	Paradise birdwing butterfly, *Ornithoptera paradisea*	Papua New Guinea, 1972		Bronze
Butterfly	Alexandra birdwing butterfly, *Ornithoptera alexandras*	Papua New Guinea, 1990	100 kina	Gold
Butterfly	Kite swallowtail, *Graphium idaeoides*	Philippines, 1990	25 sentimos	Brass
Butterfly	Poplar hawk moth, *Laothoe populi*	Poland, 2014		Brass
Butterfly	Hawk moth, *Calerio bupmoraiae*	Poland, 2014		Brass
Butterfly	Black arches, *Lymantria monacha*	Poland, 2014		Brass
Butterfly	Peacock butterfly, *Inachis io*	Poland, 2014		Brass
Butterfly	Large white, *Pieris brassicae*	Poland, 2014		Brass
Butterfly	Garden tiger moth, *Arctia caja*	Poland, 2014		Brass
Butterfly	Small tortoiseshell, *Aglais urticae*	Poland, 2014		Brass
Butterfly	Death's head hawk moth, *Acherontia atropos*	Pridnestrovian Moldavian Republic, 2011		
Cicada	Wattle cicada, *Cicadetta* sp.	Australia, 2010		Aluminum bronze

Table 1.11-9 contd. ...

...Table 1.11-9 contd.

Portrayed insects	Species (common name, Latin name)	Country, year of issue	Face value	Note
Cicada	17-year cicada	British Bermuda, 1990	2 dollars	Copper-nickel
Damselfly	Rainforest damselfly, *Diphlebia lestoides*	Cook Islands, 2000		Silver
Damselfly	Southern damselfly, *Coenagrion mercuriale*	Moldova, 2013		Silver
Dragonfly		Republic of Palau, 2010	2 dollars	Diameter: 35 mm Weight: 15.5 g (1/2 oz.) Alloy: 925/1000 Ag
Dragonfly	Emperor dragonfly, *Anax imperator*	Australia, 2010		Aluminum bronze
Dragonfly	Dragonfly	Belarus, 2012		Silver
Dragonfly	Twelve-spotted skimmer, *Libellula pulchella*	Canada, 2013	10 dollars	Silver
Dragonfly	Green darner dragonfly, *Anax junius*	Canada, 2014	10 dollars	Silver
Dragonfly	Pygmy snaketail dragonfly, *Ophiogomphus howei*	Canada, 2015	10 dollars	Silver
Dragonfly	Dragonfly	Sint Eustatius of Caribbean Netherlands, 2013		Copper + copper-nickel ring
Dragonfly	Blue-eyed hawker, *Aeshna affinis*	Palau, 2010		Silver
Dragonfly	Emperor dragonfly, *Anax imperator*	Russia, 2008		Silver
Dragonfly	Dragonfly	Tokelau Islands, 2012		Silver
Fly	Blowfly, *Protophormia terraenovae*	Australia, 2010		Aluminum bronze
Fly	Australian sheep blowfly, *Lucilia cuprina*	Australia, 2014	1 dollar	Aluminum bronze
Fly	Tsetse fly, *Glossina*	Zambia, 2010		Copper-nickel
Grasshopper		Palau, 2010	2 dollars	Diameter: 35 mm Weight: 15.5 g (1/2 oz.) Alloy: 925/1000 Ag
Grasshopper	Short-horned grasshopper, *Macrocara conglobata*	Australia, 2010		Aluminum bronze
Grasshopper	Leichhardt's grasshopper, *Petasida ephippigera*	Australia, 2014	1 dollar	Aluminum bronze
Grasshopper	Grasshopper	Sint Eustatius of Caribbean Netherlands, 2013	5 cents	Copper-nickel
Grasshopper	Ukrainian grasshopper, *Poecilimon ukrainicus*	Ukraine, 2006		Silver
Hornet	Giant hornet, Vespidae	Zambia, 2010		Copper-nickel
Ladybird	Variable ladybird, *Coelophora inaequalis*	Australia, 2015		Aluminum bronze
Ladybird	Seven-spotted ladybug, *Coccinella septempunctata*	Canada, 2011		Copper-nickel
Locust	African desert locust, *Schistocerca gregaria*	Palau, 2010		Silver

Table 1.11-9 contd. ...

...Table 1.11-9 contd.

Mantis	Praying mantis, *Mantis religiosa*	Canada, 2012	10 dollars	Silver
Mantis	Praying mantis, *Mantis religiosa*	Poland, 2011		Brass
Mantis	Praying mantis, *Mantis religiosa*	Australia, 2010		Aluminum bronze
Mantis	Giant Asian mantis, *Hierodula tenuidentata*	Kazakhstan, 2012		Copper-nickel
Mantis	Praying mantis, Mantodea	Sint Eustatius of Caribbean Netherlands, 2013	50 cents	Copper-nickel
Mosquito	Malaria mosquito, *Anopheles anopheles*	Niger, 2005		Copper-nickel + brass
Mosquito	Mosquito	Zambia, 2010		Copper-nickel
Walking stick	Goliath stick insect, *Eurycnema goliath*	Cook Islands, 2000		Silver
Walking stick	Giant prickly stick insect, *Extatosoma tiaratum*	Palau, 2011		Silver
Wasp	Cuckoo wasp, *Chrysis chrysis*	Australia, 2014	1 dollar	Aluminum bronze
Wasp	Large cuckoo wasp, *Stilbum cyanurum*	Cook Islands, 2000		Silver
Weta	Giant weta, *Deinacrida* sp.	New Zealand, 2009	1 dollar	Silver

[1] Compiled by the author with the use of numerous internet sites.

Analysis of this table shows that the beetles (40) and butterflies (34) dominate. The **bee** is the absolute winner in the "Numismatics" category. Damselflies and dragonflies (13) and ants (11) follow.

As for particular **countries**, the following use insects in coins the most often: Australia (22), Poland (15), and the Republic of Palau (11). As for **material** used to make the coins, silver (83) is the winner. Copper was used for stamping 32 coins from this list, while nickel is present in 27 coins. However, these metals are often used in combination with each other. If, however, the **bimetallic coins** are disregarded, then brass and gold were used for stamping coins 17 and 12 times, respectively.

1.11.10 Insects in philately

A vast number of postage stamps have been devoted to different insects—at least an order of magnitude more than the number of emblems related to them. In contrast with the previous sections of this chapter, the preparation of this paragraph was highly simplified due to availability of the website http://www. asahi-net.or.jp/~ch2m-nitu/indexe.htm. This site allows stamp searches both for particular countries (2,304 pieces from 150 countries) and for different kinds of insects (1,450 kinds).

Table 1.11-10 presents examples of such stamps. This information was taken from the above-mentioned site; the list of countries was limited by the first letter of the English alphabet. When two or more stamps with the same insect but different nominations were issued simultaneously by a country, they were mentioned only once.

Analysis of Table 1.11-10 shows a decided preference for **beetles** (50) and **butterflies** (36). The other insects are well behind them in frequency of occurrence.

Table 1.11-10. Insects portrayed on stamps.[1]

Portrayed insect	Species (common name, Latin name)	Country	Year Issued
Ant	Velvet ant, *Dasymutilla occidentalis*	United States of America	1999
Ant	Red bull ant, *Myrmecia gulosa*	Australia	2014
Bee	Tawny mining bee, *Andrena fulva*	British Alderney	2009
Bee	Early bumblebee, *Bombus pratorum*	British Alderney	2009
Bee	Bug mining bee, *Colletes daviesanus*	British Alderney	2009
Bee	Gooden's nomad bee, *Nomada goodeniana*	British Alderney	2009
Bee	Solitaly bee, *Halictus scabiosae*	British Alderney	2009
Bee	Western honey bee, *Apis mellifera*	British Alderney	2009
Bee	Carpenter bee, *Xylocopa aestuans*	United Arab Emirates	1998
Bee	Western honey bee, *Apis mellifera*	Austria	2009
Bee	Caucasian honey bee, *Apis mellifera caucasica*	Azerbaijan	2005
Beetle	Garden chafer, *Polyphylla fullo*	Albania	1963
Beetle	European stag beetle, *Lucanus cervus*	Albania	1963
Beetle	Gigas ground beetle, *Procerus gigas*	Albania	1963
Beetle	Northern dune tiger beetle, *Cicindela hybrida*	Albania	1963
Beetle	12-spot ladybird, *Coleomegilla maculata*	British Alderney	2004
Beetle	5-spot ladybird, *Coccinella quinquepunctata*	British Alderney	2004
Beetle	Common rose chafer, *Cetonia aurata*	British Alderney	2013
Beetle	Carrion beetle, *Nicrophorus* sp.	British Alderney	2013
Beetle	Green tiger beetle, *Cicindela campestris*	British Alderney	2013
Beetle	Common cockchafer, *Melolontha melolontha*	British Alderney	2013
Beetle	Black chafer, *Netocia morio*	British Alderney	2013
Beetle	Blister beetle, *Meloe* sp.	British Alderney	2013
Beetle	7-spotted ladybird, *Coccinella septempunctata*	British Alderney	2014
Beetle	2-spotted ladybird, *Adalia bipunctata*	British Alderney	2014
Beetle	Orange ladybird, *Halyzia sedecimguttata*	British Alderney	2014
Beetle	Multicolored Asian ladybird, *Harmonia axyridis*	British Alderney	2014
Beetle	7-spotted ladybird, *Coccinella septempunctata*	Algeria	1994
Beetle	Metallic wood-boring beetle, *Anthaxia* sp.	Algeria	1994
Beetle	Varied carpet beetle, *Anthrenus scrophulariae*	Algeria	2000
Beetle	European cockchafer, *Melolontha melolontha*	Algeria	2000
Beetle	Drugstore beetle, *Stegobium paniceum*	Algeria	2000
Beetle	Ground beetle, *Carabus* sp.	Algeria	2000
Beetle	Convergent ladybird, *Hippodamia convergens*	United States of America	1987
Beetle	Elderberry longhorn beetle, *Desmocerus palliates*	United States of America	1999
Beetle	Eyed ladybird beetle, *Anatis mali*	United States of America	1999
Beetle	Dogbane leaf beetle, *Chrysochus auratus* (Chrysomelidae)	United States of America	1999
Beetle	Eastern Hercules beetle, *Dynastes tityus*	United States of America	1999
Beetle	Bombardier beetle, *Brachinus brachinus*	United States of America	1999

Table 1.11-10 contd. ...

...Table 1.11-10 contd.

Portrayed insect	Species (common name, Latin name)	Country	Year Issued
Beetle	Dung beetle, *Phanaeus vindex*	United States of America	1999
Beetle	Spotted water beetle, *Thermonectus marmoratus*	United States of America	1999
Beetle	Branch-boring beetle, *Stephanoderes hampei*	Angola	1983
Beetle	Weevil, *Christiansenia dreuxi*	French Southern and Antarctic Lands	1972
Beetle	Antarctic rove beetle, *Antarctophytosus atriceps*	French Southern and Antarctic Lands	1972
Beetle	Weevil, *Ectemnorhinus vanhoeffenianus*	French Southern and Antarctic Lands	2015
Beetle	Anthia ground beetle, *Anthia duodecimguttata*	United Arab Emirates	1999
Beetle	Convergent ladybird, *Hippodamia convergens*	Argentina	1990
Beetle	Ground beetle, *Calleida suturalis*	Argentina	1990
Beetle	Rainbow leaf beetle, *Chrysolina aurata*	Argentina	2002
Beetle	Cocoa longhorn beetle, *Steirastoma breve*	Argentina	2002
Beetle	Ground beetle, *Procerus scabrosus fallettianus*	Armenia	2006
Beetle	Lunata ladybird, *Cheilomenes lunata*	Ascension Island	1988
Beetle	Weevil, *Alceis ornatus*	Ascension Island	1989
Beetle	Pea and bean weevil, *Neltumius arizonensis*	Ascension Island	1998
Beetle	Kiawe bruchid, *Algarobius prosopis*	Ascension Island	1998
Beetle	African ladybird, *Cheilomenes lunata*	Ascension Island	2008
Beetle	Jewel beetle, *Castiarina product*	Australia	1991
Beetle	Aenea golden stag beetle, *Lamprima aenea*	Australia	1989
Beetle	Colorado beetle, *Leptinotarsa decemlineata*	Austria	1967
Beetle	European stag beetle, *Lucanus cervus*	Austria	2007
Beetle	Blue longhorn beetle with black spots, *Rosalia alpina*	Austria	2009
Bug	Assassin bug, *Apiomerus apiomerus*	United States of America	1999
Bug	Stinkbug, *Antestiopsis lineaticollis*	Angola	1983
Bug	Cotton stainer, *Dysdercus* sp.	Angola	1994
Bug	Long-neck stinkbug, *Phtirocoris antarcticus*	French Southern and Antarctic Lands	1972
Bug	Stinkbug, *Podisus nigrispinus*	Argentina	1990
Bug	Red bug, *Nabis punctipennis*	Argentina	1990
Bug	Green stinkbug, *Edessa meditabunda*	Argentina	2002
Bug	Lace bug, *Teleonemia scrupulosa*	Ascension Island	1998
Bug	Cotton shield-backed bug, *Tectoris diopthalmus*	Australia	1991
Bumblebee	White-tailed bumblebee, *Bombus lucorum*	British Alderney	1994
Bumblebee	Buff-tailed bumblebee, *Bombus terrestris*	Azerbaijan	2005
Butterfly	Silkworm, cocoon, silk moth, *Bombyx mori*	Afghanistan	1963
Butterfly	Silkworm, *Bombyx mori*	Afghanistan	1966
Butterfly	Tiger moth, *Callimorpha principalis*	Afghanistan	1971
Butterfly	Chalcosiine moth, *Epizygaena caschmirensis*	Afghanistan	1971

Table 1.11-10 contd. ...

...Table 1.11-10 contd.

Portrayed insect	Species (common name, Latin name)	Country	Year Issued
Butterfly	Autocrator parnassius, *Parnassius autocrator*	Afghanistan	1971
Butterfly	Clouded yellow butterfly, *Colias crocea*	British Alderney	1994
Butterfly	Peacock butterfly, *Inachis io*	British Alderney	1994
Butterfly	Six-spot burnet, *Zygaena filipendulae*	British Alderney	1994
Butterfly	Common blue butterfly, *Polyommatus icarus*	British Alderney	1994
Butterfly	Small tortoiseshell, *Aglais urticae*	British Alderney	1994
Butterfly	Emperor moth, *Saturnia pavonia*	British Alderney	1994
Butterfly	Wanderer butterfly, *Danaus plexippus*	United States of America	1987
Butterfly	Luna moth, *Actias luna*	United States of America	1987
Butterfly	Wanderer butterfly (caterpillar), *Danaus plexippus* (Danaidae)	United States of America	1999
Butterfly	Wanderer butterfly, *Danaus plexippus*	United States of America	1999
Butterfly	California dogface butterfly *Colias Eurydice* (Photo 1.11-6)	United States of America	1977
Butterfly	Cotton bollworm, *Helicoverpa armigera*	Angola	1994
Butterfly	Beet armyworm, *Spodoptera exigua*	Angola	1994
Butterfly	Clothes moth, *Pringleophaga kerguelenensis*	French Southern and Antarctic Lands	1971
Butterfly	Oleander hawk moth, *Daphnis nerii*	United Arab Emirates	1999
Butterfly	Cabbage semi-looper moth, *Trichoplusia orichalcea*	Ascension Island	1988
Butterfly	Plume moth, *Trichoptilus wahlbergi* (Pterophoridae)	Ascension Island	1989
Butterfly	Cactus moth (larva), *Cactoblastis cactorum*	Ascension Island	1998
Butterfly	Long-tailed blue, *Lampides boeticus*	Ascension Island	2008
Butterfly	Hawk moth, *Langia zenzeroides* (Sphingidae)	Australia	1991
Butterfly	Red admiral, *Bassaris itea*	Australia	1976
Butterfly	Noctuid moth, *Leucania loreyimima*	Australia	1976
Butterfly	Common eggfly, *Hypolimnas bolina*	Australia	1976
Butterfly	Noctuid moth, *Austrocarea iocephala*	Australia	1976
Butterfly	Geometer moth, *Cleora idiocrossa*	Australia	1976
Butterfly	Gaudy commodore, *Precis villida*	Australia	1976
Butterfly	Crimson speckled moth, *Utetheisa pulchelloides*	Australia	1977
Butterfly	Geometer moth, *Pyrrhorachis pyrrhogona*	Australia	1977
Butterfly	Scrofa hawk moth, *Hippotion scrofa*	Australia	1977
Butterfly	Noctuid moth, *Tiracola plagiata*	Australia	1977
Butterfly	Noctuid moth, *Thrincophora aridela*	Australia	1989
Butterfly	Bee hawk moth, *Hemaris fucltormis*	Austria	2008
Cicada	Periodical cicada, *Magicicada magicicada*	United States of America	1999
Cochineal	Vordan karmir & Armenian red, *Porphyrophora hamelii*	Armenia	2006
Cricket	Field cricket, *Gryllus bimaculatus*	Ascension Island	1988
Cricket	Bush cricket, *Ruspolia differens*	Ascension Island	1988

Table 1.11-10 contd. ...

...Table 1.11-10 contd.

Portrayed insect	Species (common name, Latin name)	Country	Year Issued
Cricket	Field cricket, *Insulascirtus nythos*	Australia	1989
Damselfly	Beautiful demoiselle damselfly, *Calopteryx virgo*	Albania	1966
Damselfly	Banded demoiselle damselfly, *Calopteryx splendens*	Albania	1966
Damselfly	Blue-tailed damselfly, *Ischnura elegans*	British Alderney	1994
Damselfly	Blue-tailed damselfly, *Ischnura elegans*	British Alderney	2010
Damselfly	Ebony jewel-wing damselfly, *Calopteryx macalata*	United States of America	1999
Dragonfly	Common darter dragonfly, *Sympetrum striolatum*	British Alderney	2010
Dragonfly	Emperor dragonfly, *Anax imperator*	British Alderney	2010
Dragonfly	Brown hawker, *Aeshna grandis*	British Alderney	2010
Dragonfly	Black-tailed skimmer, *Orthetrum cancellatum*	British Alderney	2010
Dragonfly	Red-veined darter, *Sympetrum fonscolombii*	British Alderney	2010
Dragonfly	Scarlet skimmer, *Crocothemis erythraea*	United Arab Emirates	1998
Fly	Flower fly, *Milesia virginiensis*	United States of America	1999
Fly	Scorpion fly, *Panorpa nuptialis*	United States of America	1999
Fly	Sweet-potato whitefly, *Bemisia tabaci*	Angola	1994
Fly	Sub-Antarctic kelp fly, *Paractora dreuxi*	French Southern and Antarctic Lands	1971
Fly	True fly, *Amalopteryx maritima*	French Southern and Antarctic Lands	2015
Fly	Sheep maggot fly, *Lucilia sericata*	Ascension Island	1989
Grasshopper	Great green bush cricket, *Tettigonia viridissima*	British Alderney	1994
Grasshopper	Bush locust, *Zonocerus variegatus*	Angola	1983
Grasshopper	Grasshopper, *Acorypha glaucopsis*	United Arab Emirates	1999
Grasshopper	Grasshopper, *Petasida ephippigera*	Australia	1991
Hornet	European hornet, *Vespa crabro*	Azerbaijan	2005
Katydid	True katydid, *Caedicia araucariae*	Australia	1989
Katydid	True katydid, *Pterophylla camellifolia*	United States of America	1999
Locust	Red locust, *Nomadacris septemfasciata*	Angola	1963
Locust	Bush locust, *Elaeochlora viridis*	Argentina	2002
Locust	African desert locust, *Schistocerca gregaria*	Ascension Island	2008
Mantid	Praying mantid, *Blepharopsis mendica*	United Arab Emirates	1998
Mosquito	Nonbiting midge, *Microzetia mirabilis*	French Southern and Antarctic Lands	1971
Wasp	Thread-waisted wasp, *Ammophila sabulosa*	British Alderney	1994
Wasp	Golden paper wasp, *Polistes fuscatus*	Ascension Island	1989
Wasp	European wasp, *Vespula germanica*	Australia	2014
Wasp	German wasp, *Paravespula germanica*	Azerbaijan	2005

[1] Compiled by the author, using the website http://www.asahi-net.or.jp/~ch2m-nitu/indexe.htm.

Photo 1.11-6. A vast number of postage stamps have been devoted to different insects—at least an order of magnitude more than the number of emblems related to them. One of a series of US commemorative stamps featuring butterflies—Dogface—*Colias eurydice*—13-cent 1977 issue U.S. stamp.
Photo credit: By Bureau of Engraving and Printing. Designed by Stanley Galli. [Public domain], via Wikimedia Commons, 6 June 1977

1.11.11 Insects in religion, mythology, and symbolism

A vast number of myths and symbols are related to insects, and they also are reflected to a great extent in **religion**. For example, the Bible describes in detail the ten plagues of Egypt. Three of them (third, fourth, and eighth) were related to insects.

The **third plague**, black flies, refers to an attack of black flies and mosquitos, which were a common scourge in Egypt during inundations of the Nile. However, there is also alternative point of view holding that the plague involved lice or fleas. The **fourth plague**, "canine flies", involved clouds of these flies that covered people and filled the houses of the Egyptians. Most likely, "canine flies" meant gadflies, which attacked the Egyptians and their herds. The **eighth plague**, the locust invasion, was one of the deadliest. Great clouds of locusts came, and all foliage was destroyed. At the end of the day, the earth was covered by a stinking layer of locusts which was of a thickness of 12 cm (http://cogmtl.net/Articles/138.htm).

Insects are also mentioned in the Talmud. It is strictly prohibited to use insects as food, except for few kinds. The following insects are considered to **be kosher** (i.e., allowed to be used as food) (http://www.evrey.com/sitep/talm/print.php?trkt=hullin&menu=65): (1) insects with four legs, two of which are longer than other two and are used by the insect for starting out from the earth when they take flight; and (2) those with four wings covering the major part of the body. These conditions are met by the **locusts** and **grasshoppers**.

At the same time honey, which is considered to be a product of an unclean insect, is accepted as food because it (according to kashrut) is not the product of the bee's activity (https://en.wikipedia.org/wiki/Kashrut).

As for **beetles**, scarabs have been symbols of sexual vigor since ancient Egypt, and they are held to be very efficacious when used in love charms (Costa-Neto 2002). Examples of some myths, beliefs, symbols, etc., related to insects are given in Table 1.11-11.

This list is very short and incomplete. A variety of **other beliefs** and superstitions are related to insects. For example, river or lake flies have a mystic attractiveness for many indigenous peoples because they seem to appear out of nowhere and only during particular seasons. On the coast of Lake Victoria, trinkets with the chironomids *Chironomus* are believed to contribute to good fortune in business as well as to help find partners for single men and women (Macadam and Stockan 2015).

Insects also figure in books in which **dream interpretations** are given. For example, in one dream book, the appearance of butterflies in a dream is treated as a good omen. A butterfly flittering among flowers and green grass promises wealth and prosperity. If somebody sees a lot of butterflies in a dream,

Table 1.11-11. Myths, beliefs, and symbols related to insects and arachnids.[1]

Insect	People/country	Belief
Ant	China	Symbol of fairness, saintliness, and goodness
Ant	Estonia	Points at happy endings in fortune-telling
Ant	Bulgaria, Switzerland	Points at unlucky endings in fortune-telling
Ant (black)	India	Reputed as a sacred insect
Ant (white)	Buddhism	Symbol of placability and self-containment
Bee	Ancient Greece	Artemis, forever young huntress and earth goddess, was associated with the image of the sacred bee
Bee	Hittites (ancient Anatolian people)	Bee is the savior of the world from drought
Bee	Christians	Bee is considered to be a symbol of the virgin birth and emblem of the innocent Virgin Mary
Bee	Ancient Greeks, Indo-Iranians, Muslims	Bee as an allegory of pure soul
Bee (drones)	Ancient Greece	Castrated priests of the Earth goddess Artemis were identified with drones (drones are male honey bees)
Beetle (green rose chafer)	Orochs (Far East of Russia)	When a moose falls asleep, its soul takes the form of a green rose chafer (*Cetonia aurata*); when it awakens, the moose comes to the human who catches the beetle
Beetle (sacred scarab, *Scarabaeus sacer*)	Ancient Egyptians	A scarab was a divine symbol of rebirth, resurgence, and eternity of spirit
Butterfly	Ancient Greece	Psyche in Greek mythology is a personification of the soul and is represented as a butterfly or a girl with butterfly wings
Butterfly	China	Emblem of lovers
Butterfly	Japan	Emblem of women of easy virtue, geishas, while two butterflies are a symbol of happy wedded love
Butterfly	The ancients of all continents	Symbol of rebirth, reincarnation, and immortality of the soul
Butterfly	Christians	It is believed that the souls of deceased unchristened infants come into butterflies
Butterfly	Aztecs	Aztecs felt certain that butterflies were the souls of deceased brave warriors
Butterfly	Burmese	It was believed that butterflies were the souls of sleeping men; a man awakened when the butterfly-soul came back to him. For this reason, one could not awaken a sleeping man: if the soul was not able to join a body, then the startled man would surely die
Butterfly	Rural people of England	They told their fortunes for the next year by the first butterfly greeted during a season: white wings predicted a light and joyful life; golden ones, wealth; brown ones, sickness, poverty, and hunger; while black forebode an impending death
Butterfly	The Europeans in the Middle Ages	If three butterflies met together, it was considered to be an ominous sign
Butterfly	The Europeans in the Middle Ages	The encounter of the diurnal butterfly at night was the worst omen

Table 1.11-11 contd. ...

...Table 1.11-11 contd.

Insect	People/country	Belief
Butterfly	The Europeans in the Middle Ages	Meeting with a privet hawk moth (*Sphinx ligustri*) promised an unavoidable quick death
Chironomidae	Indigenous peoples around Lake Victoria	Sign of good fortune
Cicada	China	Sign of resurrection of the dead
Dragonfly	Christians	Identified with evil spirits
Dragonfly	China	Denoted elegance
Firefly	Japan	Fireflies are considered to be the souls of warriors who gave their lives for their country and merit salvation
Fly	Maasai (Kenya and Tanzania)	Souls of deceased ancestors relocate into flies
Fly	Ancient Jews	It is considered to be an unclean insect and should not appear in the temple of Solomon
Fly	Christians	Characterize fly as carrier of evil, pestilence, and iniquity
Louse	Yaghnobi people (Tajikistan)	Belief of emergence of lice in relation to thunder
Mayfly	The Europeans in the Middle Ages	Mayflies are symbols of life shortness and happiness
Mosquito	Siberian peoples (Nenets, Selkup, Nganasans, Kets) and some indigenous peoples of North America	Myth of origin of mosquitoes from sparks of fire is common
Scorpion	Ancient Rome	Mercury, the patron god of trade (commerce), was represented as a scorpion in ancient Roman mythology
Scorpion	Peoples of Africa	Scorpions were considered to be "spittle of evil spirits"
Scorpion	Peoples of western Asia	Considered to be emblem of truth, fairness, and justice
Spider	Christians	Sign of cold cruelty, greed, and malignity
Spider	Christians	Legend that a spider rescued the Christ infant from Herod's cruelty
Spider	Muslims	Legend that Muhammad was rescued from his enemies with the aid of a spider and its web
Spider	The ancients of Europe	Ominous sign of greediness and double-cross
Wasp	Siberian peoples of Russia	Belief in the ability of a shaman's soul to pass into a wasp and reach heaven and the Creator in such form

[1] Compiled by the author with the use of numerous internet sites.

it means news is coming from absent friends. For a young woman, a similar dream predicts a love that will end in a happy marriage (http://www.sonnik.ru/articles/art1398.html).

Nevertheless, traditions of different nations vary greatly, and the same dreams in different geographical zones can be explained in completely different ways.

1.11.12 Proverbs, sayings, and expressions related to insects

All the nations on the planet have proverbs concerning insects. In these proverbs, the **attitude of people** toward insects is reflected: dislike or quiet irony, admiration or contempt. Frequently, the insects are only a means to emphasize a certain side of human life, to criticize the feeble or virile features of the human character, to express a wise idea, or to represent worldly experience to young ones.

Below, some proverbs and sayings that are in circulation among the English-speaking nations of the world are given.

Table 1.11-12. English proverbs, sayings, and expressions related to insects and arachnids.[1]

Kind of arthropod	Proverb (saying, expression)
	Arachnida
Scorpion	The scorpion is sister to the snake.
Scorpion	Two scorpions in a small cave are better than two sisters in the house.
Scorpion	Even the hand of compassion is stung when it strokes a scorpion.
Spider	The spiders web lets the rat escape and catches the fly.
Spider	A spider spins his web strand by strand.
Spider	A spidery handwriting.
Spider	The spider taketh hold with her hands, and is in kings' palaces.
Tick	You are like a tick in a dog's ear.
	Insects ("bugs") in general
	The smallest insect may cause death by its bite.
	One tiny insect may be enough to destroy a country.
	The hinge of a door is never crowded with insects.
	Even a one-inch insect has a five-tenths of a soul.
	Put a bug in (someone's) ear.
	As snug as a bug in a rug.
	Bugs are bugs whether they bite or not.
	Coleoptera
Beetle	One beetle knows another.
Beetle	Beetle away.
Chafer	One chafer knows another.
Dung beetle	If you fool with the tumble-turd bug, you are apt to be spattered.
Firefly	The light of the firefly is sufficient for itself only.
Firefly	Fireflies shine only when in motion.
Firefly	It is foolish to show glowworms by candlelight.
Firefly	Glowworms are not lanterns.
	Diptera
Fly	A fly follows honey.
Fly	A fly may conquer a lion.
Fly	A person is not a fly.
Fly	A fly on the wall.
Fly	Can't kill flies with a spear.
Fly	One dead fly spoils much of a good ointment.
Fly	Dying flies spoil the sweetness of the ointment.
Fly	Die like flies.
Fly	Even a fly has its spleen.
Fly	Flies never bother a boiling pot.
Fly	You can catch more flies with molasses than vinegar.
Fly	You can catch more flies with honey than vinegar.
Fly	To a boiling pot flies come not.

Table 1.11-12 contd. ...

...Table 1.11-12 contd.

Kind of arthropod	Proverb (saying, expression)
	Diptera
Fly	A fly does not mind dying in coconut cream.
Fly	The busy fly is in every man's dish.
Fly	Make yourself all honey, and the flies will devour you.
Fly	Flies hunt (go to) lean horses.
Fly	Into a shut mouth flies fly not.
Fly	A shut mouth catches no flies.
Fly	More flies are taken with a drop of honey than a tun of vinegar.
Fly	You must lose a fly to catch a trout.
Fly	Flies come to feasts unasked.
Fly	Flies go to lean horses.
Fly	Fine fruit will have flies about it.
Fly	Almost never killed a fly.
Fly	Only a gadfly can sit on an elephant's back.
Fly	Hungry flies bite sore.
Fly	He does not harm a fly.
Fly	Even a lion must defend himself against the flies.
Fly	Flies will tickle lions being dead.
Fly	Do what we can, summer will have its flies.
Fly, hornet	Laws catch flies but let hornets go.
Fly	Laws are like cobwebs where the small flies are caught and the big ones break through.
Wasp	Waspish.
	Hemiptera
Bedbug	The best way to put an end to the bugs is to set fire to the bed.
	Hymenoptera
Ant	The constant creeping of ants will wear away the stone.
Ant	A coconut shell full of water is an ocean to an ant.
Ant	An ant can do more than an ox that is lying down.
Ant	An ant may work its heart out, but it can't make money.
Ant	Ants follow fat.
Ant	Ants live safely till they have gotten wings.
Ant	Ants never lend; ants never borrow.
Ant	Any spoke will lead the ant to the hub.
Ant	None preaches better than the ant, and she says nothing.
Ant	No worker is better than the ant, and she says nothing.
Ant	Go to the ant, thou sluggard; consider her ways, and be wise.
Ant	The ants are a people not strong, yet they prepare their meat in the summer.
Ant	She's got ants in her pants.

Table 1.11-12 contd. ...

...Table 1.11-12 contd.

Kind of arthropod	Proverb (saying, expression)
	Hymenoptera
Bee	As busy as a bee.
Bee	Old bees yield no honey.
Bee	The bee that makes the honey doesn't stand around the hive, and the man who makes the money has to worry, work, and strive.
Bee	No bees, no honey; no work, no money.
Bee	It's the roving bee that gathers honey.
Bee	If you let the bee be, the bee will let you be.
Bee	If a bee stings you once, it's the bee's fault. If it stings you twice, it's your fault.
Bee	Boys avoid the bees that stung 'em.
Bee	He who would gather honey must bear the sting of the bees.
Bee	If a bee didn't have a sting, he couldn't keep his honey.
Bee	Hit takes a bee fer ter git de sweetness out'n de hoar-houn' blossom.
Bee	From the same flower the bee extracts honey and the wasp gall.
Bee	A bee was never caught in a shower.
Bee	The three things most difficult to understand—the mind of a woman, the labor of the bees, and the ebb and flow of the tide.
Bee	Three best small things—a beehive, a sheep, and a woman.
Bee	The bee works all summer and eats honey all winter.
Bee	The drone bee dies soon after the wedding night.
Bee, spider	When the bee sucks, it makes honey; when the spider does, poison.
Bee	Where there are bees, there is honey.
Bee	While honey lies in every flower, it takes a bee to get the honey out.
Bee	A dead bee will make no honey.
Bee	Bees that have honey in their mouths have stings in their tails.
Bee	He's like the master bee that leads forth the swarm.
Bee	Every bee's honey is sweet.
Bee	When bees are old, they yield no honey.
Bee	Where bees are, there is honey.
Bee	A hive of bees in May is worth a load of hay.
Bee	Honey is sweet, but the bee stings.
Bee	The bee's knees.
Bee	She thinks she's the bee's knees.
Bee	What is good for the swarm is not good for the bee.
Bee, fly	One bee is better than a handful of flies.
	Lepidoptera
Butterfly	Butterflies store no money.
Butterfly	Bookworm.
Butterfly	Happiness is a butterfly.

Table 1.11-12 contd. ...

...Table 1.11-12 contd.

Kind of arthropod	Proverb (saying, expression)
	Lepidoptera
Butterfly	You are like the butterfly that flies from flower to flower.
Butterfly	Take not a musket and kill a butterfly.
Butterfly	What good is a red apple if it has a worm?
Butterfly	Every cider apple has a worm.
Butterfly	A social butterfly.
Butterfly	Have butterflies in one's stomach.
Butterfly	Proverbs are like butterflies—some are caught, some fly away.
Moth	Moth-eaten.
	Orthoptera
Locust	The locusts have no king, yet go they forth all of them by bands.
Grasshopper	Knee-high to a grasshopper.
	Phthiraptera
Louse	To louse something up.
	Siphonaptera
Flea	If you lie down with dogs, you'll get up with fleas.
Flea	The dog in the doghouse barks at his fleas; the dog that hunts does not feel them.
Flea	Who play wid de puppy get bit wid de fleas.
Flea	The lean dog is all fleas.
Flea	A reasonable amount of fleas is good for a dog. They keep him from broodin' on being a dog.
Flea	Even a flea can bite.
Flea	The brave flea dares to eat his breakfast on the lip of a lion.
Flea	A flea-bitten horse never tires.
Flea	Do not flay a flea for hide and tallow.
Flea	Better the wolves eat us than the fleas.
Flea	An elephant does not feel a flea bite.
Flea	With a flea in one's ear.

[1] Compiled from the following websites: http://www.webring.org/l/rd?ring=famousquotesandf;id=1;url=http%3A%2F%2Fwebspace%2Ewebring%2Ecom%2Fpeople%2Fif%2Ffiqabil%2Finsectpro%2Ehtml; http://entnemdept.ufl.edu/pubs/proverbs.htm; http://bijlmakers.com/insects/insect-proverbs-and-quotes/; http://www.seq.qc.ca/english/activites/leaflets/f2a.htm; http://intranet.puhinui.school.nz/Topics/Insects/InsectAnatomyDutch/bijlmakers/entomology/citaten_engels.htm; http://ag.arizona.edu/pubs/insects/ahb/infl3.html).

Note that in the proverbs and sayings listed, insects are mentioned that are encountered in everyday life. They include **cruel enemies** of the human (flies, fleas, lice) and those that **inspire** respect for industry and a harmonious life (bees, ants, wasps), or that **gratify** us with their beauty (e.g., butterflies).

1.12

Insect Conservation

The idea of conservation of insects is not new. Among the **first publications** on this subject was an article by J. Goldmann in 1911 that was devoted to the preservation of the Clouded Apollo (*Parnassius mnemosyne*) butterfly due to its mass capture (for sale to collectors) in Silesia, Germany (Sviridov 2011). This idea was realized in 1936 when this butterfly, together with two other species of Apollo (*Parnassius*) and Scarce Swallowtail (*Iphiclides podaliricus*), was assigned to the **first protected species** of insects. The Clouded Apollo (Photo 1.12-1) was also the first species of invertebrates included in the Convention on International Trade in Endangered Species (CITES) of Wild Fauna and Flora list of species (Naconieczny et al. 2007).

For many people, the thought of insect conservation raises eyebrows because people always see so many insects in the fields and woods that this issue seems hardly relevant. However, the data collected by entomologists show that the abundance of insects across the globe is decreasing. Many studies confirm the general **decline in insect diversity** over the last 50 years (Schuch et al. 2011).

For example, based on the national plans of monitoring of 17 species of butterflies in 19 European countries, the all-European reports referred to as **"European Butterfly Indicator"** are periodically

Photo 1.12-1. Clouded Apollo (*Parnassius mnemosyne*) butterfly is one of the first protected species of insects. Also it is the first species of invertebrates included in the Convention on International Trade in Endangered Species (CITES) of Wild Fauna and Flora list of species.
Photo credit: By Algirdas (Own work) [GFDL (http://www.gnu.org/copyleft/fdl.html) or CC-BY-SA-3.0 (http://creativecommons.org/licenses/by-sa/3.0/)], via Wikimedia Commons

published. In particular, the report of the status of populations of species inhabiting the grassy habitats (grassland butterflies) shows that the index of their abundance has declined by almost 50% since 1990, indicating a dramatic loss of grassland biodiversity (European Grassland Butterfly Indicator 1990–2011, 2013). It was also found that 71 of 576 species of butterflies inhabiting in Europe are under the threat of extinction (Swaay van et al. 2008).

The extinction of different species (including species of insects) occurs naturally and is one of the major components of the evolution of life on Earth. However, the rates of extinction accelerated greatly in recent times, and they vastly exceed the rate at which new species evolve (http://www.earthlife.net/ insects/conservation.html). Insects, then, despite their incredibly large numbers, are disproportionately at risk of extinction (Berenbaum 2009).

It should be clear from the previous section how there are numerous direct and indirect **advantages of existence of insects** for humans. Without insects most of the terrestrial life forms on this planet would slowly disappear. In addition, there are positive aspects which cannot be presented in value terms. For example, a variety of life forms is important to our mental and spiritual health and aesthetic appreciation of their beauty gives us pleasure (http://www.earthlife.net/insects/conservation.html).

Therefore, one can identify **three primary reasons** for conservation of insects: (1) our own survival and our economy depend on many insect species; (2) we find enjoyment in contemplation of wildlife; and (3) we acknowledge our moral responsibility to act as managers of the planet, taking care to maintain respect for other life forms. According to different data, insects range from 80–90% of all the animal biodiversity; it is impossible to preserve animal biodiversity if we ignore the most numerous taxon. In the past, insects have received less attention for conservation than other groups, such as birds (https://www. amentsoc.org/insects/conservation/insect.html).

For efficient conservation of insects, it is necessary to look at the ways in which human activity influences insects.

1.12.1 Human impact on insects

Human impact on insects can be direct and indirect. The **direct impact** includes the physical destruction of insects by different kinds of activity. For example, many drivers have to wash insects that stick to their vehicles after journeys into the countryside. On the roads of the United States, around 32.5 trillion insects perish every year (Messenger 2011).

Grassland farming results in the **death of many insects**. A great number of insects perish when sheep, cattle, and horses cut plants to the root as they graze. Cattle trample the grasslands and crush many insects, especially slow-moving ones. For example, in Belgium and Netherlands, the effect of sheep grazing on the hibernating caterpillars of Glanville fritillary (*Melitaea cinxia*) was studied. It was found that high-intensity pasturage of sheep resulted in the damage or destruction of 64% of the nests, whereas in low-intensity pasturage or its absence, these values were 12% and 8%, respectively (Noordwijk van et al. 2012).

Indirect influence of humans on insects is manifested in environmental changes. In comparison with this type of influence, the physical destruction of insects is absolutely insignificant. Owing to their activity, humans cause profound changes in natural complexes. Humans establish the agricultural and urban systems destroying the natural ecosystems, realize introduction of new species of plants, affect the natural ecosystems by way of melioration, fell trees, contaminate the environment with numerous substances, and so on. Human impact on the biosphere is global. It has already become one of the major factors influencing the evolution of living organisms and environmental modification (http://insecticea. ru/yekologiya_nasekomyh_i_biologicheskie_ritmy/vliyanie_cheloveka_na_nasekomyh).

Anthropogenic factors can affect insects through changes in structure and condition of the ecosystems, including abiotic and biotic factors, that is, change of the microclimatic conditions, chemical composition of the life environment, condition of plants on which the insects feed, species composition, and number of parasites and predators and other factors. These impacts can appear both in the course of the individual development of the organism and at the levels of population and ecosystem.

At the **organismic level**, the response reactions of insects are the change in the chemical composition of body and accumulation of pollutants; change in development characteristics (development time, reproductive potential, disturbances or changes of metabolism); developmental anomalies; and change in behavior. At the **level of population**, the changes in its number, reproductive structure, mean sizes of insects, etc., can occur.

Many works are published about the impact of different kinds of human **economic activity on insects**. For example, Bubova et al. (2015) did an analysis of more than 100 papers, representing most European countries, to help clarify the nature of the effect of different types of land use on butterflies. Without getting into details (influence can cardinally differ depending on species of butterfly, its life stage, season, and a variety of other factors), plant cultivation, grazing, forestry, afforestation, and drainage were acknowledged as the most crucial kinds of human activity effecting butterflies.

Character of impact is specific to each kind of activity. For example, plant cultivation is achieved mainly through the application of fertilizers, herbicides, and insecticides. The afforestation on previously cleared habitats results in fragmentation of habitats because the grassland butterflies have difficulty dispersing through forested landscapes (Nowicki et al. 2014).

Drainage has a negative effect as well; reduction of soil water levels leads to disappearance of the host plants of the moisture-loving butterflies. Many recommendations on mitigation of the effects of different kinds of economic activity (changes in intensity, time frame, applied technologies) on the particular species of butterflies at different life stages have been worked out; for example, with respect to grazing, the types of grazing animals, grazing period, and load intensity (e.g., limited to 0.2–0.5 livestock unit per hectare) (Bubova et al. 2015).

In broader terms, it is safe to say that there are **two major threats** to insects on the part of human (Swengel et al. 2011; Noordwijkvan et al. 2012; Rosin et al. 2012): (1) habitat fragmentation (i.e., division of the solid area of habitats into two and more fragments); (2) deterioration of habitat quality. Habitat **fragmentation** can lead to extinction of populations. If one large integral population can normally exist over a large area of the habitat, then often none of its fragments can provide the long-term stable existence of the subpopulation. It is related to the processes of inbreeding (closely related crossing) and gene drift (chance variation of gene frequencies). **Deterioration of habitat quality** explicitly affects the live environment that results in decrease in the number of species. These questions are considered in detail in Section 1.12.3.2.

1.12.2 Principles of insect conservation

Not all of the vast varieties of insect species require conservation in equal measures. Those most vulnerable to indirect influence and in need of conservation are insects with the following **characteristics**: (1) restricted range; (2) incapacity to migrate; (3) aggregative tendencies (i.e., propensity to form large crowds of insects); (4) restriction to an ephemeral (i.e., nondurable) habitat; (5) restriction to nutrition from a rare plant, or monophagy; and (6) an aquatic mode of life (Gornostaev 1986).

The areas of distribution of different species vary widely. There are the species that are distributed worldwide (cosmopolites), with areas covering a large part of the land, and endemic species distributed within a small geographic area. Such insects as fruit flies (*Drosophila melanogaster* and *D. mercatorum*) and red cockroach (*Blattella germanica*) can serve as examples of cosmopolites. Their widespread occurrence is largely explained by the fact that the synanthropic species are widely settled, thanks to people, and survived the difficult environments in the houses or household buildings of a human.

1. Insects whose expansion is limited by insurmountable barriers have **restricted ranges**. These are the insular or spelean forms, inhabitants of intermontane valleys or the upper zones of mountain ridges. The flightless insects have frequently the narrow ranges. For example, in the Caucasus region, ground beetles are abundant only on one or two ridges: species *Carabus polychrous* is found only in the Bzyb Range (Abkhazia), species *C. cordicollis* only on Mount Elbrus, and *C. komarovi* in Svanetia. **Similar ranges** are characteristic of the flightless gold beetles of the genus *Oreomela* in Tian Shan and Pamir-Alay. So, *O. medvedevi* lives only in one district of the Shughni Range in Pamir, *O. transalaica* in the Trans-Alay Range, and *O. bergi* in the Alpine zone of the Turkestan Range (head

of Isfara river) (http://botan0.ru/?cat=3&id=200). The Fregate beetle (*Polposipus herculeanus*) lives only on the similarly named island of the Seychelles (https://en.wikipedia.org/wiki/Polposipus_ herculeanus). Often, just such narrowly localized species suffer from human economic activity and are in need of conservation.

2. Because of human activity, development of agro-ecosystem and settlements, the range of **slowly migrating insects** evolves into a great number of small islets. Because of insufficient exchange of genetic information as well as ease of ecosystem disturbance in a small section, such earlier widely distributed species prove to be threatened with extinction. One of the most striking instances is that of the Apollo butterfly (*Parnassius apollo*) listed in the Red Book of Russia, Ukraine, Belorussia, Germany, Sweden, Norway, and Finland (Gornostaev 1986).

3. **Crowds of insects** in one place during unfavorable seasons increase the possibility of their elimination. It is known that the monarch butterfly (*Danaus plexippus*) during hibernation in Mexico and California congregates in great numbers (hundreds of thousands of individuals) in the isolated trees. There is some concern over a decrease in their number: studies carried out in 2014 showed that the population of butterflies west of the Rocky Mountains has dropped more than 50% since 1997; east of the Rocky Mountains this population has declined by more than 90% since 1995 (Xerces Society 2014). At the present time, these places of accumulations of butterflies are specially protected. Another example is the crowds of the ladybugs on the slopes of bald mountains in winter. These beetles perish sometimes in the mass at the time of forest fires or spring fires, as well as mining openings.

4. **Restriction to ephemeral (i.e., nondurable) habitats**. The examples of such biotopes include mountain steppes, seaside dunes, and dry slopes with limestone outcrops. At such places, peculiar fauna is revealed. Dunes travel continuously because of eolian processes, and intense migration and accumulation of sand deposits are characteristic to these habitats. Slopes with outcrops of limestone and chalk deposits are frequently used as grasslands and for extraction of commercial minerals, which results in the fast degradation of the vegetation cover and of the entomofauna and, more often, complexes of hemipterous insects, confined to it (Lychkovskaya 2006).

5. The possibility of extinction of monophagous insect species that obtain their nutrition from a **rare plant** is self-explanatory.

6. Many **aquatic insects** are very sensitive to chemical contamination, changes in oxygen concentration, and temperature regimen. The larvae of dragonflies and stoneflies are especially sensitive. Human economic activity resulting in changes in the condition of the water body or watercourse is inevitably reflected in the populations of insects (Gornostaev 1986).

It is known that, among vertebrate animals, the species having the **largest body sizes** become extinct first. For invertebrates, the same trend is preserved. Throughout the world is manifested a tendency to reduce the range and number of forms with the largest body size and more vivid coloring. Any distinctive feature of a creature's biology—weak capacity for migrations, restricted nutritional adaptation, low rate of reproduction, small number of generations throughout the year or long-term development cycle, and so on—can contribute to the extinction or the reduction in number of the species (Primack 2002).

Respective examples include the extinction in the Russian steppes of the wingless grasshopper *Bradyporus multituberculatus*, which earlier had a reasonably sized population (Khatukhov 2015); reduction in number of the predatory bush cricket *Saga pedo*, an extra-large grasshopper with a body size of up to 12 cm, which earlier had been all-pervasive in Europe and Asia but now is considered vulnerable globally (https://en.wikipedia.org/wiki/Saga_pedo).

The **difficulty of conservation** of insects lies in the fact that many insects use different habitats at different stages of their life (nutrition in the juvenile vs the adult form as it matures, mating season, laying eggs, hibernation stage). One can identify **two approaches** to the conservation of insects (https:// en.wikipedia.org/wiki/Insect_biodiversity): (1) allotment of large portions of land, using "wilderness preservation" as the motive for (or counteraction to) the particular processes affecting the charismatic vertebrates, which will have indirectly help with the conservation of insects (in the current environmental culture, the preservation of habitats merely for insects takes a low priority); or (2) conservation of one of the "charismatic species" of insects (butterflies or large, colorful beetles), which will have an indirect positive impact on the other species of insects (umbrella effect).

Many international, regional, and national organizations are engaged in insect conservation. These organizations can be divided into the following **categories**: (1) organizations sensitive to the conservation of specific groups of insects or invertebrates in general; (2) organizations studying insects (or their specific groups), who consider the conservation of insects as one of the principal purposes; (3) organizations more likely to support conservation of wildlife than insects in particular, but nonetheless giving attention to the latter; (4) organizations that work for the conservation of invertebrates other than insects; and (5) organizations engaging in wildlife conservation in general while recognizing the importance of insects and other invertebrates.

Examples of organizations in the **first category** include the World Save Bee Fund, British Bumblebee Conservation Trust, and Butterfly Conservation (UK); **second category**, International Union for the Study of Social Insects (IUSSI), Russian Entomological Society, and British Dragonfly Society; **third category**, the African Wildlife Foundation and Bird Life Australia; **fourth category**, the International Society of Arachnology, Xerces Society (USA), and Conchological Society of Great Britain and Ireland; and **fifth category**, the Conservancy Association (Hong Kong), Durrell Wildlife Conservation Trust, European Wildlife, International Union for Conservation of Nature, and Natural Resources (IUCN).

1.12.3 Methods of insect conservation

Insects differ from other animal as their conservations must take into account the following factors (Gorbunov and Murzin 2009): (1) strong affinity of insects to their habitats leads to the local occurrence of the majority of species; (2) their living conditions are determined by the combination of a large number of factors (light, humidity, temperature of soil and air, temperature and seasonal climatic rhythms); (3) insects require a presence of the forage plant inherent to the species under consideration for the fact that larvae and imago require different feeds; (4) size of territory should meet all the conditions.

As a consequence, the populations of insects have the following **features**: (1) high number (even minimum number of the individuals of any stable population exceeds by a factor of several times the population number of big animals); (2) sizeable fluctuations of the number caused by the vast number of the consumers of insects, frequent alteration of generations and rates of reproduction exceeding by a factor of hundreds and thousands those of mammals and birds; (3) the range of insect is often divided into the small sections with high number and the remaining part of the range where the insects are rare or absent; such sections are often isolated, that is, the species exists as the aggregate of local populations (Gorbunov and Murzin 2009).

By reference to the formal point of view, one might as well say that the conservation of insects has the centuries-old history. Here, the case is the **conservation of honey bees**. Since ancient times, the beekeeping was of great economic importance for the Slavic nations (much the largest than in the European countries). The honey (traditional drink as well as alternative to sugar and medicines) and wax (for church candles) served as the important items of the domestic trade (as well as tributes, assessments, and taxes) and export.

For example, about 800 tons of wax were exported yearly by sea from Russia only to England in the XVI century. Therefore, the measures aimed at the conservation of wild bees were taken in the country. In the course of honey harvest, it was thought that it is necessary to leave for bees a part of honeycombs (usually, about a half). At the same time, the theft of bees in the apicultural forests in Poland was even punishable by hanging (Sviridov 2011).

However, these were in essence the **measures aimed at protection of business** rather than insects. As for the insect protection proper, then, in general terms, one can identify two different methods: (1) individual protection (conservation) of species of insects; (2) conservation of insects habitats. In cases of urgency, such protective measures as transfer of populations and captive breeding of insects are applied (Poltavsky 2012).

1.12.3.1 Individual conservation of species

The **first approach** to conservation of insects is reflected in the Red Data Books and IUCN Red Lists of Threatened Animals. The Red Data Books are annotated lists of rare and endangered animals, plants and fungi; they are classified into different levels (international, national, and regional). The first edition of the IUCN Red Data Book went out in 1963. The IUCN Red List of Threatened Animals was firstly published in 1988 (https://en.wikipedia.org/wiki/IUCN_Red_List). Both the editions are off the press under the auspices of the International Union for the Protection of Nature and similar in essence. One more document of the protective purpose is the List of the insects the export, re-export, and import of which is governed by the (CITES) Convention on International Trade in Endangered Species of Wild Fauna and Flora (https://www.cites.org/eng).

The insects have not appeared in these Books in a day. For example, the insects were **first included** in the Red Book of the USSR in 1984 in number of 202 species (Nikitsky and Sviridov 1987) although its first edition came off the press in 1978. In the first instance, there were many discussions concerning the principles of attribution to the protected species. It was proposed to conserve all the useful species (pollinators, entomophages, trophic resources of useful animals, etc.), big vivid insects being of the aesthetic value, rare, relict and endemic insects, etc. (Sviridov 2011).

Now, the **degree of risk** of the species survival is taken as basis. It is assumed that, 11,200 species of insects have gone extinct since 1,600, predominantly in the tropical rainforest (Samways 2009). The IUCN Red List of Threatened Animals numbers 776 species of insects (Lemelin and Fine 2013). As on October 2, 2013, the CITES list includes 89 species of invertebrates (https://www.cites.org/sites/default/files/eng/disc/species_02.10.2013.pdf).

The Red Data Book of International Union for Conservation of Nature and Natural Resources, most Red Data Books of national level as well as the Red Lists are not juridicial (legal) documents but are exclusively **nonregulatory** ones. They cover the animal world on a global scale and contain the recommendations on conservation addressed to the countries and governments of the territories where such threatened animals have their habitat. Due to the globality of scales, these recommendations are of the most general, approximate character.

At the same time, in a number of countries (Russia, Bulgaria, Belarus, Lithuania, Ukraine, Tajikistan, Uzbekistan), an entry of the species into the Red Book automatically results in the prohibition of its acquisition independently from the species status category. Currently, insects are presented in the national Red Books by the following number of species: Russia, 95 (Tikhonov 2002); Kazakhstan, 85 (Yashchenko and Mityaev 2005); Lithuania, 123 (https://en.wikipedia.org/wiki/List_of_extinct_and_endangered_species_of_Lithuania); Ukraine, 226; and Belarus, 69 (Khoruzhik et al. 2005).

The number of insect species listed in appropriate Red Books does not reflect in many cases the real situation of their condition in nature. Such fairly dispersed species of butterflies as the Clouded Apollo (*Parnassius mnemosyne*), copper butterfly (*Neolycaena rhymnus*), and emperor moth Artemis (*Actias artemis*) turned out to be in the Russian Red Book. Often species that do not need protection and drawn arbitrarily upon the extensive list (mainly, big, beautiful, or known to the authors) are included there (Gorbunov and Murzin 2009).

Methods of conservation of insects and large animals are completely **different**. The prohibition of game hunting is enough in many cases to restore numbers of large animals; however, the prohibition of catching of insects does not provide any such effect that is confirmed by the world practice. In the early 1980s, the following experiment was carried out in the England. Three sections where the rare large blue butterfly *Maculinea arion* has ranged were exploited by two **different methods**. In one section, cattle grazing was permitted, whereas in the other one, the entomologists were allowed to catch *Maculinea arion* in unlimited numbers. The third section was used as the control, where no activities were conducted. In two years, the butterfly was almost completely destroyed in the first section while in the other two they remained in their original form (Khodorych 2002).

Science knows of no established instance of extermination of any species of insect being exterminated by way of its **collection**. This is explained by the nature of insects, which is so different from the nature of the vertebrate animals. Insects are designed by nature to be eaten by other species of animals. This does not damage insect populations because their reproduction rates take such natural extinction into account.

For example, a single colony of bats eats approximately 100 tons of insects per night in Bracken Bat Cave (http://therivardreport.com/members-night-at-bracken-bat-cave/).

Ironically, the prohibition of insect collection, on a nationwide scale, caused **damage** to the conservation of nature. In order to reach effective conservation of nature, most people in the country should appreciate nature. Collecting insects is the first, fascinating step towards communing with nature—not afflicting damage to nature, but allowing each person to make small discoveries. Nowadays, children are familiar with many models of cars, but they by no means know the natural insects. Therefore, the prohibition of insect collection is meaningless.

In Japan, with its enormous density of population, there is no the Red Book for insects. There, the equipment and instructions on collection and determination of insects can be acquired in any shop and children in schools obtain assignments for preparation of collections. Therefore, the entomofauna of Japan has been better studied than that in any other region of the planet, although entomological investigations began there more than 100 years ago (in Europe more than 250 years ago) and conservation of nature produced notable results (Gorbunov and Murzin 2009).

At times, someone's concrete **financial interests** stand behind nature protection activities. For example, in the opinion of many entomologists, the decision to enter bird wings (butterflies of the genus *Ornithoptera*) inhabiting Papua New Guinea, into the CITES list was absolutely unfounded. These butterflies were never under threat of extinction; however, they are the world's largest butterflies (e.g., Queen Alexandra's bird wing *O. alexandrae* and Goliath bird wing *O. goliath.*

Because, among other things, they are also very beautiful, their commercial value is high. Since registration of bird wings in the list of endangered species, their sale is now conducted on behalf of the special state Insect Farming and Trading Agency (IFTA). In effect, all the breeding is reduced to gathering of larvae and pupae of butterflies by local residents. IFTA purchases the "hatched" bird wings from them at a cheap rate (already in its own name) and exports the dead specimens to other countries (Khodorych 2002). The price of one specimen reaches $2000 (Muafor et al. 2012) and an annual sales total about $400,000 (Cranston 2010).

1.12.3.2 Conservation of habitats

At present, the majority of entomologists acknowledge that there are two major threats to the biodiversity of insects: (1) loss of habitats and (2) deterioration of habitat quality (Rosin et al. 2012). In the course of human economic activity, **habitat fragmentation** (i.e., their division into the smaller components) occurs permanently. Fragmentation is a result of exploitation of natural resources. Their extraction transforms the territory covering it into a network of disruptions of different sizes.

When the number of such disruptions is small or their area is not too large in comparison with native habitats, progressive succession takes place. However, upon reaching some limits of the size of disturbed sections and density of the network, the natural processes of the reproduction of aboriginal associations are blocked and original vegetation is **not restored**.

Any form of human economic activity (crop farming, cattle breeding, extraction of commercial minerals), the ecological balance is disrupted. Also, around the section themselves at which the activity is carried out (quarries, hayfields, grasslands, fields, and so on), the increase in disruptions occurs because the people performing these works also perform other actions (hunting, recreation).

Usually, the individual spots of disruptions of the natural landscape are **not isolated** from each other but rather are related to each other through any communications (e.g., paths, roads, pipelines, power lines) by which the produced resource (e.g., crop yield, timber, mineral resources) is brought out. Therefore, the fragmentation of habitats results in extinction of some species of insects.

The deterioration of **habitat quality** is another consequence of human economic activity. It takes a variety of forms (contamination of natural components owing to the application of fertilizers, pesticides, and herbicides; industrial emissions to the atmosphere and discharges of sewage waters; change in the underwater level in the course of the irrigation and drainage amelioration; soil compaction; and direct insect destruction, in the case of grassland farming and the like.

The response reactions of insects to the quality deterioration can be most diverse (e.g., developmental disabilities, drop of fertility, change in behavior). The listed threats to the biodiversity of insects can be mitigated by the change in activity intensity, duration it carried out, applied technologies, and so on.

The sufficient detailed **list of measures** needed to avoid quality deterioration of habitats of insects and other invertebrates can be found on the site of Amateur Entomologists' Society of Great Britain (https://www.amentsoc.org/insects/conservation/habitat.html). Among them there are both fairly evident (e.g., prohibition of pesticide and fertilizer application around the edges of the cultivated lands and near hedge rows; creation of uncultivated strips on the great farm fields to encourage natural enemies of pest species) and quite specific (e.g., restriction on using antiparasitic drugs such as vermectins, which make manure toxic to dung-feeding invertebrates). Examples of such measures of mitigation are also given in the article by Bubova et al. (2015).

It is evident that the real protection of insects should aid in the **conservation of ecosystems** as a whole. It is absolutely wrong to pose a question of conservation in the nature of each individual specimen of insect. Their enormous fertility shows that, in the world of insects, the population of given species and species as a whole, rather than individual specimens, is appreciated (nature intends them as the food for other creatures). For maintenance of viable populations, it is necessary to conserve their biotope with sufficient area, making it possible for the species to exist at least in minimum numbers (Gorbunov and Murzin 2009).

When using such protective measures as conservation of habitats, the particular territories are completely withdrawn from the economic turnover, and its area should be quite large. The fact is that ecosystems are subject to **successions**—the subsequent regular exchange of one plant community for another as a result of the effect of natural factors. For example, forests that were originally all birch become, in 60 to 70 years, mixed birch-fir and, in another 20 to 50 years, can be transformed into stable fir forests.

Across a large territory, the succession cannot be strictly synchronous, thanks to which the set of insects is conserved as a whole (Chenikalova et al. 2008). In addition, unfavorable **weather conditions** (drought, anomalously cold winter) during a certain year can result in conditions temporarily inapplicable for particular species, even though the state of the environment does not give rise to unfavorable criticism (https://www.amentsoc.org/insects/conservation/habitat.html).

The number of **reserves** especially established for conservation of insects is not large. One example is the Monarch Butterfly Biosphere Reserve, in Mexico, established with a view to conservation of the monarch butterflies (*Danaus plexippus*) hibernating area. On its territory, up to a billion individuals of this butterfly hibernate from November to March (https://en.wikipedia.org/wiki/Monarch_Butterfly_Biosphere_Reserve).

Another example is the reserve in the Republic of South Africa (Kwa Zulu-Natal province), established especially for conservation of Karkloof Blue butterfly (*Orachrysops ariadne*) (Samways 2005). It is recognized that, for conservation of invertebrates, such a system of specially protected natural areas is more effective when there are several smaller reserves within one natural zone rather than one reserve occupying the entire area. This is related, for example, to the fact that the individual section of invertebrate species is less than the individual section of vertebrate species (Speranskaya and Zaytsev 2011).

One of the effective measures of insect conservation is the creation of **small-scale nature reserves**. On the one hand, their small area (about 3–30 ha) does not guarantee initial entomofauna conservation. It is known that smaller reserves depend on a self-regulating system, and the conservation of insects in these territories without continuous human support has no effect (Chenikalova 2008). On the other hand, it is much easier to remove a small allotment from the commercial turnover than to withdraw the territory in order to organize a comprehensive reserve. Furthermore, for the arrangement of small-scale nature reserves, one can use different areas of inarable lands (e.g., gulleys, steep slopes).

Among the **first experiments** of this kind was the small-scale nature reserve Bumblebee Hills (Shmelinye kholmy), established in the Omsk Oblast, Russia, and designed, first of all, for the conservation of **bumblebees**. In the period of arrangement (1972), it occupied a section of the forest steppe with area of 6.5 ha (Grebennikov 2014) but now its area has increased to 284 ha.

An example of a small-scale nature reserve oriented to the conservation of **dragonflies** is the Dragonfly Kingdom, in Nakamura, Kochi Prefecture, Japan (Primack et al. 2000). There, in place of the former rice fields, with an area of 50 ha, a system of bogs and ponds was created. In this small-scale nature reserve 64 species of dragonfly live (Kadoya et al. 2004).

The Vordan Karmir State Reservation (Photo 1.12-2) was established in the Ararat plain for conservation of the **Armenian cochineal** (*Porphyrophora hamelii*); now, its area is 220 ha (https://en.wikipedia.org/wiki/List_of_protected_areas_of_Armenia).

There are also other organizational forms of insect conservation. For example, the establishment of natural monuments protected by law that are referred to as the **Sites of Special Scientific Interest** (SSSI), in the United Kingdom. The status of locality assumes the conservation of the natural and artificial processes leading to development and survival of populations. Territories awarded this status are protected by law against new construction and any interference that can damage them (https://www.amentsoc.org/insects/conservation/habitat.html).

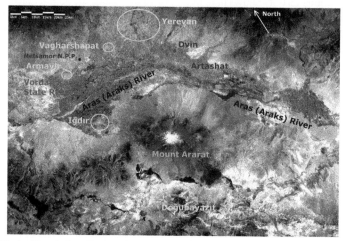

Photo 1.12-2. The Vordan Karmir State Reservation was established in the Ararat plain for conservation of the Armenian cochineal (*Porphyrophora hamelii*); now its area is 220 ha. The image shows part of its historic habitat within the Ararat Plain and Aras (Araks) River Valley northeast of Mount Ararat. The red (crimson) area marked with a gold border is the major portion of the Vordan Karmir State Reservation near the village of Arazap in Armenia, a national sanctuary for the insects and salt marsh vegetation. The purple annotations denote the historic vordan karmir-producing cities of Artashat and Dvin. Photo credit: By Ketone16 (based on a photo taken by the crew of Expedition 28 of the International Space Station) [CC0 or Public domain], via Wikimedia Commons

PART 2

Negative Aspects

The first part of this book discusses the numerous **benefits** humans derive, directly or indirectly, from insects. But there is also the other side of the coin: consider the not-infrequent situations when insects and other invertebrates create problems for people. The second part of this book focuses on some of the negative aspects of human–insect interaction.

2.1

Insects as Pests

It has been said that all creatures on Earth are intended to fulfill a certain purpose, to accomplish a certain mission. From this point of view, **there are no harmful** and **useful insects**. However, from the human viewpoint, insects are frequently the cause of economic loss. It is recognized that about 10% of species of insects fall into this category (http://slovariki.org/ekologiceskij-slovar/2667). This chapter looks at different kinds of deleterious insect activity.

2.1.1 Insects as agricultural pests

Worldwide, an estimated 67,000 different pest species attack agricultural crops: approximately 50,000 species of plant pathogens, 9,000 species of insects and mites, and 8,000 weeds (Pimentel 2009). The number of insect species damaging different agricultural crops varies widely. For example, millet is damaged by 24 insect species, whereas the grapevine is damaged by 459 insect species belonging to nine orders (Dermaptera, 2; Orthoptera, 17; Isoptera, 12; Hemiptera, 116; Thysanoptera, 34; Lepidoptera, 106; Diptera, 12; Hymenoptera, 26; Coleoptera, 134). Add to that 41 species of mites, 113 species of nematodes, 27 species of birds, 2 species of bats, 6 species of rodents, and 5 species of snails and slugs, and the total number of species detrimental to the grapevine reaches 653 (Mani et al. 2014).

2.1.1.1 Characteristics of harmfulness

The **harmfulness** of insects presents the complex biological phenomenon determined by the action force of harmful species and response reactions of the plants. The same insect can be both useful and harmful from the economic point of view. For example, pests of cruciferous vegetables are harmful when destroying the cabbage but useful when damaging the weeds. The harmfulness of insects depends, in great measure, on their selectiveness. If the pests prefer underdeveloped plants, harmfulness is less because population rehabilitation occurs. If more developed plants are primarily destroyed, the harmfulness is higher (Tansky 1988).

In certain situations, harmful insects can become useful. For example, in case of dense growth, the depression or destruction of certain plants is a favorable factor for the population as a whole because it improves the physiologic state and increases the productivity of remaining plants. For optimum seeding density, partial compensation of losses takes place. For example, when the larvae of the wheat bulb fly (*Leptohylemyia coaretata*) destroyed 50% of the wheat planted, the crop yield dropped by only 20% because the undamaged plants increased in productivity (Tansky 1988).

The **main reason** insects can have a harmful effect on agricultural plants is simple: they feed on plants (Photo 2.1-1). Insects also damage agricultural plants through nesting over planted fields or embedding eggs in plant tissues. Most commonly, plant tissues are damaged or destroyed by the feeding of gnawing pests. Damage to plant leaves leads to shrinkage of the assimilating surface area, while damage to stems slows nutrition and water delivery to other parts of the plant. Root damage disturbs the absorption of minerals and water from the soil. These negative effects weaken agricultural plants, slow their growth, and cause parts of plant tissues to rot.

In the majority of cases, **sucking insects** do not damage the apparent integrity of plants. Their presence is usually first recognized as spots on the surface of leaves, which are good indicators of pathologic changes in the affected area. Sucking pests inject their saliva gland secretions into plant tissues, which leads to biochemical changes in the plant cells and tissues. Aphids (Aphidoidea; also known as plant lice), greenflies, blackflies, and whiteflies are common sucking pests, as are mites (Trombidiformes), thrips (Thysanoptera), and true bugs (Hemiptera) (Bondarenko et al. 1991). Damage disrupts the normal physiology of plants, resulting in changes in activity of respiratory enzymes; slower transpiration rates; and decreased chlorophyll production, growth, and water storage of feeding elements, such as starch, sugars, and oils.

The outlook for an infected plant depends on **three major factors**: (1) specific type of the pest's oral apparatus (i.e., sucking or biting); (2) the part of the plant that is damaged—roots, stem, or leaves; and (3) whether or not the plant has undergone preliminary preparation by the pest. The combination of these factors results in an extensive range of potential damage and a varied outlook for the plant.

The most common damage to plants occurs without any preliminary preparation by pests; the pests generally feed on the plant as is. Damage to roots and parts of the plant that are underground could be caused by **such things** as (1) topical bites; (2) deep bites into the root's body; (3) cavities and tunnels inside of roots and root crops; and/or (4) damage inside plant bulbs.

The **stems** and **trunks** of crops might experience damage such as (1) drilled channels (e.g., holes inside of woody stems and trunks of grassy plants; (2) bite marks and deep bites at the base of stems and trunks; and/or (3) distortions and other damage.

The variable results of pest damage to **leaves** include (1) rough and random traces of eating; (2) partial eating (i.e., skeletization—the eating of tissues between veins, scrubbing of soft pulp, nibbling edges of leaves; (3) mining (i.e., the formation of mines between layers of epidermis [primary integumentary tissues]); (4) change of color in sucking spots; and (5) leaf contortion as a result of uneven growth of tissues.

Causes of pest damage to plant **reproductive organs** are also variable and may include (1) pests grazing on the surface or inside of vegetative buds; (2) injecting flower buds and fruit; (3) superficial grazing on flowers and their parts; (4) nibbling of surface or inside of healthy open flowers or flower buds; (5) nibbling of plant seeds and ovaries; and (6) mining fleshy fruits or tunneling inside of them, which can cause them to decay and drop.

The **economic consequences** caused by pests are variable. In general, the growth rate of plants is decreased due to the damage of photosynthetic leaves. The accumulation of nutrients such as starch,

Photo 2.1-1. Around 200 species of pests are known to attack strawberries both directly and indirectly. The picture shows long-jointed beetles *Lagria villosa* in Brazil (Rio Grande do Sul - Santa Maria) feeding on a strawberry.
Photo credit: Jonas Janner Hamann, Universidade Federal de Santa Maria (UFSM), Bugwood.org

sugar, fats, and proteins in seeds, fruits, and root crops can also be retarded. This damage leads to a reduction in the mass of green crop output, and the weight and quality of seeds, fruits, roots, and root crops. Damage to roots has similar consequences, resulting in crops that also have a shorter storage life because of the processes of decay. Damage to the plant's reproductive organs leads to decreased productivity, deterioration in taste quality, technical quality, seeding, and other qualities.

In general, insects are less harmful if they cause damage to parts of the plant such as the leaves, fresh sprouts, and thin roots, because these parts **recover more easily**. Damage to stems, trunk, or thick roots, on the other hand, cause **more harm**. In these cases, the damage might lead to a complete blockage of the delivery of substances to all or part of the plant.

The worst consequences economically are from damage to the reproductive organs, including damage to growth, seeds, tubers, and root crops. Root crops lose their commercial quality, even if the defects are minor (Tansky 1988).

2.1.1.2 Polyphagous insect pests

Insects are the most dangerous pests in agriculture. The numbers of plant species damaged by a particular insect can vary. Some specific insect pest species affect a very small number of plant species, whereas other insect species are omnivorous (Photos 2.1-2, 2.1-3). Most insects, however, specialize in certain groups of plants, such as cereals, vegetable crops, and fruit trees.

Photo 2.1-2. Among the most dangerous polyphagous insects are some species of locusts. Rough grazing is the type of damage to vegetation that is characteristic for them. The photo shows a desert locust (*Schistocerca gregaria*) swarm (about 3–6 square kilometers) milling over a field of harvested millet north of Bambey, Senegal.
Photo credit: M. de Montaigne, 1993. The picture is presented by FAO, image 17139.

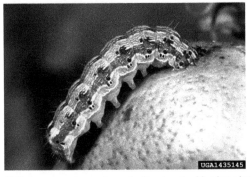

Photo 2.1-3. The corn earworm *Helicoverpa zea* is a major agricultural pest, with a large host range encompassing corn and many other crop plants. *H. zea* is the second-most important economic pest species in North America, next to the codling moth. The estimated annual cost of the damage is more than US$100 million. In the photo, larva(e) of corn earworm attacking cotton boll.
Photo credit: Clemson University, USDA Cooperative Extension Slide Series, Bugwood.org

Polyphagous insects feed on plants belonging to different botanical families. Most of them have biting mouthparts. This group of pests is represented by the order Orthoptera (Acrididae, grasshoppers/locusts; Tettigoniidae, bush crickets/katydids; and Gryllacrididae, leaf-rolling crickets), the order Coleoptera (Curculionidae, weevils; Elateridae, click beetles; Scarabaeidae, dung beetles; and Tenebrionidae, darkling beetles), and the order Lepidoptera (butterflies/months, predominantly their larvae, caterpillars). Multiplying in huge quantities, some highly polyphagous insects damage and even completely destroy agricultural crops (Agricultural entomology 1976).

Among the most dangerous polyphagous insects are some species of **locusts** (*Locusta migratoria*, migratory locust; *Dociostaurus maroccanus*, Moroccan locust; *Schistocerca gregaria*, desert locust; and *Calliptamus italicus*, Italian locust). Their mass breeding takes place near hot deserts on all continents, with the exception of those in southwestern South America (Map 2-1).

The distribution of locusts is based on the existence of **three factors**: (1) high humidity, (2) temperatures of 20°C–40°C, and (3) wind. Humidity is necessary, since locusts lay their eggs only on wet ground; temperatures of 20°C–40°C are required for the growth of grass, on which the wingless offspring of locusts feed; and wind is necessary for their flight, since locusts move in the direction of the wind (Smith 1978). A swarm of locusts is considered "tight" if more than 1 m² of the Earth's surface has greater than 100 insects (Skorer 1980). The size of swarms can be immense. There have been cases when swarms have darkened 250 km² of sky (Mavrischev 2000). In 1889, a giant swarm of locusts flew from the shores of North Africa across the Red Sea to Arabia. The movement of the insects lasted all day; their mass was 44 million tons (Gorshkov 2001).

In the autumn–winter of 1961–1962, an **anomalous locust invasion** was observed in India, Pakistan, and Iran. During this invasion, in northeast Iran, locusts infested more than 500,000 hectares (ha). The intensity of the infestation was such that a space measuring 27 x 12 km was uniformly covered with the larvae of locusts, and their concentrations ranged from 200 to 2,000 larvae per m² (Bondarev 2003).

Rough grazing is a type of damage to vegetation that is characteristic of all species of locusts. The damage caused by locusts can be colossal. In 125 BC, locusts destroyed all the crops of wheat and barley in the Roman provinces of Cyrenaica and Numidia (North Africa). As a result, almost the entire human population there—800,000 people—died of hunger (Mavrischev 2000).

The *Margarita sticticalis* **moth** is another dangerous polyphagous pest. Among its forage plants are the hundreds of species from 35 families. This moth, with a body length of 10–12 mm and wingspread

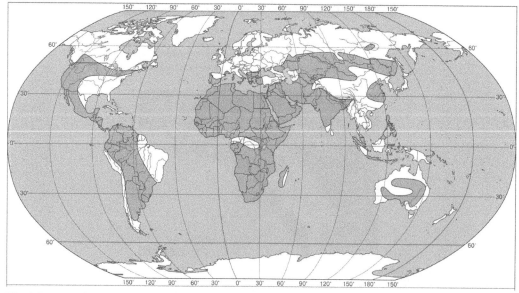

Map 2.1. Areas of mass reproduction and invasions of locusts. Adapted from (Physico-geographical atlas of the world 1964; http://www.fao.org/docrep/u8480e/U8480E0j.htm).

of 18–26 mm, is characterized by very widespread range, which includes practically all countries of the former USSR, almost the whole Europe (except for the northern Fennoscandia), Asia Minor, Iran, Mongolia, North China, Korean Peninsula, Japan, North America. The moth (*Margarita sticticalis*, beet webworm) is the primary pest of white beet and it causes the major damage to the cotton plant, sunflower, hemp, alfalfa, tobacco, many vegetables, melon, and industrial crops (Insects and ticks—crop pests, v. III, part 2, 1999).

Among the polyphagous pests is also the **turnip moth** (*Agrotis segetum*), sometimes called the north locust, with a wingspread of 40–50 mm. It is found in Russia (except for the extreme North), Belorussia, Ukraine, Moldavia, Transcaucasia, Central Asia, West Europe, Africa, Near East, Mongolia, Chine, Japan, India, and Nepal (Insects, v. III, part 2, 1999). The larvae of turnip moth can feed on the plants of more than 160 species; they damage most often white beet, sunflower, tobacco, cotton plant, maize, vegetables, melons and gourds, winter cereals, grapevine, and tee plants. The larvae destroy seeds and plantlets in the soil, eat leaves, and gnaw the plants at soil level (Bondarenko et al. 1991). One larva may destroy 10–15 plants of white beet per night and, at that, the female of turnip moth lays up to 2,000 eggs (Agricultural entomology 1976).

2.1.1.3 Highly specialized insect pests

The damage inflicted by **highly specialized pests** is often comparable with that caused by polyphagous pests. One such pest is the **grape phylloxera** (*Viteus vitifolii*). These are almost microscopic, pale yellow sap-sucking insects belonging to aphid family. The native land of the grape phylloxera is North America where it develops on the wild species of grapevine. The grape phylloxera was imported from the eastern USA to Europe from 1858 to 1862. For the first time, it was discovered in the greenhouse in London (1863) and, then, in France in 1868. From there, the grape phylloxera reached Spain, Portugal, North Africa, Switzerland, Italy, and gradually all the Balkan countries. In 1880, it appeared in Russia, on the South Coast of Crimea (http://wine.historic.ru/books/item/f00/s00/z0000001/st008.shtml).

The **two forms of phylloxera**, root and leaf, differ in external appearance and size. The way it affects the grape plant is very straightforward. Depending on the form (leaf or root), it feeds on sap sucked out the leaves or roots of grapevine. Having a long snout, the phylloxera pierces the top protective layer of the leaf or root, resulting in infection and then rot. Furthermore, the saliva of phylloxera contains agents causing tumorigenesis of the grapevine. The affected parts of roots and leaves stop sucking in nutritive substances. Nodes and swellings form on the roots of the grapevine. The more affected parts there are, the more intense the depression of the grape plant. Eventually, after chronic depression, the grapevine dies (http://www.vinograd7.ru/docs/vrediteli/filloksera.htm).

When this pest reached Europe, it caused tremendous damage to grape growing because local grape varieties were significantly less resistant than those of North America. For example, in France, it provoked a crisis and ruined many vineyards. The **most devastating effects** were recorded in 1884, when phylloxera ruined the vineyards over an area of 1.2 million ha, which corresponded to economic losses of 7.2 billion francs (Painter 1953). In this country, the total volume of wine production fell from 84.5 million hectoliters in 1875 to only 23.4 million hectoliters in 1889. Other data indicate that between two-thirds and nine-tenths of all European vineyards were destroyed (https://en.wikipedia.org/wiki/Phylloxera).

Yet another example of the highly specialized pests is **field pea weevil** (*Bruchus pisorum*). This beetle is widespread in Europe, Asia, North Africa; North, Central, and South America; and southwestern Australia (http://www.agroatlas.ru/en/content/pests/Bruchus_pisorum/). The field pea weevil attacks the fields on the edges and gradually occupies the whole area. It lays eggs on the young pea pods. The larvae eat into the young grains and live inside the grain, feeding on its flesh, until turning into pupae. The damage to grain results in reduction of its mass and quality. Because the excrement of the larvae contains the alkaloid cantharidine, the damaged grain cannot be used as fodder for cattle to be used for meat (http://www.udec.ru/vrediteli/zernovka.php).

2.1.1.4 Impact of insect pests on main branches of plant growing

The vast number of species of harmful insects makes it unreasonable to consider them in sequence, even in terms of certain families or genera. It makes sense to address the effect of pests on the individual branches of plant cultivation. Currently, there are 128 species of insect pests recorded for wheat crop, 128 species for corn, 73 for barley, 70 for rye, 42 for oats, 41 for rice, and 24 for millet. It would be erroneous to identify the damage rate of one or another agricultural crop by the number of pest species. Rice is the **most damageable** crop among the cereal grains. The rice crop showed losses of 26% on account of all insect pests, while wheat and rye losses were 5% and 3%, respectively (Bondarenko et al. 1991), which is absolutely not relevant to the number of harmful species of insects.

Cereal crops are damaged by polyphagous insects, such as Acrididae, locusts; Tettigoniidae, bush crickets or katydids; Elateridae, click beetles; Noctuidae, owlet moths; *Ostrinia nubilalis*, European corn borers, as well as by a number of specialized pests. These include cereal aphids (Aphididae), thrips (Thripidae), Eurygaster (*Eurygaster integriceps*), scarab beetles (Scarabaeidae), some owlet moths (Noctuidae), fruit flies (*Oscinella frit*), weevils (Curculionidae), European wheat stem sawflies (*Cephus pygmaeus*), amongst others. At one time, the Great Plains in the United States and Canada were losing up to 80% of the wheat harvest because of the European wheat stem sawfly (Rice 1986).

Grain legumes are damaged most frequently by polyphagous insects such as wireworms (larvae of click beetles, Elateridae), silver Y moths (*Autographa gamma*), cabbage moths (*Mamestra brassicae*), and beet webworms (*Loxostege sticticalis*). The most destructive of specialized pests are aphids (*Acyrthosiphon pisum*), Sitona weevils (*Sitona* spp.), pea weevils (*Bruchus pisorum*), dried bean weevils (*Acanthoscelides obtectus*), and pea moths (*Cydia nigricana*) (Bondarenko et al. 1991).

Sugar beet roots contain high concentration of sucrose and hence are a great attraction as a food source to many insects, including harmful pests. There are approximately 300 species of insect pests on record that damage sugar beets, including more than 130 species of beetles; 60 species of butterflies and moths; and approximately 40–50 species of aphids, true bugs, and grasshoppers. The most destructive polyphagous pests are the winter moth, meadow moth, and larvae of both click and scarab beetles. The most destructive specialized pests are beet leaf/black-bean leaf aphids (*Aphis fabae*) and sugar beet root aphids (*Pemphigus populivenae*), beet bugs (Miridae), and different types of sugar beet weevils (Curculionidae) (Govorushko 2012).

The **cotton crop** is also heavily damaged by harmful insects. There are 1,326 species of such insects on record around the world (Rice 1986). Cotton seeds and seedlings are damaged by wireworms, cutworms (larvae of the turnip moth), and false wireworms (Tenebrionidae); roots and the underground parts of stems are also damaged by wireworms (Elateridae) and mole crickets (Gryllotalpidae). Leaves are damaged by the larvae of the beet borer/armyworm/asparagus fern caterpillar (*Spodoptera exigua*), the cotton bollworm/scarce bordered straw (*Helicoverpa armigera*), the marbled clover moth (*Heliothis viriplaca*); and other cutworms, locusts, aphids, and thrips. The larvae of the small, thin, grey moth known as the pink bollworm (*Pectinophora gossypiella*) chew through cotton lint to feed on the seeds. Generative organs of cotton crops are damaged by the cotton bollworm (*Helicoverpa armigera*) larvae and also the beet borer/armyworm (*Spodoptera exigua*) (Bondarenko et al. 1991).

In the past, the most destructive worldwide pest to the cotton crop has been the **pink bollworm** (*Pectinophora gossypiella*). In Egypt, the losses were 30%–40% of its cotton; Brazil, 30%–60% (Insects, V. III, Part 2, 1999); and Mexico, 20%–25% (Agricultural entomology 1976).

Since 2000, however, yield losses have fallen sharply. **Cotton losses** to arthropod pests reduced overall yields by 2.58%. Thrips took top ranking at 0.713% loss. Lygus bugs (Hemiptera, Miridae) were ranked second at 0.614%. The bollworm/budworm complex was third at 0.486%; stink bugs were fourth at 0.371%, and fall armyworms were fifth at 0.113% (http://www.entomology.msstate.edu/resources/croplosses/2009loss.asp).

Potatoes are damaged mainly by polyphagous insects. The most dangerous pest of potatoes is the Colorado potato beetle (also known as Colorado beetle, ten-striped spearman, and potato bug) (*Leptinotarsa decemlineata*). Between 20 and 40 beetles can destroy half a potato bush (Photo 2.1-4),

Photo 2.1-4. The most dangerous pest of potatoes is the Colorado potato beetle (*Leptinotarsa decemlineata*). If the insects eat all of the leaves on the bush, the potato yield is reduced by 10 times. Damage to potato foliage by larvae of the Colorado potato beetle is shown.
Photo credit: USDA APHIS PPQ Archive, USDA APHIS PPQ, Bugwood.org

which leads to a reduction in yield by a factor of 2–3 (33%–50%). If pests eat all of the leaves on the bush, the potato crop is reduced by a factor of 10 (Bondarenko et al. 1991).

2.1.1.5 Crop losses and the economic impact of insect pests

It is understandable that the insect pests cause reduced yields of the agriculture crops and, thus, economic damage; this reduction in yields has always occurred, which is confirmed by a variety of historical examples. However, it is difficult to compare the relative quantities of losses considered for different historical periods and different regions because in most cases, the losses related to pests provided are approximate. Generally, these refer to the animal pests which include, besides insects, other arthropods, nematodes, mammals, birds, slugs, and snails. At times, an even broader interpretation occurs when the weeds and pathogens are assigned to pests.

There is a small volume of data concerning the crop and economic losses and related to only insects. Worldwide, insect pests cause an estimated 14% loss (Pimentel 2009). At the level of **particular countries**, there are recent data for the United States and Brazil. In the United States, the average annual losses of the yield reach 13%, which correspond to $26 billion (Pimentel 2009). It has been noted that in this country, the share of crops lost to insects has nearly doubled from about 7–13% in the period of 1940–1990 (Pimentel et al. 1993). In Brazil, the insect pests give rise to loss of 7.7% of yield on average. This results in loss of about 25 million tons of food, fiber, and biofuels and average annual loss of US$ 17.7 billion (Oliveira et al. 2014).

Some information regarding agricultural crops and species of insects is known for certain countries (e.g., India) (Table 2.1-1).

There are data concerning three important agricultural crops and different periods for the world as a whole (Table 2.1-2). However, we are dealing with pests of animal origin which include all of arthropods, nematodes, rodents, birds, slugs and snails.

In Russia before 1917, the total damage caused by pests (specifics unknown) amounted to 980 million gold rubles (corresponds to modern US$29.4 billion). The **amounts lost** were 10% of its gross production; horticulture, 20%; and gardening, 40% (Agricultural entomology 1955). Some global average **annual crop losses** from arthropod pest damage include potato 6.5% (Agricultural entomology 1976) and sugar beet 8.3% (Bondarenko et al. 1991). The financial loss as a result of reduction in the US cotton yield in 2009 reached $502 million (http://www.entomology.msstate.edu/resources/croplosses/2009loss.asp).

In Russia, the deleterious activity of the meadow moth (*Loxostege sticticalis*) in extreme years results in destruction of tilled and vegetable crops and 80% of sugar beets. In particular, the severe outbreak in the number and spread of the meadow moth was observed in Russia in 1986–89. At that time, the area inhabited by its caterpillar larvae covered more than 4 million ha (Finov 1997).

Table 2.1-1. Major pests and percentages of crop loss in India (Shetty and Sabitha 2009).

Crop	Major pests	Crop loss, %
Rice	Stem borer (*Scirpophaga incertulas*)	10–48
	Leaf folder (*Cnaphalocrocis medinalis*)	10–50
	Whorl maggot (*Hydrellia* spp.)	20–30
	Gall midge (*Orseolia oryzae*)	8–50
	Hispa (*Dicladispa armigera*)	5–6
Wheat	Army worm (*Mythimna separata*)	20–42
Pigeonpea	Cotton bollworm (*Helicoverpa armigera*)	14–100
	Pod webber (*Maruca testulalis*)	20–60
	Pod fly (*Melanagromyza obtusa*)	10–60
Sunflower	Cotton bollworm (*Helicoverpa armigera*)	30–60
Cotton	Spotted bollworm (*Earias vittella*)	30–40
	Cotton bollworm (*Helicoverpa armigera*)	20–80
	Pink bollworm (*Pectinophora gossypiella*)	20–95
Cabbage	Diamond back moth (*Plutella xylostella*)	20–52
Cauliflower	Diamond back moth (*Plutella xylostella*)	20–52
Brinjal	Shoot and fruit borer (*Leucinodes orbonalis*)	25–92

Table 2.1-2. Global losses of some crops caused by animal-pests (Oerke 2006).

Crop	Period	Yield, kg/ha	Loss due to animal pests, %
Wheat	1964–65	1250	5.0
	1988–90	2409	9.3
	2001–03	2691	7.9
Maize	1964–65	2010	12.4
	1988–90	3467	14.5
	2001–03	4380	9.6
Cotton	1964–65	1029	11.0
	1988–90	1583	15.4
	2001–03	1702	12.3

2.1.1.6 Mites as agricultural pests

Mites belong to the class of arachnids (Arachnida). A distinctive feature of phytophagous mites is the minute, sometimes microscopic, size of the body. Damage to plants is produced mainly during feeding. Most mites are polyphagous; that is, they are able to feed on plants of different families.

There are good examples of **polyphagy**, such as the red spider mite/two-spotted mite (*Tetranychus urticae*), garden spider mite (*Schizotetranychus pruni*), European red fruit mite/fruit tree red mite (*Panonychus ulmi*), flat scarlet mite (*Cenopalpus pulcher*), strawberry spider mite (*Tarsonemus turkestani*), amongst others. These mites can feed on and cause damage to approximately 600 different species of plants (Handbook of Agronomist on Plant Protection 1990).

Some mites are the pests of the plants belonging to a particular **family**. For example, the Tomato Russet Mite (*Aculops lycopersici*) damages the plants of the nightshade family (tomato, potato, egg plant, pepper, etc.). It is abundant almost everywhere: Australia and New Zealand, USA and Canada, South

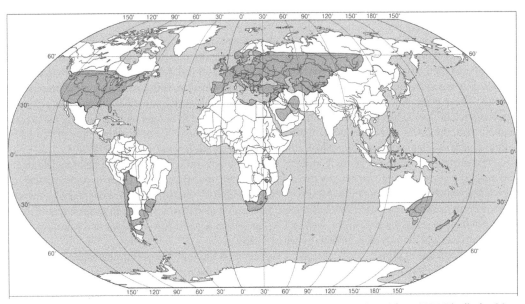

Map 2.2. Global distribution of the Pear Leaf Blister Mite (*Eriophyes pyri* Pgst.). Adapted from CABI Distribution Maps of Plant Pests, Map No. 273 (copyright CAB International, Wallingford, UK), www.cabi.org/dmpp. Reproduced with permission of CAB International. The map is partly based on the Agroecological atlas of Russia. http://www.agroatlas.ru/ru/content/pests/Eriophyes_pyri/map/.

Africa and Egypt, India and Iran, Portugal and Iceland, Argentina and Hawaii as well as many other countries (Nzonzi 2003). The wheat flower mite (*Steneotarsonemus panshini*) is transported on wheat and other cereals.

The global distribution of pear gall mite (*Eriophyes pyri*) which attacks the leaves of pear, quince, hawthorn, mountain ash, cotoneaster, is shown in Map 2-2.

There are also the **highly specialized** species. For example, the blackcurrant gall mite (*Cecidophyopsis ribis*) is a pest of gooseberry and blackcurrant and, sometimes, damages the red and white currant. The grape bud mite (*Colomerus vitigineusgemma*) and grape leaf mite (*Calepitrimerus vitis*) damage only the grapevine. The coconut mite (*Aceria guerreronis*) is a pest only on the coconut plantations.

Most mites cause harm to (1) field and vegetable crops and (2) fruit trees and berry bushes. The common spider mite is the most harmful pest to **field** and **vegetable crops**. Normally it damages cotton, soybeans, beans, potatoes, hops, and various melons. In some years, yield losses of cotton reached 25–30% in Tajikistan, and 20–60% in Uzbekistan (Bondarenko et al. 1991). The Tomato Russet Mite causes considerable damage to only tomatoes; the severely damaged plants lost from 30 to 50% of fruit yield (Nzonzi 2003).

Mites cause considerable damage to **fruit trees** and berry bushes. For example, the pearleaf blister mite (*Eriophyes pyri*) often leads to losses of up to 90% of pear crop, while the strawberry mite reduces the yield of berries by 40–70%. The brown fruit mite (*Bryobia redikorzevi*) and hawthorn spider mite (*Tetranychus viennensis*) sometimes diminish the apple crop by 65% in the United States (Bondarenko et al. 1993). Higher coconut mite *Aceria guerreronis* incidence on plants resulted in a 60% decrease in the mean number of fruits (Rezende et al. 2016).

2.1.2 Insects as forest pests

Forest pests damage or destroy trees. Insects are the most economically important forest pests, and very forest has a number of insect pests. Some insect forest pests range widely and cause considerable damage; others have less effect. Certain species cause continuous harm, while others cause occasional harm (e.g., in times of mass reproduction). Some pests cause significant damage where poor forest management exists.

2.1.2.1 *Characteristics of harmfulness*

This is not to say that trees and shrubs are defenseless against the forest pests. Comparing the plants with people, it may be said that the protective measures are generally aimed at prevention of damages and mitigation of their consequences. The **first line** lies in the chemical defense against the phytophagous insects. For example, the oak's protective mechanism against insects eating its leaves is the presence of tannins. These phenolic compounds are located in the vacuoles, which are cavities located nearby the uppersurface of leaves. Such chemical protection brings the larvae to a stand and they do not eat the oak leaves. Also, the larvae of beetles undermining the leaves overcome this protection. They pierce a leaf and consume its internal tissues, avoiding the vacuoles filled with tannin (Riklefs 1979).

The protective substance of the coniferous trees is a soft resin located in the resin canals under pressure (3–5 atm on average for healthy trees) and presenting the mixture of resin acids and essential oils (terpenes), which is toxic for insects (Maslov 1988). In addition, when the trees are damaged by insects, the soft resin floods the pests and precludes their penetration under bark. Owing to evaporation of spirit of turpentine and crystallization of the resin acids, the soft resin gets thick in the air and solidifies at the trunk's surface, and the wound is healed (Gilyarov 1986).

The **second line** of protection consists in the regeneration and restoration of the damaged or destroyed organs and tissues for which purpose many plants accumulate the nutritive substances in their roots. For example, every few years, tent caterpillars completely devour the leaves of the bird cherry tree in the northeastern United States in early spring. However, in response, this tree is able to grow new leaves because of its accumulated reserves of nutritive substances (Ricklefs 1979).

Different tree species are damaged by **different insects**, depending on the number of the insect species inhabiting them and on their persistency. For example, in the territory of the former USSR, about 850 species of insects are found on oak, about 700 species on poplars (Photo 2.1-5) and more than 300 species on the trees of elm family. Fewer pests attack maples, alder, and saxaul. Some trees, such as Amur cork tree (*Phellodendron amurense*), bird cherry tree (*Padus maackii*), honey locust (*Gleditsia triacanthos*), Chinese scholar tree (*Sophora japonica*), soap tree (*Sapindus saponaria*), shawnee-wood (*Catalpa speciosa*), are practically unaffected by the pests (Vorontsov 1982).

Photo 2.1-5. Different tree species are damaged by various insects, depending on the number of the insect species inhabiting them and on their persistency. For example, in the territory of the former USSR, about 700 species of insects are found on poplars. Different stages of development and damage inflicted by the poplar borer (*Saperda calcarata*) are presented.
Photo credit: James Solomon, USDA Forest Service, Bugwood.org

The **resistances of the tree species** to the impact of destructive insects vary widely. For example, the Siberian fir (*Abies sibirica*) and Manchurian fir (*Abies nephrolepis*) are the least resistant to needle loss. The fir stands die off frequently in case of loss of 70% of needles. Further, in the sequence of resistance, the Jezo spruce (*Picea jezoensis*), Siberian spruce (*Picea obovata*), Korean pine (*Pinus koraiensis*), Siberian pine (*Pinus sibirica*), Siberian larch (*Larix sibirica*), Dahurian larch (*Larix gmelinii*), and Sukachev's larch (*Larix sukaczewii*) follow. The reasons larches have such high resistance is because of their complete annual change of needles (Zhokhov 1975).

2.1.2.2 *Categories of forest insect pests*

According to the range of consumed tree species, forest insect pests are divided into polyphages, oligophages, and monophages. The **polyphagous** pests include gypsy moth (*Lymantria dispar*), pale tussock (*Calliteara pudibunda*), winter moth (*Operophtera brumata*), mottled umber (*Erannis defoliaria*), and chafer (*Melolontha hippocastani*), among others. The **oligophagous** pests include the black arches moth (*Ocneria monacha*) damaging the majority of coniferous species (except for juniper and yew) and some broad-leafed ones.

There are also many **highly specialized** forest pests. For example, the Oriental chestnut gall wasp (*Dryocosmus kuriphilus*) damages only chestnut trees (genus *Castanea*). The female lays her eggs in the flowerbuds of trees. The larvae hatch in 2–3 weeks after living and feeding in the buds, which causes considerable damage to the growth and fruit-bearing ability of the tree. The economic loss from the Oriental chestnut gall wasp reaches millions of dollars annually (http://www.badgersett.com/sites/default/files/info/publications/Bulletin4v1_0.pdf). It is the world's largest pest of chestnuts (Blyummer 2016). The monophages also include pine beauty (*Panolis flammea*), pine looper (*Bupalus piniarius*), birch bark beetle (*Scolytus ratzeburgi*), and others (Maslov 1988; Zhokhov 1975).

Some populations of insects maintain very low density for a long time but, after that, abruptly reach an extremely high number and extend to the immense territories. Such **outbreaks** emerge frequently in the small area places where the conditions are very favorable for reproduction and survival of pests. For example, until 1970, the population of the mountain pine beetle (*Dendroctonus ponderosae*) in the Glacier National Park (USA) was limited by area of 162 ha and, at that, from 1 to 11 trees per ha were lost there every year. However, after 1972, the population of the mountain pine beetle became quickly to increase and, by 1977, the beetles have extended over an area of about 58 thousand ha and more than 10 million trees were lost in this year alone (Berryman 1990).

The vast portion of damage to trees occurs during the process of **insect feeding**. Insects cause damage by biting off and eating parts of trees and shrubs, mostly pine needles, leaves (Photo 2.1-6), bark

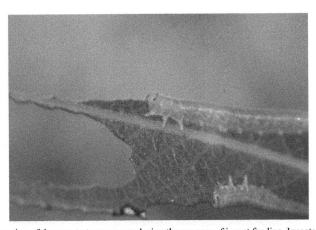

Photo 2.1-6. The vast portion of damage to trees occurs during the process of insect feeding. Insects cause damage by biting off and eating parts of trees and shrubs, mostly pine needles, leaves, bark and roots. Larvae of the willow sawfly, *Nematus oligospilus*, in Canberra, Australia is shown.
Photo credit: John La Salle (CSIRO)

and roots. They eat the seeds, damage the leaves, suck tree sap, build tunnels under the bark, burrow into the wood and form excrescence and other substances that stress trees and shrubs. These injuries in turn cause a variety of effects: distortion of stems, dying and/or felling of branches, stunting of shoots and leaves, reduced flowering, fruiting and growth. Where severe damage occurs, plants often die (Polyakov and Nabatov 1992).

Insect pests can be divided into the following **categories**: (1) pests of cones and seeds, (2) pests of buds and shoots, (3) foliage pests, (4) stem pests, and (5) root pests.

The effects of insect pests on **cones** and **seeds** are diverse. The larvae of the pinecone moth (*Dioryctria abietella*) and the tortrix moth (Tortricoidea/Olethreutidae) penetrate cones and destroy the bulk of the seeds. Leaf-footed bugs (Coreidae) pierce cones with their proboscis and suck the juice of seeds. Many beetles burrow into young cones, preventing seed development, and the cones consequently become infertile (Berryman 1990). Beetles are not characterized by outbreaks of mass reproduction, as is the case with many other insects (Vorontsov 1982). One of the most serious pests in this category is the larch fly (*Chortophila laricicola*). One larva can kill 40%–80% of larch cone seeds and 35%–40% of fir cone seeds (Maslov 1988; Berryman 2013).

Campions (Silene) are the most dangerous pests of **buds** and **sprouts**. Larvae of campions mine shoots, causing shrinkage of their buds. Caterpillars of pine-shoot moths (*Rhyacionia buoliana*) often damage the main apical stem, resulting in uneven growth of the trunk or multiple branching, which leads to the beginning of secondary apical shoot domination. Because of the pests' attacks, these trees form ugly forked trunks, so their value for commercial timber harvesting is greatly reduced or completely lost. Two of the most destructive insects of this group are the wintering pine-shoot moth (*R. buoliana*) and the white pine weevil (*Pissodes strobi*) (Berryman 1990).

Fir needle and **leaf-eating** insects destroy photosynthetic tissues of trees, resulting in a reduction of the formation of carbohydrates (Photo 2.1-7). The immediate effect of defoliation is a reduction in tree growth. A study of defoliation of a quaking/trembling aspen (*Populus tremuloides*) by leaf-eating caterpillars, in Minnesota, USA, showed that in the course of one year, when the crown of the tree was being eaten, tree growth decreased by almost 90%, and the following year it decreased by 15% (Ricklefs 1979).

The **Siberian coniferous silk moth** (*Dendrolimus sibiricus*) is among the most hazardous pests of coniferous vegetation. Outbreaks of its propagation often lead to huge losses. For example, in 1955 in the central districts of Krasnoyarsk Province, Russia, the total area of infestation was 1 million ha (Novikov 1999). The forest lost 140,000 ha and damaged timber was approximately 50 million cubic meters (Bondarev and Soldatov 1999). In outbreaks of the spruce budworm (*Choristoneura fumiferana*) in the mid-1990's in Alaska (USA) (Eaten Alive 1998) and the Douglas fir tussock moth (*Orgyia pseudotsugata*) in Oregon and Washington (USA), there were large financial losses (Berryman 1990).

Photo 2.1-7. Fir needle- and leaf-eating insects destroy photosynthetic tissues of trees, resulting in a reduction of the formation of carbohydrates. The picture shows tree mortality at La Grande, Oregon (U.S.), due to the Douglas fir tussock moth, *Orgyia pseudotsugata* (McDunnough).
Photo credit: David McComb, USDA Forest Service, Bugwood.org

The **gypsy moth** (*Lymantria dispar*) should be recognized as the most destructive leaf pest. According to various sources, gypsy moth caterpillars eat leaves from 300 (Liebhold et al. 1997) to 600 (Kuznetsov 1997) different plant species. Among its favorite trees in the Old World are the oak, fruit trees, poplar, willow, and birch (Maslov 1988). In North America, oaks and aspen are the best-known hosts (Liebhold et al. 1997). Major damage is caused by its larvae, each of which can eat from 0.6 to 3.5 grams of leaves during its development; besides that, approx. one-third of those leaves fall to the ground as leftovers (Maslov 1988). There can be several hundred caterpillars on every tree during their time of mass reproduction.

Outbreaks of the gypsy moth (*Lymantria dispar*) can lead to extremely serious consequences; the larch forests suffer the most in the eastern United States. In 1981, trees were totally devoid of leaves in an area of 12.9 million acres (about 5.2 million ha) (Liebhold et al. 1997). Global distribution of the gipsy moth (*Lymantria dispar*) is shown in Map 2-3.

Tree trunk pests include insects from the long-horned beetle family (Cerambycidae), jewel/metallic wood-boring beetles (Buprestidae), weevils (Curculionidae), horntails/wood wasps (Siricidae), cossid/ carpenter moths (Cossidae), clearwing moths (Sesiidae), amongst others. These insects live hidden under bark, or in the flesh of the trunk. They gnaw in the surface layers of the tree trunk (bast/skin fibers, cambium, sapwood), causing substantial physiological damage to the tree and their ultimate demise (Maslov 1988). Most trunk pests are oligophagous and are found in several species of trees (Concomitant of forester 1990).

Quite a lot of destructive trunk pests belong to the subfamily of bark beetles (Scolytinae), numbering more than 3,000 species (Photo 2.1-8). The peculiarity of this family is that not only the larvae, but also the adult beetles live under the bark and in the wood flesh of tree trunks. They make complex mines, relatively constant in form for each species. Bark beetles can damage the majority of forest tree species, mainly conifers (Gilyarov 1986).

Bark beetles typically undergo outbreaks. For example, a mass outbreak of a specific bark beetle, the spruce beetle (*Dendroctonus rufipennis*), was observed in Alaska, where approximately 1.3 million ha of white (*Picea glauca*) and Sitka spruce (*Picea sitchensis*) forest were impacted from 1989 to 2000 (Goodman et al. 2006). In western North America, they killed 70%–90% of large trees in some areas (Griffin et al. 2013).

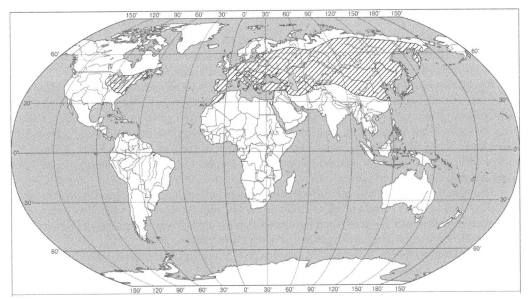

Map 2.3. Global distribution of the gipsy moth (*Lymantria dispar*) (www.fs.fed.us/ne/morgantown/4557/gmonth/world). Reproduced with permission of U.S. Department of Agriculture Forest Service.

Photo 2.1-8. The mountain pine beetle *Dendroctonus ponderosae*, is a species of bark beetle. It has destroyed wide areas of lodgepole pine forest, including more than 16 million of the 55 million hectares of forest in British Columbia. Photo shows mountain pine beetle control.
Photo credit: Mark McGregor, USDA Forest Service, Bugwood.org

Root pests include insects living in the upper soil, such as chafers, root eaters, scarab beetles, click beetles, darkling beetles, and comb-clawed beetles (Zhokhov 1975). In general, these pests do not pose a great danger to the development of trees with well-developed root systems, but they may contribute to fungal infections that cause root rot. This is mainly a problem in nurseries, in young saplings and seedlings, which have tender roots. The pests violate the normal growth of young plants by gnawing and biting them and, in some cases, causing their death (Berryman 1990).

2.1.2.3 Mites as forest pests

The number of species of mites that injure forest trees and shrubs is small. The **biggest damage** is caused by the spider mites (family Tetranychidae) and rust and gall mites. For the most, they are widespread. Among the mites, forest pests are also the host-specific pests and pantophagous species.

A good example of the **highly specialized pest** is the Kyrgyz juniper mite (*Trisetacus kirghisorum*), damaging only the seeds of the Himalayan Pencil Juniper (*Juniperus semiglobosa*). This evergreen coniferous shrub or small to medium-sized tree growing to 5–15 metres grows in the mountains of Central Asia, in northeastern Afghanistan, western most China (Xinjiang), northern Pakistan, southeastern Kazakhstan, Kyrgyzstan, western Nepal, northern Republic of India at the altitudes of 1,550–4,350 metres. The pure juniper forests are the dominant forest cover in many areas (http://survinat.ru/2010/02/mozhzhevelnik/#ixzz4HxYsVy2v).

Therefore, the periodic heavy infestations of the Kyrgyz juniper mite (*Trisetatcus kirghisorum*) have the strong adverse effect on natural regeneration. In certain years, in Kyrgyzstan and Uzbekistan, up to 90 percent of juniper seed crops of Himalayan Pencil Juniper are lost which is strongly reflected on the forest health (Food and Agriculture Organization of the United Nations 2007).

The **spider mites** (family Tetranychidae) are often pests on conifers. The most important species among them is the Spruce spider mite (*Paratetranychus ununguis*) which attacks spruce, hemlock, fir, juniper, larch, redwood, incense cedar, and pine species. This pest is widely distributed in the greater part of the Canada and the United States territory (https://en.wikipedia.org/wiki/Oligonychus_ununguis). Puncturing by the pincers (pairs of head forelimbs) the epidermis of the needles, the mites suck out the cell content. The needles turn brown, are covered with spiderwebs, and prematurely fall (Perry and Randall 2000).

The spider mite *Oligonychus coniferarum* injures mainly forest nurseries. Due to it, pine, hemlock, spruce, juniper, fir, and white-cedar suffer. The attacks of these mites result in a decrease in plant growth and yellowing of leaves. Although in most cases, these mites do not cause the mortality of seedlings, they weaken trees and predispose them to attack by insects and fungi or to damage by adverse environmental conditions (Ford and Barry 2004).

The **rust and gall mites** (superfamily Eriophyoidea) cause two kinds of damage. Rust mites feed on the leaf surface and cause discoloration. The gall mites result in formation of distinctive growths on leaves and buds. In the most cases, the consequences of their attacks consist in disfiguration of foliage and its premature fall, however, galls are not detrimental to the health of the tree. To the gall mites belong the maple bladder-gall mite (*Vasates quadripedes*) and Crimson erineum mites (*Eriophyes regulus*) (which injures the different species of maples (https://tidcf.nrcan.gc.ca/en/insects/factsheet/42)), ash flower gall mite (*Aceria fraxiniflora*) (which injures the ash), maple spindlegall mite (*Vasatesaceris crummena*), and others.

Among the forest pests are also some pests of the **fruit trees**. The major one among them is European red mite (*Panonychys ulmi*). This mite damages elm, alder, oak, sycamore, and other trees. The garden spider mite (*Schizotetranychus pruni*) can feed on different types of elm, beech, willow, maple, and others. The flat scarlet mite (*Cenopalpus pulcher*), has been observed on hornbeam, alder, goat willow and plane trees (Bondarenko et al. 1993).

Generally, the significance of mites as the forest pests is many times lower in comparison with insects. Trees are seldom killed by mites; however, seedling mortality may occur.

2.1.3 Insects and spiders as pests of fisheries

Many insects eat up of **fish juveniles** and fish roe, causing damage to the fish industry. Among them are water scorpions (Nepidae), giant water-bugs (Belostomatidae), backswimmers (Notonectidae), water boatmen (Corixidae), diving beetles (Dytiscidae), scavenger water beetles (Hydrophilidae), some species of caddis worms (for example, caddisfly *Plectrocnemia conspersa*, family Polycentropodidae), nymphs of dragonflies (for example, Pacific spiketail (*Cordulegaster dorsalis*), common aeschna (*Aeschna juncea*), etc.).

Among the **semi-aquatic spiders**, there is a considerable number of carnivores eating up the baby fishes. They mainly inhabit the latitudes between 40 degrees S and 40 degrees N. The phenomenon of fish predation was registered in more than a dozen spider species from the superfamily Lycosoidea (families Pisauridae, Trechaleidae, and Lycosidae), two species of the family Ctenidae, and in one species of the family Liocranidae. More often (more than 75% of cases), the fish is caught by the Araneomorphae spiders of genus *Dolomedes* better known as the fishing spiders. As a rule, the fish caught by these spiders has a length of 2–6 cm and belongs to the most common species characteristic of the given geographic area (for example, mosquitofish [*Gambusia* spp.] in the southeastern USA, killifish [*Aphyosemion* spp.] in Central and West Africa, fish of the genera *Galaxias, Melanotaenia,* and *Pseudomugil* in Australia (Nyffeler et al. 2014)).

Information on the juvenile fish eaten by different insects is presented in Table 2.1-3.

Table 2.1-3. Number of juvenile fishes eaten by insects under experimental conditions (Nikolsky 1961).

Insect Species	No. of Fish Eaten	No. of Days	Average No. of Fish Eaten per Day
Great diving beetle (*Dytiscus marginalis*), adult	102	22	4.5
Great diving beetle (*Dytiscus marginalis*), larva	91	12	7.6
Lesser diving beetle (*Acilius sulcatus*), adult	18	8	2.2
Lesser diving beetle (*Acilius sulcatus*), larva	131	24	5.5
Backswimmer (*Notonecta glauca*), adult	45	9	5
Backswimmer (*Notonecta glauca*), larva	103	15	7
Water scorpion (*Nepa cinerea*), adult	17	8	2.1

For example, significant damage to aquaculture farms is caused by the **great diving beetle** (*Dytiscus marginalis*) and, to a larger extent, by its larvae. This insect is widespread and found in most of the stagnant water bodies of the middle latitudes (http://beneficialbugs.org/bugs/Diving_Beetle/diving_beetle.htm). It paralyzes its prey by injecting secretions from a venom gland, located in the base of its upper jaw, which dissolves the prey's body tissues, after which the beetle sucks up the liquid. The diving beetle catches, first of all, the suppressed and slowly swimming fishes. A larva may suck up 10 fishes, each 3 cm (1.5 in) long, per day (http://www.bibliotekar.ru/7-rybovodstvo/68.htm). The predation of adult large diving beetles was studied by Frelik (2014), who ascertained that the juvenile fishes are eaten by the beetles of the species *D. circumcinctus* most often, by *D. marginalis* slightly less often, and by *Cybister lateralimarginalis* the least often.

The larva of a species of **backswimmer** (*Notonecta glauca*) kills its prey, juvenile fish, by puncturing the skin by the proboscis and injecting digestive liquid. After a time, the insect sucks out the killed prey (Nikolsky 1961). Generally, the damage caused to the fishing industry by the insects is not large, in comparison with the impact of insects on the other types of human activity.

2.1.4 Biodeterioration by insects

Biodeterioration is any adverse change in the properties of materials caused by the vital functions of organisms—in our situation, insects. The **categories of properties** of materials affect the damage that can be caused by insects: (1) organoleptic, (2) antibiotic, and (3) constructive (Zhuzhikov 1983).

Organoleptic properties of a material have an impact on visual, chemical, and tactile receptors in insects. Attractiveness of a material to an insect depends on a defined set of characteristics and properties that it prefers when searching for food, oviposition sites, shelter, and so forth. Material stability against insect damage depends on its organoleptic properties, which can be of **two types** (Zhuzhikov 1983). The *first type* of material stability is due to its lack of one or more attracting properties. The *second type* is due to the presence of repelling properties.

Antibiotic properties of a material depend on characteristics such as its toxicity or lack of nutrients. The most impregnable materials are those on which a population of certain insect species are unable to sustain their existence.

Structural properties are also important. Because materials are damaged only by insects with biting mouth parts, the hardness of a coating or of the material itself will contribute to its preservation (Zhuzhikov 1983).

Insects may cause mechanical damage to materials, mainly in their **search for food**. If a product is suitable for insects to colonize and has cavities, then internal contamination is possible. Damage to materials also occurs when insects develop a colony in the cavities of a material and use its particles for their construction activities.

The **most significant** insects in this category are beetles, butterflies, biting lice, termites, and bristletails. Based on the way they feed, they are described as keratophagous or xylophagous. **Keratophagous insects** can digest and assimilate keratins—proteins that form the basis of the horny layer of skin and hair of mammals and the feathers of birds. **Xylophagous insects** feed on plant materials.

The biodeteriorations caused by insects result in significant economic loss. For example, the depreciation caused by keratophagous insects approximates to $1 billion dollars annually in the United States alone (Arnault et al. 2012). From the economic point of view, the pests of stored food, fabrics, and wood-destroying insects are most significant.

2.1.4.1 Insects as stored food pests

Some insects live in places of food storage or processing and actually feed on the stored food. The **food reserves** include breadstuffs and derived products, bean- and oil-bearing crops, dried fruits and vegetables, medicinal herbs, confectionery products (chocolate and sweetmeats), different kitchen herbs, tobacco, and various products of animal origin (cheese, meat, fish, smoke products). The stored products are *major food* for some insects and only a *habitat* for other insects, where they find other foods such as mold fungi, other insects, arachnids, and so on. Throughout the course of human history, some insect species have been constant competitors for food resources. When humans began to grow and store grains, insects became attracted by this plentiful supply, and destroyed a lot of food meant for humans.

Damage by insect pests to stored foods is enormous and caused by the following **factors** (Burakova 2005): (1) destruction of a part of a stock; in this case, in spite of the insignificant sizes of insects, the losses can be considerable because the number of pests can be enormous; (2) deterioration of edible, technological (flour-grinding, baking, etc.), and seed (reduction in germination ability) qualities of grain; (3) contamination of food reserves with bodies and waste products of insects: cobwebs, excrement, molting skins, and so forth, which requires additional cleaning costs; (4) creation of conditions (rise in temperature and humidity), contributing to the reproduction of different microorganisms (bacteria, mold fungi), which in turn attracts other species of insects feeding on these microorganisms; and (5) stopping production equipment (for example, the larvae of Indian mealmoth (*Plodia interpunctella*) gnaw through the sieves and plug up the pipes that move the products with their webby nests).

Besides economic losses, these groups of insects cause **health problems**, including the following **factors** (Mason et al. 2012; Cherepnev et al. 2013): (1) ability to disseminate pathogenic microorganisms (many storage insect pests have hairs and indentations on their exoskeletons that contribute to the mechanical transfer of pathogens); (2) allergic responses (when infected grain is handled, dermatitis, conjunctivitis, and dermatitis develop); (3) emergence of other medical problems: bread strongly infected with pests is unfit for consumption, because the widely distributed grain pests, the grain weevil (*Sitophilus granaries*) and the rice weevil (*Sitophilus oryzae*), contain cantharidin, resulting in not only skin and mucous membrane irritation but also vomiting, headache, and convulsions).

Destruction of food reserves may be considerable. For example, one of the Ukrainian elevators held 217 tons of buckwheat. When this lot was tested for presence of pests after 40 days, contamination of the lot with the grain weevil was discovered. When the beech wheat was passed through the clean-up separator, its weight decreased by 8.5 tons (Cherepnev et al. 2013). In Benin, after 6 months of storage, the average losses of rice reach 5.47% in the southern region, 4.07% in the central region, and 1.64% in the northern region (Togola et al. 2013).

The number of food storage insect pests is enormous. From the economic point of view, among the **major pests of insects** worldwide are the angoumois grain moth (*Sitotroga cerealella*), rice weevil (*Sitophilus oryzae*), granary weevil (*Sitophilus granarius*) (Photo 2.1-9) maize weevil or greater rice weevil (*Sitophilus zeamais*), lesser grain borer (*Rhyzopertha dominica*), larger grain borer (*Prostephanus truncates*) (Hansen et al. 2004), and brown flour mite (*Gohieria fusca*). The most malicious pests of **stored legumes** are the common bean weevil (*Acanthoscelides obtectus*), cowpea weevil (*Callosobruchus maculates*) (Mason et al. 2012), and flour mite (*Aleurobius farinae*).

Among the major pests of **stored fruits and nuts** are the Indian mealmoth (*Plodia interpunctella*), almond moth (*Cadra cautella*), tobacco moth (*Ephestia elutella*), raisin moth (*Ephestia figulilella*), red flour beetle (*Tribolium castaneum*), and sawtoothed grain beetle (*Oryzaephilus surinamensis*) (Burks et al. 2012). The detailed descriptions of different pests (biology, behavior, and ecology) of the food reserves are given in the appropriate publications (Sallam 2000; Hagstrum et al. 2012).

Photo 2.1-9. The granary weevil (*Sitophilus granarius*) occurs all over the world and is a common pest in many places. It can cause significant damage to harvested stored grains.
Photo credit: CSIRO

The **germinating ability** of seeds is reduced according to the **developmental stage of the insects**. A study of the germinating capacity of maize grains in Brazil showed that allowing the maize weevil or greater rice weevil (*Sitophilus zeamais*) to lay in eggs in them lowers this index by 13%; due to the laying of the eggs of angoumois grain moth (*Sitotroga cerealella*), the index is lowered by 10.9%. When reaching the adult stage, the germinating capacity drops by 93% and 85%, respectively (Sallam 2000). Decline in germinating capacity depends also on the species of insect pest. So, the germinating capacity of the wheat seeds drops by 92%, 75%, 75%, and 53% in case of damage by the larger grain borer (*Prostephanus truncates*), rice weevil (*Sitophilus oryzae*), sawtoothed grain beetle (*Oryzaephilus surinamensis*), and confused flour beetle (*Tribolium confusum*), respectively (Cherepnev et al. 2013).

2.1.4.2 *Insects as pests of fabrics, fur, and leather*

Insect that damage fabrics, fur, and leather can be divided into **three groups**: (1) carpet beetles (Dermestidae), (2) hide beetles (Trogidae), and (3) clothes moths (Tineidae). Among the **carpet beetles,** the four most commonly encountered species include the black carpet beetle (*Attagenus megatoma*), varied carpet beetle (*Anthrenus verbasci*), furniture carpet beetle (*Anthrenus vorax*), and common carpet beetle (*Anthrenus scrophulariae*). The larval stage of all carpet beetles is responsible for the damage (Klein 2008; Querner 2015).

The larvae concentrate in dark, undisturbed locations and, in the course of their feeding, damage such objects as wool, silk, leather, fur, natural bristle hairbrushes, pet hair, and feathers. They do not feed on synthetic fibers. The **distinctive feature** differentiating moth larvae from carpet beetle larvae is that, in general, carpet beetles are more likely to damage a large area on one portion of a garment or carpet, whereas moth damage more often appears as scattered holes. Also, carpet beetle larvae leave brown, shell-like, bristly-looking cast skins when they molt. These skins and a lack of webbing are usually good clues that carpet beetles are the culprits (http://ipm.ucanr.edu/PMG/PESTNOTES/pn7436.html).

As for the **hide beetles**, the most damage is caused by three species: larder beetle (*Dermestes lardarius*), hide or leather beetle (*D. maculatus*), and black larder beetle (*D. ater*). These beetles reside in imperceptible holes (with diameter 3 mm [⅛ inch]) in fabrics and leather to pupate, however, in large infestations, this causes structural damage (https://www.catseyepest.com/pest-library/pantry-pests/beetles/hide-beetle). Both the adults and the larvae damage materials.

Among **moths**, the common clothes moth or webbing clothes moth (*Tineola bisselliella*), case-bearing clothes moth (*Tinea pellionella*), and carpet moth (*Trichophaga tapetzella*) are the most harmful pests of materials (Ebeling 2002; Querner 2016). These small unattractive butterflies fly from dusk to the first half of the night and live an average of 7–10 days. During this time, the females lay approximately 60–120 eggs. The adult moths do not feed; all damage is done by the larvae.

Food-related damage is inflicted by larvae of all ages. The larvae can feed on clothing (Photo 2.1-10), carpets, rugs, furs, fabrics, blankets, stored wool products, upholstery, piano felts, brush bristles, and felt gaskets in devices (Brimblecombe and Querner 2014). The amount of material eaten by larvae depends on the moth species, material quality, and temperature and relative humidity of the air. The mass of food eaten by the moth larvae over a period of their development has been found to be 3,600 times their own body mass (http://www.biochemi.ru/chems-1103-2.html).

The moth larvae gnaw through the thin woolen cloth at a rate of one hole per day. During the larvae's mass reproduction, the unprotected material of keratinous origin can be fully destroyed. The digestive juice released by the moth larvae, with a pH of 9.9, is alkaline, and keratin has low chemical resistance to the alkaline attack. The moth larvae may also damage hybrid fabrics, needing to consume more of these materials because the synthetic thread is indigestible with lower nutritive value. The amount of damage caused by moth larvae increases, to a great degree, when fur or wool is severely contaminated with excrement, larval skins, and webby nests.

An example of **non–food-related damage** is the damage to materials when larvae construct their webby passages and larval indusia, the walls of which are encrusted with gnawed-off bits of material, during their migration in search of food—or a convenient place for pupation, if damaged material gets in their way. The hungry larvae may damage materials that are nonedible but accessible for their jaws,

Photo 2.1-10. Clothes moths cause damage when they feed on furs, wools, felt pads in devices, heat- and sound-insulating felt material, leather bindings of old books, woolen fabrics, and zoological collections. The total amount of wool destroyed by a single moth brood is 50 grams per year. A sheepskin that was practically eaten by moths is shown.
Photo credit: S.M. Govorushko, Pacific Institute of Geography, Vladivostok, Russia, 21 February 2004

including but not limited to paper, cardboard, cotton, linen, synthetic fabrics, polyvinylchloride and polyethylene films, and insulation of telephone wires (http://lib4all.ru/base/B3161/B3161Part11-46.php).

From an **economic** viewpoint, the common clothes moth or webbing clothes moth (*Tineola bisselliella*) is the pest that causes the most damage. It differs in capacity for mass year-round reproduction; it inhabits many human living quarters, clothing and fur shops, warehouses, and enterprises that process wool, furs, and leather. Throughout their lives, its females lay small amounts of eggs on the surfaces of different keratin-containing materials and products affected by molds and fungi; however, they may also lay eggs on a nonfood surface. The larvae hatched from eggs have a sensation of hunger lasting more than 1 month. In search of food, the microscopic larvae crawl into cupboards, suitcases, and commodes through the smallest cracks and eyelet holes. This feature of the common clothes moth explains unexpected damage to well-packed-up things.

The second most harmful insect in moderate climates is the case-bearing clothes moth (*Tinea pellionella*) (http://lib4all.ru/base/B3161/B3161Part11-46.php).

2.1.4.3 *Wood-destroying insects*

The enormous number of insects associated with timber fall generally into the category of forest pests. However, the number of species able to develop in wood not in contact with the soil and with humidity lower than the hygroscopic limit is sufficiently small. From the viewpoint of taxonomy, they can be divided into **three groups**: (1) termites (Isoptera); (2) carpenter ants (genus *Camponotus*); (3) beetles (Coleoptera).

The nature of damaged objects can be divided into pests of fresh-wood timber and pests of dry-wood materials, buildings, and furniture. **Harvested logs** are usually damaged by bark beetles (Scolytinae) and long-horned beetles (Cerambycidae), red palm weevil (Curculionidae; *Rhynchophorus ferrugineus*); they are occasionally damaged by jewel beetles (Buprestidae), ship timber beetles (Lymexyidae), and horntails (Siricidae). Beetles sometimes create deep wormholes (for example, certain types of long-horned beetles penetrate 6–12 cm into wood), which reduces the yield of intact material from logs. In the northern part of Minnesota (USA), capricorn beetles destroy about 5% of harvested timber, turning it into the category of firewood (Karasev 1983).

The carpenter ants damage also both the **harvested logs** and **wood structures** contacting with soil (posts, piles, cellars, etc.). They do not feed on the wood but pierce in it tunnels to their nests. The carpenter ants prefer the wood that is naturally soft or has softened by decay. Mounds of sawdust indicate their presence (https://en.wikipedia.org/wiki/Carpenter_ant).

Dry timber and **wood products** are vulnerable to certain capricorn beetles or long-horned beetles, death watch beetles (Anobiidae) (Photo 2.1-11), true powder post beetles (Lyctinae), auger beetles (Bostrichidae), and termites (Isoptera). Of the beetles, the most damage is caused by death watch beetles, which damage coniferous wood in damp and poorly ventilated spaces beneath buildings (Ebeling 2002). The most destructive insects for buildings and wood are **termites**. They are widely distributed (Map 2-4) and are divided into **three groups**.

1. Subterranean termites: build nests in the ground, forming huge mounds; in search of food, they cover great distances, building closed tunnels between their nests and the food source.

Photo 2.1-11. Dry timber and wood products are vulnerable also to certain capricorn beetles, death watch beetles (Anobiidae), true powderpost beetles (Lyctinae), auger beetles (Bostrichidae), etc. A wooden picture frame infested by furniture beetles *Anobium punctatum* in a museum in Austria is shown.
Photo credit: P. Querner (University of Natural Resources & Life Sciences, Vienna, Austria), 23 June 2015

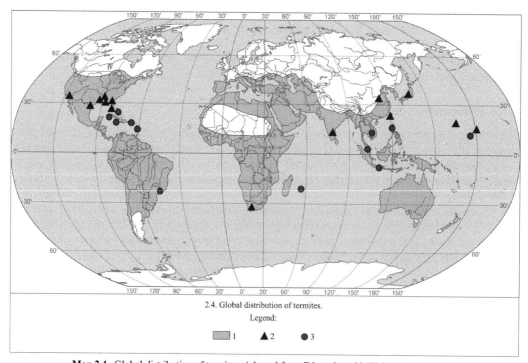

2.4. Global distribution of termites.
Legend:

▓ 1   ▲ 2   ● 3

Map 2.4. Global distribution of termites. Adapted from Edwards and Mill (1986); Su (2001).
Legend: (1) global distribution; Points of detection of most destructive subterranean termite species: (2) Formosan subterranean termite *Coptotermes formosanus*; (3) Asian subterranean termite *Coptotermes gestroi*

2. Damp-wood termites: their nests are mainly located underground, close to wood, because they need contact with the soil but may build their nests in wood if necessary

3. Dry-wood termites: do not depend on the soil and build their nests in the dry wood (Photo 2.1-12), they can also settle in other materials containing cellulose such as plywood, cardboard, fiberboard, and chipboard.

In some regions, termites became a true **scourge** of wooden buildings. For example, 42.7% of buildings are infested with termites in Brazil (Milano et al. 2002), 80%–90% are infested in China (Zhong et al. 2002), 53.2% in Spain (Gaju et al. 2002); and 70% (residential) and 20% (industrial) in Malaysia (Lee 2002). In most cases, termites consume the wood entirely, leaving only a thin outer layer of sound wood to protect them from the atmosphere. Because the surface of the wood looks intact, the infestation is difficult to discover until it reaches a grand scale. There is also the fear that the accidental transport of termites to apartments extends their range to regions that they do not usually inhabit because of climate.

Of 3,106 known species of termites, the significant damage to the wood constructions is caused by **83 species**. In North America, there are 9 species threatening building structures (https://en.wikipedia.org/wiki/Termite); 20 species in Australia (Termite risk management 2002); 26 species in the Indian subcontinent, 24 species in tropical Africa, 17 species in Central America and the West Indies (Su and Scheffrahn 2000); and 9 species in Sri Lanka (http://pestcontrol.lk/termites/).

Each region focuses on its own specific **high-priority** pest species. For example, in Southeast Asia the most economically important species is the Asian subterranean termite (*Coptotermes gestroi*). This insect attacks mainly wooden structures like cabinets, parquet floors, windows, door frames, and roofs. In Malay Peninsula, the Asian subterranean termite is responsible for approximately 85%–90% of termite infestations (Neoh 2013). In the United States, the Formosan subterranean termite (*Coptotermes formo*) was recognized as most harmful (Yates et al. 1997). On the Azores archipelago, the major economic loss is caused by the West Indian dry-wood termite (*Cryptotermes brevis*) (Guerreiro et al. 2014).

Data on global **losses** related to damage by termites to wooden structures are unknown. Su (2002) estimated the annual economic losses due to termites at US$11 billion for the United States and US$22 billion worldwide. At the same time, they cause more than $3 billion worth of damage to wooden structures annually throughout the United States (Verma et al. 2009). If this proportion is true worldwide, then the damage by wood-destroying termites can be roughly estimated at $6 billion. The control cost for termite pests in the United States was estimated at $1.5 billion annually in 1994 (Su and Scheffrahn 2000).

Photo 2.1-12. Termites are most destructive insects for buildings and wood. Photo shows termite *Coptotermes acinaciformis* nest in pole.
Photo credit: Forestry and Forest Products, CSIRO

2.1.4.4 *Insects as pests of museum collections*

Long recognized worldwide, and confirmed by the Committee of Conservation of ICOM (International Council of Museums), insect pests can be very destructive to museum collections (Photos 2.1-13, 2.1-14). Any museum (or individual collector) cannot fully evade damages related to deterioration of them. A list of insect pests that threaten museum collections is given in Table 2.1-4.

The most commonly encountered and malicious pests of the **zoological collections** are the skin beetles: more than 20 species of them cause damage in museums of Russia (Golub et al. 2012). The common carpet beetle (*Anthrenus scrophulariae*), varied carpet beetle (*Anthrenus verbasci*) and museum beetle (*Anthrenus museorum*) are most harmful of them. They inflict irreparable damage to museum specimens: in a short time, they can destroy an entire collection of stuffed animals and birds (Linnie et al. 2000).

The fauna of insects that **herbarium specimens** is not specifically diverse. In any case, the genes of the predominant pests, more or less widely distributed in the herbariums of the world, are not numerous. Universally, the most significant pests are beetles (Coleoptera), the larvae of which do the damage. The major pests of herbarium specimens are the biscuit beetles or drug-store beetle (*Stegobium paniceum*) and the cigarette or tobacco beetle (*Lasioderma serricorne*), which feed immediately on dried vegetative material (Pinniger et al. 2015).

Photo 2.1-13. Insect pests can be very destructive to museum collections. Any museum (or individual collector) cannot fully evade damages related to deterioration of them. The ethnographic object of gingerbread infested by biscuit beetle *Stegobium paniceum* in a museum in Austria is shown.
Photo credit: P. Querner (University of Natural Resources & Life Sciences, Vienna, Austria), 14 April 2014

Photo 2.1-14. There are a few types of dermestids which commonly attack insect collections, including those in the genera Anthrenus, Attagenus, and Reesa. Dermestids will likely make their way into a collection as eggs on a specimen which has been incorporated. Signs of dermestid damage include holes in pinned specimen, a black "dust" under specimen (the frass of the beetles), and empty exuviae on or around a specimen. Infested insect drawer (beetle collection) in a natural history museum, Vienna is shown.
Photo credit: P. Querner (University of Natural Resources & Life Sciences, Vienna, Austria), 04 July 2014

Table 2.1-4. Insect pests of museum collections (Pinniger et al. 2015).

Specimen Type	Pest	Signs of Damage
Dried insect collections, particularly large Lepidoptera and Coleoptera	*Anthrenus, Attagenus, Ptinus, Reesa*	Loose wings, fatty bodies eaten, frass, and larval skins in cases
Dried arachnids and Crustacea, particularly poorly cleaned decapods	*Anthrenus, Attagenus*	Loose legs, frass, and larval skins in cases
Bird skins	*Anthrenus, Attagenus, Tinea, Tineola*	Stripped feathers, webbing, and frass
Mammal skins	*Anthrenus, Attagenus, Tinea, Tineola*	Loose hair, webbing, cast skins, and frass
Freeze-dried mammals and birds	*Anthrenus, Attagenus, Ptinus, Stegobium, Lasioderma Tinea, Tineola*	Loose hair, round emergence holes, webbing, cast skins, and frass
Reptiles, particularly poorly cleaned	*Anthrenus, Attagenus*	Cast skins and frass
Bones and horns, poorly cleaned	*Anthrenus, Attagenus, Dermestes*	Cast skins and frass

Lesser damage is caused by the insects feeding on mold or detritus. If their population is not large, such insects present more of a hindrance than a danger; however, in large numbers, they can inflict serious damage to the collections. Pests of this category include the spider beetles, silverfish, firebrats, and booklice (*Liposcelis* spp.) (Ashby 2007). Among this group of insects are the Australian spider beetle (*Ptinus tectus*) and the plaster beetle (*Cartodere filum*), which, to a greater extent, constitutes a serious problem for mushrooms. The only leather beetle damaging the herbarium specimens is the varied carpet beetle (*Anthrenus verbasci*), which feeds on seeds and pollen, especially those of the sunflower family Asteraceae (Compositae) (Byalt et al. 2014).

A threat to the **herbarium specimens** is highest in the tropical and subtropical regions where the high temperature and humidity afford an opportunity for the pests to quickly grow and reproduce; numerous local pests penetrate easily into these herbariums. In moderate climates, most herbarium pests do not live away from heated spaces, so reinfestation takes place rarely. However, some pests typical of moderate conditions (for example, silverfish [*Lepisma saccharina*] appear in the present tropical herbariums where air-conditioning creates suitable conditions for them to live (Byalt et al. 2014)).

Some plants are not as damaged by insects as others in the same family: cruciferous (Cruciferae), aster (Compositae), umbelliferous (Umbelliferae), and liliaceous (Liliaceae). Even in the same family, not all genes are equally destroyed by insects. For example, in the aster family, the representatives of such genes as *Scorzonera, Centaurea, Crepis, Taraxacum, Chondrilla, Carduus,* and *Cirsium* are damaged to the maximum extent; at the same time, insects are not found on the herbarium specimens of *Artemisia* and *Lappa*. The plants collected in the salt marshes or chalk deposits are very rarely damaged (Pavlov et al. 1976).

2.1.4.5 *Other insect pests*

The damages of the food reserves, fabrics, wood, and museum collections are most significant from the viewpoint of extent of caused loss (Querner et al. 2013, 2017). However, the insects damage also several other materials, such as paintings, paper, and silk cocoons.

The highest hazard for **paintings** is related to all-pervasive common housefly (*Musca domestica*). On any surface, including those of artistic paintings, it leaves the stains of excrements, which contain corrosive acids. It is impossible to wash off these stains without traces. Within 24 hours, the fly leaves about 50 stains of excrement, contaminating about 3 cm^2 of the surface (Toskina 1998). The paintings in watercolors are damaged by silverfish (*Lepisma saccharina*), small wingless insects of the bristletail order. Silverfish eat watercolors and some dyes (for example, blue inks) (Uniform rules 2009).

The insects can have an **indirect** effect on preservation of the pictures. For example, during the restoration of a series of Brussels tapestries preserved in the Abbey of Sacromonte (Granada, Spain), large losses of material and color alterations were discovered on one of them. The studies showed that this biodeterioration effect was caused by the remains of beetles (*Pyrrhalta luteola*). Although the beetles did not consume the textile material, they took part directly in the destruction of the tapestry: their presence resulted in higher mold growth, which in turn caused the color alterations (Sánchez-Piñeroa and Bolívar 2004). Silverfish can also cause damage to tapestries (http://carnivoraforum.com/topic/9754556/1/).

A number of insects damage **paper** and anything made from paper (books, wallpaper, labels, postage stamps, paper currency). For example, silverfish damages all of the items listed here (Pinniger et al. 2015). This insect eats the upper layers of paper (Photo 2.1-15), especially coated paper, and gnaws out the holes in cigarette paper and filters (Uniform rules 2009). Cockroaches gnaw cigarette paper roughly, breaking the edges and leaving the characteristic excrements in the form of short black rods about 1.5 mm long (Toskina 1998).

Termites are malicious pests which destroy paper and paper products (Photo 2.1-16). Once having penetrated inside the house, they are not limited to wood: anything containing high cellulose content serves as potential food. Sometimes, the consequences of this are catastrophic; for example, in South America, it is seldom that a book older than 50 years can be found out due to continuous presence of termites (Akimushkin 1975). Because the books are made not only of paper, but also of a number of other materials, then the number of their potential pests is much more (Photo 2.1-17). The kinds of damages by the drugstore beetle and spider beetles are similar: the fly pages, flag pages, and several first and last pages as well as book cover through which the beetles gnaw the flight holes are damaged (Toskina et al. 2007).

Of all insects that damage **book bindings** containing flour paste, the psocids (*Liposcelis bostrychophila*), also known as booklice, are most widely known. The colonies of these insects contribute to the degradation of precious library collections (Green et al. 2015). The paper bindings are also damaged by silverfish and firebrats (Klein 2008).

Leather beetle larvae damage leather, silk, and velvet bindings; animal and wheaten adhesives in the book block; and, sometimes, paper. They eat holes and dig small trenches in the bindings, erase the

Photo 2.1-15. Silverfish damage paper and anything made from paper (books, wallpaper, labels, postage stamps, paper currency). Paper damaged by Silverfish *Lepisma saccharina* is shown.
Photo credit: P. Querner (University of Natural Resources & Life Sciences, Vienna, Austria), 21.07.15

outer layer of skin until it becomes rough, and destroy the sizing compounds in the book back. Clothes moth larvae can destroy antique cardboard; the wool fibers from the cardboard enter into its composition and damage the plush on the book binding if it has a wool nap (Pekhtasheva 2012).

Skin/carpet beetles in the cocoon enterprises badly hurt the **cocoons of mulberry silkworms**, gnawing holes in the cocoons and making them inadequate for rewinding. Sometimes these insects even damage products that are inedible by making holes in them. Such damage can be quite dangerous for complex radioelectronic plants and expensive equipment, where even the minor disturbances of insulation can make the whole system inoperative. There have been cases of damage to electric power cables in a steel mill, to the automatic lock system for a high-speed railway, and to radars at an airport; these incidents sometimes led to accidents with fatalities (Ilyichev et al. 1987).

Photo 2.1-16. Termites are malicious pests of paper and paper products. Dry wood termites damaging a book in a library in Sri Lanka are shown.
Photo credit: P. Querner (University of Natural Resources & Life Sciences, Vienna, Austria), 11 November 2013

Photo 2.1-17. The Tobacco Beetle is worldwide in distribution. It is mainly damage stored tobacco (e.g., tobacco leaves, cigars, chewing tobacco, cigarettes). Tobacco Beetles can also be found in food storage areas and are know to chew through books, manuscripts, furniture fabrics and other organic materials. Damage of historic books by tobacco beetle *Lasioderma serricorne* in infested library in Sri Lanka is shown.
Photo credit: P. Querner (University of Natural Resources & Life Sciences, Vienna, Austria), 15 November 2013

2.2

Insects as Vectors of Plant Diseases

Plant disease may be defined as a disturbance by phytopathogens of a plant's normal metabolism, resulting in reduced productivity of the plant or in its total loss. Phytopathogens can be transmitted in a lot of ways: wind, animals, etc. Insects also play a certain part in the transfer of disease. The phytopathogens penetrate the tissues of plants through air pores (pores of leaves), lenticels, honeycups, water glands, epidermis cells, wounds, cracks from sunscalds, frost cracks, and so forth. Often, the "gateway" through which agents of disease can penetrate is caused by plant damage inflicted by insects (http://belboh.com/magflori/115-mag45.html).

Disease agents can be transferred by unspecific (arbitrary) or specific ways (Panteleev 2012). In an **unspecific transfer**, pathogens may be carried on the insect's outer covering, attaching to the cuticle of different parts of the insect's body, or may enter the insect's digestive system when the attacked plant tissues are eaten. In a **specific transfer**, throughout their life cycle, insects "cultivate" microscopic pathogens on the vegetable substrate that serves as their nutrient source. Therefore, insect phytophages are potential vectors in transfer of disease agents.

2.2.1 Insects as vectors of agricultural plant diseases

The vectors of diseases are fungi, virus, bacteria, and mycoplasma. When comparing the frequencies and spreading of different diseases, fungal diseases are most common and diverse: 80% of all plant diseases fall into this category (http://belboh.com/magflori/115-mag45.html). Fungal diseases are followed by viral, bacterial, and mycoplasmal diseases. However, insects' role in spreading disease is most significant in the transfer of **virus diseases:** 76% of plant viruses are spread by insects (http://g.janecraft.net/virusy-rastenijj/). This is not to say that the viruses are spread solely by insects; they can be propagated also in many other ways (such as via seeds, sap of diseased plants in the process of pricking-out, pruning, contact between diseased and healthy plants, and soil nematodes).

There are several **types of transmission of plant viruses** by insects (http://www.meddiscover.ru/newmed-520.html):

1. External transmission by a stylet, without virus persistence in the transmission vector organism; in this case, virus settles on the insect stylet top when it feeds on the infected plant and can be immediately transferred to the healthy plant. The capacity for the virus transmission can be lost in a moment or can last for a few days;

2 Regurgitative transmission, when a virus remains in the upper intestine of insects (aphids and beetles) over a long period and is transmitted to a healthy plant by way of regurgitation of the intestinal contents;

3. Circulative transmission, when a virus is transmitted to another plant only upon completion of the certain latent period (from several hours to several days) rather than after the vector feeds on the diseased plant;

4. Propagative transmission, when a virus reproduces in the tissues of the insect before it reaches the mouthparts of this insect; the length of the latent period is determined by the time needed for virus replication, and, eventually, the virus enters the salivary glands of the insect and it transmitted with saliva to the plant (leafhoppers, a large class of vectors of plant viruses—provide the virus transmission to plants almost always by this mechanism).

The top 10 list of **economically important plant viruses** (affecting not only the plants by which they are named) is as follows (Rybicki 2015):

1. Tobacco mosaic tobamovirus (TMV)

2. Tomato spotted wilt tospovirus (TSWV)

3. Tomato yellow leaf curl begomovirus (TYLCV)

4. Cucumber mosaic cucumovirus (CMV) (only cucumbers)

5. Potato virus Y (potyvirus, PVY)

6. Cauliflower mosaic caulimovirus (CaMV)

7. African cassava mosaic begomovirus (ACMV)

8. Plum pox potyvirus (PPV)

9. Brome mosaic bromovirus (BMV)

10. Potato virus X (potexvirus, PVX)

Most of them can be transmitted by insects and mites. For example, aphids, true bugs, and mites can be the vectors of TMV (Starukhina and Bondarchuk 2014). The TSWV is transmitted by thrips (Photo 2.2-1) (Sin et al. 2005). TYLCV and ACMV are transmitted by sweet potato whitefly (*Bemisia tabaci*) (Glick et al. 2009; Dubern 1994). CMV, CaMV, and PPV are transferred by aphids (https://en.wikipedia.org/wiki/Cucumber_mosaic_virus) (Bak et al. 2013; Kegler and Hartman 1998). BMV is transmitted by the cereal leaf beetle (*Oulema melanopus*) and their larvae (Vlasov and Teploukhova 1993). Only PVY and PVX cannot be transferred by insects (https://en.wikipedia.org/wiki/Potato_virus_X).

Photo 2.2-1. The western flower thrips *Frankliniella occidentalis* is the major vector of tomato spotted wilt virus, a serious plant disease. This species of thrips is native to North America but has spread to other continents including Europe, Australia, and South America via transport of infested plant material.
Photo credit: By Frank Peairs, Colorado State University, Bugwood.org [CC BY 3.0 (http://creativecommons.org/licenses/by/3.0)], via Wikimedia Commons

The **bacterial diseases** transmitted by insects are also numerous. One can identify the special group of bacterioses that are propagated most exclusively by insects. Among them are bacterial wilt of corn (infectious agent, *Erwinia stewarti*), bacterial wilt of cucurbits (infectious agent, *Erwinia tracheiphila*), and tuberculosis of olive (infectious agent, *Pseudomonas savastanoi*).

Bacterial wilt of corn is transmitted by the corn flea beetle (*Chaetocnema pulicaria*) and the toothed flea beetle (*C. denticulata*), as well as the larvae of western spotted cucumber beetle (*Diabrotica duodecimpunctata*), which gnaw on the corn's roots in the soil and in that way transfer the infectious agents from one plant to the other. The digestive tract and salivary glands of these insects contain *Erwinia stewarti*, which infects new plants through gnawing the roots.

Olive knot is transmitted exclusively by the olive fruit fly (*Dacus oleae*). Females' ovipositors always contain *Pseudomonas savastanoi*; each time a female lays her eggs into the olive tree bark, the tree receives a portion of the *P. savastanoi* bacteria. This infection results in the formation of tuberculous tumors (http://agro-portal24.ru/bolezni-rasteniy/5078-istochniki-infekcii-i-sposoby-rasprostraneniya-bakteriozov-chast-5.html).

In other bacterial diseases, it is difficult to identify the groups in which the insects would play the prevailing part. A significant role is played by bees, flies, and other insects in propagation of the **bacterial blight (fireblight) of fruit cultures** caused by the bacterium *Erwinia amylovora* (https://en.wikipedia.org/wiki/Fire_blight). Different insects gnawing on the roots can also be an important factor in spreading different soft rots caused by bacteria *E. carotovora*, *E. aroideae*, and *E. phytophthora*. For example, the bacterial soft rot caused by the bacteria *Erwinia carotovora* (or *Pectobacterium carotovorum*) affects the banana, beans, cabbage, carrot, cassava, coffee, corn, cotton, onion, other crucifers, pepper, potato, sweet potato, and tomato. Seedcorn maggots (*Delia platura*) play an important role in the dissemination and development of bacterial soft rot in potatoes. Generally, the bacteria appears in a potato field with the infected seed material, but it can also live in all developmental tages of the insect, including the pupae (Agrios 2002).

Bacterial ring rot is a very serious disease of potatoes. The pathogens are transported by the Colorado potato beetle (*Leptinotarsa decemlineata*) and green peach aphid (*Myzus persicae*) (http://phytopath.ca/wp-content/uploads/2015/03/DPVCC-Chapter-16-potato.pdf). Other bacterioses spread by insects include bacterial bean blight, citrus canker, cotton boll rot, crown gal, bacterial spot, and canker of stone fruits.

Insects also play an essential role in transmission of **fungal diseases** of plants. They transfer the pathogenic fungi from diseased to healthy plants, breaking the epidermis and other protective tissues of plants with their mouthparts or with their ovipositor, and thereby allowing the fungus to enter in the process of feeding and egg laying. The infection can also be transferred during pollination.

Weevils (*Calendra parvula* and *Anacentrus deplanatus*), Hessian fly (*Phytophaga destructor*), spotted cucumber beetle (*Diabrotica undecimpunctata howardii*), and northern corn root worm (*Diabrotica longicornis*) take part in the root infections caused by such fungi as *Pythium*, *Fusarium*, and *Sclerotium*. There are known cases of infection of the corn stems with *Gibberella*, *Fusarium*, and *Diplodia* through the European corn borer *Pyrausta nubilalis* (Agrios 2002).

Mycoplasma diseases (phytoplasma diseases) were first diagnosed in 1967 by Japanese virologists. More than 300 diseases, formerly considered to be viral infections, are now known to be phytoplasma diseases (Bertaccini and Duduk 2009). Phytoplasma diseases are spread mainly by insects belonging to the families Cicadellidae (leafhoppers), Fulgoridae (planthoppers), and Psyllidae (psyllids), which feed on the phloem tissues of infected plants. A pathogen is transferred to the other plant only upon completion of a particular latent period rather than after the vector feeds on the diseased plant (http://agro-archive.ru/zaschita-rasteniy/141-fitopatogennye-mikoplazmy.html). The phytoplasma enters the insect through its stylet, after which it moves through the intestine and is absorbed into the hemolymph. Sometimes the transfer of phytoplasma is transovarial (i.e., through the offspring). For example, such insect–disease combinations were recorded as the American grapevine leafhopper (*Scaphoideus titanus*)–aster yellows; leafhopper (*Hishimonoides sellatiformis*)–mulberry dwarf; leafhopper (*Matsumuratettix hiroglyphicus*)–sugarcane white leaf; and psyllid (*Cacopsylla melanoneura*)–apple proliferation disease (Bertaccini and Duduk 2009).

2.2.2 Insects as vectors of forest plant diseases

The situation with insect vectors of forest plant diseases is the same as with agricultural plant diseases. In most cases, the insects carry the infectious agents on the surface of their bodies. More rarely, infectious agents enter their digestive systems when the insect feeds on the tissues of the infected plant and, later on, are transmitted to the healthy trees.

Fungal diseases of plants (mycoses) are most abundant and economically significant. The number of pathogenic plant fungi species reaches the tens of thousands (Krutov and Minkevich 2002). However, insects' role is the largest in spreading viral diseases. In the vast majority of cases, insects are only one of the ways infection can be transmitted to healthy trees; the other sources including wind, rain, birds, rodents, and nematodes.

Fungal spores are spread by various species of insects. For example, many bark beetles transmit the spores of the wood-staining and wood-destroying fungi (Kuz'michev et al. 2004). Different species of flies and beetles, the larvae of which develop in the fruit of pileate fungi and Polyporaceae, also help spread the fungi spores. Some insects (particularly, flies) contribute to the transmission of rust fungi (Forest Encyclopedia 1986). Various fungal disease agents are transmitted by insects, including the vascular mycosis of oak, nectria canker of pines, beech bark canker, scleroderris canker of pine and spruce, and different kinds of vascular wilts (persimmon wilt, mango wilt, oak wilt, and Dutch elm disease).

The **vascular mycosis of oak** affects the vascular system of trees and is manifested in the decay and drying-up of leaves, crown thinning, as well as in the gradual or sudden drying-up of some branches or the whole tree. Its causative agent is a fungus *Ceratocystis fagacearum*, which develops in the vascular system of the tree and results in its clogging. The infection of trees occurs when the spores of the causative agent are spread by stem pests including the European oak bark beetle (*Scolytus intricatus*), longhorn beetle (*Mesosa myops*), and oak borer (*Agrilus angustulus*).

The cicadae, gallflies, and mites are also of some importance in this regard. The infection penetrates the tree tissues in **two ways**: (1) through the fresh wounds inflicted by the stem pests; (2) through insect damage to foliage along the vascular system to the branches and stem of trees (http://lsdinfo.org/gribnye-bolezni-lesnyx-nasazhdenij-i-borba-s-nimi/).

The nectria canker of pines is transferred by spittlebug (*Aphrophora saratogensis*). The beech bark canker is caused by two species of fungi, *Neonectria faginata* and *Neonectria ditissima* (Cale et al. 2015). **Their vectors** are beech scale insects, *Cryptococcus fagisuga,* and, to a lesser degree, scale insect *Xylococcus betulae* (Houston 1994).

The **vascular wilts** are caused by different fungi affecting the vascular system and tissues. The **Dutch elm disease** is the best known disease of this kind. It is caused by the fungus *Ophiostoma ulmi,* which is spread from diseased to healthy trees by the bark beetles. In North America, these include the native elm bark beetle (*Hylurgopinus rufipes*), European elm bark beetle (*Scolytus multistriatus*), and banded elm bark beetle (*Scolytus schevyrewi*); in Europe, these are the European elm bark beetle (*Scolytus multistriatus*), large elm bark beetle (*Scolytus scolytus*), and bark beetle (*Scolytus pygmaeus*). In the expansion of the pathogen, the leaf-eating insects, for example, the elm-leaf beetle (*Xanthogaleruca luteola*) can also participate.

The **persimmon wilt** is the wasteful disease caused by the fungus *Cephalosporium diospyri.* Its causative agent is transmitted by the powder-post beetle (*Xylobiops basilaris*) and the twig girdler beetle (*Oncider cingulatus*). The **mango wilt** is caused by two species of fungi, *Diplodia recifensis* and *Ceratocystis fimbriata.* The first of them is transmitted by the beetle *Xyleborus affinis* (Carrillo et al. 2013), while the second, by a scolytid beetle (*Hypocryphalus mangiferae*).

The **oak wilt** is caused by fungus *Ceratocystis fimbriata* and is an essential disease of the forest trees. The fungus is transmitted by nitidulid beetles such as *Carpophilus lugubris, Colopterus niger,* and *Cryptarcha ample.* In addition to the nitidulid beetles, the scolytid beetles (*Monarthrum fasciatum* and *Pseudopityophthorus minutissimus*), brentid beetle (*Arrhenodes minuta*), buprestid (*Agrillus bilineatus*), and flat-headed borer (*Chrysobothrys femorata*) are involved in this process. The spores are transmitted both externally on their bodies and internally through their digestive tracts (Agrios 2002).

When insects spread **phytopathogenic bacteria**, mechanical transmission occurs, accompanied sometimes by damage to the plant and pathogens being introduced into it. Nevertheless, "biological" transmission of bacteria by insects is also possible. For example, the bacterium *Erwinia amylovora*, transmitted by aphids, beetles, ants, wasps, and bees, is preserved in the intestine of the transmitting insect and can be transferred with its eggs.

A close biologic relationship exists between the bacterium causing the **"tuberculosis" of ash** and the **olive fruit fly** (*Dacus oleae*), the organism in which the bacterial growth takes place. In the absence of bacteria, the flies do not develop and, in the absence of flies, there is no effective diffusion of bacteria (Forest Encyclopedia 1986).

The bacterium *Xylella fastidiosa* is one of the most dangerous plant bacteria worldwide, causing a **variety of diseases**. These bacteria inhabit the xylem (vascular tissue of plants, which conducts water) in the tropical and subtropical regions of the world. In nature, they are only transmitted by xylem-feeding insects, such as the sharpshooter leafhoppers (Cicadellinae) and spittlebugs (Cercopidae) (Agrios 2002). Examples of insects transmitting the bacterioses within a temperate climate include longhorn beetles (*Saperda populnea* and *S. carcharias*), which transmit the vector of **bacterial canker of poplar**, and the ash bark beetle (*Hylesinus varius*), transmitting the vector of **bacterial ash cancer** (http://givoyles.ru/articles/uhod/tainye-puti-infekcii/).

The **viral diseases** are spread most often by insects with the sucking mouthparts (aphids, cicadas, thrips, true bugs). Depending on the nature of the effect of virus on the plant, features of the pathologic process and external appearance, **two basic types** of viral diseases of plants are identified: mosaics and yellows.

The **mosaics** are mainly manifested in change of the leaves and flowers color and their deformation. The diseases of this group include the mosaics of leaves of the ash-leaved maple (*Acer negundo*), elm, ash, and other wood species. The **yellows** are characterized by more diverse disturbances in metabolism, growth, and development of affected plants. Some elements of the plant are deformed, related to the virus's suppression or stimulation of the growth process. The total underdevelopment, sharp stagnation (dwarf), different degenerative changes in plants, and tissue overgrowth (histomas, galls) can be observed.

The witches' broom disease serves as a good example (http://forest.geoman.ru/forest/item/f00/s00/e0000403/index.shtml). As another example, aphids transmit the causative agents of the mosaics of white hazel, apple, chestnut, and maple (Kuz'michev et al. 2004).

The diseases caused by phytoplasma are called **phytoplasma diseases** and are manifested in the pathological changes of the conductive tissues of plants (more often, in inner bark). The typical symptoms of phytoplasma diseases are chlorosis, stooling, cutting-back, virescence and proliferation of flowers, reduction of reproductive organs, and general dwarfism.

The **most dangerous** phytoplasma disease of the trees in North America is the **elm phloem necrosis**. The cork of the affected elm species—American elm (*Ulmus americanus*) and Texas cedar elm (*U. crassifolia*)—becomes yellowish-brown, and leaves abruptly turn yellow and wilt. The death of a tree can occur in as little as several weeks after the onset of initial symptoms of disease or after 1–2 years or more. It is caused by phytoplasma that infects the phloem (inner bark) of the tree. This disease is transmitted from infected to healthy trees by the whitebanded elm leafhopper (*Scaphoideus luteolus*), meadow spittlebug (*Philaenus spurarius*), and another species of leafhopper (*Allygus atomarius*) (https://en.wikipedia.org/wiki/Elm_yellows).

Yet another example of **phytoplasma disease** is the **ash yellows**. The symptoms of these diseases occur over a period of up to 3 years after the mycoplasma enters the inner bark of the tree. They are expressed in growth slowdown, leaf color loss, progressive decline, and witches' broom on trunks. White ash (*Fraxinus americana*) is the most resistant to this mycoplasmosis; green ash (*F. pennsylvanica*) has medium resistance; and black ash (*F. nigra*) is the most susceptible to the disease (Cloyd 2013).

2.3

Insects as Vectors of Disease in Animals and Humans

Insects are vectors of many animal and human diseases. The transmission of pathogens can be mechanical and specific. In **mechanical transmission,** the causative agents received by the vector maintain their viability and disease-inciting power only for some time on the surface of its body or in its digestive tract. In **specific transmission**, the pathogen continues to parasitize the vector organism, continuing to develop inside the vector organism and possibly reproducing inside one of its organs or tissues. The transmission of the infectious agent may not be realized until after some time, which provides an interval for these events to occur (Tarasov 2002).

2.3.1 Mosquitoes as vectors of disease

More than 3,400 species of mosquitoes (Culicidae) are known, and only 10% of them are medically meaningful (Goddard 2012). Mosquitoes transmit many causative agents. The general way it works is this: a pathogen develops in a mosquito, which becomes in this case the intermediate, or definitive, host. The pathogen penetrates the mosquito's saliva, which is injected into a human when the mosquito bites. A mosquito's proboscis has a valve-like device, through which the blood can be transferred in one direction only (inward); a mosquito cannot release the blood of the bitten person outward.

The causative agents of diseases transmitted through the bites of mosquitoes belong to **various groups**: protozoa, viruses, bacteria, and parasitic worms. These diseases are widespread in subtropical and tropical areas of Africa, Asia, South and North America, and southern Europe. Every year, more than 1 million people die of virus infections resulting from mosquito bites and several hundreds of millions have health problems caused by infections transmitted through mosquito bites (http://www.medicinform.net/ infec/infec62.htm). The **most important diseases** include malaria, dengue fever, yellow fever, lymphatic filariasis (several types), encephalitis viruses (e.g., West Nile virus, eastern equine encephalitis, La Crosse encephalitis, Japanese encephalitis), and other arboviruses (e.g., Zika fever, chikungunya, Ross River fever).

Malaria can be transmitted by several species of mosquitoes, but it is usually transferred by those commonly known as malarial mosquitoes, which inhabit mainly tropics and subtropics. The causative agents of malaria are apicomplexan protozoa (hemamoebas of five species: *Plasmodium knowlesi*, *P. vivax*, *P. ovale*, *P. malariae*, and *P. falciparum*). Each species causes its own type of malaria. To actively transfer the hemamoebas, the mosquitoes should live under conditions that match those in which the hemamoeba reproduces. This is possible nearer to the equator and within the subequatorial zone (http://www.medicinform.net/infec/infec62.htm).

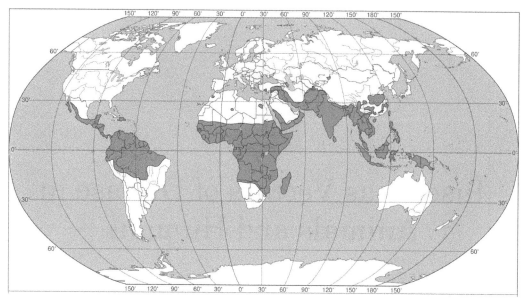

Map 2.5. Global distribution of malaria (http://www.tankonyvtar.hu/hu/tartalom/tamop412A/2010-0010_10_Climate_change_climate_protection_climate_policy/4711/index.scorml; https://en.wikipedia.org/wiki/Malaria).

Malaria is the **most common disease** transmitted by mosquitoes in the world (Map 2-5). According to World Health Organization estimates, 124 to 283 million events of infection with malarial hemamoebas and 367,000 to 755,000 deaths from malaria take place each year. Also, 85–90% of the infection events occur in the regions of Africa to the south of the Sahara (World Health Organization 2010). Worldwide, 3.3 billion people are currently at risk of being infected with malaria (Delil et al. 2016). In 2015, there were more than 214 million new cases of malaria and approximately 438,000 deaths from malaria (Ntonifor and Veyufambom 2016). The economic damage caused by this disease in Africa is estimated at $12 billion annually because of increased health care costs, lost ability to work, and negative effects on tourism (Greenwood et al. 2005).

Dengue fever is found in more than 110 countries, predominantly in Southeast Asia, Africa, Oceania, and the Caribbean (Ranjit and Kissoon 2011). The yearly incidence is about 50 million people (Simmons et al. 2012) and about 10,000 to 20,000 deaths (Stanaway et al. 2016). Infected humans, monkeys, and bats can serve as sources of infection. Mosquitoes transmit the infection from the sick human or monkey (*Aedes aegypti* from humans and *Ae. albopictus* from monkeys) (https://en.wikipedia.org/wiki/Dengue_fever).

Yellow fever originated in Africa, from which it spread in the 17th century to South America through the slave trade. Its causative agent is spread primarily by one mosquito species, *Ae. aegypti* (Photo 2.3-1) (Goddard 2012). Currently, yellow fever is commonly found in tropical areas of South America and Africa but not Asia (Map 2-6). About 900 million people reside in the 32 African and 13 Latin American countries where this disease is widespread (Agampodi and Wickramage 2013). Yellow fever causes 200,000 infections and 30,000 deaths every year (https://en.wikipedia.org/wiki/Yellow_fever); about 90% of all cases of infection occur in Africa (Tolle 2009). Since the 1980s, the number of cases of yellow fever has been increasing (Barrett and Higgs 2007).

The **filariatoses** are a group of helminthic invasions of humans and animals, transmitted by the nematodes of the order Spirurida. Humans and vertebrate animals are definitive hosts; different species of mosquitoes (*Culex*, *Aedes*, *Mansonia*, and *Anopheles*) are intermediate hosts and also vectors of the parasite. The basic filariatoses that infect humans are wuchereriasis (causative agent, *Wuchereria bancrofti*), Malayan filariasis (causative agent, *Brugia malayi*), loaiasis (causative agent, *Loa loa*), and onchocerciasis (causative agent, *Onchocerca volvulus*) (http://www.eurolab.ua/diseases/508/).

Filariatoses are spread in tropical and subtropical countries in Africa, tropical Asia, and Central and South America. About 1.1 billion people live in regions where filariatoses are spread, and about

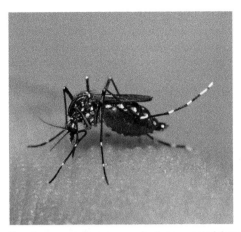

Photo 2.3-1. *Aedes aegypti*, the yellow fever mosquito, is a mosquito that can spread dengue fever, chikungunya, Zika fever, Mayaro and yellow fever viruses, and other diseases. *Aedes aegypti* feeding in Dar es Salaam, Tanzania is shown.
Photo credit: By Muhammad Mahdi Karim (Own work) [GFDL 1.2 (http://www.gnu.org/licenses/old-licenses/fdl-1.2.html)], via Wikimedia Commons, 2009

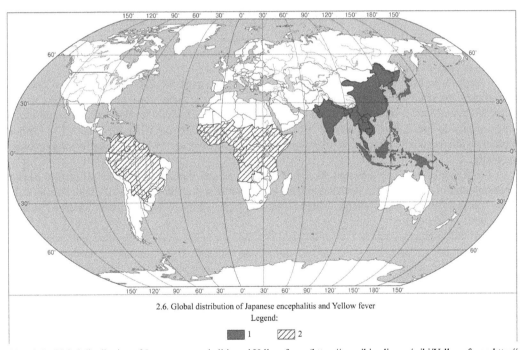

Map 2.6. Global distribution of Japanese encephalitis and Yellow fever (https://en.wikipedia.org/wiki/Yellow_fever; http://folk.uib.no/mihtr/Fredskorpset/FK082006.html; https://en.wikipedia.org/wiki/Japanese_encephalitis).
Legend: (1) Japanese encephalitis; (2) Yellow fever

120 million of them are afflicted with these diseases (Das et al. 2002). The **economic loss** related to filariatoses is nearly $1 billion (US) in India only (Ramaiah et al. 2000). Domestic animals such as cattle, sheep, and dogs are also subject to the filariatoses (https://en.wikipedia.org/wiki/Filariasis).

   The **West Nile virus** is transmitted to humans by mosquitoes of the genus *Culex* that suck the blood of birds and mammals. The virus was first found in humans in 1937 near the West Nile region, Uganda, Africa. Since then, the virus has spread and caused severe epidemics of West Nile virus in North Africa, Europe, Asia, and North and South America: for example, there was a large-scale epidemic of 1996 in Romania (Marka et al. 2013) and the virus appeared in the United States in 1999 (Nash et al. 2001).

During the last two decades, the occurrence of the West Nile virus has increased, and today West Nile is estimated as one of the most common arboviruses in the world (Vlckova et al. 2015).

La Crosse encephalitis (virus first identified near La Crosse, California, USA) is an acute zoonotic virus infection that occurs in summer. The virus source is chipmunks and tree squirrels; the virus vectors are mosquitoes *Aedes triseriatus*. The disease signs and symptoms include fever, headache, vomiting, mental impairment, convulsions, and focal neurologic deficits. La Crosse encephalitis is relatively rare: in the United States, about 80 to 100 disease cases are recorded each year, mostly from the midwestern states (Minnesota, Wisconsin, Iowa, Illinois, Indiana, and Ohio) (https://en.wikipedia.org/wiki/La_Crosse_encephalitis).

Zika fever is caused by viruses of the family Flaviviridae. The basic route of Zika virus transmission to a human is through the bite of *Aedes* species mosquitoes that are active in the daytime: *Aedes aegypti* and *A. albopictus* (Korzeniewski et al. 2016). The virus was first detected in 1947; for the next 60 years, only 15 cases of Zika fever were described in sub-Saharan Africa and Southeast Asia (Enserink 2015). In 2007, the Zika fever occurred for the first time outside the limits of Africa and Asia and caused a major outbreak on Yap Island (Federated States of Micronesia). In October 2013, an even larger outbreak was observed in French Polynesia (affecting more than 30,000 people). From there, the infection spread to other Pacific Islands.

The **first cases** of Zika fever in people in the Western Hemisphere were noted on Easter Island, (a Pacific territory of Chile), in February 2014. The epidemic continued to spread quickly; over the next 2 years, cases of Zika fever were recorded in many countries of Latin America. The most seriously affected country was Brazil, where between 500,000 and 1,500,000 cases of Zika fever were recorded in 2015 (Chan et al. 2016). In March 2015, Zika fever was recorded in 22 of Brazil's 26 states (de Goes Cavalcanti et al. 2016).

Zika fever in itself is of no serious hazard to people; however, the Zika virus can cause a serious **hereditary deformity** (microcephaly) in developing fetuses. A baby thus affected is born with a disproportionally small head and reduced brain size; later, these children develop various psychomotor impairments. In Brazil, from November 2015 through the end of January 2016, there were 4,182 babies born with microcephaly, whereas from 2010 through 2014, an average of 163 such cases were recorded annually.

Furthermore, men can transmit the virus to their sexual partners (Sikka et al. 2016). **Sexual transmission** of Zika virus has been recorded in nine countries: Argentina, Canada, Chile, France, Italy, New Zealand, Peru, Portugal, and the United States (https://en.wikipedia.org/wiki/2015%E2%80%9316_Zika_virus_epidemic/).

The **chikungunya virus** is transmitted through the bite of the same mosquitoes that transmit the dengue fever and malaria. The symptoms of chikungunya include high temperature; headache; fatigue; nausea; muscular aches; hives on the body, face, hands, and legs; rheumatism (joint pains); and rarely death (http://www.medicinform.net/infec/infec62.htm). Mosquito vectors of the virus infection can be found not only on the Indian subcontinent, Africa, and Asia, but also currently Europe and North America.

The **major outbreak** of disease was recorded in 2005 on Reunion Island, Indian Ocean, when 266,000 people (out of 770,000 total population) caught the disease (Roth et al. 2014). In an outbreak in India in 2006, 1.25 million cases were recorded (Muniaraj 2014). The chikungunya virus was recently reported (2013–2014) in the Western Hemisphere, where 1,118,763 suspected cases and 24,682 confirmed cases have been reported (http://en.wikipedia.org/wiki/Chikungunya).

2.3.2 Ticks and mites as vectors of disease

Ticks and mites belong to the class of arachnids (Arachnida). Ticks (including hard and soft ticks) belong to the order Parasitiformes, while mites (e.g., harvest mites, food mites) belong to the order Acariformes. They differ in size: ticks may reach 5 mm in length, whereas mites usually measure 0.2 to 0.4 mm (rarely, reaching 3 mm).

Ticks, the sanguivorous external parasites of mammals, birds, and reptiles, comprise about 850 species worldwide (http://www.livescience.com/46117-ticks-lyme-disease.html). Ticks are the **most**

important transmitters of disease in the United States, and second most important worldwide. They transmit the following **pathogens** (https://en.wikipedia.org/wiki/Tick-borne_disease): bacteria (Lyme disease or borreliosis, relapsing fever, Rocky Mountain spotted fever, Helvetica spotted fever, ehrlichiosis/anaplasmosis, tularemia), viruses (tick-borne meningoencephalitis, Colorado tick fever, Crimean-Congo hemorrhagic fever, severe febrile illness), protozoa (babesiosis, cytauxzoonosis), and toxins (tick paralysis).

Lyme disease or borreliosis is caused by spirochete bacteria of the genus *Borrelia*. It got its popular name from the town where it was first encountered in 1975: Old Lyme, Connecticut (US). The natural hosts of Lyme disease agent include white-tailed deer (in the United States), rodents, dogs, sheep, birds, and black cattle. The major transmitters of the disease agent are hard-bodied ticks of the genus *Ixodes*: bacteria are transmitted to a human through a bite of an infected ixodic tick. In Europe, these are sheep tick or castor bean tick (*I. ricinus*) (de Mik et al. 1997); in Asia, the taiga tick (*I. persulcatus*) (Sun and Xu 2003). In North America, the agent is transferred by black-legged ticks or deer ticks (*I. scapularis*) on the East Coast and by western black-legged tick (*I. pacificus*) on the West Coast (https://en.wikipedia.org/wiki/Lyme_disease).

Lyme disease is the **most common disease** transmitted by ticks in the Northern Hemisphere. This disease affects 300,000 people a year in the United States (Shapiro 2014) and 65,500 people a year in Europe (Leeflang et al. 2016). The early manifestations of the disease can include fever, arthritis, neuroborreliosis, erythema migrans, cranial nerve palsy, carditis, fatigue, and flulike illness. Delay in treatment can result in development of chronic Lyme disease, which is difficult to cure and can end in disability or even death (https://en.wikipedia.org/wiki/Lyme_disease).

Tick-borne meningoencephalitis is a virus infection characterized by fever, intoxication, and cerebral affectation. The disease can be found across a practically continuous belt through the forest zone of Eurasia from the Atlantic Ocean and Mediterranean Sea to the Pacific Ocean (Map 2-7). The disease can lead to neurological and psychiatric complications and even death. Between 1990 and 2007, 157,584 cases of tick-borne encephalitis were recorded (i.e., 8,755 cases per year on average). At that, 5,950 cases were found in Russia and 2,805 cases in the other European countries (Süss 2008). From the late 1980s, the incidence of tick-borne encephalitis grew tenfold and mortality increased from 0.3–2.7%.

The **natural sources** of the virus include more than 130 species of different warm-blooded wild and domestic animals and birds, in particular, wild ungulates, ruminants, birds, rodents, carnivores, and

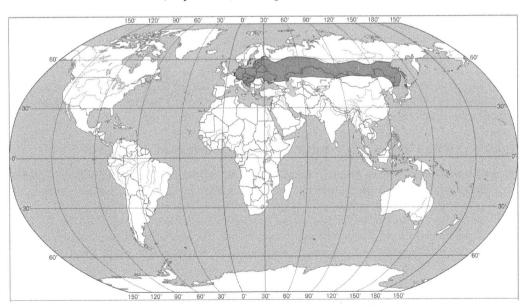

Map 2.7. Global distribution of Tick-borne encephalitis
(https://www.travmed.com/pages/health-guide-chapter-9-insect-borne-diseases;   https://en.wikipedia.org/wiki/Tick-borne_encephalitis).

horses. The vectors of the causative agent are ixodic ticks—the taiga tick (*Ixodes persulcatus*) and sheep tick or castor bean tick (*I. ricinus*) (Medico-geographical Atlas of Russia 2015).

Rocky Mountain spotted fever (bacterium *Rickettsia rickettsi*) was named after the Rocky Mountains where it was first identified in 1896 (in Idaho, US). However, the disease is widespread the United States, Canada, Brazil (São Paulo, Rio de Janeiro, Minas Gerais states), Colombia (departments of Cundinamarca and Santander) (Abarca and Oteo 2014), as well as in Mexico (states of Sonora, Sinaloa, Durango, Coahuila) and Panama.

Human infection results from bites of infected ticks. In the United States, the **major vectors** of the causative agent of Rocky Mountain spotted fever are the American dog tick (*Dermacentor variabilis*), Rocky Mountain wood tick (*D. andersoni*), and lone star tick (*Amblyomma americanum*). In the United States, 1,000 to 2,000 cases are reported each year (Goddard 2012). Maximum disease incidence and death rates are recorded in the American Indian reservations in Arizona, where it occurs in 7% of the people, compared with about 1% occurrence in other regions of the United States (Regan et al. 2015).

The **Crimean-Congo hemorrhagic fever** was for the first time identified in 1944 in Crimea. An analogous disease was discovered in the **Congo** in 1956, and the studies of this virus have established its absolute identity with the virus found in Crimea. The natural reservoir of the causative agent includes rodents, cattle and small ruminants, birds, and wild species of mammals, as well as the ticks themselves. Ticks can transmit the virus to their offspring through their eggs, vectors becoming lifetime virus carriers. The source of causative agent is the infected human patient or infected animal.

The virus has been isolated from at least 31 different species of ticks from the genera *Haemaphysalis* and *Hyalomma* (Mehravaran et al. 2012); however, its **major vectors** are ticks *Hyalomma marginatum*, *Dermacentor marginatus*, *Ixodes ricinus*. The disease occurs in Russia (Astrakhan and Rostov Regions, Krasnodar, and Stavropol Krais), in southern Ukraine; Central Asia; China; Bulgaria; the former Yugoslavia; Pakistan; central, eastern, and southern Africa (e.g., Congo, Kenya, Uganda, Nigeria) (Butenko and Trusova 2013). The **mortality rate** from this disease is 10–40% (https://en.wikipedia.org/wiki/Crimean–Congo_hemorrhagic_fever).

The larvae of **harvest mites** in the family Trombiculidae, sometimes called chiggers, or red bugs, are of global importance from the medical viewpoint because they cause dermatitis and transmit the causative agent of **scrub typhus**, bacterium *Orientia tsutsugamushi*. Scrub typhus was first reported in Korea in 1951 (Shin et al. 2014); currently it occurs throughout Asia, Africa, Chile, and the Middle East. In Southeast Asia, 1 million cases occur per year (Taylor et al. 2015).

Of great medical importance is the parasitic **itch mite** (*Sarcoptes scabiei*). Not a vector of diseases, the itch mite is the intradermal parasite causing a strong itch sensation in humans and many other mammals (dogs, cats, ungulates, wild boars, bovids, wombats, koalas, and great apes). The mites burrow passages in the skin of the host and feed on its blood, causing a strong itch that worsens at night. People infested with these mites scratch the affected spots, enabling penetration of bacterial infection and resulting in purulence and inflammatory processes. Sometimes these infections lead to serious life-threatening complications (Swe et al. 2014).

2.3.3 Sand flies, tsetse flies, and black flies as vectors of disease

Sand flies, small blood-sucking dipteran insects of the family Psychodidae, are found mainly in tropical and subtropical regions of the globe. It should be noted that different taxa are assigned, depending on the country; comprehensive information on this subject is contained in the article by Ready (2013).

Sand flies are vectors of leishmaniases and bartonellosis (Carrion's disease) (Beati et al. 2004). The **leishmaniases** are a group of transmissible diseases found in tropical and subtropical countries and caused by parasitic protozoa of the genus *Leishmania*. In the Western Hemisphere, the leishmaniases are spread by sand flies of the genus *Lutzomyia*; whereas in the Old World these diseases are spread by sand flies of the genus *Phlebotomus*. In total, 350 million people are at risk of infection in 98 countries around the world (Ballart et al. 2016). Yearly, about 2 million new cases of leishmaniases (Photo 2.3-2) are recorded (Barrett and Croft 2012) along with 20,000 to 50,000 deaths (https://en.wikipedia.org/wiki/Leishmaniasis). Global distribution of Leishmaniasis is shown in Map 2-8.

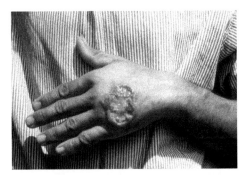

Photo 2.3-2. Leishmaniases are spread by sand flies. In total, 350 million people are at risk of infection in 98 countries around the world. Yearly, about 2 million new cases of leishmaniases are recorded and 20,000 to 50,000 deaths are registered. Skin ulcer due to leishmaniasis in the hand of a Central American adult is shown.
Photo credit: By CDC/Dr. D.S. Martin [Public domain], via Wikimedia Commons

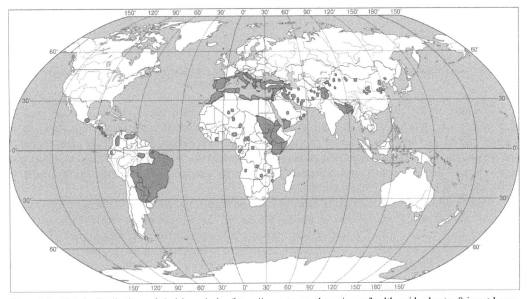

Map 2.8. Global distribution of Leishmaniasis (https://www.travmed.com/pages/health-guide-chapter-9-insect-borne-diseases; http://www.austincc.edu/microbio/2993u/ld.htm).

More than 30 species of sand flies of the genus *Lutzomyia* are significant vectors of bartonellosis, or **Carrion's disease**, a tropical infection caused by the bacterium *Bartonella bacilliformis*. The disease is found in Peru, Ecuador, and Colombia (Maguina et al. 2001), with several cases recorded in Bolivia and Chile (Gomes et al. 2016). In Peru, where the area of its distribution is 145,000 km^2, 1.7 million people are exposed to risk of disease (Huarcaya et al. 2004). Children also are affected by this disease; fatality rates of approximately 10% have been recorded (Rolain et al. 2004).

Tsetse flies belong to genus *Glossina* of family Glossinidae (Diptera); 23 species of this genus are known, differing according to the strict confines of certain landscapes and biotopes. They inhabit tropical Africa between 15°N and 20°S latitude, and their area of range is 10,000,000 km^2. Tsetse flies feed on the blood of vertebrate animals and are vectors of trypanosomiases, which are diseases of animals and humans caused by the parasitic protozoa, trypanosomes.

The **most important species** of these protozoa are *Trypanosoma vivax* and *T. congolense*, which mainly infect livestock, and *T. brucei*, which infects both humans and animals (Yaro et al. 2016). The principal diseases transmitted by tsetse flies are sleeping sickness or human African trypanosomiasis (in humans) and nagana (in animals). Global distribution of human African trypanosomiasis is shown in Map 2-9.

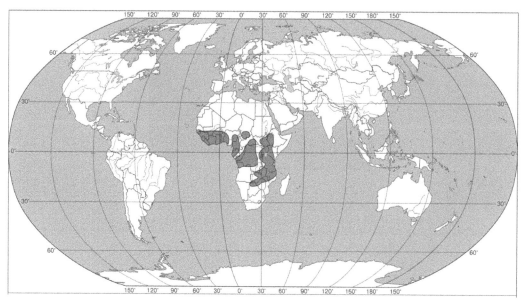

Map 2.9. Global distribution of human African trypanosomiasis (http://lifecenter.sgst.cn/para/Trypanosoma.html; http://www.ideluxe.org/trypanosomiasis.html).

Exposure to the risk of **sleeping sickness** affects 70 million people living over an area of 1.55 million km^2 in 20 countries (Simarro et al. 2012). The populations of the Democratic Republic of Congo, Angola, and Sudan are the most affected by this disease (Goddard 2012); 84% of deaths occur in the Democratic Republic of Congo (Wamwiri and Changasi 2016). In recent decades, the mortality rate due to human sleeping sickness has decreased gradually: for example, mortality was 34,000 people in 1990 and 9,000 people in 2010 (Lozano 2012).

Nagana disease results in dramatic death of cattle and loss in productivity of livestock farming. To avoid this disease, only 45 million heads of black cattle out of 172 million heads are situated at present in the tsetse-infested areas. The rest of the animals are managed in less-productive areas such as the highlands or the semiarid Sahel zone, leading to overgrazing by animals and overuse of lands for foodstuff manufacturing by people. The economic losses in cattle production alone are $1.0 to $1.2 billion; economic losses in agricultural production losses reach $4.75 billion per year (Eshetu and Begejo 2015).

Onchocerciasis (river blindness) is a disease caused by infection with the nematode worm *Onchocerca volvulus*. It is characterized by formation of skin nodules, severe itching, and ocular lesions that can progress to partial or total blindness (Photo 2.3-3). The nematodes are transmitted by the bites of **black flies** (Simulidae) of the genus *Simulium* (also called buffalo gnats, turkey gnats, and Kolumbtz flies). These flies live nearby the rivers that explains the second name of disease. In order for the infection to happen, the multiple bites are needed. There is no vaccine against this disease.

According to varying data, people with this disease may number from 17 million (https://en.wikipedia.org/wiki/Onchocerciasis) to 37 million (Hoerauf 2011). Global distribution of Onchocerclasis is shown on Map 2-10. Most infections happen in sub-Saharan Africa, although they are also recorded in Central and South America. Individual cases were registered in Yemen (https://en.wikipedia.org/wiki/Onchocerciasis).

Most significant in onchocerciasis infection are black flies of the **species** *Simulium damnosum sensu lato*, which is widespread in Africa; in east Africa, the disease is also transmitted by black flies *S. neavei*. In Central and South America, vectors of onchocerciasis are *S. ochraceum* and *S. metallicum* (Goddard 2012). The people of Africa are the most affected by this disease: 99% of affected people are African (Noormahomed et al. 2016). In total, approximately 270,000 cases of microfilarial-induced blindness and another 500,000 people with severe visual impairment were recorded (Hoerauf 2011).

Photo 2.3-3. Onchocerciasis is a disease caused by infection with the nematode worms *Onchocerca volvulus* which are transmitted by the bites of black flies (Simulidae) of the genus *Simulium*. In certain regions of West Africa, onchocerciasis is a most important cause of blindness. In some villages it is common to see young children leading blind adults; in highly endemic areas the blindness rate in men over 40 years old may be 40% or higher.
Photo credit: World Health Organization, 17 November 2006

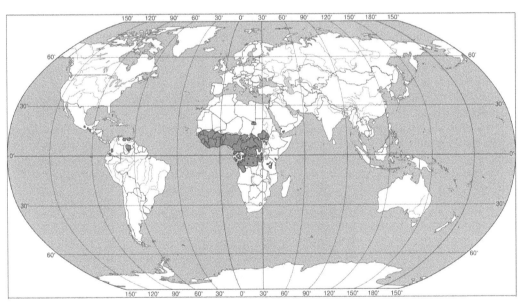

Map 2.10. Global distribution of Onchocerclasis (http://medpdffinder.com/d/divergence.com1.html; http://www.actionagainstpoisoning.com/page53/page499/page499.html).

2.3.4 Biting midges as vectors of disease

Biting midges (genus *Culicoides*) as vectors of diseases are of importance to cattle breeding. The principal diseases which are distributed by biting midges include bluetongue, Schmallenberg virus, African horse sickness, and Eastern equine encephalomyelitis.

Bluetongue is a virus disease of ruminants (most often affecting sheep) characterized by the inflammatory-necrotic lesions of tunica mucosa of the mouth, particularly the tongue, digestive tract, and hooves, as well as dystrophy leading to lameness. Sheep either die (after 8–10 days) or improve over a long period of time, with hair loss, infertility, and growth retardation (Inzhuvatova et al. 2016).

The **origin of** the bluetongue virus is in south Africa; the first information on it is dated the latter half of the 18th century. Although for a long time, this disease was recorded only in this region of Africa, the range of the spread of bluetongue has gradually expanded. Moving through northwestern Africa, bluetongue reached the European countries on the Mediterranean coast and since then has been recorded in Asia, North and South America, Australia, and New Zealand (Bouchemla and Agoltsov 2014).

The major **vectors** of bluetongue among biting midges are *Culicoides imicola* in Africa, Middle East, and southern Europe; *C. sonorensis* in North America; and *C. brevitarsis* (Photo 2.3-4) in Australia (Sprygin et al. 2015). In central and northern Europe, the vectors of bluetongue virus are biting midges from the *C. obsoletus* complex (*C. obsoletus, C. scoticus, C. montanus*) (Melhorn 2007) and the *C. pulicaris* complex (Purse et al. 2005).

The **financial losses** caused by bluetongue are related to death of the animals (decline in sheep) or, in the case of survivors, to the lowering of the animals' productivity, impairment of their reproductive function, and losses in export trade.

Loss of animals due to bluetongue can be very significant. For example, because of the outbreak of bluetongue in 1956, more than 130,000 sheep were lost in Spain over a period of four months (Inzhuvatova et al. 2016).

In recent years, bluetongue has affected mostly Mediterranean countries. From 2000 through May 2014, 63.08% of sickness cases were recorded in the countries of this region. The average **morbidity rate** for total number of sheep was 3.57%, while the mortality rate (ratio between the number of deaths from this disease and the number of animals living with the disease) was 31.78%, and death rate (ratio between the number of deaths from this disease and the total number of animals) was 1.13% (Bouchemla and Agoltsov 2014).

Black cattle are not protected against bluetongue; however, its consequences are much weaker in these animals. During the same time period (2000–May 2014), the incidence of bluetongue disease in black cattle in the Mediterranean countries was 1.91% on average, morbidity 0.24%, and death rate practically absent (Bouchemla and Agoltsov 2014).

Worldwide losses in the export trade related to bluetongue are estimated at $3 billion annually and, to livestock industries of some countries, these losses may be more distinct than the disease itself (Schmidtmann et al. 2011).

The disease caused by the **Schmallenberg virus** was initially reported in Germany. The new disease got its name from the location of its discovery (Schmallenberg, in North Rhine-Westphalia) in August 2011. Since then, the disease has been found in the Netherlands, Belgium, France, Luxembourg, Italy,

Photo 2.3-4. Scanning electron micrograph of a biting midge *Culicoides brevitarsis*. This insect is a vector of bluetongue virus, a viral infection of ruminants. Image produced by Electron Microscopy Unit, Australian Animal Health Laboratory. Photo credit: Electron Microscopy Unit, AAHL (CSIRO)

Spain, the United Kingdom, Switzerland, Ireland, Finland, Denmark, Sweden, Austria, Norway, Poland, and Estonia (https://en.wikipedia.org/wiki/Schmallenberg_virus).

Ruminant cloven-hoofed animals (cattle, sheep, goats) are susceptible to the virus. The disease is characterized by the gastrointestinal distresses, body temperature rise, deaths of animals, and birth of young stock with developmental defects (http://lefortvet.ru/bolezn-shmallenberga). The **vectors** of Schmallenberg virus are such species of biting midges as *C. dewulfi*, *C. scoticus*, *C. obsoletus sensu stricto*, and *C. chiopterus* (De Regge et al. 2014; Elbers et al. 2013a,b).

African horse sickness is an extremely contagious and fatal virus disease characterized by fever and abnormalities of the blood circulatory and breathing systems. It affects horses, mules, and donkeys. Its major vector is the biting midge *Culicoides imicola*. In addition, this pathogen is also transmitted by some species of mosquitoes including *Culex*, *Anopheles*, and *Aedes* and species of ticks such as *Hyalomma* and *Rhipicephalus* (https://en.wikipedia.org/wiki/African_horse_sickness).

The disease is endemic to the part of Africa located to the south of the Sahara Desert and was **first recorded** in the mid-1600s when the horses were delivered to Africa (https://en.wikipedia.org/wiki/African_horse_sickness). In 1930, the disease was first noted in the Middle East (Yemen) and then, in the 1940s and 1950s, in Palestine and Egypt. Reemerging in the countries of the Middle East in 1959, the disease in the form of panzooty has penetrated Iran, Pakistan, Turkey, Lebanon, Syria, Iraq, Jordan, Afghanistan, India, and Cyprus; in 1965 to 1966, the countries of the North Africa (Algeria, Tunisia, Morocco, Libya) (http://zooresurs.ru/horse/horse-zb/748-afrikanskaya-chuma-loshadej.html).

The **economic damage** caused by the disease is considerable. Horses are the most susceptible to this disease (about 90% of deaths of diseased animals), followed by mules (50%) and donkeys (10%) (https://en.wikipedia.org/wiki/African_horse_sickness). In 1959 to 1961, more than 300,000 horses died of the African horse sickness in the countries of the Middle East. An even greater number of horses, donkeys, and mules died off in 1965 to 1966 in north Africa. Carrying out the animal epidemic countermeasures (expenses for the vaccination, desinsection, restriction of horse pasturing, and so on) was related to considerable material expenses (http://zooresurs.ru/horse/horse-zb/748-afrikanskaya-chuma-loshadej.html).

2.3.5 Bedbugs and kissing bugs as vectors of disease

Bedbugs are widespread blood-sucking insects feeding on the blood of animals and humans. The two species of bedbug that feed on human blood are **common bedbug** (*Cimex lectularius*) found in most parts of the world, and **tropical bedbug** (*C. hemipterus*) found mainly in tropical countries (World Health Organization 2011).

Whether bedbugs are **vectors of communicable disease** is still an open question. A search of MEDLINE and EMBASE databases (1960–October 2008) for articles using the keywords bedbugs, *C. lectularius*, humans, parasitology, pathogenicity, and drug effects has found 53 articles meeting these criteria. In summary, transmission of more than 40 human diseases is attributed to bedbugs. Nevertheless, the authors of the study arrived at the conclusion that, for now, there was too little evidence that bedbugs are vectors of communicable disease (Goddard and de Shazo 2009). More recently, it was considered most highly likely that bedbugs transmit hepatitis B virus (World Health Organization 2011).

Bedbugs cause harm to humans by their bites interrupting the normal rest and sleep patterns and thereby reducing efficiency in the workplace. Additionally, in some cases, the bite can result in skin rashes, psychological effects, and allergic symptoms (Doggett Russell 2009).

Triatomine bugs, or **kissing bugs** (Triatominae), are widespread in North America and South America; several species also inhabit Asia, Africa, and Australia. Some kissing bugs are vectors of **Chagas disease** (American trypanosomiasis), a tropical parasitic disease caused by protozoan species *Trypanosoma cruzi*. This disease is spread in 18 countries from northern Argentina to southern United States (Map 2-11).

Its **most important vectors** in Central and South America are species *Panstrongylus megistus*, *Rhodnius prolixus*, *Triatoma infestans*, and *T. dimidiata* (Goddard 2012). Some believe that Charles Darwin suffered for the most of his life from Chagas disease, which he acquired when visiting Chile in 1835, as recorded in his *Voyage of the Beagle* (Clayton 2010).

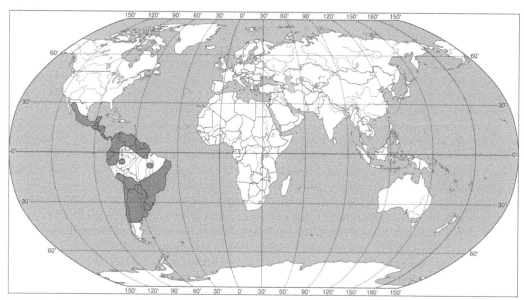

Map 2.11. Global distribution of Chagas' disease (https://revpana.wordpress.com/2012/08/31/chagas-one-bite-now-causes-heart-failure-20-years-later/; http://itg.author-e.eu/Generated/pubx/173/chagas_disease/distribution.htm).

The **reservoir** of infection are humans and some domestic and wild mammals (e.g., dogs, cats, rabbits, pigs, monkeys, rodents, and armadillos). Kissing bugs attack their sleeping victims, biting the labial mucous membranes, eyelids, and other parts of the face (which accounts for the bug's name). Infection of the bug takes place at the time of first blood-sucking, after which the trypanosomes develop in its intestinal tract. Within 1 to 2 weeks of the bite of the sick person, the bug is able to transmit the agent of the Chagas disease.

Most often, the trypanosomes are absorbed into the human organism through the feces of bugs (defecating close to place of bite) which can be instinctively wormed into the place of bite, eye conjunctiva, labial mucous membrane, or into any fissure or scratch on the skin. Transmission of trypanosomes is also possible through food infected with trypanosomes; when transfusing blood drawn from infected donors; from infected mother to infant through pregnancy or childbearing; when transplanting from infected donors; and during laboratory incidents.

According to different data, from 8 million people (Rassi et al. 2010; Bermudez et al. 2016) to 10 million people (https://en.wikipedia.org/wiki/Chagas_disease) living in endemic Latin American countries and an additional 300,000 to 400,000 living in non-endemic countries including many European countries and the United States suffer from Chagas disease. Every year, about 41,200 new cases of infection in the endemic countries are recorded and 14,400 infants are born with congenital Chagas disease (https://en.wikipedia.org/wiki/Chagas_disease). The **mortality rate** for Chagas disease was 10,300 deaths in 2010, and 9,300 in 1990 (Lozano 2012).

2.3.6 *Fleas as vectors of disease*

Fleas are blood-sucking insects (order Aphaniptera) that occur in all continents (even in Antarctica, where they parasitize birds) (Steele et al. 1997). Frequently, they are vectors of agents of different diseases of humans and animals, 2,086 species of fleas are known (Zhang 2013). The following species are the most important from the medical point of view (World Health Organization 2011): rat flea (*Xenopsylla cheopis*), human flea (*Pulex irritans*), and cat flea (*Ctenocephalides felis*). The fleas can transmit more than 25 diseases, of which the plague and murine or endemic typhus are most dangerous.

The **plague** is the extremely dangerous and very infective disease caused by the bacterium *Yersinia pestis*. In nature, the sources and reservoirs of infection are rodents (e.g., marmots, ground squirrels, gerbils, rats) while the fleas are their vectors. The natural foci of disease lies within the broad belt between the 55°N and 40°S (https://en.wikipedia.org/wiki/Plague_(disease). The rats are the major sources in the inhabited localities. The reasons for the plague outbreaks are high numbers of rats and fleas. When the rats die, the fleas leave them and search new hosts; therefore, human epidemics often follow in the tracks of epizooties with high death rate among the rats.

The most common forms of plague are bubonic plague and pneumonic plague. The **bubonic plague** is so named because it is accompanied by formation of buboes (swelling of the lymph nodes due to inflammation). It develops in case of penetration of the causative agent through the skin (as a rule, when biting the infected flea and the "rat-flea-human" scheme is considered to be the common path of its propagation). However, during the large-scale epidemics of the past, a human carrier of bubonic plague can be a source of infection that was transmitted according to the "human-flea-human" scheme. Progression of inflammation in the lungs is characteristic of **pneumonic plague**; the disease develops as a result of airborne infection of the human's respiratory organs. Earlier, the mortality rate due to bubonic plague reached 95%, while that due to pneumonic plague reached 98–99%. At present, if a treatment is correct, the mortality rate is 5–10% (Supotnitsky and Supotnitsky 2006).

The number of people who have died of plague has reached many hundreds of millions. In the past, the plague was called the Black Death, and it caused the disastrous epidemics. Three pandemics of plague are best known. The first of them was called the **Plague of Justinian**. It began in Egypt in AD 541–542 and, from there, the infected rats (and fleas) were brought on board grain ships to Constantinople. During the outbreak in Constantinople, about 5,000 people died every day. In AD 588, the second major wave of plague began and spread this time from the Mediterranean coast of France. This epidemic was nearly worldwide in scope, affecting central and south Asia, north Africa, and Arabia, and practically all of Europe (Lotfy 2015). In the period between AD 541 and 700, about 100 million people died across the world (https://en.wikipedia.org/wiki/Plague_(disease)).

The **Second Pandemic** (Black Death) began in China between 1347 and 1351 and spread by way of the Silk Road or by ship through Asia, Europe, and Africa. It reduced the world population from 450 million to between 350 and 375 million in 1400 (Lotfy 2015). It is estimated that China had lost half of its residents (from 123 million to 65 million), Europe, about one third (from 75 to 50 million), and Africa, one eighth (from 80 million to 70 million). The Second Pandemic was particularly widespread in the following years: 1360–1363; 1374; 1400; 1438–1439; 1456–1457; 1464–1466; 1481–1485; 1500–1503; 1518–1531; 1544–1548; 1563–1566; 1573–1588; 1596–1599; 1602–1611; 1623–1640; 1644–1654; and 1664–1667 (https://en.wikipedia.org/wiki/Second_plague_pandemic).

The **Third Pandemic** began in China's Yunnan province in 1855 and has been recorded on all continents. Bubonic plague spread as a result of transportation by ships of cargoes with infected rats. The more contagious pneumonic plague was predominantly concentrated in Manchuria and Mongolia (Lotfy 2015). As a result of this pandemic, more than 12 million people died in India and China alone (https://en.wikipedia.org/wiki/Plague_(disease)). This pandemic ended only in 1959 when worldwide casualties dropped to 200 per year.

At present, the spread of this disease is essentially smaller. From 1989 through 2004, about 40,000 cases were recorded in 24 countries. In a number of countries in Asia (Kazakhstan, China, Mongolia, and Viet Nam), Africa (the Congo, Tanzania, and Madagascar), and South America (Peru), the cases of infecting humans are registered practically every year. In the period of 2001 to 2010, the plague affected 21,725 persons, with 1,612 deaths; at that, the fatality rate of 7.4% was fixed (Butler 2013).

The **murine or endemic typhus** is caused by bacteria *Rickettsia typhi*. Its most common vector is the rat flea (*Xenopsylla cheopis*) (https://en.wikipedia.org/wiki/Murine_typhus). The infection of humans takes place when worming the feces of the infected fleas into the skin in the scratched points, as a result of contact of infected excrements with eye mucosa as well as through the products contaminated with the urine of sick rodents. The endemic regions include northern and southern Africa, Southeast Asia, Australia, India, coastal regions of the North, Baltic, Caspian, and Black Seas (http://diseases.academic. ru/1070/ТИФ_СЫПНОЙ_ЭНДЕМИЧЕСКИЙ). From 1913 to mid-1940, more than 5,000 cases of murine typhus were recorded in the United States (Civen and Ngo 2008).

Such cases in the United States are unique. However, the outbreaks of the disease arise periodically in the **less developed countries**. In 2001, there was a crowd disease (756 disease cases) in Nepal

(Zimmerman et al. 2008). The mortality rate of the endemic murine typhus is currently fairly low: with antibiotic treatment, mortality is 1%; without treatment, mortality is 4% (Civen and Ngo 2008).

2.3.7 Lice as vectors of disease

The blood-sucking lice (order Anoplura) are parasites of humans, many other mammals, and birds widespread across the globe. More than 3,000 known species of lice are known (http://tolweb.org/Phthiraptera). **Three species of lice** living on humans are head louse (*Pediculus humanus capitis*), body louse (*Pediculus humanus humanus*), and crab or pubic louse (*Pthirus pubis*) (https://en.wikipedia.org/wiki/Sucking_louse). Lice are vectors of such diseases as epidemic typhus, trench fever, and epidemic relapsing fever. Body lice have the maximum **epidemiological consequences**; head lice are of less importance, and pubic lice are not known to transmit disease organisms (Goddard 2012).

Epidemic typhus is caused by the bacterium *Rickettsia prowazekii*. It is transmitted predominantly through body lice and, rarely, through head lice. The bite of the infected louse does not result directly in ingress of infection; rather infection is introduced through the worming of louse feces into the skin lesions. The epidemic typhus is an anthroponosis (i.e., only humans get sick with it). After feeding on the blood of the sick, the louse becomes contagious after 1 to 2 weeks, continuing until the end of its life (i.e., 30–40 days). The onset of disease is unexpected and characterized by severe headache, a sustained high fever, cough, rash, severe muscle pain, chills, and falling blood pressure (https://en.wikipedia.org/wiki/Epidemic_typhus).

Epidemic typhus has been known for several centuries; **its first descriptions** were made in Germany in the 16th century. The unsanitary conditions and congestion of population contributed to the disease's propagation. Typhus once had a global reach and its epidemics were observed initially during times of war, famine, and deprivation. In the history of wars, epidemic typhus was sometimes the deciding factor because the death toll from this disease often exceeded that from the war. Some examples are provided by the Thirty Years' War (1618–1648) and Crimean War (1853–1856) (https://ru.wikipedia.org/wiki/Сыпной_тиф).

The **best-known mass cases** of epidemic typhus occurred from 1577 to 1579, when about 10% of the English population were lost (https://en.wikipedia.org/wiki/Epidemic_typhus); 1812 (Napoleon invasion of Russia); and the period of the World War I and civil war in Russia (1917–1921). For example, at the beginning of the military activities against Russia in 1812, Napoleon's army numbered 500,000 to 700,000 men, but only 3,000 men returned. It is assumed that epidemic typhus was responsible for 20% of these losses (Bechah et al. 2008).

In Russia after World War I, during civil war between the White and Red armies, 3 million people **died** of epidemic typhus out of an estimated population of 20 to 30 million people with the disease.

In many cases, a disease was a major cause of death for those nursing the sick. During World War II, the epidemic typhus resulted in the death of hundreds of thousands of prisoners in Nazi concentration camps (https://en.wikipedia.org/wiki/Epidemic_typhus).

This disease spread mostly in **Europe**, but it has also caused serious problems elsewhere. For example, in Mexico 22 epidemics were recorded from 1655 to 1918. For 19 of the 22 epidemics, it was found they were related to droughts and low crop yields, resulting in famine (Burns et al. 2014).

Gradually, the incidence of epidemic typhus, frequency of epidemics, and death rate during them has lowered. The last great outbreak of this disease was recorded in 1997 in Burundi during the civil war. Then, about 100,000 people caught the disease, with case fatality of 15% (Bechah et al. 2008). In the 21st century, mass incidences of epidemic typhus are possible in the refugee camps during wars or natural disasters (https://en.wikipedia.org/wiki/Epidemic_typhus).

Trench fever is a disease caused by bacterium *Bartonella quintana*. The vector is the body louse and a human becomes infected as a result of the bite of the infected louse (bacteria are contained in saliva and excrements) (Goddard 2012). The focus of infection is a diseased human. The disease is characterized by high fever, severe headache, pain on moving the eyeballs, and soreness of the muscles of the legs and back. During World War I, more than 1 million people had trench fever. Among British troops, the disease was recorded in one fifth to one third of troops, while among the German and Austrian troops, incidence of the disease was about 20% (https://en.wikipedia.org/wiki/Trench_fever).

The epidemics of this disease were also recorded in the course of World War II, both among troops and prisoners living in crowded and dirty conditions. **Now**, the incidence of trench fever is rare, but such cases are still registered in Bolivia, Burundi, Ethiopia, Mexico, Poland, and north Africa (World Health Organization 2011).

The **epidemic relapsing fever** is caused by bacterium *Borrelia recurrentis.* The vectors of the causative agent, body lice (*Pediculus humanus humanus*), can transmit the infection within 4 to 8 days of feeding on the sick. Transmission occurs when the louse is crushed and the infected hemocoel is released onto the human skin (http://ecdc.europa.eu/en/healthtopics/emerging_and_vector-borne_diseases/louse-borne-diseases/pages/louse-borne-relapsing-fever.aspx).

In the first half of the 20th century, the **major outbreaks** of the disease were registered in Eastern Europe, the former Soviet Union, and Africa, and they were related to wars and famine. With improvements in living standards, the spread of epidemic relapsing fever dropped off. Now, it is characteristic mainly of Ethiopia, Somalia, and Sudan, with some cases recorded in the rural Andean community in Peru and in northern China. In the absence of appropriate treatment, death occurs in 10–40% of cases; if treatment is given, only 2–5% of patients die (World Health Organization 2011).

2.3.8 Mechanical disease transmission by insects

Many insects (ants, beetles, cockroaches, flies) take part in the **mechanical transmission** of the causative agents. It happens when the insects physically transport the pathogens from one place to the other with the use of parts of the body, that is, there are no biological relations between transmissive agent and vector. For example, ants, beetles, and cockroaches, feeding on dead animals or excrement and moving afterwards to another place, transfer in their hairs, spines, setae, and so on, the causative agents to the food products and everyday objects (Goddard 2012). Therefore, house flies can transmit the agents of typhoid fever, dysentery and cholera; gadflies can transfer the bacteria of Siberian plague; and cockroaches can transmit diphtheria, tuberculosis, salmonellosis, infectious hepatitis, and tetanus.

As an example of a disease transmitted by insects, the **surra disease** can be mentioned. This chronic wasting disease is caused by a hemoflagellated parasite *Trypanosoma evansi* (Banerjee et al. 2015). Surra is a crucial disease of domestic animals, wild herbivores, and carnivorous animals in tropical and subtropical countries. A spreading of this disease is considered in detail in a recent paper (Desquesnes et al. 2013).

Affecting mostly cattle, buffalo, camels, and horses in Southeast Asia, the disease is mainly **transmitted mechanically** by horseflies (family Tabanidae). Clinical symptoms include severe anemia, fever, hypoglycemia, weight loss, poor weight gain, poor drinking ability, infertility, miscarriage, and even death (Ligi et al. 2016). People can also become sick with this disease, sometimes leading to death (Truc et al. 2013).

The mechanical transmission of the causative agents is also characteristic of louse flies (family Hippoboscidae). The transmission of *Bacillus anthracis* by louse flies (*Hippobosca rufipes*, *H. equina*) and sheep ked (*Melophagus ovinus*) is possible (Doszhanov 2003). The transmission of the West Nile fever virus was registered for the louse flies *Icosta americana* (Ganez et al. 2002; Farajollahi et al. 2005) and *I. ardeae* (Matyukhin et al. 2013). The louse fly *Pseudolynchia canariensis* is a vector of trypanosomiasis (*Trypanosoma hannae*) transmitted to healthy pigeons from diseased ones (Matyukhin et al. 2012).

Blood-sucking Insects and Ticks

The common feature among members of this category of arthropods is that they all feed permanently or temporarily on the blood of humans and animals. Sections 2.3.1 to 2.3.7 in the previous chapter considered the role these parasites play in the transmission of causative agents. In this chapter, we will discuss their effect on humans and animals as it relates to the physiological harm caused by the pricking of skin and sucking of blood.

2.4.1 General characteristics of blood-sucking insects and ticks

Despite their commonality, blood-sucking insects and ticks belong to different biological taxa and thus differ from each other in lifestyle, nature of effect, and so on. With respect to degree of association with the host, these parasites are divided into **permanent** (lice) and **temporary** (bugs, fleas, midges, mosquitoes, biting midges, horse flies, blood-sucking flies). By type of attack, these insects can be either active attackers (e.g., mosquitoes, black flies, horse flies) or insidious insects (e.g., fleas, bugs) (Tarasov 2002). Insects in this category differ from each other in many ways, characteristics of which are included in the Table 2.4-1.

Gnats are a group of blood-sucking two-winged insects that comprise many of the parasites under consideration here. This group includes the representatives of different families: mosquitoes, black flies, biting midges, sand flies, gadflies or deer flies, and some species of blood-sucking flies (e.g., tsetse flies, stable flies).

The diversity of groups and species of blood-sucking gnats is closely related to the **geographical zones** they inhabit. Generally, mosquitoes and black flies predominate in the tundra zone; their total number is very high, despite a limited number of species. In the taiga and mixed forest zone, there is more diversity in gnats: the number of species of mosquitoes and black flies increases, and biting midges, gadflies (e.g., horse flies), and blood-sucking flies appear. Farther south, the diversity declines again: the number of species of mosquitoes, black flies, biting midges, and gadflies decreases, but sand flies appear.

Many species of insects have specific places of mass breeding that are unique to them. The most favorable sites for these breeding locations are river flood plains and delta fans, as well as stagnant and low-flow reservoirs, bogs, and water-logged lands (in the middle and high latitudes) and areas of the artificial irrigation and paddy culture (in the lower latitudes) (Great medicinal encyclopaedia V. 6, 1977).

Factors influencing people's susceptibility to insect bites are listed in Table 2.4-2.

Table 2.4-1. Some characteristics of the blood-sucking insects and ticks (Tarasov 2002; Abuladze, et al. 1990; Newman 1989; Gilyarov 1986; Great medicinal encyclopaedia, v. 11, 15, 19; Goddard 2012; numerous Internet sites).

Common name	Latin name	Sizes, mm	Lifespan of adult	Sex	Duration of blood-sucking, min	Quantity of blood meal at a time, mg	Duration of digesting, days	Duration of starvation	Travel speed	Travel distance, km
Diptera										
Mosquitoes	Culicidae	4–14	Female: 42–56 days; male: 10 days	Only females	Up to 3	Up to 3	–	–	3.2 km/hour	2–3, to 35 by means of wind
Black flies	Simuliidae	1.5–6	Females 2 months; males 1–2 weeks	Only females	1–3	1–3	–	–	–	4–10, up to 200 by means of wind
Biting midges	Ceratopogonidae	0.5–3.0	–	Only females	–	0.005	4–5	–	–	0.2–1.0
Sand flies	Phlebothiminae Phlebotomus, Sergentomyia, Lutzomyia, Brumptomyia, Wfarileya	1.3–3.5	2 weeks	Only females	1–2	0.4–0.5	3–10	–	–	1.5
Horse flies or deer flies	Tabanidae	6–30; 10–20 on average	20–30 days	Only females	2–20	40–300	–	–	Up to 60–70 km/hour	3–5 to 10
Stable flies	*Stomoxys calcitrans*	5.5–7.0	2–3 weeks	Males and females	–	–	–	–	Up to 8 km/hour	Up to 10 by means of wind
Tsetse flies	Glossinidae	9–14	1–6 months	Males and females	1	15–156	–	–	–	–
Louse flies	Hippoboscidae	1.2–12; 4–8 on average	5–8 months for females	Males and females	–	0.2–1.5	–	–	–	–

Table 2.4-1 contd. ....

...Table 2.4-1 contd.

Common name	Latin name	Sizes, mm	Lifespan of adult	Sex	Duration of blood-sucking, min	Quantity of blood meal at a time, mg	Duration of digesting, days	Duration of starvation	Travel speed	Travel distance, km
Anoplura										
Lice	*Rhynchophthirina, Ischnocera, Amblycera*	0.5–5	1–2 months	Males and females, but males of lice suck blood much less than females	5–30	0.5–3.0	–	2–3 days; up to 10 days	–	–
Siphonaptera										
Fleas	*Pulicidae*	1–6, sometimes up to 16 (after blood-sucking)	100–500 days, up to 5 years	Males and females	1 min–several hours	Up to 100 (per day)	–	Up to 12–18 months	–	1–60 m
Hemiptera										
Bedbugs	*Cimicidae*	4.2–6.5	12–14 months	Males and females	–	Up to 7 g	–	In warm rooms 2–6 months; in cold rooms up to 1 year	1 m/min	0.02
Kissing bugs	*Reduviidae*	12–36	Several months–1 year		11–28	–	–	–	–	–
Acari										
Ticks	*Ixodidae Argasidae Nuttalliellidae*	0.2–5.0; in fullness condition, up to 30	0.5–5 years	Males and females but the males of ticks suck much less blood than females	Some min–2 h	–	–	1–6 months; argasid ticks up to 11 years	–	–
Mites	*Acarina*	0.2–0.4	10–70 days	Males and females	–	–	–	–	–	–

Table 2.4-2. Variables influencing susceptibility of individuals to insect bites (Singh and Mann 2013).

	Who are susceptible?	Insects attracted	Mechanism(s) involved
Environmental factors	Persons living in tropical areas, summer in nontropical areas	Any	Fewer clothes; exposure of larger areas
	Spending time in garden	Any	Increased exposure to insects
	Overcrowding and poor hygiene	Lice, fleas, bedbugs	Increased reinfestation
	Moving to a house in which previous owner kept pets	Cat and dog fleas	Cocoons hatch and attack humans in scarcity of natural host
	Dilapidated housing	Bedbugs	Source of crevices in which bedbugs multiply
	Tourism destinations, trains, cinemas, hospital wards and clinic waiting rooms, staff and student accommodation, hotels	Bedbugs	Increased transfer of insects
Host factors	Body heat and carbon dioxide in exhaled air	Mosquitoes, fleas, bedbugs	Increased attractiveness to insects
	Vibrations caused by host	Fleas	Displacement of air attracts insects
	Human sweat	Mosquitoes	Increased attractiveness to insects
	Human skin flora	Mosquitoes	Microflora produces compounds attracting insects
	Human body odor	Mosquitoes, sand flies	Increased attractiveness to insects
	Pregnancy	*Anopheles gambiae* complex	Increased heat and increased release of volatile substances from the skin surface
	Alcohol and beer	Mosquitoes	Unknown
	Lipoatrophy in patients on antiretroviral therapy	Mosquitoes	Lipoatrophic subcutaneous tissue may present a more accessible capillary network and increased release of volatile substances from the skin surface

2.4.2 Distribution of different blood-sucking insects and ticks

Blood-sucking mosquitoes (Culicidae) are widespread across the globe and inhabit all continents except Antarctica. They are also absent in Iceland and on some other northern islands (https://en.wikipedia.org/wiki/Mosquito). The range of the common house mosquito (*Culex pipiens*) is widest; it occurs wherever a human—its main prey—exists.

About 3,000 types of mosquitoes are known, and 2,400 of them live in tropical and subtropical areas. The **maximum species** diversity of mosquitoes is in Central and South America, where one third of the known species lives. Another third lives in tropical Asia and on the Malay Archipelago. In the Ethiopian region (Africa, 20° S latitude) live one sixth of mosquito species. In the Palearctic region (Europe, temperate Asia, and north Africa including the Sahara), less than 9% of species are found; in the Nearctic region, about 7% of species (Daniel 1990).

Usually only the **female** mosquitoes are blood-suckers; however, the male of the species *Opifex fuscus* living in New Zealand is an exception (Daniel 1990). About 90 species of blood-sucking mosquitoes reside in the former Union of Soviet Socialists Republic. Within the zone of tundra and northern taiga,

Photo 2.4-1. Among the most numerous bloodsucking insects are mosquitoes, numbering about 3,000 species. In general, they dominate in the Arctic prairie (tundra) zone; their total numbers are very high there, while the number of species is not. The picture was taken in the lower part of the Indigirka River (north-eastern Yakutia, Russia). Photo credit: S.M. Govorushko (Pacific Geographical Institute, Vladivostok, Russia), July 1975

their diversity is low (8–16 species), but the number is extremely high (Photo 2.4-1). Moving southward, the diversity increases (22–24 species), but the total number is lower. In semidesert and desert zones, the number of species of blood-sucking mosquitoes falls to 4 to 6 (Great medicinal encyclopaedia V. 11, 1979).

Sand flies (Phlebotominae) inhabit predominantly tropics and subtropics; however, the northern boundary of their range passes slightly to the north of latitude 50° N in Canada and slightly to the south of 50° N in northern France and Mongolia. Sand flies can be found as far south as latitude 40° S. Sand flies are not found in New Zealand or on the Pacific islands. Their habitat ranges from below sea level (Jordan Valley and Dead Sea) to 3,300 m above sea level (Afghanistan). In Russia, sand flies live near Sochi and in Crimea.

Sand flies are biologically fairly close to mosquitoes. **Differences** between them include development of sand fly larvae in moist soil rather than water and the sand fly's noiseless flight. Several genera are assigned to sand flies: in particular, *Phlebotomus* and *Sergentomyia* in Africa, Europe, and Asia and *Lutzomyia* in North and South America, with a combined total of more than 700 species (Sun et al. 2009).

Notably, "sand flies" appears to be a very imprecise term. It may indicate different species in different countries. For example, in the United States, sand fly refers to members of the family Tabanidae (e.g., greenheads) or Ceratopogonidae (e.g., sand gnat, sandflea, no-see-em); outside the United States, sand fly may refer to members of the subfamily Phlebotominae within Psychodidae (https://en.wikipedia.org/wiki/Sandfly).

Stable flies (genus *Stomoxys*) are insignificant in number, from the viewpoint of biodiversity. Just 18 species are known, and of those, the stable fly *S. calcitrans* is most significant, being the only species that occurs worldwide and the only species that is synanthropic (http://entnemdept.ufl.edu/creatures/URBAN/MEDICAL/Stomoxys_calcitrans.htm).

Louse flies (Hippoboscidae) feed on the blood of some mammals and birds and do not attack humans. There are 780 species of louse flies (Pape et al. 2011). Hippoboscidae occur throughout the world, with most abundance in the tropics of Africa, Europe, and Asia (Hutson 1984).

Tsetse flies (Glossinidae) live in 37 countries of tropical Africa and comprise 30 species and subspecies (http://www.fao.org/docrep/009/p5178e/P5178E05.htm). All are two-winged blood-sucking flies of *Glossina* genus. Tsetse flies are most abundant in wet locales, mainly, moist tropical forests and lands along the river shores; however, tsetse flies may also live in the savannah (Map 2-12).

Black flies (Simuliidae) inhabit all the continents except Antarctica. In the remaining parts of the land, they occur everywhere except for certain remote islands and deserts. In practice, black flies inhabit wherever water reservoirs with circulating water exist (Photo 2.4-2). Their number depends more on breeding place availability and height above sea level; temperature conditions are of lesser importance. The maximum

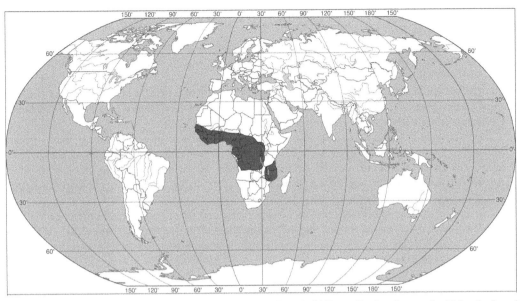

Map 2.12. Global distribution of tsetse fly (https://en.wikipedia.org/wiki/Tsetse_fly; http://www.microbiologybook.org/lecture/trypanosomiasis.htm).

Photo 2.4-2. Black flies are active during daylight hours only. Under the conditions of the polar day in the high latitudes, they attack 24 hours a day. A black flies attack in the Canadian arctic, Dubawnt River Nunavut is shown.
Photo credit: By en:user:NicolasPerrault [CC0], via Wikimedia Commons, July 2015

abundance of black flies is recorded within a zone of taiga and deciduous forests (Tarasov 2002). At present, 2,204 living species of black flies were recorded around the world (Adler and Crosskey 2016).

Horse flies (Tabanidae), the largest blood-sucking two-winged insects, live on all the continents except Antarctica. Horse flies are also absent in Greenland, Iceland, Hawaii, and some other oceanic islands (http://entnemdept.ufl.edu/creatures/livestock/deer_fly.htm). They are most numerous in the tropical regions. There are about 4,400 species of horse flies worldwide (https://ru.wikipedia.org/wiki/Слепни).

These heat- and light-loving insects attack in the afternoon and evening (before nightfall). They prefer temperatures of 19°C to 30°C for flying (Abuladze et al. 1990). The **highest total number** and number of species of horse flies (up to 20 in each locality) occurs in water-logged areas, on the boundaries of different ecotopes, in the places of cattle-grazing. Their altitude range is from sea level to at least 3,300 m above (https://en.wikipedia.org/wiki/Horse-fly).

Biting midges (Ceratopogonidae) are the smallest two-winged flies among the gnat group. They are widespread in almost any aquatic or semiaquatic habitat throughout the world, as well as in mountainous areas (https://en.wikipedia.org/wiki/Ceratopogonidae). Sometimes, they are the predominant component of the gnat group. Thus, in some regions of Siberia and the Far East, biting midges represent more than 90% of all blood-sucking insects attacking a human (Abuladze et al. 1975; Tarasov 2002). Worldwide, there are more than 1,000 species of biting midges (Sprygin et al. 2015).

Blood-sucking biting midges comprise five genera (Abuladze et al. 1990). Of these, genera *Leptoconops*, *Culicoides*, and *Forcipomyia* have the **most significant** effects on human activities: in *Leptoconops*, all the species are blood-sucking; in *Culicoides*, the majority of species are blood-sucking; and in *Forcipomyia*, about 10 species are blood-sucking. Most biting midges are nonhazardous insects that feed on plant moisture (Tarasov 2002).

The insects considered below are not part of the gnat group. Their fundamental difference from the earlier-considered blood-suckers rests in the fact that they are "born to creep and **cannot fly**". However, an inability to fly does not mean that they are less harmful. The group of blood-sucking insects considered below includes lice, bugs, and fleas.

Lice (Anoplura) are spread across every continent; wherever their victims (birds and mammals) inhabit, lice can be found (Boutellis 2014). For example, 15 species of lice were found on penguins in Antarctica (Banks and Paterson 2004). Nearly 5,000 species of louse are known; slightly over 4,000 species parasitize birds and 800 species parasitize mammals (https://en.wikipedia.org/wiki/Louse).

The geographic distribution of **bedbugs** (Cimicidae) is not limited to human dwellings; they can also parasitize hens, pigeons, cats, dogs, bats, and small rodents. Two species are most significant: bedbug *Cimex lectularius* (cosmopolitan in distribution) and tropical bedbug *C. hemipterus* (distributed in the tropics and subtropics). Humans are also attacked by other species, such as martin bug (*Oeciacus hirundinis*), but the role of these bugs is not significant.

According to varying data, 130 to 140 species of **kissing bugs** (Triatominae) occur. One species, *Triatoma rubrofosciata*, is found throughout the world. It is assumed that these bugs traveled the world on ships, feeding upon the blood of the ships' rats. Several species inhabit in Southeast and East Asia and Africa, but the most species (about 125) are found only in the Western Hemisphere (Schofield and Galvno 2009).

Worldwide diversity of **fleas** (Siphonaptera) includes 2,000 species and 550 subspecies belonging to 18 families. Forest foothills with temperate or subtropical climate present the most favorable conditions for fleas (https://www.zin.ru/Animalia/Siphonaptera/distr.htm).

The following group of blood-suckers—**ticks**—do not fall into the category of insects. Ticks are small arthropods of the class of arachnids. Of the three families of ticks, families Ixodidae and Argasidae are most significant. The 702 species of Ixodidae (Guglielmone et al. 2010) are also known as "hard ticks" because of their hard chitinous exteriors. None of the more than 200 species of Argasidae have such covering, so they are known as "soft ticks" (Allan 2001). Ticks are widely distributed all over the world, even in Antarctica (tick *Ixodes uriae* is parasitic on penguins and other birds) (https://ru.wikipedia.org/wiki/Иксодовые_клещи).

Mites are also widely distributed across the globe; however, in contrast to ticks, mites are microscopic. A total of 48,200 species of mites have been described (Halliday et al. 2000). Only the larvae of the *Trombicula* species feed on blood; they suck the blood of mammals, reptiles, birds, and amphibians. The process of blood-sucking lasts two or more days (http://stopvreditel.ru/parazity/perenoschiki/krasnotelkovyje-kleshhi.html).

2.4.3 Impact of blood-sucking insects on humans

Blood-sucking two-winged insects of the gnat category affect humans and animals collectively, so that sometimes it is difficult to identify the specific effects. Generally, the damage gnats have done to humans and animals is significant. During a mass attack, humans lose their capacity for normal rest, their working efficiency decreases by 25–30%, and their occupational trauma level increases. During the construction of the Bratsk Hydroelectric Power Station, the work was interrupted on some days because of the high gnat activity (Volovnik 1990). Next, we look at the effects of specific categories of blood-sucking insects on humans.

Mosquitoes (Culicidae)

Mass attacks of mosquitoes take place during the morning and evening twilight. In their search for a host, the insects are attracted to the odor of body sweat, as well as to temperature, emission of carbon dioxide, and color and surface of the clothes (Daniel 1990). The female plunges her piercing-sucking probe into the skin. If an appropriate capillary is not found, then she repeats her attempt. The moment of skin piercing is practically **painless**.

However, the saliva injected by the mosquito to prevent the clotting of blood causes the **painful skin** (itchiness, burning) and systemic reactions. Sometimes the bites are accompanied by formation of blisters and hives or local festering because of microbes entering the bite wound. The degree and character of body reaction depends largely on the species of mosquitoes, vulnerability of the recipient, and number of bites (Great medicinal encyclopaedia, vol. 11, 1979).

Sand flies (Phlebotominae)

Similar to biting midges, sand flies are twilight insects. Attacks against humans happen both outdoors and indoors. Before the blood-sucking, the female makes from 3 to 30 short piercings of the skin in search of the convenient place. At the point of the bite, scratching causes a blister to arise and redden; then a vesicle covering with a crust forms at its center. The places of scratching are easily contaminated and often become covered with small pustules (Tarasov 2002). Effects in humans include loss of sleep and appetite, temperature increases, and chronic ulcerated dermatitis. All of this has a significant impact on human health and results in performance degradation; after a sand fly bit, the itchiness can last 1 to 2 weeks (Chebyshev 2012).

Stable flies (genus Stomoxys)

The influence on a human is exerted, first of all, by the (autumn) stable fly *Stomoxys calcitrans.* Although these flies attack predominantly animals, sometimes they fly into human living space and bite humans. Both females and males feed on blood. Humans usually get bitten on the legs, behind the knees, and on the elbows (http://entnemdept.ufl.edu/creatures/URBAN/MEDICAL/Stomoxys_calcitrans.htm).

Louse flies (Hippoboscidae)

These insects mostly attack animals, but in high numbers some of them (e.g., deer fly [*Lipoptena cervi*]) bite humans and feed on their blood (http://http-wikipediya.ru/wiki/Оленья_кровососка).

Tsetse flies (Glossinidae)

All species of tsetse flies (females and males) feed on the blood of humans and animals. Roughly every 3 days a tsetse fly needs a blood meal in order to maintain its vital functions (Hoppenheit 2013). It is unclear whether tsetse flies prefer animal or human blood. Studies of nutrition of two species of tsetse flies carried out in the Sikasso region, southeast Mali, showed that human bites accounted for more than 66% of *G. p. gambiensis* blood meals, whereas for *G. tachinoides* human and cattle DNA were found in equal parts (Hoppenheit et al. 2013).

Black flies (Simuliidae)

Similar to mosquitoes, **only female** black flies suck blood, though not in all of the species. Depending on living conditions of the larvae, the same species may or may not be a blood-sucker. Sometimes, the stock of nutrients gained by larvae is sufficient to support females in their laying of eggs without blood sucking (Great medicinal encyclopaedia, vol. 15, 1981).

Black flies suck the blood of mammals, birds, and humans. In any locality, approximately **one third of species** attack human and domestic animals. Alighting on their victim's skin, black flies creep for some time before striking their probe into the skin, which makes their attacks importunate and especially fatiguing. The duration of blood sucking is 1 to 3 minutes. Black fly attacks typically happen outdoors; attacks against humans and animals do not occur indoors.

Black flies are active during daylight hours only. Under the conditions of the polar day in the high latitudes, they attack 24 hours a day. At temperatures below 6°C, in rain, and at wind speed of more than 3 m/s, black flies do not fly. Most attacks occur at temperatures of 17°C to 20°C (Tarasov 2002).

Swelling (and therefore pain) is greater from the bites of black flies than from the bites of mosquitoes. Black flies also inflict an actual bite taken out of the flesh, whereas mosquitoes pierce the skin with their thin, stylet-like mouthparts (https://ru.wikipedia.org/wiki/Мошки).

Horse flies (Tabanidae)

Most often, these insects attack humans and animals when the **body is moistened** with water or sweat. In contrast with mosquitoes and sand flies, horse flies do not bite indoors. The painfulness of their bites is due to the fairly large wound size and the injection of saliva containing anticoagulants and toxic substances. These bites may lead to edema and, sometimes, symptoms of allergic reaction may occur.

Biting midges (Ceratopogonidae)

These are, generally, twilight insects that attack outdoors and are especially aggressive before rain. However, some species can attack during the day and in fine weather; the females of some species fly into houses and livestock buildings where they exhibit high activity. The number of biting midges is higher during years with moist summers (Tarasov 2002). Their activity is highest at temperatures of 7°C to 20°C and in the absence of wind (not more than 0.5 m/s).

The painfulness of these bites is related to the introduction of **toxic saliva**. The site of the piercing becomes itchy, and hives appear. The intensity of the biting midges' attacks against humans can reach 4,000 times/hour (Great medicinal encyclopaedia, vol. 15, 1981). In some areas of the forest zone, biting midges represent more than 90% of all blood-sucking insects attacking humans. Mass attacks from these insects result in the lost productivity of humans (Chebyshev 2012).

Lice (Anoplura)

The parasites of a human include **three species** of lice: (1) head lice (*Pediculus humanus capitis*); (2) body lice (*P. h. vestimenti*); (3) pubic lice (*Phthirus pubis*). Lice spend their entire lives on the body of the host; separated from their host, they die within a few days. They may resettle by means of creeping across contacts between hosts (Abuladze et al. 1990). Lice feed frequently and a little at a time, they cannot feel hungry for a long time. The females suck more blood than the males. After piercing of skin with their probes, lice introduce their saliva, containing toxins, into the wound. Creeping on the body and piercings cause the skin irritation and itchiness.

People in history with a high infestation of lice have died as the result of their bites: Roman general and statesman Lucius Cornelius Sulla Felix (138–78 BCE), Roman king of Judea Herod the Great (73–4 BCE), and King of Spain Philip II (1527–1598). Death was actually caused by massive septic fever brought on by multiple scratches on the body becoming infected with pathogen bacteria (usually staphylococci) and covered with pyogenic crusts (Talyzin 1970).

Bedbugs (Cimicidae)

Most often, these bugs attack around 3:00 A.M., but if hungry, some individuals are active even in the light. Attacking sleeping humans, the bugs bite them mainly around the eyes (where the skin passes into the eye mucosa), on the lips, and on other parts of face (Tarasov 1996). When piercing the skin, the bugs introduce into the wound their **anticoagulant-containing saliva**.

Both females and males suck blood. Often the bites cause itch, blisters, and scratches. In their mass reproduction, these bugs disturb humans and deprive them of normal sleep and rest. Nevertheless, about 20% of humans are hyposensitive to their bites. Blood loss from parasitizing of these bugs is insignificant, although long-term, mass attacks result in a decrease in the hemoglobin level (Insects and ticks of the Far East 1987).

Kissing bugs (Triatominae)

These insects feed on blood during the night and prefer to bite humans around the mouth or eyes, which explains their name. This is because just the face is usually uncovered at bedtime. The typical kissing bug bite leads to a small raised skin lesion with a central punctum and an inflammatory infiltrate in the subcutaneous tissue. Swelling of tissue at the bite site may last up to 7 days (Stevens et al. 2011).

Fleas (Siphonaptera)

Both female and male fleas suck the blood. In contrast with lice, they can go hungry for a long time, living without food for **up to 18 months** (Abuladze et al. 1975). The volume of blood sucked by different species of fleas varies depending on the biological features of the given species, sizes of flea, air temperature, specificity of the host, and so on. Generally, the females suck a larger amount of blood than the males (Tarasov 1996). Repeatedly, the flea pierces the skin, sucks the blood, and injects droplets of saliva containing toxins.

The sites of flea bites become itchy, and hives and allergic reactions arise. Subsequent scratching contributes to intrusion of microorganisms into the wounds. Mass attacks result in considerable **blood loss**. From the medical point of view, the females of the chigoe flea (*Tunga penetrans*), inhabiting the tropical zones of both hemispheres are most significant. Most often, humans encounter the chigoe flea by walking barefoot. Attacking a human, these fleas penetrate the skin (usually between the toes) almost entirely and stay there for the rest of their lives. During their course of growth, they reach the size of a pea and inflict terrible pain (https://en.wikipedia.org/wiki/Tunga_penetrans).

Ticks

Ticks are different with respect to nutrition, even though a considerable portion of the known 25,000 species of ticks feeds on the blood of humans and animals. From this perspective, the ixodic (Ixodidae family) and argasid (Argasidae family) ticks are **most important** (Insects and ticks of the Far East 1987). Twelve species of argasid ticks (*Argas* and *Ornithodoros* genera) attack humans, their bites injecting toxic saliva and causing itch and hives. The bite of the pajaroello tick (*Ornithodoros coriaceus*) is most painful (https://ru.wikipedia.org/wiki/Аргасовые_клещи).

The females of all species of these ticks feed on blood while it is **less characteristic of the males**: not all of species suck blood and blood is only supplementary ration for many species consuming it. Cutting off the skin covering of its host, the tick begins to suck blood and periodically injects to the wound the saliva which anaesthetizes the bite, prevents blood clotting, and increases the permeability of the blood vessel walls (Tarasov 1996).

Trombiculid mites (Trombiculidae)

The larvae of some species of the trombiculid mites attack humans, and the considerable number of their bites can cause intense skin irritation (Ljamin and Pakhorukov 2009). Some trombiculid mite species that normally parasitize birds (e.g., *Dermanyssus gallinae*) will also bite poultry farm workers (https:// en.wikipedia.org/wiki/Mites_of_livestock). These mites continue to feed in the place of their bite for 2 to 4 days, and then they will drop off (http://bezvrediteley.ru/kleshhy/chem-opasen-krasnyj-kleshh).

2.4.4 Impact of blood-sucking insects on animals

Mosquitoes (*Culicidae*) inflict **significant damage** on the health and productivity of animals. In June in the forest zone of central Russia, during a 5-minute interval, a horse attracted up to 1,500 mosquitoes, a cow up to 600, and a calf up to 200. During the intense flight of mosquitoes, milk yields were reduced by 20–30%, and body weight gain of growing stock was reduced by 20–40% (http://www.ronl.ru/lektsii/ biologiya/847004/).

In case of an attack of 5,000 to 7,000 mosquitoes, they can suck about 100 to 120 g of blood per hour. Over a summer, mosquitoes suck about 7 to 9 L of blood from one animal (Abuladze et al. 1975). The mass bites result in anemia. Penetration of saliva into the wounds causes toxicosis, swelling, and dermatitis. Cases are known in which large numbers of animals **die** as a result of mass attacks by mosquitoes (Abuladze et al. 1990).

Sand flies (*Phlebotominae*) attack about 300 species of animals (mammals, birds, reptiles, and amphibians). Attacks may occur outside as well as indoors. The painfulness of the bites is due to their saliva containing toxins. The effect on animals is expressed in reduction in weight gain, loss of productivity, and so on. The bites of sand flies are overcome with most difficulty by horses, dogs, and birds (Chebyshev 2012).

As for **stable flies** (genus Stomoxys), both females and males feed on blood. There are several species of stable flies: biting house fly *Stomoxys calcitrans*, stable fly *Liperosia irritans*, stable fly *Haematobia atripalpus*, and others. The biting house fly lives both outside and indoors. Other species of stable flies inhabit only the open biotopes. Attacking animals in the grasses outdoors, they suck blood for a long time. Cattle and horses are mainly afflicted (flies accumulate on the legs and abdomen), but stable flies also feed on goats, sheep, swine, donkeys, cats, and dogs; in case of small animals, they feed around the ears because of the superficial blood vessels there, and on the head and legs (http://entnemdept.ufl.edu/ creatures/URBAN/MEDICAL/Stomoxys_calcitrans.htm).

In addition, stable flies are mechanical irritants of the nerve terminals in the skin, mucosa of eyes, nose, mouth, genitals, and wounds. All of this results in atrophy and reduction in **productivity of animals** (Akbayev et al. 1992). Stable flies cost the U.S. livestock industry $2.2 billion every year (http:// entnemdept.ufl.edu/creatures/URBAN/MEDICAL/Stomoxys_calcitrans.htm).

For the most part, **louse flies** (*Hippoboscidae*) are specific parasites. For example, the sheep ked (*Melophagus ovinus*) feeds on sheep, forest fly (*Hippoboscidae equine*) on horses, deer fly *Lipoptena cervi* on moose and deer, camel fly *Hippoboscidae camelina* on camels, louse fly *Hippoboscidae canis* on dogs, and species of *Ornithomyia* genus on birds (Abuladze et al. 1990).

Some species have **many hosts**. For example, the deer fly (*Lipoptena cervi*) feeds mainly on moose, deer, roe deer, elk, and cattle; however, these flies have also been found on wild boars, badgers, foxes, wolverines, bears, dogs, sheep, and goats. The number of louse flies on one animal sometimes reaches 5,000; more often, this number ranges from about 200 to 300. Deer flies (*Lipoptena cervi*) feed 15 to 20 times a day, sucking at each feeding from 0.2 to 1.5 mg of blood (http://http-wikipediya.ru/wiki/Оленья_ кровососка).

As with many other blood-sucking two-winged insects, louse flies intensely bother animals by their biting and creeping. After the biting subsides, inflammation develops, with intumescence, redness, and intense itchiness. Mass attack by great numbers of flies can cause abrupt atrophy, growth retardation, and, in many cases, death.

The **maximum damage** to animal breeding is caused by sheep ked. One sheep can be parasitized by 250 to 3,700 louse flies of this species simultaneously (Insects and ticks of the Far East 1987). The animals attempt to disturb the flies, rub against different objects, and scratch itchy skin areas with their teeth. Effusions of blood, loss of hair, and dermatitis arise. Because wool has been contaminated with dead insects, its quality drops. Louse flies often cause the deaths of animals, particularly lambs (Abuladze et al. 1975).

Both females and males of all species of **tsetse flies** (Glossinidae) feed on the blood of animals. The amount of absorbed blood varies from 65 to 156 mg. The flies attack moving prey—most often, animals going toward the watering place (Great medicinal encyclopaedia, vol. 16, 1981). Cattle, sheep, goats, dogs, pigs, camels, horses, and most wild animals may be attacked.

All species of domestic animals, but mostly black cattle and horses, are subject to attacks of **black flies** (Simuliidae). Before attacking the animal, the flies form a cluster around it. This produces agitation in the animal, which increases rises when black flies get into the eyes, ears, and upper respiratory tract. Bites from black flies are painful and often cause a severe reaction because the saliva contains strong hemolytic poisons. At the place of the bite a pinprick of blood appears, and, in a few minutes, edema develops, accompanied by extreme itchiness and a loss of skin sensibility. In mass attacks of black files, the affected animals show general weakness, temperature increase, and loss of appetite (Great medicinal encyclopaedia, vol. 15, 1981).

Many regions **record reductions** in milk yields and in weight gain of young stock as a result of intense black fly attacks. In the Irkutsk oblast (Russia), the milk yields decreased on certain days by 53.4% (Fedorova 2009). Sometimes, the cause of death of these animals is flies clogging the upper respiratory tract (Insects and ticks of the Far East 1987). Black cattle deaths were observed in a number of regions of Russia, Belarus, Ukraine, Germany, Denmark, and other countries (Fedorova 2009).

The painfulness of the bites of **horse flies** (Tabanidae) is related to the fairly large wound size and the injection of saliva containing anticoagulants and toxic matters into the wound. Sites of bites show edema and sometimes an allergic reaction. Mass attacks against domestic cattle result in reduction of weight gain, decrease in milk yield, and a lowered resistance to disease because of total intoxication (Great medicinal encyclopaedia, vol. 23, 1984). Animals affected the most by horse flies include horses, reindeer, and black cattle, especially young stock; and in the south, camels (Abuladze et al. 1990). During mass flight, 200 to 800 horse flies may attack one animal within 15 minutes (Tarasov 1996).

Biting midges (Ceratopogonidae) attack mainly outdoor animals; however, the females of certain species fly into livestock houses where activity is high. The painfulness of the bites is related to the injection of the toxic saliva. Once the piercing is done, itchiness and hives develop. The intensity of attacks of the biting midges against the animals reaches 10,000 specimens per hour (Great medicinal encyclopaedia, vol. 15, 1981). Black cattle, camels, and birds are the animals **most susceptible** to the bites of biting midges. The mass attacks result in development of dermatitis, edema of subcutaneous tissue, and partial loss of hair (Abuladze et al. 1975). The consequences of the bites of biting midges are practically the same as in the case of mosquitoes: weight loss, reduction in milk productivity, and so on (Chebyshev 2012).

Lice (Anoplura) are specialized parasites in that they feed on a particular species of animals. For example, *Haemotopinus macrocephalus* lives on horses, *H. curysternus* on black cattle, *H. suis* on swine. The strong infestation with lice causes agitation in the animals and affects their dietary regimen. The animals lose weight and may even starve to death. Reductions occur in the animals' working capacity, milk productivity, and quality of wool (in sheep and goats) (Insects and Ticks of the Far East Being of Medico-veterinary Importance 1987).

Bedbugs (Cimicidae) have the most significant effect on poultry breeding and rabbit breeding. Bedbugs make these animals and birds anxious; they scratch the disease sites on their body and they lose weight. Hens produce fewer eggs. Chicks 1 to 6 days old are affected the most; in case of mass attacks of bedbugs, chicks will sit with feathers raised, eyes semi-closed, and wings drooping. Rabbits often develop dermatitis as a result of scratching (Abuladze et al. 1975).

Kissing bugs (Triatominae) may feed on animals like raccoons, opossums, and wood rats; these bugs do not affect cattle breeding.

Piercing the skin, **fleas** (Siphonaptera) suck blood while continuously releasing into the wound droplets of saliva containing toxins. The sites of piercings become itchy, and hives and allergic reactions may arise. The subsequent scratching contributes to microorganisms penetrating the wounds. In the course of mass attacks, significant blood loss takes place. For example, the female flea *Vermipsylla* *alacurt* absorbs up to 100 mg blood a day and 700 to 7,000 of them can be simultaneously parasitic on one sheep. The sheep's hemoglobin level decreases by 20%, and wool losses reach 40%. Flea attacks on hens has led to reductions in egg production by 5–50% (Abuladze et al. 1990).

Among **ticks**, the ixodic (Ixodidae family) and argasid (Argasidae family) ticks are **most important** (Insects and ticks of the Far East 1987). Cutting the skin covering of the host, the tick begins to suck blood, periodically injecting into the wound the saliva that anesthetizes the bite, prevents blood clotting, and increases the permeability of blood vessel walls. The characteristic feature of ticks is that they can absorb an enormous mass of blood that sometimes exceeds hundreds of times the hungry tick's body weight. The linear dimensions of ixodic ticks increase by 10 or more times and volumes by up to 300 times by the end of blood sucking (Tarasov 1996).

The effect of ticks can be subdivided into **three constituents**: (1) mechanical breach of the skin coverings; (2) toxic influence of tick saliva; and (3) withdrawal of blood. All of this may cause anemia, local inflammation, itchiness, and, consequently, a general sense of aggravation. For example, during mass parasitizing of ixodic ticks on cows in Belarus, the average milk yield decreased to 2 to 3 L per animal. In the same area, the hemoglobin content in sheep blood was found to be reduced by 12–15% (Balashov 1967).

Trombiculid mites (Trombiculidae) inflict damage predominantly to poultry breeding. For example, red poultry mites (*Dermanyssus gallinae*) reproduce in poultry houses in vast numbers, especially in summer. Commonly, they attack the birds at night, while during the day they hide in the cracks of roosts and walls as well as in the dry poultry manure. The birds get chewed up and have poor growth; the growing birds frequently die (http://paukoobraznye.ru/books/item/f00/s00/z0000010/st016.shtml). Similar problems are also created by northern fowl mite *Ornithonyssus sylviarium* and tropical fowl mite *Ornithonyssus bursa* (Mullens 2009).

2.5

Venomous Insects and Arachnids

Toxicity is a fairly widespread phenomenon among arthropods. Generally, venomous animals need toxins for protection and predation. In the case of insects, spiders, and scorpions, toxicity finds an acceptable explanation. The reason for toxicity in mites has not so far been clarified.

2.5.1 Venomous insects

Among insects, 800,000 species use venom, or so-called chemical defense (Izmailov 2005). By **way of venom introduction**, insects are divided into **several groups** (Gelashvili et al. 2015). The **first** group introduces the venom into the body of the victim through a specialized stinging apparatus. This group includes different species of wasps, bees, bumblebees, ants, and the like.

Insects of the **second group** use the mouthparts for venom introduction. Among these are some species of true bugs, horsefly larvae, robber flies, and ant lions. The insects of the **third group** have no wound-inflicting apparatus and the toxic substances are released from openings in legs (blister beetles, lady beetles), anal glands (some ground beetles, Bombardier beetles), and so on. The **fourth group** of insects have wound-inflicting devices, but their venomous secretion is passively released because the gland cell generating the toxins has no compressor muscles. This group includes some butterflies as well as their larvae.

Three types of **responses to insect venom** are possible: local, general toxic, and allergic. A local response is characterized by redness, edema, pain, itching or burning near the site, and local lymphoid nodular hyperplasia. A **general toxic response** is characterized by chills, fever, nausea, vomiting, pain in the joints, and headache; it is commonly observed in cases of multiple stings. An **allergic response** arises in persons prone to it (Section 2.7 discusses this subject).

Venomous insects are mainly concentrated in the taxa of hymenopterans (Hymenoptera), beetles (Coleoptera), butterflies (Lepidoptera), true bugs (Hemiptera), and dipterans (Diptera).

2.5.1.1 Venomous hymenopterans

Practically all venomous hymenopterans introduce their venom by stinging. Among them are bees, bumblebees, wasps, hornets, ants, and ichneumon wasps.

Bees are the most significant in this regard. For a human, the sting of even one bee is extremely painful, while mass stings (e.g., attacks by a swarm of bees) can sometimes be lethal. The pattern of intoxication depends on locations of stings (the stings of head, neck, and throat are most dangerous); number of stings; and functional state of organism. The median lethal dose (LD_{50}) for adults is reached upon receipt of 500 stings. Numerous cases of loss of life have occurred upon receipt of 500 to 1,000 stings (Gelashvili et al. 2015). Nevertheless, a case is known in which an affected person survived after being stung by 2,500 bees (Mueller 1990).

In most fatalities as a result of multiple stings, death comes in a **few days**, when the general toxicity results in impaired kidney function, blood-clotting disorder, or brain necrosis (Modern Medical Encyclopaedia 2002). In case of allergic responses, death usually occurs within the **first hour** and nearly always within the first day after the sting.

At regular intervals, deaths of people by single stings not related to allergic responses are recorded. That happens very often when a person is stung in the tongue, throat, or roof of the mouth (e.g., when eating honey containing a live bee). Tongue inflammation is a very serious symptom, but edema of the throat mucosa, which tumefies and blocks the opening of the throat, is more serious, causing suffocation of the individual.

Sensitivity of various species to bee venom is much different. The horses are most sensitive, but cases of sheep and goats being killed by bee stings are also known. On ingestion of bees, geese and ducks die after several minutes. At the same time, bee venom is safe for a number of animals, such as bears, hedgehogs, badgers, martens, shrews, mice, rats, lizards, frogs, toads, and some birds (herons, woodpeckers, sparrows, tits, swallows) (Lipnitsky and Piluy 1991).

The **Africanized honey bee**, known colloquially as "killer bee", is far more dangerous. It is a hybrid of the western honey bee and the African honey bee. By mistake, the queens of a new hybrid were set free in Brazil in 1957. Since then, the bees have advanced northward with an average speed of 270 km/year and now their range occupies a considerable part of the United States (Map 2-13). More than 1,000 people fell victim to the Africanized honey bee (https://en.wikipedia.org/wiki/Africanized_bee).

We have no information on the distribution of victims according to the cause of death (toxicity or anaphylactic shock). However, considering the fact that the loss of life in most cases was related to multiple stings, one can assume a predominance of toxicity as compared with allergic reactions among the victims.

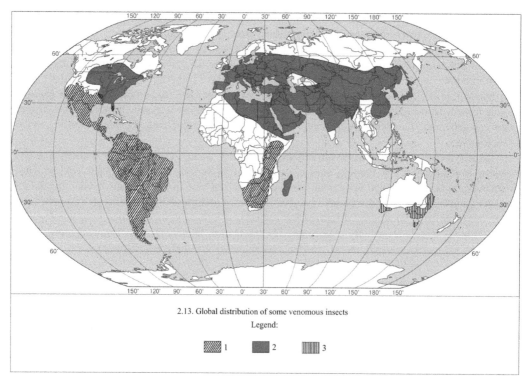

2.13. Global distribution of some venomous insects

Legend:

    1   2   3

Map 2.13. Global distribution of some venomous insects (http://www.vespa-crabro.com/hornets.htm; http://entnemdept.ufl.edu/creatures/misc/bees/AHB.htm; https://ru.wikipedia.org/wiki/Чёрный_муравей-бульдог).

Legend: (1) Africanized honey bee *Apis mellifera scutellata*; (2) hornets (*Vespa* spp.); (3) jack jumper ant *Myrmecia pilosula*

This species of bee also causes damage to cattle breeding. For example, 1,071 cases of stings of domestic animals had 715 lethal outcomes in Trinidad from 1979 through 1990 (Rodriguez-Lainz et al. 1999).

Bumblebees are less aggressive than bees. In contrast with bees, bumblebees can sting multiple times without consequences to their health. The symptoms of bumblebee stings are similar to those of intoxication by bee venom—pain and edema (Tsurikov 2005). As for wasps, the most significant species for humans are mainly concentrated in the families of social wasps (Vespidae), sand wasps (Sphecidae), and spider wasps (Pompilidae).

Social wasps include the common wasp *Vespula vulgarus*, German wasp *Vespula germanica*, and different species of hornets (e.g., *Vespa crabro*, *Vespa mandarinia*, *Vespa insularis*). A representative of sand wasps is the European bee wolf *Phylantus triangulum*. Spider wasps include the spider wasp *Cryptocheilus annulatus* and the wasps *Pepsis femoratus* and *Pompilus ciliatus*.

The wasp's sting is very painful. When a sting occurs to the neck or to the throat, if a live wasp in fruit drink, jelly, or soft fruit is accidentally swallowed, the most urgent measures should be taken because the delay can result in death by asphyxiation due to edema of the breathing passages. Deaths have occurred in people drinking water from a teapot's neck where a wasp was hiding. Deaths as a result of mass attacks of wasps have also been reported (Tsurikov 2005).

In most cases, people are subjected to stings by wasps when the wasps turned out to be inside the clothes or entangled in the hair. The risk is highest for people stirring up a nest of wasps. Many publications are devoted to stings by wasps. The most commonly encountered **clinical signs** are pain at the site of the sting (86.8%), edematic-inflammatory response to venom (75.0%), general uneasiness (63.2%), pruritus and urticaria (50.5%), Quincke's edema away from the sting site (47.1%), temperature rise (45.6%), dryness of the mouth (44.1%), headache (27.9%), heart palpitations (26.5%), and faintness and darkening of vision (25.0%) (Orlov and Gelashvili 1985).

It is recognized that stings by wasps are more dangerous than those of the domestic bee (Mueller 1990); however, because of less-frequent occurrence of wasps, the incidents with them happen more rarely. Nevertheless, in Brazil alone, about 2,000 individuals check into hospitals per year for intoxication related to wasp stings (Silva et al. 2016).

The **most dangerous representatives** of social wasps are hornets, which comprise 23 species. Best known is the European hornet (*Vespa crabro*), with sizes from 20 to 30 mm, living throughout Europe and northeast Asia. The hornets of southeastern Asia are larger and more dangerous. Their sizes reach 40 to 50 mm, and the Asian giant hornet (*Vespa mandarinia*) is considered to be most venomous. The danger from its stings is related to the great quantity of venom (up to 20 mg) contained in its venomous bubble; they can be lethal to a human (Gelashvili et al. 2015). Stings to the tongue or roof of the mouth are especially dangerous, and such cases are not infrequent; death following these kinds of stings have been described many times in the medical books.

Stings by any hornets are very painful. They cause **local** symptoms (pain, edema, inflammation) and **general** symptoms (headache, faintness, heart palpitations, temperature rise) of toxicity. It is highly dangerous to stir up a nest of hornets, which are distinguished by high aggressiveness. The danger of some hornets lies also in the fact that they can shoot venom into the eyes of their human victims and cause severe burns (Uzhegov 2007).

The best-known representative of sand wasps is the **European beewolf** (*Phylantus triangulum*). It is a medium-sized wasp, reaching a length of 13 to 17 mm. European beewolf stings do not constitute a serious risk for people; skin redness disappears after 1 to 2 days (Orlov et al. 1990).

Stinging ants belonging to the families Myrmicidae and Poneridae have a rather strong venom. Fire ants *Solenopsis*, bulldog ants, and jumper ants *Myrmecia* and bullet ants *Paraponera* are most significant. Bullet ants are most venomous of these (Haddad et al. 2015).

In some regions of the world (e.g., southern United States, Australia, Israel, Taiwan), these ants present a serious epidemiological problem. In the southern United States, in addition to three endemic species, two species of ants imported from South America have occupied this space in recent decades. The biggest problems are caused by the red imported fire ant (*Solenopsis invicta*), inhabiting the southern United States. Every year, about 33,000 people seek medical advice for stings by these ants (Photo 2.5-1),

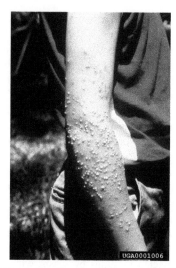

Photo 2.5-1. Stinging ants have rather strong venom. They constitute a serious epidemiological problem in Australia, Israel, and southern states of the United States. In the United States, 33,000 people ask for medical assistance every year after stings by red imported fire ants (*Solenopsis invicta*). The photo shows numerous stings of red imported fire ants in the United States. Photo credit: Murray S. Blum, University of Georgia, Bugwood.org

and some severe cases require hospitalization. Deaths are rare and are mainly related to allergic responses (Davis and Grimm 2003).

The manifestations of the stings by different species of ants **can vary**. The venom of red imported fire ants is characterized by its dermatopathic action. The skin's response to stings by the jack jumper ant (*Myrmecia pilosula*) (widespread in southern Australia), is expressed by erythema leading to edema. Pain and itching persist sometimes a few days (Orlov and Gelashvili 1985). The venom of the ichneumon wasp is only dangerous to the insects they parasitize by laying their eggs in them.

2.5.1.2 *Other venomous insects*

Some beetles have venomous properties. Venomous beetles are concentrated generally in the families of blister beetles (Meloidae), leaf beetles (Chrysomelidae), and rove beetles (Staphylinidae). The toxic substances produced by these beetles can be released from openings on legs, anal orifices, and so on.

The hemolymph (liquid circulating in the vessels) of all **blister beetles** is venomous and, in the event of emergency, they release it from the openings located between the lower legs and upper legs (blood-letting). Blister beetles cause dermatitis when crushed on the skin. Most often, skin of uncovered body parts (hands, neck, and face) is affected. Wounds, scratches, or wetting of the skin increase the venom's absorption and further development of general symptoms of toxicity (Tsurikov 2005). The mucosa of nose, lips, tongue, and conjunctiva are the most sensitive to these toxins, which have a serious effect on these areas (Orlov and Gelashvili 1985).

The **most venomous** in the blister beetle family are oil beetles (*Meloe*), Spanish flies (*Lytta*), blister beetles of the genus *Mylabris*, and blister beetles of the genus *Epicauta*. The venomous properties of all these beetles have been known for a long time. The effective agent of the venomous hemolymph of the blister beetles is cantharidin. When crushing the beetles, the venom enters the skin through openings of hair follicles, which results in emergence of big blisters. The burning and itching symptoms of envenomation by *Paederus* beetles are more intense than those caused by *Lytta* and *Epicauta* beetles (Haddad et al. 2015).

The **maximum danger** is food toxicity that occurs when cantharidin (or the whole beetle) penetrates into the digestive tract. The venom is intensely absorbed by the mucosae of the digestive tract and causes quick intoxication, sometimes with a lethal outcome. The LD_{50} of cantharidin for a human by ingestion is believed to be 20 to 30 mg; however, others find the LD_{50} to be only 10 mg (Barbier 1978). Typical consequences of poisoning with cantharidin include glomerulonephritis and cystitis.

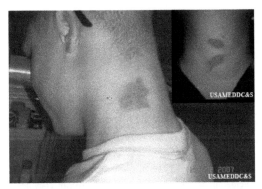

Photo 2.5-2. Paederus dermatitis is skin irritation resulting from contact with the hemolymph of certain rove beetles, a group that includes the genus *Paederus*. Paedarus dermatitis, two photographs, one showing "kissing" lesion where pederin was transferred to two skin surfaces that touch each other.
Photo credit: US Army Public Health Command (US Army Center for Health Promotion and Preventive Medicine USACHPPM), 28 July 2011

The best-known **beetle** of the Chrysomelidae family is the dangerous pest of potatoes, the Colorado potato beetle (*Leptinotarsa decemlineata*). Its hemolymph is poisonous for some invertebrate and vertebrate animals (Gelashvili et al. 2015); however, it has low toxicity for people. At the same time, toxicity of some other members of the Chrysomelidae family is extremely high, including diamphidia beetles (*Diamphidia locusta* and *D. nigro-ornata*) widespread in Africa. A toxin produced from the larvae of *Diamphidia* was named diamphotoxin. The LD_{50} for mice is 0.000025 mg/kg and, hence, it is one of the strongest natural toxins (Orlov and Gelashvili 1985).

Most species of the **rove beetles** (Staphylinidae) family are little plain beetles inhabiting various decaying substances, fungi, and manure and can be found under stones, in anthills, and in soil and forest litter. Among rove beetles, the members of *Paederus* are most venomous. When crushed (a beetle creeping on uncovered body parts is commonly crushed), they cause Paederus dermatitis, destroying the deep layers of the skin (Photo 2.5-2). At least 20 of the more than 600 species of *Paederus* beetles have been associated with Paederus dermatitis (Mullen and Durden 2009). Mass infestation of Paederus dermatitis caused by these beetles was observed in Russia (Lower Povolzhye), Brazil, India, and Algeria (Gelashvili et al. 2015).

The toxin of rove beetles is called **pederin**. Similar to blister beetles, rove beetles are most dangerous when ingested, causing food intoxication, which results in development of enteritises (inflammatory diseases of the small gut). These enteritises have often been observed in residents of the Marshall Islands, from using palm oil infected with rove beetles (Pigulevsky 1975).

Trouble for people can also be caused if a beetle gets in their **eye**. If caught by the eyelids, the beetle immediately releases a venomous hemolymph with strong irritating qualities, causing conjunctivitis and inflammation of eyelids, cornea, and iris. The uncontrolled dacryops can last for up to three hours, especially if the beetle is not removed quickly (Tsurikov 2005).

Lepidopterans, or **butterflies**, use toxic substances as a chemical defense against predators. Adult insects contain toxic compounds in their bodies, and larvae have sharp venomous stinging hairs, as protection against enemies. These hairs are hollow and filled with venomous secretion. The tips of hairs are very thin and are easily broken, resulting in the toxin flowing out. All lepidopterans are passively venomous insects. Although some larvaes' venom apparatus itself is painful, the toxic substance flows out passively because the gland producing the venom has no compressor muscle (Orlov et al. 1990).

The **pattern of intoxication** depends on the toxicity of the species of larvae considered, on intensity of the effect as determined by the number of entering hairs, and, finally, on localization of their penetration. As a rule, uncovered body parts (face, neck, arms) are affected (Photos 2.5-3, 2.5-4). More serious suffering may be caused by hairs penetrating an eye. Penetration of the stinging hairs in the digestive tract (e.g., with unwashed fruits) and breathing passages is also known. The most typical symptoms inflicted by the lepidopterans are dermatitis and conjunctivitis. Intoxication by the lepidopterans may be random

Photo 2.5-3. Stink bugs (family Pentatomidae) are common around the world. They derive their name from an unpleasant scent from a glandular substance released from pores in the thorax when disturbed. The stink bug in Brazil is shown.
Photo credit: Vidal Haddad Jr. (Boticatu School of Medicine, São Paulo State, Brazil)

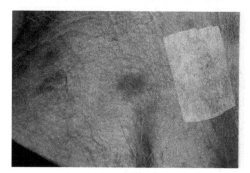

Photo 2.5-4. The chemicals of secretion involved include aldehydes, making the smell similar to that of coriander. In some species, the liquid contains cyanide compounds and a rancid almond scent, used to protect themselves and discourage predators. The photo shows erythematous and oedematous plaques caused by contact with a stink bug on the neck of a human being.
Photo credit: Vidal Haddad Jr. (Boticatu School of Medicine, São Paulo State, Brazil)

or could mostly affect professionals (gardeners, cultivators of the mulberry silkworm). Further, the most typical representatives of the venomous lepidopterans are involved.

Moths of the genus *Zygaena*. These vivid insects have a wingspan of 3 to 3.5 cm. They release a strongly smelling liquid from the front of the head. When it comes into contact with human blood, this liquid causes a violent reaction: if a drop of this liquid is applied to a scratch in the skin, in 6 minutes the person turns pale, perspires, feels a choking sensation, and experiences a quickening pulse, attaining 120 to 130 beats per minute. In an hour, the symptoms of poisoning disappear.

Garden tiger moth (*Arctia caja*). These insects with a wingspan of 50 to 80 mm are named for the appearance of their larvae: shaggy, covered with long hairs, they resemble tiny bear cubs. The hairs of larvae cause conjunctivitis; however, the adults are also venomous.

Siberian silk moth (*Dendrolimus sibiricus*). This moth has a wingspan of 8 cm and coloring of light yellowish brown or light gray to dark grey verging on black. Its larvae are up to 7 cm in length. In the period of mass reproduction, the 1.7 mm-long stinging hairs of the larvae are very dangerous. In the forest, in breeding grounds of the Siberian silk moth, people experience inflammation of the eye mucosa and dermatitis of various severities. A.S. Rozhkov (1965) described 168 patients with eye lesions from the stinging hairs of the Siberian silk moth larvae from 1946 to 1949 in Irkutsk Oblast. Penetration of numerous hairs into the skin can cause inflammation of finger and wrist joints.

Processionary caterpillars, larvae of the oak processionary moth (*Thaumetopoea processionea*) and the pine processionary moth (*Thaumetopoea pityocampa*). These have thin, easily broken, serrated hairs. These hairs contain formic acid in such high concentration that it causes a burn when it contacts human skin. If the hairs are ingested by grazing cattle, they can cause serious inflammation and even death (Gelashvili et al. 2015).

Brown-tail moth (*Euproctis chrysorrhoea*). This medium-sized moth has a wingspan of 26 to 40 mm; these are members of the tussock moth family (Lymantriidae). It is widespread in the forest and forest-steppe zones of European Russia. The stinging hairs on it cause skin irritation.

Cabbage White (*Pieris brassicae*). The larvae of the Cabbage White butterfly have a venom gland on the lower surface of the body between the head and the first body segment. To protect themselves, they eructate green gruel from the mouth to which the secretions of the venom gland are also added. People collecting these insects with bare hands have sometimes been hospitalized because of the skin inflammation and swelling of their hands. Ingested cabbage butterfly larvae have resulted in the death of domestic ducks.

Saturniid moths (Saturniidae). Some members of this family of insects, including the giant silk worm moth (*Lonomia obliqua*), found in Argentina, Uruguay, and southern Brazil, are also dangerous. The poisoning (which may cause death) results from careless contact with the larvae (Haddad et al. 2012). For example, in Rio Grande do Sul, more than 1,000 people suffered from such poisoning from 1997 through 2005 (Gelashvili et al. 2015).

In Uruguay, among venomous lepidopterans, the larvae *Megalopyge urens*, commonly known as "bicho peludo", is extremely dangerous. The effects of its venom include symptoms of general intoxication, such as restlessness, death anxiety, brachycardia, spasms, and vomiting, in addition to the strongest local pain. In the United States, the puss caterpillar (*Megalopyge opercularis*), encountered in all southeastern states, is the most dangerous of 50 species of venomous lepidopterans. Every year (from June through September), it causes epidemics of dermatitis, affecting thousands of people (Gelashvili et al. 2015).

2.5.2 Venomous spiders

2.5.2.1 Distribution of venomous spiders and the patterns of their bites

The order of spiders is divided into **three infraorders**: Liphistiomorphae, Mygalomorphae, and Araneomorphae. The species able to bite through the skin are considered to be dangerous for humans and homeothermic animals. Approximately 0.5% of these can inhabit all parts of the globe, except the coldest regions; all of them belong to the second and third infraorders.

Spiders of the infraorder Mygalomorphae include species of genera *Acanthoscurria*, *Phormictopus*, and *Avicularia*, living in South America; species of the genus *Pterinochilus*, in East Africa; and species of the genus *Atrax*, in Australia.

As for spiders of infraorder Araneomorphae, the species of the genus *Ctenus*, family Ctenidae (wandering spiders), widespread in the tropical and subtropical areas of the world, are very venomous. Species of families Araneidae (orb-weaving spiders) living in the tropics, Sparassidae (huntsman spiders) living in tropics and subtropics, Theridiidae (tangle-web spiders) living in all parts of the world, Lycosidae (wolf spiders) found in Asia and Europe, Salticidae (jumping spiders) in all parts of the world, and Sicariidae (six-eyed brown spiders) living in tropics and subtropics, are also dangerous (Gelashvili et al. 2015). Spider bites mainly affect people in the Americas, Australia, and Africa (Laustsen et al. 2016). Distribution of the some venomous spiders is shown on Map 2-14.

As a rule, spiders **do not attack** people and bite them only in self-protection. Female spiders are far more dangerous; male spiders of many species do not bite at all or are incapable of inflicting essential damage to a human.

The **distribution of bites** by time of year and time of day varies greatly. As for spiders inhabiting in nature within middle latitudes, the majority of their bites occur in the warmer months. For tropical and subtropical residents and spiders living in the house, the intra-annual distribution of bites is more uniform. For spiders living predominantly outdoors, the peak of bites is during the daylight hours. For example, 76.5% of bites of spiders of the genus *Phoneutria* in Brazil occur in the daylight hours (Bucaretchi et al. 2000). On the other hand, bites of *Lampona cylindrata* and *L. murina* spiders in Australia were found to peak between 8 P.M. and 12 A.M. (Isbister and Gray 2003).

The **sex composition** of the bitten people depends also on the spiders' habitat; victims of house spider bites are predominantly women. Thus, women constituted 54% of 130 people bitten by spiders

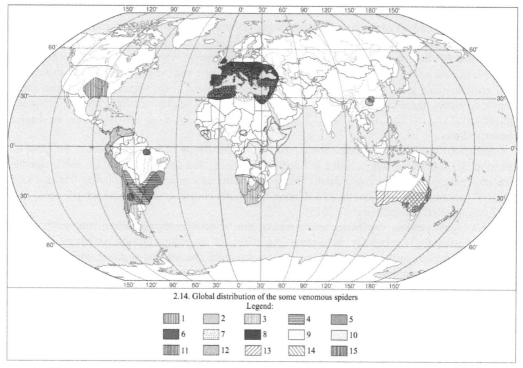

2.14. Global distribution of the some venomous spiders
Legend:

Map 2.14. Global distribution of some venomous spiders (Nieuwenhuys 2008; Binford et al. 2008; https://en.wikipedia.org/wiki/Segestria_florentina).

Legend: (1) Brown recluse spider (*Loxosceles reclusa*); (2) Chilean recluse spider *Loxosceles laeta* group (24 spp.); (3) Spiders of genus *Sicarius*; (4) Recluse spider *Loxosceles spadicea* (3 spp.); (5) Recluse spider *Loxosceles gaucho* group (4 spp.); (6) Recluse spider *Loxosceles amazonica*; (7) Mediterranean recluse spider *Loxosceles rufescens*; (8) European segestriid spider *Segestria florentina*; (9) Recluse spider *Loxosceles spinulosa* group (3 spp.); (10) Other African *Loxosceles* (9 spp.); (11) Recluse spider *Loxosceles aphrasta*; (12) Recluse spider *Loxosceles lacta*; (13) South Island Whitetail spider *Lampona cylindrata*; (14) North Island Whitetail spider *Lampona murina*; (15) Australian funnel-web spiders (genus *Hadronyche*)

Lampona cylindrata and *L. murina* in Australia (Isbister and Gray 2003). In Brazil, 56% of 515 people bitten by spiders of Lycosidae family were men (Diaz 2004). As for body parts bitten, the extremities are to a greater degree subject to bites; of 422 patients in Brazil bitten by spiders of the genus *Phoneutria*, 40.9% of bites were on the feet and 34.3% were on the hands (Bucaretchi et al. 2000).

As for the **kinds of activities** occurring when the person is bitten, bites may occur when the person is going to bed or sleeping (32%), dressing (20%), lying on the floor or sitting in an armchair (17%), or toweling off after shower or bath (11%) (Isbister and Gray 2003). In other words, the situations constituting a threat are when a spider is between the fabric (clothes, shoes, bed sheet, towel) and the human skin.

Spiders can **control the dose** or quantity of venom released on each bite. Because the production of venom is fairly energy-consuming, this property is of great importance. Spider bites are either attacking or defensive. When attacking, the quantity of released venom depends on the size of the prey. In case of defense, the sole purpose of the bite is to prevent against damage as a result of being pinched or crushed. If the spider estimates a risk to it as negligible or the size of prey is small, then it makes a so-called dry bite, without venom release (Spider bites 2009).

Information on the venom productivity and toxicity of some spiders is given in Table 2.5-1.

Table 2.5-1. Distribution, venom productivity and venom toxicity of some species of spiders (compiled by author using the following materials: Barbaro et al. 1996; Bayram et al. 2007; Herzig et al. 2002; Hung and Wong 2004; Liang et al. 1993; Manzoli-Palma et al. 2003; Ori and Ikeda 1998; Sheumack et al. 1984; Sutherland et al. 1980; Vetter and Visscher 1998; http://www.fauna-toxin.ru/15.html).

Scientific name	English name	Distribution	Sizes, mm	Venom productivity, mg	LD$_{50}$ mg/kg
Lycosa singoriensis	Russian tarantula, black widow	Central Asia, southern Russia and Ukraine, Turkey, Iran, southern Europe	27–37 (females); 25 (males)	–	15
Latrodectus tredecimguttatus	European or Mediterranean black widow, karakurt	Southern Europe, North Africa, western Asia	11–13 (females); 4–7 (males)	0.02–0.3	0.04
Latrodectus hasselti	Redback spider	Australia	10 (females); 3–4 (males)	–	–
Latrodectus mactans	Southern black widow	Southeastern US, Mediterranean countries	8–13 (females); 4–6 (males)	0.02–0.3	0.002
Latrodectus variolus	Northern black widow	Northeastern US, southeastern Canada	9–11 (females); 4–5 (males)	–	–
Latrodectus hesperus	Western black widow	Mexico, western US, southwestern Canada	14–16 (females); 7–8 (males)	–	–
Loxosceles reclusa	Brown recluse	North-central and southern US	6–20	0.13–0.27	–
Eratigena agrestis	Hobo spider	Pacific coast of US, southwestern Canada, western and central Europe	11–15 (females); 8–11 (males)	–	–
Loxosceles laeta	Chilean recluse	North, Central and South Americas, Australia, Finland (introduced)	8–30	–	–
Loxosceles intermedia	Brown spider	Brazil, Argentina	–	–	–
Latrodectus geometricus	Brown widow, brown button spider	Northeastern and southern US, Australia, South Africa, Cyprus, Japan	25–32	–	–
Cheiracanthium punctorium	Yellow sac spider	From central Europe to Central Asia	15 mm	–	–
Cheiracanthium mildei	Northern yellow sac spider	US, Canada, Argentina, Europe, Israel	60–70	–	–
Lithyphantes paykulliana	Steatoda	Black Sea region, Crimea, Caucasus, Central Asia, and Kazakhstan	–	–	–
Araneus diadematus	European garden spider, diadem spider, cross spider	Europe, US, Canada, up to extreme north	Up to 25 (females)	–	–

Table 2.5-1 contd. ....

...Table 2.5-1 contd.

Scientific name	English name	Distribution	Sizes, mm	Venom productivity, mg	LD$_{50}$ mg/kg
Argiope lobata	Lobed argiope	South of European part of Russia, Caucasus, Kazakhstan, Central Asia	Up to 15 (females)	–	–
Segestria florentina	Tube web spider	Crimea, adjacent areas of Black Sea region and Azov Sea region, Caucasus	15–22	–	–
Pterinochilus murinus	Orange baboon tarantula, Mombasa golden starburst tarantula	Central, East, and South Africa	100–150	–	1.5
Atrax robustus	Sydney funnel web, northern tree funnel web spider, southern tree funnel web spider, blue mountains funnel web spider	Australia (within a radius of 160 km of Sydney)	24–32	0.25 (females); 0.81 (males)	0.16
Missulena occatoria	Red-headed mouse spider	Australia	10–35	–	–
Lampona murina	White-tailed spider	Southern and eastern Australia, New Zealand	Up to 15 (females); up to 10 (males)	–	–
Phoneutria nigriventer	Brazilian wandering spider	Brazil, northern Argentina, Uruguay	17–48	1.1–2.15	0.00061–0.00157
Phoneutria fera	Armed spider, banana spider	Ecuador, Peru, Brazil, Surinam, Guyana, Venezuela	17–48	–	0.0009
Phidippus audax	Daring jumping spider, bold jumping spider	Southeastern Canada, US, north of Central America	13–20	–	–
Sicarius hahnii	Six-eyed sand spider	Sandy deserts of southern Africa	17	–	–

2.5.2.2 *Medical and veterinary importance of spiders*

According to different data, 100 (Bayram et al. 2007), 180 (Ori and Ikeda 1998), or 200 (Diaz 2004) species constitute a **threat to humans**. In order of their danger, they can be divided into four **categories**:

1. Human deaths by spiders of this group are numerous and well documented; this group includes the genera *Phoneutria, Atrax, Latrodectus*, and *Loxosceles.*
2. Spiders of this group are suspected in causing human deaths, but data are not reliable; this group includes genera Tagenaria and *Haplopelma.*
3. Spiders of this group have a venom potentially dangerous to human life; this group includes the genera *Hadronyche, Missulena*, and *Sicarius.*
4. Bites from spiders in this group cause diverse sensations of pain; this group includes 11 genera.

In addition, serious complications and even death can occur due to **secondary infection**, a wound, or anaphylactic shock. Therefore, in 0.9% of 750 bites by spiders in Australia, the introduction of infection into wounds was registered (Isbister and Gray 2003).

According to its effect, spider venom is divided into two **categories**: neurotoxic (affecting the nervous system) and necrotic (affecting the tissues around the bite and, sometimes, influencing the vital organs and systems of the organism). **Necrotic** venom is characteristic of spiders of genera *Latrodectus, Atrax*, and *Phoneutria*, whereas **neurotoxic** venom is typical of spiders of genera *Loxosceles, Tegenaria, Cheiracanthium*, and *Sicarius.*

In mass media and in scientific literature, the question is frequently asked about which spider species is most dangerous. The answer is not simple, and the following considerations should be taken into account. The toxicity of spiders' venom is determined for mice; however, the susceptibility of a human to it is often very different. The significance of some species is great because of the high number of their population. The venom productivity is also of a certain importance. For example, the quantity of venom released in a bite by a Brazilian wandering spider (*Phoneutria nigriventer*) is tens times the analogous value for the black widow (*Latrodectus tredecimguttatus*). It must be noted that spider venom as a whole is underinvestigated compared with scorpion venom and, even more, snake venom.

From the viewpoint of aggressiveness and toxicity of venom, the Brazilian wandering spider (*Phoneutria nigriventer*) in South America and the Sydney funnel-web spider (*Atrax robustus*) in Australia (Haddad et al. 2015) can be considered **most dangerous**. Taking into account the other factors (e.g., high number, frequent occurrence in houses, confinedness of ranges to areas with high population density), one can add to them *Loxosceles laeta, L. Intermedia*, and *Phoneutria fera* in South America; *Loxosceles reclusa* (Photo 2.5-5) and *Latrodectus mactans* in the United States; and *Sicarius hunnii* in South Africa (Diaz 2004).

The **mortality** rate from spider bites (Table 2.5-2) is absolutely negligible at the present time. Before formulation of antivenom, the mortality rate was 4–5% of the number of bitten people. In comparison, in

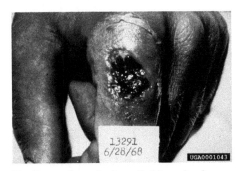

Photo 2.5-5. The brown recluse spider (*Loxosceles reclusa*) usually bites only when pressed against the skin, such as when tangled up within clothes, bath towels, or bedding. Most bites are minor with no necrosis. However, a small number of bites do produce severe dermonecrotic lesions (i.e., necrosis), and an even smaller number of bites produce severe systemic symptoms. The photo shows consequences of a brown recluse spider bite.
Photo credit: Center for Disease Control Archive, Centers for Disease Control and Prevention, Bugwood.org

292   *Human–Insect Interactions*

Table 2.5-2. Bites by spiders and their consequences, according data of different authors.

Country	Average number of bitten/year	Average number of fatalities/year	Fatality rate, %	Period	Dominant species	Source
Asia						
Russian Empire (Kazalinsky district, Syrdarya Region), now Kazakhstan	394	16	4	1896	*Latrodectus tredecimguttatus*	Tarnani 1907
Europe						
Russia (Samara Governorate)	48	2	4	1898	*Latrodectus tredecimguttatus*	Pavlovsky 1950
North America						
United States	–	0.4	–	1984–2004	*Loxosceles reclusa*	Spider bites 2009
	–	–	5	1900–1940	*Latrodectus hesperus*	Marikovskii 1947
	13,000	0	0	1997	*Latrodectus mactans*	http://www.fauna-toxin.ru/15.html
	–	–	5	Till 1956	*Latrodectus mactans*	Spider bites 2009
	–	0.63	–	1950–1989	*Latrodectus mactans*	Miller 1992
	4,699 (2,463 + 2,236)	0	0	2005	*Latrodectus mactans* *Loxosceles reclusa*	http://www.city-data.com
	–	6,5	–	1950–1959	*Latrodectus mactans*	Harves and Millikan 2008
South America						
Continent as a whole	–	Several	–	–	*Loxosceles laeta*	Spider bites 2009
	–	–	3–4	–	*Loxosceles laeta*	Schenone et al. 1989
Brazil (Santa Catarina)	267	–	1.5	–	*Loxosceles laeta* *Loxosceles intermedia*	Sezerino et al. 1998
Brazil	–	6	–	–	*Phoneutria nigriventer*	https://en.wikipedia.org/wiki/List_of_medically_significant_spider_bites
Australia and Oceania						
Australia	–	0.3	–	1920–1956	*Latrodectus hasselti*	Miller 1992
	30–40	0.25	–	1927–1980	*Atrax robustus*	Brown 2001
Worldwide	–	–	5–10	1966–1996	*Latrodectus* spp.	Bucaretchi et al. 2000

1896, 394 persons bitten by black widow spider were registered in Kazalinsky district (Syr-Daria Oblast, now Kazakhstan) and 16 of them (i.e., about 4%) died (Tarnani 1907). The mortality rate due to bites by spider *Latrodectus mactans* in the United States was 5% (Marikovskii 1947; Spider bites 2009).

After creation of antivenom, the situation changed dramatically and loss of life due to spider bites became **exceptionally rare**. For example, in 2005, none of the 4,699 persons bitten by spiders *Latrodectus mactans* and *Loxosceles reclusa* died in the United States (http://www.city-data.com). True, G.A. Jelinek (1997) says that there are published data of the present-day mortality rate of 5–10% due to bites by spiders of genus *Latrodectus*; however, he does not give any duration of time.

Therefore, the **present-day mortality** rate from spider bites does not exceed 10 people/year (except for cases of anaphylactic shock). Generally, this number is related to deaths of people in Brazil, where six people die from spider bites every year, mostly from the Brazilian wandering spiders of the genus *Phoneutria* (https://en.wikipedia.org/wiki/List_of_medically_significant_spider_bites). Among other spiders in Brazil that are significant medically are spiders belonging to the genera *Loxosceles* and *Latrodectus* (Cordeiro et al. 2015).

Nevertheless, the medical problem of spider bites remains operative. The number of people bitten by them is fairly high and the painfulness of the bites causes disability, which has a considerable social and economic impact. In 1994, in the United States, 9,418 spider bites were registered (Litovitz et al. 1998). In Australia, 1,474 bites were recorded for 27 months (Diaz 2004).

Apart from the effect on public health, spider bites affect labor productivity in some areas (e.g., in plant cultivation). The Sri Lanka ornamentals (*Poecilotheria fasciata*) inhabiting the coconut palms in Sri Lanka frequently bite the nut gatherers. In South America, vineyard workers suffer frequently from bites by spiders *Mastophora gasteracanthoides* spinning their cobweb in vines (Gelashvili et al. 2015).

The **symptoms of bites** by different spiders vary, depending substantially on the type of venom. Spiders with **necrotic** venom grossly affect the skin. Symptoms of general toxicity include anemia, intravascular blood coagulation, and impaired kidney function. The bites by spiders having the neurotoxic venom paralyze the central and peripheral nervous systems. They result in hallucinations, rigor, breathing difficulty, nausea, and vomiting (Great Medicinal Encyclopaedia 1986).

The **sensations of pain** are also very different. The bites by spiders possessing necrotic venom are accompanied with burning pain persisting for a prolonged period (about 1 day) and, at the site of the bite, extensive swelling of tissues is observed. If the venom is neurotoxic, pain at the moment of the bite is weak and disappears quickly, and there is no swelling at the point of the bite; however, intense pain arises in all organisms after a while (Gelashvili et al. 2015).

As a means of emergency medical aid in cases of spider bites, antivenom is administered. Information on the existing antidotes is presented in Table 2.5-3.

Other medical measures include an immobilization of the affected body part, cold compress to the bite point, increased fluid intake, and administration of medicinal preparations which are discussed in the literature (Great Medicinal Encyclopaedia 1986; Modern Medical Encyclopaedia 2002). Bite prevention measures can include wearing gloves and clothes with long sleeves, cleaning cobwebs from visiting spots (e.g., lavatories, tents, hangars), moving beds away from corners and walls, treating rooms with insecticides, and inspecting shoes, socks, hats, gloves, and towels before use. In farming, it is effective to drive or graze some animals not susceptible to bites by spiders; for example, sheep and swine devour karakurts with impunity (Pavlovsky 1950).

Spiders exert a great influence on **livestock farming**. In 1896, in Kazalinsky district, 1,045 different domestic animals were bitten by karakurt (black widow) spider (Syr Daria Oblast); 276 of 738 camels, 39 of 192 horses, 5 of 39 heads of black cattle, 20 of 85 sheep and goats died, and the total number of deceased animals reached 340 (Tarnani 1907). In 1869, in the Kirghiz steppes, 87 of 173 camels, 36 of 218 horses, and 4 of 11 black cattle, respectively, died after being bitten by karakurt (Pavlovsky 1950). As we can see, camels and horses are most susceptible to the bites. Even when the animals overcame their illness, they were left unable to work (Marikovskii 1947). At the same time, hedgehogs, dogs, bats, amphibians, and reptiles have low sensitivity to karakurt bites (Gelashvili et al. 2015).

Table 2.5-3. Antivenoms on the market for treatment of spider bites (Laustsen et al. 2016).

Product name	Producer	Country	Spiders
Funnel web spider antivenom	CSL Ltd.	Australia	Hexathelidae family (funnel-web spiders)
Red Back Spider antivenom	CSL Ltd.	Australia	*Latrodectus hasselti* (redback spider)
Aracmyn	Instituto Bioclon	Mexico	*Latrodectus mactans* (black widow spider), *Loxosceles* spp. (recluse spiders)
Reclusmyn	Instituto Bioclon	Mexico	*Loxosceles* spp. (recluse spiders)
Soro antiescorpionico	Instituto Butantan	Brazil	*Loxosceles* spp. (recluse spiders), *Phoneutria* spp. (Brazilian wandering spiders)
Anti-Latrodectus antivenom	Instituto Nacional de Biologics A.N.L.I.S.	Argentina	*Latrodectus mactans* (black widow spider)
Suero antiloxoscélico Monovalente	Instituto Nacional de Salud, Perú	Perú	*Loxosceles* spp. (recluse spiders)
Soro Antilatrodéctico	Instituto Vital Brazil	Brazil	*Latrodectus mactans* (black widow spider)
Antivenin (*Latrodectus mactans*)	Merck Sharp and Dohme International	USA	*Latrodectus mactans* (black widow spider)
SAIMR Spider Antivenom	South African Vaccine Producers	South Africa	*Latrodectus indistinctus* (black button spider)

2.5.3 Venomous scorpions

Strictly speaking, all scorpions are venomous but to different extents. A sting from a scorpion is dangerous for humans, especially if no one is nearby. The **ratio** of sizes of pincers and the stinger on the tail of the scorpion serves as a visual indicator of its envenoming capacity: the most venomous species have small pincers as compared with a larger stinger on the tail, while the nonvenomous species are distinguished by larger pincers and a comparatively small stinger.

2.5.3.1 General characteristics of scorpions

There are over 2,000 species of scorpions belonging to about 100 genera of 15–17 families. The territory of their distribution encircles the globe with a belt on both sides of equator between 48° N and 54° S. The southern boundary of the scorpions coincides with the southern limits of continents, however, in Australia, it also includes Tasmania. On the Pacific islands, imported species are generally widespread; however, some islands (New Caledonia, Fiji) contain endemic species. In Japan and New Zealand, there are no scorpions (Gelashvili et al. 2015).

The **northernmost** isolated place of scorpion habitation is Sheerness on the Isle of Sheppey in the United Kingdom (Kent County) located at 52° N. A small colony of scorpions (*Euscorpius flavicaudis*) has existed there since the 1860s. It is suggested that they were delivered there with imported fruits (http://en.wikipedia.org/wiki/Scorpiones). The **southernmost** habitat of scorpions is the Brunswick Peninsula in Chile. Distribution of the most dangerous scorpions is shown in the Map 2-15.

With respect to **some features of biology**, within the bounds of their range, scorpions live in sandy deserts, on high rocky highlands, and in places with a humid climate. The number of scorpions is not infrequently found to be enormous. In some sandy deserts, there is one scorpion to every square meter. In this connection, the question arises: on what do they feed in such a scanty ecosystem? Two considerations help them to survive. **First**, scorpions are not fussy/picky and eat all and sundry. **Second**, they eat very little. Because of their extremely slow metabolism, they can be restricted to one cricket per month (Fet 2006).

Scorpions are exclusively **nocturnal**. In daylight hours, they hide under stones, in forest litter or bark, in the cracks of structures, and so on. They feed on different invertebrates (e.g., insects, spiders, millipedes), or sometimes on small lizards or mice. If the prey is large and actively resists, the scorpion immobilizes it by using its venom (Orlov et al. 1990).

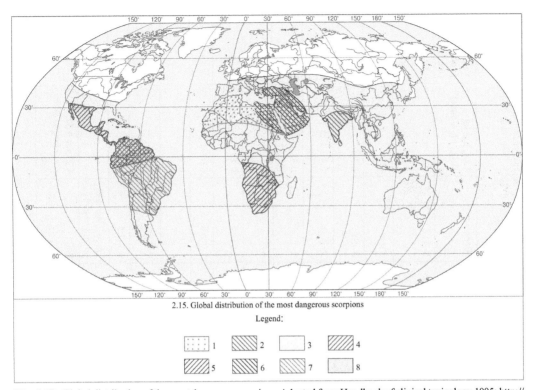

2.15. Global distribution of the most dangerous scorpions

Legend:

Map 2.15. Global distribution of the most dangerous scorpions. Adapted from Handbook of clinical toxicology 1995; http:// scorpion.amnh.org/page3/page3.html). Reproduced with permission of Prof. Julian White.
Legend. Genera: (1) *Androctonus*; (2) *Buthotus*; (3) *Buthus*; (4) *Centruroides*; (5) *Parabuthus*; (6) *Leiurus*; (7) *Tityus*; (8) Global distribution

The stinger of the scorpion is located in the end of the abdomen ("tail") and is supplied with a thin curved needle. At the base of the stinger are two vesicles with secretions, which contain the venom components. Near the needle point, there are two openings through which the venom is injected. Each bag is connected with one opening. Constricting the muscles of these bags, the scorpion injects the venom, combining a mixture of dozens of toxins, into the prey. It is interesting to note that scorpions use their venom sparingly. They control the venom spray and measure the quantity of venom that should be forced through each opening. Generally, the quantity of venom is from 0.1 to 0.6 mg. Sometimes, they sting their prey without venom release. The scorpions make their attack forward, above the head (Fet 2006).

There is no unanimity of opinion relative to the number of **species lethal to humans**. According to different data, their number can reach 25 species (de Roodt et al. 2003), 30 to 40 species (Fet 2006), 50 species (Ozkan et al. 2006; Cheng et al. 2016), or 75 species (Langley 1999). Virtually all species dangerous to humans belong to the family Buthidae and only two of them are included in the family Hemiscorpiidae. In terms of venom lethality, the venom of *Androctonus australis* and *Leiurus quinquestriatus* are the most toxic (Cheng et al. 2016; Deal 2006). Information on the majority of species dangerous to humans is presented in Table 2.5-4.

A zoning of the globe by frequency of scorpion stings is presented in Map 2-16. In the territories where there is a risk to be stung by scorpions, about 2.3 billion people live (Chippaux and Goyffon 2008). In each country, there exists its own set of medically **significant species** of scorpions. In some places, one species dominates. For example, in Tunisia, 80% of all people stung and 90% of deaths are from the scorpion *Androctonus australis*. In Iran, the scorpion *Hemiscorpius lepturus* is responsible for only 12% of known stings but, at the same time, it is the cause of 95% of the deaths from scorpions (Radmanesh 1990). In Zimbabwe, 75.4% of incidents were related to the Transvaal thick-tailed scorpion (*Parabuthus*

Table 2.5-4. Distribution, venom productivity, and venom toxicity of some species of scorpions (compiled by author with the use of the following materials: Gelashvili et al. 2015; Fet 2006; Flindt 1992; Cheng et al. 2016; Diaz 2004; Goyffon et al. 1982; Jarrar and Al-Rowaily 2008; Radmanesh 1990; de Roodt et al. 2003; Otero et al. 2004; Ozkan and Filazi 2004; Uawonggul et al. 2006; http://en.wikipedia.org/wiki/Scorpiones).

Scientific name	English name	Distribution	Sizes, cm	LD_{50} mg/kg (LD_{50} µg/mouse)	Venom productivity (mg/bite)
Leiurus quinquestriatus	The death stalker, Omdurman scorpion, Israeli desert scorpion, Palestine yellow scorpion	Deserts of north and southwest Africa, Near East, Pakistan	9–11.5	0.16–0.50 (5.1)	0,62
Centruroides noxius	Mexican scorpion, Scorpion of Nayarit, Mexican bark scorpion	Everywhere in Mexico (mountains, plateaus, near shore plains, deserts)	Up to 5.0	0.26	–
Androctonus australis	Yellow fat tailed scorpion	Deserts of north Africa, Near East, Pakistan, India	4–10	0.3–0.7 (7.0)	1,5–2,0
Androctonus crassicauda	Arabian fat-tailed scorpion, black fat-tailed scorpion	North and southeast Africa, Near East, Iran, Afghanistan, Pakistan, Armenia, Azerbaijan	9–10	0.31–11.5 (8.0)	0,3
Androctonus mauretanicus	Moroccan fat tail scorpion	Deserts of Morocco and Mauritania	9–11	0.32 (6.3)	–
Tityus serrulatus	Brazilian yellow scorpion, Yellow scorpion	Brazil	6–7	0.43 (8.6)	–
Centruroides suffusus	Durango Scorpion, Mexican bark scorpion	Mexico	5.0–6.5	0.43	–
Centruroides limpidus	Mexican scorpion	Central and southern Mexico	5.0–6.0	0.69–1.56	–
Androctonus amoreuxi	Fat tail scorpion, Egyptian Yellow Fat Tail Scorpion	North Africa, Near East, Iran, Afghanistan, Uzbekistan	11–12	0.75	–
Buthus occitanus	Mediterranean yellow scorpion	France, Spain, Malta, countries of north Africa	3–5	0.9 (17.9)	–
Buthus tunetanus	Common European scorpion	Tunisia, Algeria, Egypt, Libya	5.5–7.5	0.90	–
Centruroides sculpturatus	Arizona bark scorpion	Southwestern USA, Mexico	Up to 7 (female); up to 8 (male)	1.12	–
Androctonus bicolor	Black fat tail scorpion	Algeria, Egypt, Eritrea, Libya, Morocco, Tunisia, Israel, Jordan	8–9	1.21 (or 0.31, according to other data)	–

Species	Common name	Distribution	Size	Toxicity
Tityus bahiensis	Brown scorpion, Brazilian scorpion	Southeastern Brazil, Paraguay, Uruguay, northern Argentina	6–7	1.38
Mesobuthus eupeus	Lesser Asian scorpion, yellow scorpion	Balkan Peninsula countries, south Asia (from Turkey to Korea)	6.0–7.5	1.45
Parabuthus granulatus	Granulated thick-tailed scorpion	Angola, Botswana, Namibia, Republic of South Africa, Zimbabwe	12–16	1.56
Hottentotta tamulus	Eastern Indian scorpion, Indian red scorpion	India, Pakistan	7–8	2.25
Scorpiomaurus	Large-clawed scorpion, Israeli gold scorpion	North Africa, Near East	6.5–7.5	4.25
Parabuthus transvaalicus	Black spitting thick-tailed scorpion, South African giant fat tail	Botswana, Republic of South Africa, Zimbabwe, Mozambique	12	4.25
Vaejovis spinigerus	Stripe-tailed scorpion, "devil" scorpion	Arizona (US)	7–8	5.37
Hemiscorpius lepturus	–	Iran, Iraq, Pakistan, Yemen	Up to 5 (female); up to 8 (male)	5.81
Centruroides vittatus	Striped bark scorpion, Texas bark scorpion	Southwestern US, northern Mexico	6–9	7.7
Hottentotta judaicus	Judean/Israeli black scorpion	Egypt, Israel, Jordan, Lebanon, Syria, Turkey	7–8	7.94
Hadrurus arizonensis	Giant desert hairy scorpion	Southwestern North America	10–12	168
Opistophthalmus glabrifons	Yellow legged creeping scorpion	Southern Africa	10–11	600
Opistophthalmus boehmi	Golden scorpion	Southern Africa	5	625
Nebo hierichonticus	Common black scorpion	Israel, Jordan	10–13	1,000
Pandinus imperator	Emperor scorpion, imperial scorpion	West Africa (Congo, Cote d'Ivoire, Ghana, Guinea, Guinea Bissau, Nigeria, and Togo)	20–23	–
Hadrurus spadix	Black hairy desert scorpion	Southwestern North America	10–11	–

Note: Species of scorpions are ranked by venom toxicity.

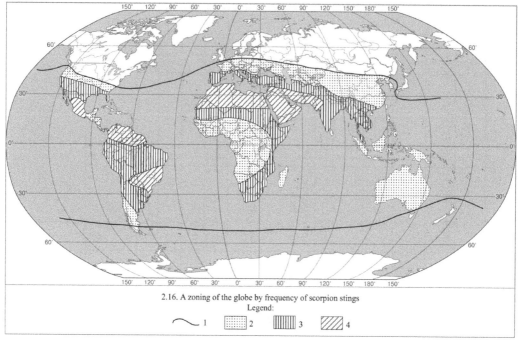

2.16. A zoning of the globe by frequency of scorpion stings
Legend:

~~~1   2   3   4

Map 2.16. A zoning of the globe by frequency of scorpion stings (Chippaux and Goyffon 2008; Lourenço et al. 1996). Legend: (1) Limits of distribution; Incidence per 100,000 inhabitants: (2) <1; (3) 1 to 100; (4) >100

Photo 2.5-6. The Arizona bark scorpion (*Centruroides sculpturatus*) is a small (7–8 centimetres in length) light brown scorpion common to the south-western United States. The bark scorpion is the most venomous scorpion in North America, and its venom can cause severe pain (coupled with numbness and tingling) in adult humans, typically lasting between 24 and 72 hours. Close-up (macrograph) of the barb of an Arizona bark scorpion is shown.
Photo credit: J. Adams, 19 September 2009

transvaalicus) (Bergman 1997). In the United States, the Arizona bark scorpion (*Centruroides sculpturatus*) is the only dangerous species (Photo 2.5-6).

In some countries, **two priority species** are known; these are the death stalker *Leiurus quinquestriatus* and fat-tail scorpion *Androctonus crassicauda* in Saudi Arabia (Jarrar and Al-Rowaily 2008), Eastern Indian scorpion *Mesobuthus tamulus* and *Heterometrus swammerdami* in India (Bawaskar and Bawaskar 1998), and *Tityus bahiensis* and *Tityus serrulatus* in Brazil (von Eickstedt et al. 1996). There are countries with a wider choice. For example, in Colombia, 39% of stings are related to *Tityus pachyurus*, 24% to *Centruroides gracilis*, 22% to *Tityus fuehrmanni*, and 5% to *Tityus asthenes* (Otero et al. 2004).

2.5.3.2 *Stings of scorpions*

As for the **times of the increased risk** of stings by scorpions, these are commonly the morning and evening hours. In the United States, 49% of stings occur from 6 to 12 A.M. and 30% from 6 to 12 P.M. (Cheng et al. 2016). This is quite explainable because the peaks of activities of both human and scorpions fall during these hours. The most common times seasonally for scorpion stings to occur are during the warmest months. For example, in Saudi Arabia, stings usually take place from May through October (Jahan et al. 2007) and the maximum frequency (21.5%) is in June while minimum frequency (1.5%) is in December (Jarrar and Al-Rowaily 2008). In Iran (Khuzestan province), the maximum frequency of stings (16%) is also in June and minimum frequency (0.6%) is in February (Shahbazzadeh et al. 2009).

The **parts of the human body** on which the stings occur can be most diverse. To a greater degree, the extremities are exposed to risk. So, in Colombia, 27.9% of stings fall on the hands and 26.4% on the feet (Otero et al. 2004). A similar situation is also observed in Iran: 39.3% stings occur on the hands, and 37.3% stings occur on the feet (Shahbazzadeh et al. 2009). At the same time, in Colima, Mexico, most stings occurred on the hands/arms (64.2%), followed by stings on the legs (25.3%), thorax (7.4%), head (1.9%), and neck (1.2%) (Chowell et al. 2006).

From the standpoint of the **sex**, males are most often affected: for example, of all stung people 77.3% of cases in Saudi Arabia (Jarrar and Al-Rowaily 2008), 72.9% in Qatar (Alkahlout et al. 2015), and 55% in Colima, Mexico (Chowell et al. 2006) were male. The mean age of stung people in different countries is virtually constant; in Saudi Arabia, on the basis of analysis of 6,465 cases it was 23 ± 17 years (Jahan et al. 2007). In Colima, Mexico, on the basis of 13,223 cases it was 27 ± 19 years (Chowell et al. 2006). In Brazil, the mean age was 30 years (Coelho et al. 2016).

A scorpion sting is very painful. At the point of the sting, the smarting pain arises, the skin reddens and swells quickly, and sometimes blisters appear. The most common symptoms presented included local pain (94.7%), local paresthesia (67.2%), pruritus (54.3%), sensation of a lump or hair in the throat (47.3%), sialorrhea (27.7%), and restlessness (26.4%) (Chowell et al. 2006).

Data on the annual average **number of the people stung** by scorpions and **number of casualties** in different countries are given in Table 2.5-5. Generally, there is little such information. Data on the mortality rate (i.e., percent of deceased persons of total number of stung ones) are also not numerous. In the old days, scorpion stings in some regions took the form of a disaster. For example, 1,600 people died in town of Durango (Mexico) with population of 40,000 people in the period of 1890 through 1926 with bites by scorpions belonging to genus *Centruroides* (Barbier 1978). Early in the 21st century, the annual number of people stung by scorpions across the globe reached 1.2 million people and 3,250 of these people (0.27%) died (Chippaux and Goyffon 2008). Later, the number of stings increased to 1.5 million; however, the death toll decreased to 2,600 people (Chippaux 2012). Such a considerable reduction in mortality rate resulted from the formulation and wide use of antivenoms. Information on antivenoms is given in Table 2.5-6.

The other **methods of protection** against scorpions are as follows. It is necessary to shake out clothes and shoes before putting on because scorpions frequently crawl into them on nights. When staying overnight in the desert, it is also important to inspect the territory around a tent before bedtime using an ultraviolet torch. The skins of all scorpions fluoresce in the ultraviolet radiation, shining in the green-blue spectrum. Using a good-quality torch, one can see a large scorpion at 5 to 10 m (Fet 2006).

For **mitigation** of a sting that has just occurred, an ice cube should be applied to the site for pain relief or a pain-relieving ointment containing antihistaminic, anesthetic, and corticosteroid preparations may be used (Modern Medical Encyclopaedia 2002). In the event of severe poisoning, intensive therapy and antivenom is necessary. In a number of countries, local scorpions' venom-specific sera are widely used. A serum is obtained from goats, and its use is sometimes complicated by the anaphylactic shock and serum disease (http://medbiol.ru/medbiol/env_fact/000688e2.htm).

Table 2.5-5. Stings by scorpions and their consequences according to data of different authors.

Country	Average number bitten/year	Average number of decreased/year	Fatality rate, %	Period	Dominant species	Source
Asia						
India, hospital, Mahad (Maharashtra state)	46.1	2.57	5.57	1984–1990	*Hottentotta tamulus (Mesobuthus tamulus)*	Cheng et al. 2007
Israel (hospital in Jerusalem)	10.2	0.4	3.9	1980–1984	–	Amitai et al. 1985
Iran	100,000		–	–	*Androctonus crassicauda, Mesobuthus eupeus, Hemiscorpius lepturus*	Cheng et al. 2007
Iran (Khuzestan province)	329	5	1.52	1984	*Hemiscorpius lepturus*	Radmanesh 1990
Saudi Arabia	14,500	–	–	–	*Leiurus quinquestriatus, Androctonus crassicauda*	Jarrar and Al-Rowaily 2008
Saudi Arabia (Qassim province)	1,293	–	–	1999–2003	–	Jahan et al. 2007
Africa						
Algeria	1,186.1	22.7	1.91	1942–1958	*Androctonus australis*	Balozet (1971)
Egypt	36,000			1933	–	Shteinmann (1984)
Tunisia (hospital in Sfax)	717	–	0.35	1980	–	Goyffon et al. (1982)
North America						
United States	14,950	–	–	2004	–	http://medbiol.ru/medbiol/env_fact/000688e2.htm
	13,642	–	–	1999	*Centruroides exilicauda*	Bush 2007
Mexico	200,000	–	–	–	*Centruroides sculpturatus*	Uawonggul et al. 2006
	100,000	800	–	1980–1989	*Centruroides sculpturatus*	http://en.wikipedia.org/wiki/Scorpiones
	–	1,000	–	–	–	Cheng et al. 2007
	–	1,933	–	1946	–	Diaz 2004
Mexico (Durango, population 40,000)	–	43.25	–	1890–1926	*Centruroides sculpturatus, Centruroides suffusus*	Barbier 1978

South America					
Brazil	3,858	31.3	1988–1990	*Tityus serrulatus, Tityus bahiensis*	von Eickstedt et al. 1996
Brazil (hospital in Belo Horizonte)	241.6	0.68	1977–1992	*Tityus serrulatus*	Fleire-Maia et al. 1994
Colombia (10 hospitals)	129	–	2002	*Tityus pachyurus, Centruroides gracilis, Tityus fuehrmanni*	Otero et al. 2004
Trinidad	139.6	6.6	1929–1933	*Tityus trinitatis*	Waterman 1938
	42.9	0.4	1940–1949	*Tityus trinitatis*	Waterman 1960
Globe as a whole	–	More than 5,000	–	–	http://medbiol.ru/medbiol/env_fact/000688e2.htm
	1.2 million	3,250	–	–	Chippaux and Goyffon 2008
	1.5 million	2,600	–	–	Chippaux 2012

Table 2.5-6. Antivenoms on the market for treatment of scorpion sting envenomings (Laustsen et al. 2016).

Product name	Producer	Country	Scorpions
Suero antialacran	BIRMEX	Mexico	*Centruroides* spp. (bark scorpions)
Suero antiescorpiónico	Centro de Biotecnologia de la Universidad central de Venezuela	Venezuela	*Tityus* spp. (thin-tailed scorpions)
Le serum antiscorpionique (monovalent)	Institut Pasteur d'Algerie	Algeria	*Androctonus australis* (fat-tailed scorpion), *Buthus occitanus* (common yellow corpion), *Androctonus crasicauda* (Arabian fat-tailed scorpion)
Scorpion antivenom	Institut Pasteur du Maroc	Morocco	*Buthus occitanus* (common yellow scorpion), *Androctonus mauritanicus* (Moroccan fat-tailed scorpion)
Le serum antiscorpionique	Refik Saydam Hygiene Center	Turkey	*Androctonus crassicauda* (Arabian fat-tailed scorpion), *Leiurus quinquestriatus* (Israeli yellow scorpion)
Alacramyn	Instituto Bioclon	Mexico	*Centruroides* spp. (bark scorpions)
Soro antiescorpionico	Instituto Butantan	Brazil	*Tityus* spp. (thin-tailed scorpions)
Soro antiescorpionico	Instituto Butantan	Brazil	*Tityus bahiensis* (black scorpion), *Tityus serrulatus* (Brazilian yellow scorpion)
Soro antiescorpionico	Instituto Vital Brazil	Brazil	*Buthus occitanus* (common yellow scorpion)
Polyvalent scorpion antivenom	National Antivenom and Vaccine Production Center	Saudi Arabia	*Leiurus quinquestriatus* (Israeli yellow scorpion), *Androctonus crassicauda* (Arabian fat-tailed scorpion), *Buthacus arenicola*, *Hottentotta minax*, *Buthus occitanus* (common yellow scorpion), *Androctonus amoreuxi* (fat-tailed scorpion)
Le serum antiscorpionique	Pasteur Tunis	North Africa	*Androctonus australis* (fat-tailed scorpion), *Buthus occitanus* (common yellow scorpion)
Monovalent scorpion antivenom	Razi Vaccine and Serum Research Institute	Iran	N/A
Polyvalent scorpion antivenom	Razi Vaccine and Serum Research Institute	Iran	*Androctonus crasicauda* (Arabian fat-tailed scorpion), *Hemiscorpius lepturus*, *Hottentotta saulcyi*, *Hottentotta schach*, *Mesobuthus eupeus*, *Odontobuthus doriae*
Scorpifav	Sanofi Pasteur	North Africa and Middle East	*Androctonus australis* (fat-tailed scorpion), *Leiurus quinquestriatus* (Israeli yellow scorpion), *Buthus occitanus* (common yellow scorpion)
SAIMR Scorpion Antivenom	South African Vaccine Producer	South Africa	*Parabuthus transvaalicus* (dark scorpion)
Scorpion antivenom Twyford	Twyford Pharmaceuticals	North Africa	*Androctonus australis* (fat-tailed scorpion), *Buthus occitanus* (common yellow scorpion), *Leiurus quinquestriatus* (Israeli yellow scorpion)
Purified Polyvalent antiscorpion serum	VACSERA	Egypt	*Leiurus quinquestriatus* (Israeli yellow scorpion), *Scorpio maurus* (large-clawed scorpion), *Androctonus crasicauda* (Arabian fat-tailed scorpion), *Buthus occitanus* (common yellow scorpion)
Scorpion Venom Antiserum	Vins Bioproducts Ltd.	India	*Leiurus quinquestraitus* (Israeli yellow scorpion), *Androctonus amoreuxi* (fat-tailed scorpion)
Soro Antiescorpiônico (FUNED)	Fundação Ezequiel Dias	Brazil	*Tityus serrulatus* (Brazilian yellow scorpion)
Antiscorpion Venom Serum	Haffkine Bio-Pharmaceutical Corporation Ltd.	India	*Mesobuthus tamulus* (red scorpion)

2.5.4 Venomous ticks

Ticks are known to a greater degree as the vectors of a wide variety of diseases of humans and vertebrate animals. However, some species of ticks result in pathological changes in the host organism due to the transfer of noninfectious toxic components in the process of feeding. Toxins have been identified in whole tick extracts, salivary gland secretions, salivary gland extracts, and tick eggs. Toxicity may be either inherent per se or due to the transformation of tick secretory components in the host. Alternatively, it may be the result of breakdown of the host tissue or be ascribed to a product of a symbiotic organism in the tick (Mans et al. 2002).

Most often, the toxic agents are contained in the **saliva** injected into the wound at the moment of suction to the host; this provides an anesthetic effect (Cordeiro et al. 2015). In case of prolonged blood sucking (not less than 4 to 5 days), a considerable amount of neurotoxin is injected into the human or animal. **Tick paralysis** is the best-known toxicosis in human and veterinary medicine. It is characterized by the blocking of neuromuscular transmission, resulting in paraplegia. In some cases, respiratory system disorder also occurs. If the tick is not removed, this can cause death.

Tick paralysis can be caused by 69 of the 869 tick species, that is, approximately 8% of all tick species. Among them are 55 hard tick species (Ixodidae) and 14 soft tick species (Argasidae) (Cabezas-Cruz and Valdés 2014). They belong to all major tick genera, except *Carios* and *Aponomma* (Mans et al. 2004). In most cases, tick paralysis is recorded in South America and Australia.

Most often, tick paralysis is caused by the following species of **hard ticks** (Hall-Mendelin et al. 2011; Cabezas-Cruz and Valdés 2014): the castor bean tick *Ixodes ricinus*, widespread predominantly in Europe and Asia; paralysis tick *Ixodes holocyclus*, in Australia; deer tick *Ixodes scapularis*, American dog tick *Dermacentor variabilis*, Rocky Mountain wood tick *Dermacentor andersoni*, lonestar tick *Amblyomma americanum*, in North America; cayenne tick *Amblyomma cajannense*, in South America; karoo paralysis tick *Ixodes rubicundus* and red-legged tick *Rhipicephalus evertsi evertsi*, in South Africa; as well as the brown dog tick *Rhipicephalus sanguineus*, widespread across the globe.

Following are the soft ticks for which paralysis has been described, demonstrated, or suspected (Mans et al. 2002): fowl tampan *Argas (Persicargas) walkerae*, North American bird argas *A. (P.) radiatus, Ornithodoros capensis*, sand tampan *Ornithodoros savignyi*, spinose ear tick *Otobius megnini* (South Africa), and *O. lahorensis* (Eurasia).

The **castor bean tick** *Ixodes ricinus* having a wide distribution in Europe and Asia occurs also in North Africa and North America. Its common habitats include deciduous and mixed forests, bushes, and grasslands. The larvae and nymphs are parasitic on small mammals, birds, and lizards; the adult ticks parasitize cattle, dogs, hares, and, not infrequently, humans. The study of response of skin tissues at the points of bites by ticks *I. ricinus* has shown that, in combination with the mechanical action of the mouthparts, the tick's saliva causes epidermis necrosis, that is, deadening of epidermis cells. In addition, as a result of interaction between the toxic saliva of the tick and host tissues, the pathological products are produced that penetrate the blood and, for multiple bites, cause symptoms of systemic poisoning (Gelashvili et al. 2015).

The **Australian paralysis tick** *Ixodes holocyclus* is distributed mainly in the forests along coastal eastern Australia. The toxic effect of the venomous secretion of this species of ticks poses a serious danger for humans and domestic and livestock animals. The bite by *I. holocyclus* results in development of flaccid paralysis, sometimes ending in death. In the 20th century, at least 25 people died from its bites (Angelutsa 2001). The tick can be parasitic on many animals; however, it is found most often on small Australian animals (bandicoots) and dogs.

Toxic substances are contained in the **eggs of hard ticks**. The toxic substances released from the eggs of different species of ticks (*Amblyomma hebraeum, Boophilus decoloratus, B. microplus, Rhipicephalus evertsi evertsi*) cause similar histopathological changes in the tissues of animals (Gelashvili et al. 2015).

Besides tick paralysis, **other types of toxicoses** caused by ticks are also known. Among them are sand tampan toxicosis, caused by bites by *Ornithodoros savignyi*, found in the northwestern parts of South Africa and also in the semidesert regions of north and east Africa, the Middle East, India, and Ceylon (Mans and Neitz 2004). Two forms of toxicosis occurring in Africa are described as mhlosinga and magudo. They are caused by bites by the tick *Hyalomma truncatum* and afflict cattle, sheep, and pigs (Ristic and McIntyre 1981). In South Africa, the spring lamb paralysis is widespread (Mans et al. 2002).

Venomous ticks inflict serious damage to **cattle breeding**. In South Africa, the karoo paralysis tick *Ixodes rubicundus* causes mass mortality of sheep and other domestic animals. For example, from 1983 to 1986, 29,000 animals (91% sheep; 9% black cattle and goats) were killed by this tick.

In North America, the Rocky Mountain wood tick *Dermacentor andersoni* is a significant species. From 1900 to the early 1970s, it was responsible for the deaths of more than 3,800 head of black cattle and sheep in British Columbia, Canada, and Washington, Idaho, and Montana in the United States (Mullen and Durden 2009).

In Australia, the Australian paralysis tick *Ixodes holocyclus* bites up to 10,000 companion animals (dogs, cats, and so on) and up to 100,000 head of livestock every year (Hall-Mendelin et al. 2011). In South Africa, the karoo paralysis tick *Ixodes rubicundus* inflicts serious damage to sheep stock (Petney and Fourie 1990). Bites from the sand tampan *Ornithodoros savignyi*, also in South Africa, cause mass mortality in young calves and lambs (Mans et al. 2002).

The **number of animals lost** due to the toxicoses caused by ticks is much larger than that of humans who died from the same cause. There is a simple explanation for this. Because several days are required for development of toxicosis, a human bitten by a venomous tick usually succeeds in finding the biting tick or notifying others of his or her discomfort. In most cases, ticks on companion animals are revealed by their owners. Tick-caused toxicoses in livestock animals are observed more frequently because their number is higher while they receive less attention. As for wild animals, because no help is forthcoming, tick toxicoses in these animals must be controlled by their body's own defenses.

2.6

Invasions of Insects

Invasions are migrations of species beyond their native range. The concept of "biological invasions" includes all cases of such invasions caused by **human activity** (introductions) or by **natural dispersal** of species beyond the areas of their general distribution. One can identify the following **types** of invasions: (1) natural invasion—independent penetration of insects into a new territory; (2) intentional introduction—deliberate resettlement of an insect species by humans to localities not inhabited earlier by the insects; and (3) accidental introduction—casual importing of an insect for any reason related to human activity.

In most cases, migrating insects do not survive in new places, and perish. However, some species penetrate quickly and successfully into the local ecosystems. To accomplish this requires a combination of such conditions as the favorable climate, availability of the appropriate food, adaptive capacities, and absence of natural enemies controlling their numbers within the native range. In such cases, the species are given the unique opportunity to uncontrollably reproduce and expand into new territories. The invasive species are considered to be second after destruction of habitats in order of importance of **risk to biodiversity** (Early et al. 2016).

2.6.1 General characteristics of insect invasions and introductions

The desire to **expand their range** along with aiming for **reproduction** are attributes of living organisms. Because of insects' vast variety, insignificant size, high flying ability, and often hidden mode of life, insects, more often than the other invertebrates, are transferred to new regions, penetrating far beyond their original ranges. Not all species penetrating into new areas survive, and not all of the settling alien species (i.e., breeding new generations) constitute a threat. Every year, there are many thousands of invasions (natural and anthropogenic). At that, hundreds of species settle down, but only dozens of these species become economically important and constitute a danger for human activities (Maslyakov and Izhevsky 2011).

The intensities and global patterns of introduction and disturbance are changing more rapidly today than at any other time in human history (Early et al. 2016). For example, from 1930 to 1990, a new destructive alien species was found within the Soviet Union territory every 22 months on average (Izhevsky 1990). In the 1990s, the **intensity of invasions** increased: a new species was detected in the European part of Russia every 18 months on average; from 2000 to 2010 a new species was detected every 12 months (Maslyakov and Izhevsky 2011).

A similar situation is also characteristic of other regions of the globe. For example, the average rate of introduction of species in Europe reached 10.9 species per year for the period 1950 to 1974, while it reached 19.6 species per year for 2000 to 2008 (Roques et al. 2016) (i.e., the rate has roughly doubled over 30 to 40 years).

The majority of insect species are introduced accidentally. For example, in Europe, only 14% of introductions were intentional (most of these for biological control); the remaining 86% were accidental (Hajek et al. 2016). A similar situation has been observed elsewhere around the world (Kumschick et al. 2016).

When comparing data on the **taxonomic composition** of adventitious species in such great regions of the world as Russia, United States, and western Europe, a certain similarity is found. In all three regions, the representatives of the order Homoptera (aphids, scale insects, cicadas, and leafhoppers) predominate in the composition of adventitious species. Their share far exceeds that of the orders Coleoptera (beetles) and Lepidoptera (moths and butterflies). Meanwhile, the taxonomic composition of adventitious species of the Korean Peninsula is very much different (Table 2.6-1).

The highest numbers of invasive alien species in the world, the strongest management efforts, and the greatest knowledge about the extent of invasions are found in economically developed countries (McGeoch et al. 2010). This is quite explainable. A considerable number of imported species is based on trade relations. Careful attention to this problem is related to vast economic loss. Information on the volume of financial loss in the agriculture and forestry in some countries is presented in Table 2.6-2.

The directionality of invasion of insects to one or another country depends significantly on the existing trade and economic relations. Under modern globalization, this means that they can arrive practically from any direction. For example, among the alien species settling down in European Russia are representatives of North American fauna: Colorado potato beetle (*Leptinotarsa decemlineata*), fall webworm (*Hyphantria cunea*), and western flower thrips (*Frankliniella occidentalis*). The potato tuber moth (*Phthorimaea operculella*) and tomato leafminer (*Tuta absoluta*) originated in Central America and South America. The Oriental fruit moth (*Grapholita molesta*), San Jose scale (*Diaspidiotus perniciosus*), and many pests of subtropical cultures have arrived from Asia. The cottony cushion scale (*Icerya purchasi*) is native of Australia (Maslyakov and Izhevsky 2011).

Table 2.6-1. Taxonomic composition of adventitious species of insects in the separate regions of the world (Maslyakov and Izhevsky 2011; Honga et al. 2012; Sailer 1978).

Taxon	USSR/Russia Number of species	%	USA Number of species	%	Europe Number of species	%	Korea Number of species	%
Homoptera	120	62.7	327	25.9	55	44	—	—
Coleoptera	26	13.6	295	23.5	14	11.2	67	39.2
Lepidoptera	22	11.1	120	9.5	19	15.2	21	12.3
Thysanoptera	5	2.6	80	6.4	17	13.6	8	4.7
Hymenoptera	10	5.3	272	21.5	3	2.4	15	8.8
Diptera	7	3.7	86	6.8	13	10.4	15	8.8
Orthoptera	1	0.5	27	2.1	—	—	—	—
Hemiptera	1	0.5	55	4.3	4	3.2	35	20.5
Other	—	—	—	—	—	—	10	5.8
Total	192	100	1262	100	125	100	171	100

Data for USSR/Russia over the period 1870 to 2010; data for USA over the period 1640 to 1977; data for Europe over the period 2002 to 2006; and data for Korean Peninsula 1900 to 2010.

Table 2.6-2. Economic losses due to introduced arthropod pests in crops and forests in some countries (billion dollars per year) (Pimentel 2002).

	United States	United Kingdom	Australia	South Africa	India	Brazil
Crops	15.9	0.96	0.94	1.0	16.8	8.5
Forests	2.1	—	—	—	—	—

In most cases, after the settling of a species to a new place, an "extermination war" is declared on it. However, as a rule, the invasions are **irreversible** and the attempted control measures, at best, only restrict the species' distribution speed. For example, the Colorado potato beetle (*L. decemlineata*), under constant siege, has since 1953 increased the area of its range in Russia 12,190 times (Kuznetsov 2005). This species appeared in China from Kazakhstan in 1993 and, in 2010, its geographic range reached 277,000 km^2 (Liu et al. 2012).

Maslyakov and Izhevsky (2011) have analyzed long-term data for results of the **phytosanitary control** of plant products imported to Europe, from January 2002 through March 2004, from 45 countries. During this period, about 4,100 different species of phytophagous insects and mites were revealed as a result of examination of imported plant products by quarantine organizations. On average, 5.2 potential invaders were found every day at the European borders. Considering that only a very small portion of all delivered plants are subjected to customs examination, this number is actually much higher.

National plant quarantine services are not in a position to fully preclude the invasion of alien insects. It is not realistic to expect quick and careful supervision of all imported plant cargo. This cargo is delivered to a country from all continents in railway cars, aboard ships, in containers, and on heavy-duty trucks and aircrafts. Only a very small percentage gets examined, and this usually occurs within the borders of the importing country—in loading areas, storage warehouses, and. It is more difficult to reveal the species of insects hidden inside bark, fruits, seeds, leaves, and needles.

Most often, insects in the following groups become **invaders** (Maslyakov and Izhevsky 2011): (1) phytophagous insects (pests of live plants); (2) xylophagous insects (wood pests); (3) pests of food and nonfood stores; and (4) synanthropic (i.e., endophilic) insects.

There is a global invasive species database (http://www.iucngisd.org/gisd/). As of January 2017, 13 insects were counted in the list of 100 of the world's worst invasive alien species (http://www.iucngisd.org/gisd/100_worst.php). Summarized information on these invasive alien species is given in Table 2.6-3. It is not difficult to see that **ants and termites** have the maximum representation (5 of 13) in this list.

2.6.2 Natural invasions of insects

Natural invasions are migrations of insects and arachnids that are carried out without human input. Migration of insects to new regions is frequently inhibited by **natural barriers**. These can be water areas, mountain ridges, and so on (Caplat et al. 2016). Different natural factors can help bypass these barriers; **three such factors** are (1) wind, (2) currents, and (3) climate changes.

Wind. Wind is the acknowledged factor of insect migration. The airborne dispersal (transport) is termed *ballooning*. To active migrations of insects, **three categories** of winds are relevant: (1) global (monsoons, trade winds) and numerous local winds; (2) tropical cyclones (typhoons); (3) jet streams.

As a rule, the speed of **monsoons, trade winds**, and local winds does not exceed 20 to 30 km/hour; therefore, a migration with wind assistance is effective over short distances. The American false webworm (*Hyphantria cunea*) was passively carried by wind in 1952 from Hungary to the Zakarpattia Oblast of Ukraine. The transport of the Colorado potato beetle (*Leptinotarsa decemlineata*) by wind from Poland to the Kaliningrad Region of Russia was recorded for the first time in 1953 (Maslyakov and Izhevsky 2011). In China, the Colorado potato beetle was dispersed over 115 km in 16 days with the help of wind (Liu et al. 2012).

Transportation of insects by **tropical and extratropical cyclones** is also quite a well-known phenomenon (Gillespie et al. 2012). Air-swirling motions promote the ascent of invertebrates, and strong winds (often exceeding 100 km/hour) allow them to be carried great distances quickly (de la Giroday et al. 2012).

For example, early in September 2016, the subtropical noctuid moth *Risoba yanagitai* was first detected in the Vladivostok suburbs (south of Russian Far East). The closest points of the native habitat of this moth are 1,150 km away (Oita Prefecture, Kyushu Island, Japan) and 1,000 km (Namhae Island, South Korea). It is reasonable to assume that this moth was delivered there via the strong Lionrock typhoon from Japan. The typhoon passed across Japan and the Sea of Japan on August 30, 2016, just

Table 2.6-3. Insects from list of 100 of the world's worst invasive alien species (compiled by author from numerous Internet sites).

Common names	Latin name	Native range	Alien range	Means of introduction	Impact
Asian tiger mosquito, forest day mosquito, tiger mosquito	*Aedes albopictus*	Southeast Asia, the islands of the Pacific and Indian Oceans, northern China, Japan, and Madagascar	Brazil, Guatemala, El Salvador, Honduras, USA, Mexico, Cuba, 18 European countries, Madagascar, Nigeria, Cameroon, Gabon, Equatorial Guinea, New Guinea	Via the international tire trade (due to the rainwater retained in the tires when stored outside) and bamboo plants	Transmission of many human diseases, including dengue virus, West Nile virus, and Japanese encephalitis.
Common malaria mosquito	*Anopheles quadrimaculatus*	Eastern North America, El Salvador, Panama	Zimbabwe ?	Aboard ocean ships; floating plants; in cockpits and cabins of aircrafts and cars, with military cargoes	Primary vector of malaria in North America and vector for other diseases, including Cache Valley virus, West Nile virus
Yellow crazy ant, crazy ant, gramang ant, long-legged ant, Maldive ant	*Anoplolepis gracilipes*	West Africa	Northern Australia, Caribbean islands, some Indian Ocean islands (Seychelles, Madagascar, Mauritius, Réunion, the Cocos Islands and the Christmas Islands), and some Pacific islands (New Caledonia, Hawaii, French Polynesia, Okinawa, Vanuatu, Micronesia, and the Galapagos archipelago)	In the shipping containers; in holds of cargo ships	Present a severe hazard to existence of the red land crab (*Gecarcoidea natalis*) populations on Christmas Island, endangered birds such as Abbott's booby (*Sula abbotti*), through habitat alteration and direct attack by the ants. They prey on, or interfere in, the reproduction of a variety of arthropods, reptiles, birds, and mammals on the forest floor and canopy.
Asian long-horned beetle, starry sky beetle	*Anoplophora glabripennis*	Japan, Korea, and China	United States, Canada, Austria, France, Germany, Italy, Poland, Slovakia	Arrived in United States in solid wood packing material from China	Threatens 30%–35% of trees in urban areas of eastern USA. The economic, ecological and aesthetic impacts on the United States will be devastating if the beetle continues to spread. Potential losses have been estimated in the tens to hundreds of billions of US dollars.
Cotton whitefly, sweet potato whitefly, sweetpotato whitefly	*Bemisia tabaci*	India	Global, except Antarctica	It has been spread throughout the world through the transport of plant products that were infested with whiteflies	Over 900 host plants have been recorded for *B. tabaci* and it reportedly transmits 111 virus species. Once established, *B. tabaci* quickly spreads and through its feeding habits and the transmission of diseases, it causes destruction to crops around the world.

Cypress aphid	*Cinara cupressi*	Europe and the Middle East: eastern Greece to Iran	North America: United States, Canada; Europe: Italy, Belgium, Netherlands, France, Bulgaria, Portugal, Germany, Italy, Lithuania, Poland, Slovakia, Russia, Switzerland, United Kingdom; Middle East: Israel, Jordan, Yemen; Asia: China, Pakistan; Africa: Malawi, Tanzania, Burundi, Rwanda, Uganda, Kenya, Congo Democratic Republic, Zimbabwe, South Africa, Libya, Morocco, Mauritius; South America: Argentina, Chile, Colombia, Brazil	Transported on live plant materials	It feeds on various trees. *C. cupressi* sucks the sap from twigs, causing yellowing to browning of the foliage on the affected twig. The overall effect on the tree ranges from partial damage to eventual death of the entire tree. This aphid has seriously damaged commercial and ornamental plantings of trees around the globe.
Formosa termite, Formosan subterranean termite	*Coptotermes formosanus*	Southern China to Taiwan	Guam, Japan, South Africa, Sri Lanka, Hawaii, and the continental United States	Arrived accidentally on ships from the Pacific	Competes with native species; causes structural damage to buildings. *C. formosanus* has its greatest impact in North America. It is currently one of the most destructive pests in the United States, estimated to cost consumers over US$1 billion annually for preventive and remedial treatment and to repair damage caused by this insect.
Argentine ant	*Linepithema humile*	Parana River drainage (Argentina, Brazil, Paraguay, Uruguay)	Australia, Bermuda, Chile, Cuba, France, Italy, Japan, Mexico, New Zealand, Peru, Portugal, South Africa, Spain, Switzerland, United Arab Emirates, United Kingdom, United States	The primary mode of introduction is probably shipping. It is believed the ant came to Louisiana through Argentine shipments of coffee or sugar. It spread across the southern states, most likely by train, and eventually moved into California. The ant was probably introduced to Hawaii by way of goods shipped from California	It is the greatest threat to the survival of various endemic Hawaiian arthropods and displaces native ant species around the world (some of which may be important seed-dispersers or plant-pollinators), resulting in a decrease in ant biodiversity and the disruption of native ecosystems.

Table 2.6-3 contd. ...

...Table 2.6-3 contd.

Common names	Latin name	Native range	Alien range	Means of introduction	Impact
Asian gypsy moth, gypsy moth	*Lymantria dispar*	All of Europe, Asia Minor, Turkestan, Caucasus, all of Siberia, Japan, southern Europe, northern Africa, Asia, and Pacific	North America	The gypsy moth was introduced into North America in 1869 from Europe by a French naturalist, Etienne Leopold Trouvelot, who imported the moths with the intent of interbreeding gypsy moths with silk worms to develop a silkworm industry. The moths were accidentally released from his residence in Medford, Massachusetts	One of the most destructive pests of shade, fruit, and ornamental trees throughout the Northern Hemisphere. It is also a major pest of hardwood forests. Asian gypsy moth caterpillars cause extensive defoliation, leading to reduced growth or even mortality of the host tree.
Big-headed ant, brown house ant, coastal brown ant, lion ant	*Pheidole megacephala*	Cameroon	Throughout the temperate and tropical zones of the world.	Sailing ships in the 18th and 19th centuries; general freight and household movements from infested areas	It is a serious threat to biodiversity through the displacement of native invertebrate fauna and is a pest of agriculture as it harvests seeds and harbors phytophagous insects that reduce crop productivity. *P. megacephala* are also known to chew on irrigation and telephone cabling as well as electrical wires.
Red imported fire ant, RIFA	*Solenopsis invicta*	Tropical areas of South America	United States, Mexico, Caribbean Islands, Australia, Malaysia, Singapore, Taiwan, Philippines, China	Possibly introduced in ships' ballast	It has become a major pest of humans and animals; it readily stings humans, producing local swelling and pruritus with development of a pustule at the site of the sting and, in rare cases, it can cause anaphylactic shock with death from respiratory or cardiac arrest. Can attack and cause painful stings on humans, pets, and livestock.

Common name	Scientific name	Native range	Distribution	Pathway	Impact
Khapra beetle	*Trogoderma granarium*	Tropics and subtropics of India	Western European, Central Asian, southern American and New Zealand countries, number of Mediterranean, Middle Eastern, Asian and African countries	Usually found in imported cargo at airports and maritime ports	Destructive pest of grain products and seeds.
Common wasp, common yellow jacket	*Vespula vulgaris*	Northern Hemisphere	New Zealand and Australia	Queen wasps stow away in human goods and accidentally transported	They are economic pests of primary industries such as beekeeping, forestry, and horticulture. They damage electrical insulation, disabling electric motors, pumps, telephone relays, fire systems, traffic lights, air conditioners, and other equipment. In addition to causing painful stings to humans, they compete with other insects and birds for insect prey and sugar sources. They will also eat fruit crops and scavenge around rubbish bins and picnic sites.
Cocoa tree-ant, little fire ant, little introduced fire ant, little red fire ant, small fire ant, West Indian stinging ant	*Wasmannia auropunctata*	Amazon basin, Argentina, Barbados, Bolivia Brazil, Colombia, Costa Rica, Cuba, Dominican Republic, French Guiana, Grenada, Guadeloupe, Guatemala, Guyana, Haiti, Honduras, Jamaica, Martinique, Mexico, Nicaragua, Panama, Paraguay, Peru, Puerto Rico, Saint Lucia, Suriname, Trinidad and Tobago, Uruguay, Venezuela	Australia, Bahamas, Bermuda, Cameroon, Canada, Cook islands, Costa Rica, Ecuador, Fiji, French Polynesia, Gabon, Guam, Israel, New Caledonia, New Zealand, Papua New Guinea, Solomon islands, Tuvalu, United Kingdom, United States	This species is commonly associated with and distributed by humans; nurseries, fruit tree orchards, and ornamental plants are all potential habitat for this ant	Reducing species diversity, reducing overall abundance of flying and tree-dwelling insects, and eliminating arachnid populations. It is also known for its painful stings.

before the date of the specimen collection. The wind speed over the Sea of Japan reached 50 to 70 km/hour, enabling the moth to reach Vladivostok for 15 to 20 hours (Beljaev and Velyaev 2016).

The long-distance wind-borne dispersal of insects was also noted in other regions. Such cases were recorded with the moth *Cornifrons ulceratalis* in the northern Mediterranean (Dantart et al. 2009). The wind displacement of the noctuid moth *Helicoverpa armigera* with the speed of 24 to 41 km/hour across the Bohai Sea (North China) was recorded, and, at that, the duration of displacement was 8 to 11 hours (Feng et al. 2009). Other examples of the distant migrations of the noctuid and other moths by force of the wind are given in the literature (Mikkola 1986; Drake and Gatehouse 1995).

Jet streams are fairly narrow air flows situated at the altitude of over 6 km, moving at speeds of 50 to 180 km/hour (Govorushko 2012). However, the high altitude, low temperature, decreased atmospheric pressure, and direct solar radiation dramatically reduces the probability of insects' survival (Bell et al. 2005). Therefore, among the different winds, the tropical and extratropical cyclones are **more probable dispersal vectors** because of their speed, strength, and variable altitude (Gillespie et al. 2012).

For migration to new regions via wind, some insects use **webs**. Broken web strips can hover for a long time in the flows of cold and warm air. In such manner, some mites and some caterpillars migrate (https://en.wikipedia.org/wiki/Ballooning_(spider)). However, this mode of transportation is mostly characteristic of spiders.

Migration of spiders by passive flight **(ballooning)** is a typical natural phenomenon (Krichevsky 2011). It is widespread, even in middle latitudes; within warm climates, especially in South America, this phenomenon occurs on a vast scale. A detailed description of the flight preparation and all stages of travel (creation of spider web, flying-off, altitude change, alighting) is presented on the website (http://www.8lap.ru/section/interesnye-fakty/vsye-o-paukakh-kak-letayut-pauki/).

Spiders have worldwide distribution, which is to a large extent related to their capacity of "ballooning" for long distances (Photo 2.6-1). Seamen have reported discovering spiders on the cloths of vessels when distance to nearest land was more than 1,600 km (990 mi) (Hormiga 2002). Spiders have also been observed and reported in the middle of the ocean (Hayashi et al. 2015).

Spiders are among the first arthropods to invade **newly appearing islands**. For example, the island of Anak Krakatau (Indonesia) arose in 1929 as a result of a submarine volcano eruption (Photo 2.6-2). Less than two years later, the linyphiid *Maso krakatauensis* was detected on this island (Hormiga 2002). They have even been detected in atmospheric data balloons collecting air samples at slightly less than 5 km (16,000 ft) above sea level (https://en.wikipedia.org/wiki/Ballooning_(spider)).

During migration, spiders can live without food for a long time. For example, the American house spider *Parasteatoda tepidariorum* can resist starvation for 25 days or longer (Valerio 1977). Not all

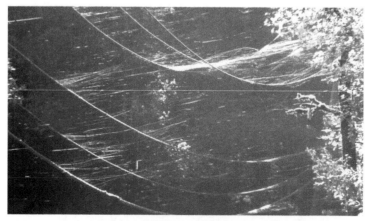

Photo 2.6-1. Spiders have worldwide distribution, which is to a large extent related to their capacity for "ballooning" for long distances. Spiderlings ballooning in the Santa Cruz Mountains of the San Francisco Peninsula is shown.
Photo credit: By Little Grove Farms (https://www.youtube.com/watch?v=5_QJV9kqOzY) [CC BY 3.0 (http://creativecommons.org/licenses/by/3.0)], via Wikimedia Commons, 17 August 2013

spiders balloon; this is characteristic of crab spiders, a few species of tangle-web spiders, and wolf spiders (Hayashi et al. 2015).

Taking advantage of **water currents**, the second most common method of invasion is **rafting on drifting tree trunks** or **mats of vegetation**. This way of dispersal is also characteristic, to a greater degree, of spiders. For example, the trap-door spider (*Titanidiops canariensis*) has reached the Canary Islands by rafting from northern Africa (Opatova and Arnedo 2014). The arrival of linyphiids (*Orsonwelles*) to Hawaii was also accomplished by this rafting method (Hormiga 2002). It is assumed that anyphaenid sac spiders (*Anyphaenidae*) of genus *Amaurobioides* reached New Zealand in this way (Opell et al. 2016). On vegetation (macrophyte) rafts in the ocean mygalomorphs, salticids, spiders of genus *Dysdera*, and intertidal spiders (*Desis marina*) were observed (Ceccarelli et al. 2016).

Independent movement across the water surface is characteristic of a number of semiaquatic spiders (e.g., *Dolomedes* raft spiders) (Hayashi et al. 2015). These spiders also use vegetation (macrophyte) rafts for traveling on **rivers**. Raft spiders were found in the lower course of the Solimões River (stretch of upper Amazon River) not far from the city of Manaus (Schiesari et al. 2003).

Invasions of **insects** frequently occur in a similar manner. All islands in the Pacific Ocean were colonized by insects that came there by rafting on drifting trees or mats of vegetation. An evident example of such insects is the weevils (*Rhyncogonus*) occurring on oceanic islands throughout the Pacific (Gillespie et al. 2012). Cases were recorded of the Colorado potato beetle invasion by sea into the territories of neighboring states from Poland. Beginning in 1954, their coming ashore was observed many times in Lithuania, Latvia, and Kaliningrad Region (Russia) (Maslyakov and Izhevsky 2011).

Climate changes are the third factor contributing to the natural dispersal of insects beyond their ranges. The distribution of insects, as much as all living beings, depends on a wide range of factors, the most important of which are climatic characteristics. Primary factors of climate changes are temperature and humidity. Changes in these factors, depending on the trend of process, results in expansion or reduction of the range. Climate changes occurring in recent decades, consisting largely in increases in average annual temperature of surface air and changing the level of humidity of territories, result in

Photo 2.6-2. Spiders are among the first arthropods to invade newly appearing islands. For example, the island of Anak Krakatau (Indonesia) arose in 1929 as a result of a submarine volcano eruption. Less than 2 years later, the linyphiid *Maso krakatauensis* was detected on this island.
Photo credit: By Raul Heinrich (Own work) [CC BY-SA 3.0 (http://creativecommons.org/licenses/by-sa/3.0) or GFDL (http://www.gnu.org/copyleft/fdl.html)], via Wikimedia Commons, 21 September 2007

displacement of the boundaries of the ranges and zones of mass reproduction of insects (Popova and Popov 2013).

Let us consider the effect of this factor by the example of the **mountain pine beetle** (*Dendroctonus ponderosae*). This bark beetle is native to the forests of western North America from Mexico to central British Columbia. It is the major insect pest of pine forests (Keeling et al. 2012) and most dangerous forest pest of the North America among bark beetles (Weed et al. 2015). From 1997 to 2010, 63% and 94% of total bark beetle–caused tree mortality in the United States and Canada, respectively (Early et al. 2016). Formerly, the distribution of this beetle was restricted by the minimum winter temperatures. Winter air temperatures dropping below –40°C resulted in 100% beetle mortality. However, since the 1960s, the extreme winter temperatures across the western US have warmed by about 4°C (Weed et al. 2015). This allowed the mountain pine beetle to overcome the **geoclimatic barrier** of the northern Rocky Mountains and to extend over the other macroslope (de la Giroday et al. 2012).

Many **other cases** of the dispersal of insects due to climate changes are known. For example, in recent years, various previously absent lepidopterans were recorded in the southern Russian Far East, including *Acosmeryx naga, Ananarsia lineatella, Grapholita dimorpha, Apocheima Ypthima multistriata, Spilarctia alba, Acherontia styx, Ambulyx tobii, Idaea trisetata, Thinopteryx crocoptera*, and *Rhamnosa angulate* (Beljaev and Velyaev 2016).

In most cases, these flights do not lead to the foundation of populations within a new territory. However, in some cases, their naturalization, sometimes rather quick, takes place. Hawk moths (*Ambulyx tobii* and *Acosmeryx naga*) flying in from the territory of eastern China and South Korea have formed very fast, for years, the populations in the south of Primorsky Krai (Russia). In addition, the boundaries of their ranges advance gradually farther to the north at the present time (Koshkin and Bezborodov 2013; Koshkin et al. 2015).

Other examples in Russia and neighboring countries include the Colorado potato beetle (Yasyukevich et al. 2011; Popova and Semenov 2013); castor bean tick (*Ixodes ricinus*) (Popov et al. 2016); and Italian locust (*Calliptamus italicus*) (Popova et al. 2016). Global warming has allowed the Asian long-horned beetle to spread northward in China (Wang et al. 2011).

2.6.3 Introductions of insects

Because the second part of this book considers the **dark side** of the interaction between humans and insects, we discuss here the cases when the introduction of insects in order to address a problem resulted in the emergence of new **problems**. Decisions to introduce a species of insects into a new region are preceded by studies aimed at revealing possible negative consequences of such actions. Therefore, very few examples exist of how an introduced species led to ecological, economic, or other forms of damage.

The overwhelming majority of insect species was introduced for the purpose of **biological control**. In addition, the introductions were carried out for the **following purposes** (Kumschick et al. 2016): (1) silk production; (2) laboratory animals for scientific experiments; (3) habitat and soil improvement and restoration; (4) human food; (5) animal feed; (6) vivarium; (7) model taxa for schools; (8) live exhibits; (9) pollination; (10) managed relocation; (11) waste processing; (12) dye production; (13) pet trade; (14) food for pets; (15) ornamental trade; (16) cultural practices; (17) bait for fishing; (18) medicinal use; and (19) bioweapons.

Some examples of introductions causing negative consequences have been considered in the previous sections of this book (multicolored Asian lady beetle *Harmonia axyridis* [1.7.1.1 Control/regulation of agricultural pests]; African honey bee *Apis mellifera scutellata* [2.5.1.1 Venomous hymenopterans]). Here, we turn our attention to the negative consequences of insects' dispersal for the purposes of pollination and silk production.

To a greater or lesser degree, the resettlement of the following species can be classified as unsuccessful introductions of insects-pollinators: western honey bee (*Apis mellifera*) and its subspecies, buff-tailed bumblebee (*Bombus terrestris*) and its subspecies; solitary multivoltine bee (*Megachile apicalis*), alfalfa leafcutter bee (*M. rotundata*) (Kenis et al. 2009; Goulson 2010). Let us consider the consequences of the buff-tailed bumblebee introduction.

Bumblebees are used for the pollination of over 20 different crops (Murray et al. 2013). At that, they are frequently resettled into new areas to countries where this species was previously absent. For example, the buff-tailed bumblebee (*B. terrestris*) inhabits temperate Eurasia, but it has been actively displaced to different regions of the globe since the 1800s. The dispersal process of *B. terrestris* became most intense in the 1980s, when the artificial rearing and bulk delivery of bumblebees for greenhouse pollination services began (Dafni et al. 2010).

The buff-tailed bumblebee was imported to 57 countries, and 16 of them were situated outside its native range (Ings et al. 2010). It has quickly emerged that this species is invasive and it may settle down in new locations and disturb local ecosystems (Photo 2.6-3). At present, the buff-tailed bumblebee has become established in the wild in Japan, Chile, Argentina, and New Zealand. It was recognized as a pest in Israel, Japan, and Tasmania (Dafni et al. 2010).

The consequences of its introduction into Japan were well studied. It was imported there in 1991 and widely used mainly for pollination of tomato plants. By 2004, the number of its colonies had increased until it reached almost 70,000 (Goka 2010).

The environmental threats related to the introduction of this species in Japan are limited to the following (Murray et al. 2013; Dafni et al. 2010; Goka 2010): (1) negative interactions with local bee fauna (depletion of floral resources, displacement of local bee species possible through resource competition); (2) competition for nesting sites (*B. terrestris* competes and also occupies nesting sites of local *Bombus* species); (3) genetic contamination of local *Bombus* species (*B. terrestris*) can copulate with local *Bombus* species; (4) invasion and spread of parasites and pathogens (parasitic protozoans and endoparasitic mites were already found in the introduced colonies of *B. terrestris*, such as endoparasitic mite *Locustacarus buchneri*); (5) disturbances of reproduction of local flora (less effective pollination and reduction of seed set when *B. terrestris* is a main visitor as well as fruit quality, reducing pollination, piercing and robbing flowers, possibility to interfere with the plants' reproductive success, disruption of native plant-pollinator systems may precipitate reduced seed set in native plants, the presence of bumblebees could decrease the density of these natural pollinators, through competition for nectar and direct displacement; in the long term, this could jeopardize fruit and seed availability); and (6) rapid invasion rates reaching 90 km/year in New Zealand and 12.5 km/year in Tasmania.

Damage from the introduction of the western honey bee *Apis mellifera* was related to introducing mites to new regions; for example, mites were introduced into the Western Hemisphere from Japan and eastern Russia through the transportation of European honey bee colonies (Goka 2010).

Photo 2.6-3. Global trade in bumblebee colonies probably exceeding 1 million nests per year. The buff-tailed bumblebee was imported to 57 countries, and 16 of them were situated outside its native range. At present, the buff-tailed bumblebee is recognized as a pest in Israel, Japan, and Tasmania. Buff-tailed bumblebee (*Bombus terrestris*) in Oeiras, Portugal is shown. Photo credit: By Paulo Costa (Own work) [CC BY-SA 3.0 (http://creativecommons.org/licenses/by-sa/3.0)], via Wikimedia Commons

From among insects introduced for silk production, the following species were recognized as harmful (Kumschick et al. 2016): Chinese tasar moth (*Antheraea pernyi*), Japanese silk moth (*A. yamamai*), ailanthus silkmoth (*Samia cynthia*), and gypsy moth (*Lymantria dispar*).

The gypsy moth (*Lymantria dispar*) is widespread in Europe and Asia; however, it was initially absent in North America. The eggs of the gypsy moth were delivered to the United States in 1869 by French naturalist Etienne Leopold Trouvelot (Photo 2.6-4), who wanted to interbreed this species with the mulberry silkworm (*Bombyx mori*). Experiments were carried out at the individual lot of Trouvelot in Medford, Massachusetts; however, the larvae spread quickly in the surrounding forests. Trouvelot contacted colleagues in order to try to eliminate the problem, but measures were not taken at the time. In 1889, the gypsy moth was recognized as a pest in the United States. From the date of introduction, the gypsy moth now spread over 17 states and the District of Columbia (Brewer 2008).

At the larva stage, the gypsy moth damages more than 500 species of trees and bushes. In North America, the trees most susceptible to damage are oak, cherry, white birch, maple, alder, willow, elm, and trembling aspen. In some years, gypsy moth larvae defoliated as much as 5.2 million hectares of forest in the United States (Wayland et al. 2015). The potential area that is climatically suitable for the gypsy moth is estimated to be 595 million hectares (Global Invasive Species Database 2017b).

Yet another example of unsuccessful introduction for the purpose of rearing silkworms is related to the ailanthus silkmoth (*Samia cynthia*), originally from China and Korea. From the middle 19th century, the ailanthus silkmoth was introduced for the purpose of manufacturing silk in many regions of the globe: Japan, India, Australia, Canada, United States, Venezuela, Uruguay, Brazil, Tunisia, France, Austria, Switzerland, Germany, Spain, Bulgaria, and Italy (https://en.wikipedia.org/wiki/Samia_cynthia). However, in most cases, the production of silk was not established. Nevertheless, the ailanthus silkmoth was successfully naturalized and is often a forest pest (Frank 2015).

Photo 2.6-4. Etienne Leopold Trouvelot (1827–1895) was a French artist, astronomer, and amateur entomologist. He is noted for the unfortunate introduction of the Gypsy Moth into North America. In the U.S., silk-producing moths were being killed off by various diseases. Trouvelot believed that Gypsy Moth could potentially be used for silk production. He brought some Gypsy Moth egg masses from Europe in the mid-1860s and was raising gypsy moth larvae in the forest behind his house. Unfortunately, some of the larvae escaped into the nearby woods. He immediately realized the potential problem he had caused and notified some nearby entomologists, but nothing was done.
Photo credit: Author is unknown—http://www.na.fs.fed.us/fhp/gm/trouvelot/images/trouvelot_portrait.jpg, Public domain, https://commons.wikimedia.org/w/index.php?curid=8913602

2.6.4 Accidental invasions of insects

Examples of accidental introductions of insects are numerous. Most introductions do not result in invasions and become apparent only to competent specialists (Zenni and Nuñez 2013). However, the dispersal of some insects is accompanied by such serious economical, ecological, medical, and other consequences that they become apparent to everyone.

Accidental insect invasions can occur in the **following ways**: (1) planting materials (e.g., cuttings, tubers, rhizomes, bulbs); (2) fruits; (3) ornamental plants in pots, containers, and the like; (4) cut flowers; (5) seeds and products of their processing; (6) timber (under bark, in cracked wood); (7) soil, dirt, and other substrates; (8) packaging and packing materials; (9) transport; and (10) directly by people (Govorushko 2016).

As examples, we will consider insects belonging to different taxonomic groups, the invasions of which happened in a variety of ways. Among ants, the **red imported fire ant** (*Solenopsis invicta*) is the most notorious. The native habitat of this little (2–6 mm) ant is South America. In 1930, a ship transported the fire ants from Brazil to the port of Mobile, Alabama (US). Presumably, they were in ground ballast (Lockey 2007). The fire ants gradually spread north and west, despite attempts to exterminate them or at least to slow their dispersal. Since then, they have spread to 18 southern states (US) and to northeast Mexico. The dispersal of fire ants is continuing; they migrate from one area to another in turf, root balls of nursery plants, and other agricultural products.

In 2001, these fire ants were accidentally introduced to Queensland, Australia (McCubbin and Weiner 2002). It is assumed that the ants were transported there in shipping **containers** arriving at the port of Brisbane, most likely from North America (Henshaw et al. 2005). In December 2014, a nest of ants was found in suburban Sydney, New South Wales. In September 2015, the fire ants were found at a Brisbane airport (https://en.wikipedia.org/wiki/Red_imported_fire_ant).

These ants were also found in New Zealand in 2001; however, several years later, their population was destroyed (Wetterer 2013). Red imported fire ants have been reported in India (Rajagopal et al. 2005), Malaysia (Na and Lee 2001), Cayman Islands, Singapore (Wetterer 2013), and Philippines (https://en.wikipedia.org/wiki/Red_imported_fire_ant). In 2004, the ants were found in Hong Kong and mainland China (Wang et al. 2013), from there, in 2005, they entered Macau and Taiwan (Zhang et al. 2007). In Europe, the only nest was found in the Netherlands in 2002 (Noordijk 2010).

Around 1980, red imported fire ants became to extend across the **West Indies** and were for the first time found in Puerto Rico (Buren 1982) and the US Virgin Islands (Wetterer and Snelling 2006). Between 1991 and 2001, these ants were found in Trinidad and Tobago, the Bahamas, British Virgin Islands, Antigua, and Jamaica (Davis et al. 2001), as well as a number of other islands in the Caribbean. In most cases, the way these invasions can be only assumed. However, it is well established that their migration to Jamaica happened with sod for golf courses, imported from Florida (Wetterer 2013).

Red imported fire ants live mainly in **urban areas**; for example, in courtyards of private houses, on school grounds, or on roadsides. They damage electrical insulation, disabling electric motors, pumps, telephone relays, fire systems, traffic lights, air conditioners, and other equipment (Davis and Grimm 2003).

In the process of building their nests and mounds, these ants can break so much ground that they can cause problems for roadways and retaining walls. In agriculture, red imported fire ants cause damage to many species of plants; their activity in gardens often creates difficulties for workers collecting fruits. Numerous ant mounds (usually approximately 100 and sometimes 400 per hectare) interfere with hay-making and transportation on farmland.

Approximately 33,000 people ask for **medical assistance** in the United States each year after being bitten by imported red fire ants; some cases are severe and require hospitalization. Death can occur in people with allergic reactions (Lockey 2007). Red imported fire ants cause US\$3 billion of losses in the United States annually (Davis and Grimm 2003). Accidental introduction of the Argentine ant *Linepithema humile* is also widely known. Its natural habitat and current distribution are shown on Map 2-17.

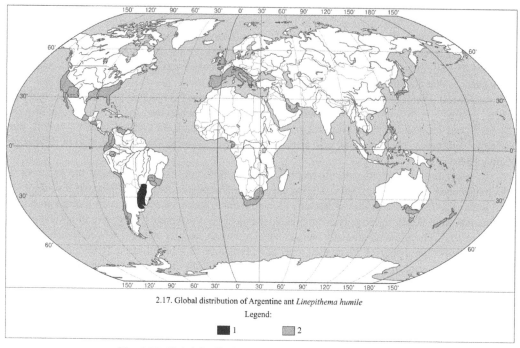

2.17. Global distribution of Argentine ant *Linepithema humile*
Legend:

▮ 1   ▮ 2

Map 2.17. Global distribution of Argentine ant *Linepithema humile*
http://scienceblogs.com/myrmecos/2008/04/how-to-identify-the-argentine-ant-linepithema-humile.php. Reproduced with permission of Dr. A. Wild.
Legend: (1) Native; (2) Introduced

The native habitat of the **Asian tiger mosquito** (*Aedes albopictus*) includes Southeast Asia, the islands of the Pacific and Indian Oceans, northern China, Japan, and Madagascar (Carvalho et al. 2014). Its dispersal across the globe is related mainly to the **tire trade**. There are two main reasons of spreading of Asian tiger mosquito: (1) discarded tires trade; (2) bamboo trade. In last 30–40 years Asian tiger mosquito became most quickly spreading species of mosquitoes (Bonizzoni et al. 2013).

A. albopictus has been identified in 20 countries in the Americas, 27 in Europe, and at least 5 in Africa—and **continues to expand** its range (Carvalho et al. 2014). The first efficient establishment of *A. albopictus* in a temperate country outside its native area was recorded in Albania in 1979 (Adhami and Reiter 1998). It began to colonize in the United States in 1985, arriving from Japan (Kuno 2012). The first detection of this mosquito in Brazil was in 1986 in the states of Rio de Janeiro and Minas Gerais (Carvalho et al. 2014). The mosquito was detected in Nicaragua in 2002 (Belli et al. 2015) and in Catalonia, northeastern Spain, in August 2004 (Lucientes-Curdi et al. 2014).

As a rule, arrival in a country usually took place **aboard ships**, while further spread was realized by other means of transport. For example, arrival in Brazil happened through the ports located in the southeast of the country and further dispersal was probably effected though the **railway network** (Carvalho et al. 2014). It is assumed that the moving of mosquitoes over a distance of more than 400 km in Spain from San Cugat del Vallés (Catalonia) to Orihuela (Alicante Province) happened **by car** (Lucientes-Curdi et al. 2014).

In most cases, after the establishment of *A. albopictus* in a country, further spread is realized at a **very high rate**. For example, between January 1995 and March 2016, 1,241 counties from 40 states and the District of Columbia reported occurrence of *A. albopictus* (Hahn et al. 2016). The same situation was also observed in Brazil. Now, this mosquito inhabits at least 59% of the Brazilian municipalities and 24 of 27 federal units and one federal (metropolitan) district (Carvalho et al. 2014). Although the migration was initially from the native range, the countries in which *A. albopictus* was introduced became the source of invasion. For example, this mosquito migrated in 1990 from the United States to western Europe, especially Mediterranean areas (Kuno 2012).

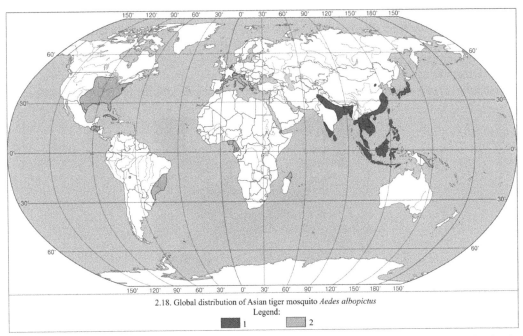

2.18. Global distribution of Asian tiger mosquito *Aedes albopictus*
Legend:
1   2

Map 2.18. Global distribution of Asian tiger mosquito *Aedes albopictus* (https://ru.wikipedia.org/wiki/Aedes_albopictus).
Legend: (1) Native; (2) Introduced

The risk of *A. albopictus* is related to the fact that it is a propagation vector of a number of diseases, specifically chikungunya and Zika viruses (Vanlandingham et al. 2016). Its natural habitat and current distribution is shown on Map 2-18.

Another insect from the list of 100 of the world's worst invasive alien species is the **Asian long-horned beetle** (*Anoplophora glabripennis*). Its native range includes Japan, Korea, and China; however, with untreated wood packaging materials, as well as with forest products and planting material, this beetle has spread widely in recent years and has been found in the United States, Canada, and at least 11 countries in Europe (Meng et al. 2015). Later, the Asian long-horned beetle was found established at Paddock Wood in Kent, in southern England (Straw et al. 2016) and in the extreme southwestern Primorsky Krai (Maritime Territory), in Russia, where it was detected in August 2014 (https://www.zin.ru/animalia/coleoptera/rus/anoglamd.htm).

The Asian long-horned beetle was first discovered in the **United States** in Brooklyn, New York, in August 1996. In order to get rid of beetle's breeding ground, in this case, it was necessary to cut down 6,262 trees (Haack et al. 1997). Information on its further dispersal across the United States is given on the website, https://ru.wikipedia.org/wiki/Anoplophora_glabripennis; the number of the trees that had to be cut down is indicated in an article by Shatza et al. (2013). In the United States, as in Europe, the Asian long-horned beetle feeds and develops in both young and old trees of more than 24 hardwood species growing mainly in urban parks and gardens (Faccoli and Gatto 2016).

Damage is caused by both adult insects and larvae of this species. The former feed on bark and leaves, causing the extensive damage to fresh shoots, which are lost in such cases. The basic damage to the plants is afflicted by larvae, which gnaw through the channels (from 10 to 30 cm) inside the trunk and branches. Infestation from the Asian long-horned beetle carries a strong visible impact and often dictates the need for removal and replacement of highly valued ornamental trees along streets and in parks and gardens (Straw et al. 2016).

This is associated with considerable expenses. In 2008, total **expenditures** for eradication of the Asian long-horned beetle reached US$373 million for the United States, €464,000 in Austria, €55,000 in France, €65,000 in Germany, and CAN$23.5 million in Canada (Faccoli and Gatto 2016). It was evaluated that, in the event of distribution of this beetle across the potentially available territory, it will be capable of destroying 30.3% of the urban trees in the United States, resulting in a financial loss of US$669 billion (Meng et al. 2015).

Another famous example of unintentional invasion is the spread of the Colorado potato beetle (Photo 2.6-5) *Leptinotarsa decemlineata*. Its natural habitat and current distribution is shown on Map 2-19.

Photo 2.6-5. Colorado potato beetle penetrated into Europe after World War II. In 1997 Mihály Búzás and József Szolnoki erected the bronze statue for the Colorado potato beetle in Hédervár, Hungary in connection with 50th anniversary of this event.
Photo credit: By Own work (Own work) [GFDL (http://www.gnu.org/copyleft/fdl.html) or CC BY-SA 3.0 (http://creativecommons.org/licenses/by-sa/3.0)], via Wikimedia Commons

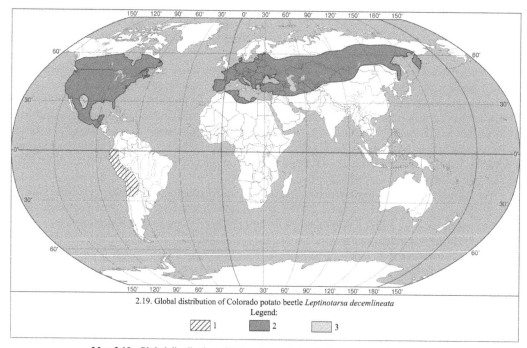

2.19. Global distribution of Colorado potato beetle *Leptinotarsa decemlineata*
Legend:
1   2   3

Map 2.19. Global distribution of Colorado potato beetle *Leptinotarsa decemlineata*
(https://en.wikipedia.org/wiki/Colorado_potato_beetle; Liu et al. 2012; Matsishina 2011).
Legend: (1) Native; (2) Introduced; (3) Origin of potato

2.7

Insect Allergy

An insect's allergens enter the human organism in **several ways**: (1) through venom from the stings of Hymenoptera (wasps, bees, hornets, bumble bees, ants); (2) through the saliva on bites of insects (mosquitoes, sand flies, black flies, biting midges, bugs, fleas, lice), as well as arachnids (ticks, spiders); (3) through inhalation of particles of insect bodies and metabolites, being a part of domestic dust; (4) through direct contact with the skin; (5) through the digestive tract, upon ingestion of either particles of insect bodies and metabolites, or of foods infected with insects.

2.7.1 Insect sting allergy

Stinging insects belong to the order Hymenoptera. From the viewpoint of their allergenicity, the most critical groups are the sphecoid wasps (Sphecoidea); bees (Apoidea; 20,000 species); vespine wasps (Vespoidea; about 15,000 species); and ants (Formicidae; about 15,000 species) (Fitzgerald and Flood 2006). More than 10% of all cases of anaphylactic shocks are caused by stinging insects (Tankersley and Ledford 2015).

2.7.1.1 Most important stinging insects and types of reactions to their stings

The **most significant** allergenic stinging insects on the global scale include the following species (de Graaf et al. 2009):

1. Wasps (genus *Vespula*): common wasp (*V. vulgaris*), German wasp (*Vespula germanica*), eastern yellow jacket (*V. maculifrons*)

2. Arboreal wasps (genus *Dolichovespula*): bald-faced hornet (*D. maculata*), median wasp (*D. media*)

3. Paper wasps (genus *Polistes*): paper wasp (*P. gallicus*), European paper wasp (*P. dominula*), paper wasp (*P. annularis*), social wasp (*P. exclamans*)

4. Hornets (*Vespa* genus): European hornet (*V. crabro*)

5. Bees (genus *Apis*): western honey bee (*A. mellifera*), Asiatic honey bee (*A. cerana*), giant honey bee (*A. dorsata*)

6. Bumble bees (genus *Bombus*): large ground bumble bee (*B. terrestris*), American bumble bee (*B. pennsylvanicus*)

7. Ants (genus *Solenopsis*): red imported fire ant (*S. invicta*), tropical Asian fire ant (*S. geminata*)

8. Jumper ants (genus *Myrmecia*): jack jumper ant (*M. pilosula*)

Some of these taxa are fairly rare. There are about 250 species of bumble bees (William and Osborne 2009). According to various estimates, there are 211 (http://vespabellicosus2008.narod.ru/polistes.html), more than 300 (http://en.wikipedia.org/wiki/Polistes), or 395 species of paper wasps (http://zipcodezoo.com/Animals/P/Polistes_annularis/); 25 species of hornets (http://en.wikipedia.org/wiki/Hornet); 23 species of wasps of genus *Vespula* (http://en.wikipedia.org/wiki/Vespula); 18 species of wasps of genus *Dolichovespula* (http://ru.wikipedia.org/wiki/Dolichovespula); and 11 species of bees of genus *Apis* (http://what-when-how.com/insects/apis-species-honey-bees-insects/). The distribution of the some significant allergenic stinging insects is shown on the Maps 2-20–2-21.

Most often, allergic reactions arise from the stings of **flying stinging insects**—mostly bees and wasps. In many regions, a large proportion of allergic reactions is from ants. In this regard, imported fire ants *Solenopsis invicta* and *S. geminata* are most significant. They create the most problems in the United States, where they sting about 1 million people annually. As a result, tens of thousands of people seek medical advice for allergic reactions (Hile et al. 2006). In Australia, the jack jumper ant (*Myrmecia pilosula*) (Photos 2.7-1, 2.7-2), which is responsible for more than 90% of Australian ant venom allergy, is the most dangerous species (Brown et al. 2003). The rest of cases involve bull ants *M. pyriformis* (Brown and Heddle 2003).

Allergic reactions are also caused by stings of other ants, although their danger is far less. They include the **following species** (Klotz et al. 2005; Brown et al. 2011): wood ant *Formica rufa* (Europe) in the subfamily Formicinae; inchman ant *Myrmecia forficata* (Tasmania) in the subfamily Myrmeciinae; Samsum ant *Pachycondyla sennaarensis* (United Arab Emirates); Asian needle ant *Pachycondyla chinensis* (Korea, Japan); trap-jaw ant *Odontomachus bauri* (Venezuela); *Hypoponera punctatissima* (worldwide), in the subfamily Ponerinae; green head ant *Rhytidoponera metallica* (Australia) in the subfamily Ectatomminae; southern fire ant *Solenopsis xyloni* (Mississippi, Arizona, California, Texas [USA]); desert fire ant *S. aurea* (California [USA]); tropical fire ant *Solenopsis geminata* (Okinawa [Japan], Guam [USA], Oahu [Hawaii, USA]), red harvester ant *Pogonomyrmex barbatus* (Texas, Oklahoma [USA]), Maricopa harvester ant *Pogonomyrmex maricopa* (Arizona [USA]), rough harvester

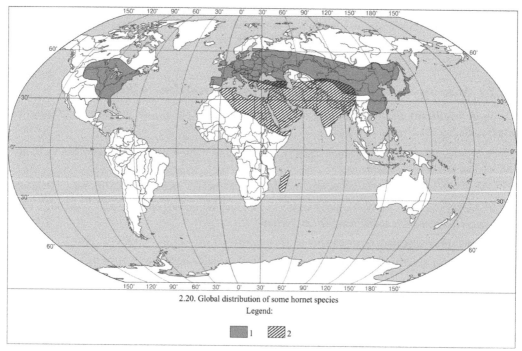

2.20. Global distribution of some hornet species
Legend:

1   2

Map 2.20. Global distribution of some hornet species (http://www.muenster.org/hornissenschutz/weitere/distribution.jpg; http://www.hornissenschutz.de/oriental-hornet.htm).
Legend: (1) European hornet *Vespa crabro*; (2) Oriental hornet *Vespa orientalis*

ant *Pogonomyrmex rugosus* (Arizona [USA]) in the subfamily Myrmicinae, and twig or oak ant *Pseudomyrmex ejectus* (Georgia, Florida [USA]) in the subfamily Pseudomyrmecinae.

Commonly, a sting will cause the **following types** of allergic reactions (Golden 2015; Lee et al. 2016): (1) normal; (2) local allergic; (3) systemic allergic; (4) anaphylactic shock; (5) toxic. Mild systemic reactions are usually generalized skin symptoms such as flush, urticaria, and angioedema. Typically, dizziness, dyspnea, nausea are symptoms of a moderate sting reaction, whereas anaphylactic shock, asthma, loss of consciousness, or cardiac or respiratory arrest indicate a severe sting reaction (Dhami et al. 2016). Data on percentage of population experiencing allergic reactions to the stings of hymenopterans is given in Table 2.7-1.

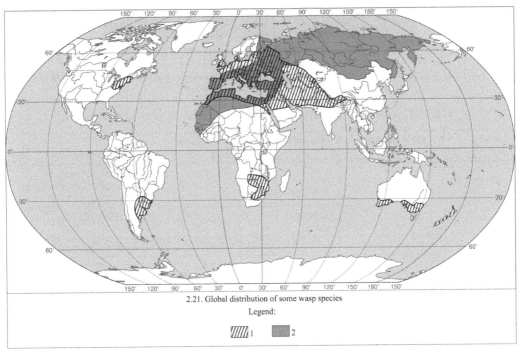

2.21. Global distribution of some wasp species

Legend:

Map 2.21. Global distribution of some wasp species (http://en.wikipedia.org/wiki/German_wasp; http://www.latoxan.com/VENOM/ARTHROPOD/Polistes-gallicus.php).
Legend: (1) German wasp *Vespula germanica*; (2) paper wasp *Polistes gallicus*

Photo 2.7-1. Photo shows a Jack Jumper ant (*Myrmecia pilosula*) on it's nest in Tasmania.
Photo credit: By en:User:Ways [GFDL (http://www.gnu.org/copyleft/fdl.html) or CC-BY-SA-3.0 (http://creativecommons.org/licenses/by-sa/3.0/)], via Wikimedia Commons, 19 October 2004

Photo 2.7-2. The Royal Hobart Hospital in Hobart, Tasmania is shown. It offers a desensitisation program for people who are prone to severe anaphylactic reactions to jack jumper ant stings.
Photo credit: By Wiki ian (Own work) [CC BY-SA 3.0 (http://creativecommons.org/licenses/by-sa/3.0)], via Wikimedia Commons, 27 December 2013

Table 2.7-1. Percentage of people suffering from allergic reactions to stings of hymenopterans.

Country	Local reactions	Systemic reactions	Anaphylactic shock	Source
USA	—	—	0.5–5.0	http://www.wellness.com/reference/allergies/insect-sting-allergy
	—	1 (children), 3 (adults)	—	Klotz et al. 2009
	—	—	1 (fire ants)	Prahlow and Barnard 1998
Australia	—	—	2.7 (bees)	O'Hehir and Douglass 1999
	—	2.4 (ants)	—	
Turkey	—	7.5	2.2	Karakaya et al. 2011
Sweden	—	1.5 (bees, wasps)	—	Björnsson et al. 1995
Greece	4.6 (air force officers)	3.1 (air force officers)	—	Grigoreas et al. 1997
Spain	—	0.4–3.3	—	Fernandez et al. 1999
South Korea	23	—	1 (Asian needle ant *Pachycondyla chinensis*)	Shek et al. 2004
Europe and North America	—	Up to 5	—	Karakaya et al. 2011
Europe	—	0.5–3.3 (adults)	—	Pesek and Lockey 2014
Europe	—	0.15–0.8 (children)	—	Pesek and Lockey 2014
Worldwide	—	3 (adults), 0.34 (children)	—	Golden 2015
	—	—	Up to 3	Bilo et al. 2005
	Up to 10 (adult population)	—	—	Golden et al. 2009
	—	0.4–0.8 (children)	—	Bilo and Bonifazi 2008
	—	0.8–5.0	—	de Graaf et al. 2009
	—	0.4–4.0	—	Schuetze et al. 2002

Allergic reactions to stings can be early or late. Early reactions begin immediately or during the first hour after the sting and are responsible for 95–98% of cases. Late reactions develop within 6 to 12 hours after the sting and represent 2–5% (Pukhlik 2002).

The rate of appearance and development of symptoms serves as a rough indicator of the severity of the patient's reaction: reactions developing 1 to 2 minutes after a sting are the most severe and delayed reactions are less severe (Fedoskova 2007). The more severe the initial reaction, the more dangerous are subsequent stings. So, after a local allergic reaction, a systemic reaction after subsequent stings develops in less than 5% of cases; after a moderate systemic reaction in 15–30% of cases; and after a severe reaction in more than 50% of cases (Annila et al. 1996).

The duration of allergic reactions can be very different. The great local reactions last from several days to 2 weeks. Anaphylactic shock symptoms usually do not last longer than several hours; however, death can occur from 5 to 30 minutes up to 1 to 2 days. In the case of hypoxic brain damage, mental and physical problems may persist for the rest of the person's life.

Little information exists on the percentage distribution of the frequency of the symptoms' occurrence. Data (Perez-Pimiento et al. 2005) show that the most commonly encountered symptoms are pruritus (77.8%), urticaria (57.5%), edema (54.8%), erythema (52.2%), faintness (51.3%), and respiratory difficulty (49.5%); however, these percentages are based on an investigation of only 113 patients and it is unlikely they can be recognized as representative.

2.7.1.2 *Populations at risk of allergic reactions*

The question of whether the use of beekeeping products by people who are intolerant to venom of bees is urgent. Because the honey and venom of bees have no common antigens, the honey is acceptable. On the other hand, the propolis (an adhesive substance collected by bees from the spring buds of trees for caulking cracks in beehives) has antigens in common with the venom and therefore should not be used (Artomasova 2011). There is no connection between use of any ethnic foods and development of allergic reactions (Rueff et al. 2009).

The probability of coming in contact with a stinging insect is determined by its specific habitat. For example, because yellow jackets (Vespidae) usually build their nests in the ground, the risk of getting stung is greater for people working in the garden or around the yard. The nests of hornets are located, as a rule, on trees or shrubs and, therefore, the probability of encountering them is higher for people taking walks in the woods. Special care should be taken when cutting branches of trees and shrubs. Digger wasps (Sphecidae) often build their nests under the cornices of houses and sheds, which increases the risk of being stung for people near these structures. An additional risk factor is eating food in the open air because food and beverages attract stinging insects (Bilo and Bonifazi 2008).

People in certain professions may have a higher frequency of allergic reactions from various sting insect species. Bee stings occur more frequently in professional movers and, naturally, in beekeepers; bumble bee stings in people involved in flower-growing; and wasp stings in salesmen of fruits, juices, sweets, and meats and in garbage collectors. Some concept of the occupational risk is presented in Table 2.7-2.

Increased allergy to stings among beekeepers is most known. It is recognized that allergy to the venom of bees among this population is observed in 15–43% of cases (Annila et al. 1996). The investigation of 852 beekeepers in the United Kingdom has shown that 28% of them had great local reactions to the bee venom, while 21% were subject to systemic reactions (Richter et al. 2011). In Finland, 191 beekeepers were investigated; similarly, 28% of them were found to have great local reactions to the bee venom, while 38% were subject to systemic reactions. In addition, local and systemic reactions to wasp stings were found to be 2% and 13%, respectively (Annila et al. 1996). In an investigation of 246 beekeepers on the Canary Islands (Spain), it was found that 128 of them (52%) suffered from great local and systemic reactions (de la Torre-Morin et al. 1995).

There is also a different degree of proclivity for allergy related to ethnicity. In Israel, 10,000 junior high school students were studied with regard to their reactions to insect stings. Compared with their Jewish counterparts, Arab children had an increased tendency toward allergic reactions (Graif et al. 2006).

Table 2.7-2. Proportion of representatives of different professions in the total number of patients having an allergy to stinging insects (Bonadonna et al. 2008).

Type of susceptibility and occupation	Proportion of patients having allergic reactions, %	Proportion of total population, %
No risk	68.3	91.3
Incidental risk	8.3	0.9
High risk	23.4	7.8
Farmer	12.4	4.9
Brick layer	4.3	2.2
Truck driver	4.0	0.7
Gardener	1.5	—
Beekeeper	0.8	—
Garbage collectors	0.2	—
Others	0.2	—

2.7.1.3 Mortality from allergic reactions

Death from anaphylactic shock related to contact with stinging insects is fairly common. Table 2.7-3 shows the proportion of deaths due to anaphylactic shock.

Therefore, approximately **one-fifth** of cases of anaphylactic shock are related to stinging insects. Fatality due to anaphylactic reactions to their venom is 64–84% (http://medvuz.info/load/allergologija/insektnaja_allergija/48-1-0-566).

The **earliest recorded case** of death related to anaphylactic shock was the death by hornet sting of the first ruler of two Nilotic kingdoms, Egyptian pharaoh Menes, in 2641 B.C.E. (Fassahov et al. 2009). More advanced data on this mortality rate are given in Table 2.7-4. The available rate of mortality from allergy to stinging insects has been essentially underestimated because many so-called unexplained deaths are actually connected to this allergy (Boz et al. 2003). **Global mortality** from allergy to stinging insects is estimated to be 1,000 to 1,200 people annually.

The question of the most dangerous species of stinging insects depends on the type of venom and the frequency of stings, which depends in turn on the aggressiveness of the insects, their number, density of human population, and so on. It is recognized that, in conjunction with these factors (more severe reactions to stings, high aggressiveness, attraction to human food), wasps of genus *Vespula*: common wasp (*V. vulgaris*), German wasp (*V. germanica*), and Eastern yellow jacket (*V. maculifrons*) are on the whole **more dangerous** than other representatives of stinging insects (Rueff et al. 2009).

If only the severity of allergic reaction to sting is taken into account, then **hornets** would be recognized as **most dangerous**. For example, the clinical records of 157 patients, 97 of whom were stung by various species of wasps (*Vespula*), 35 by European hornet (*Vespa crabro*), and 25 by western honey bee (*Apis mellifera*), were examined. The examination showed that the relative danger to life was three times higher for those with European hornet stings than for those with wasp and western honey bee stings (Antonicelli et al. 2003).

Table 2.7-3. Proportion of deaths due to anaphylactic shock caused by stinging insects.

Country	Period	Total number of fatal cases related to anaphylactic shock	Number of fatalities due to stinging insects (%)	Source
Great Britain	1992–2001	202	47 (23)	Pumphrey 2004
New Zealand	1985–2005	18	4 (22)	Low and Stables 2006
Florida (USA)	1996–2005	44	9 (20)	Simon and Mulla 2008
Australia	1997–2005	112	20 (18)	Liew et al. 2009

Nevertheless, the significance of some species in different regions of the globe can be very different. For example, the overriding priority **in Australia** belongs to the western honey bee (*A. mellifera*) because more than 5,000 Australians have severe allergic reactions to its sting (Angelutsa 2001). In areas where plants are pollinated by hornets, the significance of hornets is very high (Bucher et al. 2001). This is the case in the **Netherlands**, where tomato crop farm workers have increased frequency of allergy to their hornet venom (de Groot 2006). Data on the comparative mortality rates from bees and wasps in different countries are given in Table 2.7-5.

Table 2.7-4. Mortality caused by stinging insects.

Country, Period	Population at that time, million people	Total fatalities	Mortality, people/year	Number of deaths per million residents/year	Source
USA, 1950–1959	164	229	22.9	0.14	Parrish 1963
USA, 1962–1982	200	677	32.2	0.16	Nall 1985
USA, 1991–2001	270	533	48.5	0.18	Langley 2005
England and Wales, 1959–1971	50	61	4.7	0.09	Müeller 1990
West Germany, 1979–1983	60	53	10.6	0.18	Müeller 1990
Denmark, 1960–1980	5	26	1.24	0.25	Mosbech 1983
Switzerland, 1961–1983	6	61	2.7	0.45	Müeller 1990
Australia, 1960–1981 (only bees)	14	27	2.25	0.16	Harvey et al. 1984
Australia, 1979–1998 (wasps and bees)	16	45	2.25	0.14	McGain and Winkel 2002
Australia, 1980–1999 (ants)	16	6	0.3	0.02	Brown et al. 2003
France, 1981–1991	55	—	16–38	—	Charpin et al. 1994
Sweden, 1981–1990	8	20	2.0	0.25	Johansson et al. 1991
The Netherlands	15	—	2.0	—	Müller et al. 1990
Europe and America			Some hundreds		Müller et al. 2008
Europe	—	—	200		Müller 2010

Table 2.7-5. Regional differences in mortality rate due to allergy to venom of bees and wasps (death from other insects was not taken into account).

Country	Mortality rate due to bees, %	Mortality rate due to wasps, %	Source
Sweden	10	90	Johansson et al. 1991
Great Britain	12	88	Pumphrey 2004
Denmark	37	63	Mosbech 1983
USA	50	50	Barnard 1973
Florida [USA]	50	50	Simon and Mulla 2008
Australia	84	16	McGain and Winkel 2002
	93	7	Liew et al. 2009

The available information on **simultaneous allergy** to several species of stinging insects is contradictory. It is reported that approximately one in four with hypersensitivity to the venom of wasps and one in six with hypersensitivity to the venom of bees can have an allergy to venoms of other insects (http://medvuz.info/load/allergologija/insektnaja_allergija/48-1-0-566). On the other hand, Richardson (2004) notes that the persons having an allergy to venom of wasps are rarely allergic to the venom of bees. When polling 1,399 residents of three areas of Sweden, the allergy to stings of wasps or bees was found in 130 respondents. In this study, allergy to both bees and wasps was recorded in only eight residents (6%) (Björnsson et al. 1995).

Thus, an allergy to the venom of stinging insects is a serious problem for the world public health service. It makes a considerable contribution to the global mortality rate due to different representatives of wild life. The **number of fatalities** from allergy to the venom of stinging insects exceeds 2–3 times that caused by the general toxic reaction as a consequence of multiple stings. Compared with mortality due to other venomous arthropods, mortality from allergy to the venom of stinging insects is many times **higher** than the mortality from venomous spiders (tens of people/year) and is considerably **less** than mortality from venomous scorpions (about 2,600 people/year) (Chippaux 2012).

2.7.2 Insect bite allergy

Apart from the allergic reactions caused by the stings of insects, the allergy to their bites is also often observed. Stinging insects (mainly, hymenopterans) inject venom into their victims, whereas biting (predominantly, blood-sucking) insects inject their saliva into the wound. The anticoagulants contained in that saliva often result in an allergic reaction. Allergic reactions are mainly manifested as skin damage. A list of insects causing allergic reactions is presented in Table 2.7-6.

Table 2.7-6. Insects of dermatological significance (Burns 2010).*

Order Diptera
Suborder Nematocera
Family Culicidae (mosquitoes)
    Family Psychodidae (sandflies)
        Genus *Phlebotomus*
        Genus *Lutzomyia*
    Family Simuliidae
        Simulium damnosum complex (black flies)
        Simulium posticatum (Blandford fly)
    Family Ceratopogonidae (biting midges, punkies, "no-see-ums")
        Genus *Culicoides*
        Genus *Leptoconops*
Suborder Brachycera (circular-seamed flies, muscoid flies, short-horned flies)
    Family Tabanidae
        Genus *Tabanus* (horse flies)
        Genus *Chrysops* (deer flies)
        Genus *Haematopota* (clegs)
    Family Rhagionidae (snipe flies)
    Family Chloropigae (eye flies, frit flies)
    Family Muscidae (house flies, lesser house flies, stable flies, tsetse flies)
    Family Hippoboscidae (flat flies, louse flies)
    Family Calliphoridae (blow flies)
    Family Sarcophagidae (flesh flies)
    Family Oestridae

Table 2.7-6 contd. ...

...Table 2.7-6 contd.

Order Siphonaptera (<u>fleas</u>)
  Family Tungidae
  Family Pulicidae
    Pulex irritans (<u>human flea</u>)
    Ctenocephalides canis (<u>dog flea</u>)
    Ctenocephalides felis (<u>cat flea</u>)
    Xenopsylla cheopis (<u>tropical rat flea</u>)
  Family Ceratophyllidae (<u>bird fleas</u>)

Order Hymenoptera (bees, wasps and ants)

Order Phthiraptera (<u>lice</u>)

Suborder Anoplura
  Family Pediculidae
    Pediculushumanus humanus (<u>body louse</u>)
    Pediculushumanus capitis (<u>head louse</u>)
  Family Pthiridae
    Pthiris pubis (<u>crab louse</u>)

Order Hemiptera (true bugs)
  Family Cimicidae
    Genus *Cimex*
      Cimex lectularius (<u>common bed bugs</u>)
      Cimex pipistrelli (<u>batbug</u>)
      Cimex hemipterus (<u>tropical batbug</u>)
    Genus *Leptocimex*
    Genus *Oeciacus* (<u>martin bug, swallow bug</u>)
    Genus *Haematosiphon* (<u>Mexican chicken bedbug</u>)
  Family Anthrocoridae
  Family Pantatomidae
    Palomena prasina (<u>green shield bug</u>)
  Family Reduviidae (<u>kissing bugs, assassin bugs, cone-nosed bugs</u>)

Order Thysanoptera (<u>thrips</u>)

Order Coleoptera (beetles)

Order Dictyoptera (cockroaches)

Order Orthoptera (locusts, grasshoppers)

Order Lepidoptera (butterflies, moths)

*Biting insects are underlined.

  As in the case of stings, allergic reactions to bites can be local, systemic, or anaphylactic. **Local** allergic reactions commonly develop immediately after the bite occurs and appear as edema, intumescence, and redness. **Systemic** reactions appear as hives (red spots throughout the body), accompanied with intense itchiness, Quincke's edema (laryngeal edema and lung spasm), and bronchospasm. Development of **anaphylaxis** is manifested by nausea, vomiting, tachyarrhythmia, breathlessness, drop in blood pressure, and loss of consciousness. Without emergency medical assistance, anaphylaxis can result in death.

  Symptoms of allergic reaction to bites of different insects are covered in the following articles: Lee et al. 2016; Walter et al. 2012; Jarisch and Hemmer 2010. **Anaphylactic shock** upon the bites of insects is rarely recorded. Nevertheless, there have been case reports of anaphylaxis due to bites from kissing bugs (Klotz et al. 2010), tsetse flies (Singh and Mann 2013), mosquitoes (Ramirez and Zeoli 2015), horse fly bites, and ticks (Lee et al. 2016).

Some animals also may have allergies to insect bites. In the United Kingdom, from 3–11.6% of **horses** have insect bite hypersensitivity; in some regions of Germany 37.7%; in parts of Queensland, Australia, 10–60%; and in parts of the Netherlands, 71.4% (Martib et al. 2015). The most serious problem for domestic carnivorous animals is flea bite allergy. The symptoms include pruritus, excoriation, and alopecia and their manifestations increase in the warmer months of the year (Noli and Beck 2007).

Allergic reactions to bites of some insects have their **own names**. For example, the collection of symptoms to mosquito saliva is called **skeeter syndrome**; to saliva of black flies, **simuliidosis**; to saliva of sand flies, phlebotodermia or **mosquito dermatosis**. As a whole, the seriousness of allergic reactions to insect bites is **much lower** than allergic reactions to stings.

Mites can also contribute to this type of allergy. Bites from some species—for example, the terrible grain itchy mite (*Pediculoides ventricosus*), oak leaf gall mite (*Pyemotes herfsi*), and grain itch mite (*Pyemotes tritici*)—cause the grain itch. These mites inhabit grain, straw, grass, cotton, dust of flour bag dust, spoiled cereal products, and tare. The mites bite the skin of people in contact with the infested grain and, more rarely, with its dust. The disease lasts 1 to 2 weeks and is accompanied by the strong itch and headache, nausea, urticaria, fever, and sleep loss. Large blisters appear at the locations of the mite bites (James et al. 2016).

2.7.3 Inhalational insect allergy

Inhalational allergy to insects occurs when a person inhales the body scales and metabolites of insects and arachnids forming part of the domestic dust on immediate contact with insects, thus allowing the allergens to enter the human body.

The most important source of allergens from inhalation is the **house dust mite** (*Dermatophagoides*) inhabiting human dwellings. Their sizes range from 0.1 to 0.5 mm and they are distributed around the world. These mites inhabit books, upholstered furniture, carpets, mattresses, cushions, coverlets, indoor shoes, and so on. They feed on flakes of human skin, which are shed by humans every day in the quantity of 1.5 g. House dust everywhere contains mites of the *Dermatophagoides* genus. House dust mites are one of the most common causes of bronchial asthma, atopic dermatitis, allergic rhinitis, and conjunctivitis worldwide (https://en.wikipedia.org/wiki/House_dust_mite).

The house dust mites are the most prevalent **source of indoor allergens**. Their low number is characteristic only of regions with a few months of relative humidity below 50% (Thomas et al. 2010). By now, about 150 species of mites are found in house dust. The most important species is the worldwide European house dust mite (*Dermatophagoides pteronyssinus*). The American house dust mite (*Dermatophagoides farinae*) is the second in order of importance; it inhabits drier areas. The mites *Blomia tropicalis* and *Euroglyphus maynei* are also of great importance (Patel and Meher 2016); they are distributed in humid geographical regions and are often present in concentrations of more than 100 individuals per 1 g of dust. Many species of mites are present in dwellings simultaneously; Thomas et al. (2010) provide details on co-inhabiting species and their proportions in different regions of the world.

The **major cause** of allergic reactions is the feces of mites; however, almost all body parts of the mites (exoskeleton, gut, cuticles, and eggs) are allergens, triggering allergy in 85% of asthmatics (Gregory and Lloyd 2011). Around the world, house dust mites are the most common cause of perennial allergic rhinitis with or without conjunctivitis. They are also responsible for 15–30% of cases of atopic dermatitis in childhood (Bae et al. 2016).

Another important source of allergens entering a human organism by inhalation is the **cockroach**. The two species found to induce most cases of allergies are the German cockroach (*Blatella germanica*) and the American cockroach (*Periplaneta americana*). Other species also cause allergies: the brown-banded cockroach (*Supella longipalpa*), brown cockroach (*Periplaneta brunnea*), Australian cockroach (*Periplaneta australasiae*), and harlequin roach (*Neostylopyga rhombifolia*). The cases of allergy were recorded even for the Madagascar hissing cockroach (*Gromphadorhina portentosa*), which is often kept in the house as a pet (Patel and Meher 2016).

Besides mites and cockroaches, allergic reactions related to inhalation of **insect body, secretions**, excrement, and other remains are caused by the lepidopterans (butterflies, moths), caddisflies, orthopteroid insects (grasshoppers), hymenopterans, and so on. More often, these allergic reactions are characterized by respiratory symptoms (rhinitis, asphyxia) and, more rarely, by dermatitis (https://ru.wikipedia.org/wiki/Инсектная_аллергия).

Publications devoted to allergy caused by inhalation of the particles or metabolic byproducts of different insects abound. For example, the cases of the allergic rhinitis caused by crushed Muscadomestica adults, hatched eggs, contaminated nests and sand, as well as fly feces were described (Tas et al. 2007). On the coast of the Black Sea region of Turkey, an allergy is caused by the larvae and adult individuals of the varied carpet beetle (*Anthrenus verbasci*) (Idge et al. 2009).

The number of allergenic species of insects increases often at the expense of the deliberately or accidentally **introduced** species. For example, the Asian lady beetle (*Harmonia axyridis*) was introduced into Europe and North America for control of aphids and scales. In recent decades, it became the common cause of the inhalant allergy (allergic rhinitis, asthma, and urticaria). These diseases peak in the autumn months when large swarms descend on human dwellings. The ladybug hemolymph is the primary source of allergens (Goetz 2009). The brown marmorated stink bug (*Halyomorpha halys*), was first discovered in the United States, in Allentown, Pennsylvania, in September 1998 (Mertz et al. 2012). By now, it has spread to the territories of more than 40 American states. The allergy is caused by the bug's odor (http://www.thefullwiki.org/Halyomorpha_halys).

2.7.4 Other kinds of insect allergy

In addition to earlier considered types of allergy, there is a number of **other ways** of delivery of allergens of insects and arachnids to a human organism. First, it is a contact way (when allergens penetrate into organism through the skin covering). Second, the allergens are delivered through the digestive tract. It is possible in case of unintended ingestion of metabolites and parts of bodies of insects, use of products infected with insects as well as insects themselves as food.

An allergy due to **contacts** with insects is also widespread. As an example of contact forms of allergy, the eczema of the hands in beekeepers which develops in the summer period while working with bees as well as beekeeping products, mainly, with bee glue, can serve. As is known, propolis is a resinous substance produced by bees for caulking of cracks and regulation of the entrance block passability. It possesses a mass of positive properties, but, at the same time, it contains 26 allergens. The allergic reactions to propolis are commonly expressed as contact dermatitis, however, the manifestations can be in form of rhinitis, conjunctivitis, inflammation of the mucous membranes of the mouth and ulcers, etc. (Basista-Sołtys 2013).

A study by Münstedt and Kalder (2009) showed that, among German beekeepers, 3.6% of respondents were allergic to **propolis** (37 of 1,051 beekeepers). In addition to beekeepers, the group of risk includes also musicians and people who make string musical instruments. In the world, from 0.76–4.04% of beekeepers afflicted with allergic reactions to propolis, while this index varies from 0.64–1.3% in the rest of the healthy population (Basista and Filipek 2012). Review of the world literature indicates that 25% of people allergic to propolis are beekeepers (Basista-Sołtys 2013). Some people also are allergic to honey bee royal jelly (Jarisch and Hemmer 2010).

As an illustration of contact allergic reaction, let us consider **lepidopterism**, which is dermatitis with erythema (redness) and itch. This contact dermatitis is caused by irritating caterpillar or moth hairs coming into contact with the skin or mucosa (https://en.wikipedia.org/wiki/Lepidopterism).

Mites also make their contribution to contact allergy. The **grocer's itch** is a well-known disease. Its causative agents are mites, such as fruit mite (*Carpoglyphus passularum*) and common house mite (*Glyciphagus domesticus*). They inhabit food products such as figs, dates, prunes, grain, cheese, and other dried foods. Moving to a human, they cause dermatitis (James et al. 2016). The flour mite (*Tyroglyphus siro*) can cause the **cheese mite dermatitis** (occupation disease of vanilla sorters). When sorting the frowsy vanilla, it is subjected to brushing, which causes the dust of mites and mold to float in the air.

The dust enters the eyes and settles on the face and hands. The face swells, a papular rash appears, and the skin peels from the face and is replaced with new skin (http://dommedika.com/laboratoria/474.html).

Penetration of allergens through the **digestive tract** is also widespread. Intact insects, parts of bodies, and their metabolites, are abundant in green plants, mushrooms, fruits, and products containing starch. For example, **chocolate** contains a fair amount of chitin. When grinding the insufficiently refined cacao beans, the tropical cockroaches are unavoidably ground and it is fairly problematic and expensive to dispose of them. The higher is the grade of chocolate, the more careful is the cleaning of cacao beans (Photo 2.7-3) from cockroaches and the lesser is the percentage of chitin in it (Mosov 2015).

According to the standards of the US Food and Drug Administration, 100 g of chocolate can contain up to 60 **fragments of insects**. An especially great number of insect fragments is found in kitchen herbs. For example, an average of 925 or more insect fragments are permitted per 10 g of ground thyme, and an average of 400 or more insect fragments is acceptable per 50 g of ground cinnamon (https://en.wikipedia.org/wiki/The_Food_Defect_Action_Levels).

Often, many insect fragments are contained in **flour**. It is produced as a result of milling of grains of different agricultural plants, predominantly cereal grains. Residing in these grains are insects and mites, which are unavoidably found in the composition of the end product. A consumption of contaminated products often causes allergic reactions in people prone to them. Sometimes the consequences of consumption of culinary products composed of contaminated flour can be very serious, leading to anaphylactic shock (Suesirisawad et al. 2015).

An allergic reaction to food products infected with mites is called the **oral mite anaphylaxis**, or pancake syndrome. More often, allergens of the European house dust mite *Dermatophagoides pteronyssinus*, and other mites such as *Blomia tropicalis* and *Suidasia pontifica* result in similar cases (Mangodt et al. 2015). The disease appears more often in tropical and subtropical climates. Among the other mites causing allergic reactions upon entering the digestive tract with food are the cheese mite (*Tyrolichus casei*), mould mite (*Tyrophagus putrescentiae*), mite *Caloglyphus rodionovi*, and others.

Allergy to insects is also found in countries where insects are traditionally used for food. For example, in Laos, **giant water bugs** (Belostomatidae) are widely used in traditional cuisine. However, their consumption causes an allergic reaction in some people (van Huis et al. 2013). There have also been cases of allergic reaction following ingestion of **carmine** (a natural food coloring made from female cochineal insects) (Jarisch and Hemmer 2010).

Photo 2.7-3. Penetration of allergens through the digestive tract is also widespread. For example, chocolate contains a fair amount of chitin. When grinding the insufficiently refined cacao beans, the tropical cockroaches are unavoidably ground. According to the standards of the US Food and Drug Administration, 100 g of chocolate can contain up to 60 fragments of insects. Cocoa beans drying in the sun are shown.
Photo credit: Irene Scott/AusAID [CC BY 2.0 (http://creativecommons.org/licenses/by/2.0)], via Wikimedia Commons

Conclusion

Well, you have read the book. It must be said that scientific and popular science books (in contrast with fiction books) are rarely read from the beginning to the end, and even more rarely read without interruption. Rather, readers usually address first the sections of interest to them. This book is likely no exception.

The trends of scientific development are such that increasing the **depth of exploration** takes place permanently against the background of narrowing its field. The first scientists were **encyclopedists**—they examined a wide variety of issues. For example, **Aristotle** (384–322 B.C.) was the first and greatest of the polymaths succeeding in many fields of science. He is considered to be the founder of logic. Aristotle left behind writings on a wide variety of themes, concerning geology, physics, optics, meteorology, biology, and medicine (https://en.wikipedia.org/wiki/Aristotle).

The works of Abu Arrayhan Muhammad ibn Ahmad Al-Biruni (973–1048), Persian scholar-encyclopedist, covered many different branches of knowledge—astronomy and geography, mathematics and physics, geology and mineralogy, chemistry and botany, history and ethnography, and philosophy and philology (https://en.wikipedia.org/wiki/Al-Biruni).

The Russian scholar Mikhail **Lomonosov** (1711–1765) made a major contribution to the development of chemistry, physics, mineralogy, geology, astronomy, meteorology, history, philology, and art (https://en.wikipedia.org/wiki/Mikhail_Lomonosov). The German scholar-encyclopedist Alexander **von Humboldt** (1769–1859) became renowned as physicist, meteorologist, geographer, botanist, and zoologist (https://en.wikipedia.org/wiki/Alexander_von_Humboldt).

However, as scientific study advanced, it gradually became subdivided into **certain disciplines**. With respect to the subject of this book, biology came first, followed by zoology. After that came the subdivision of zoology investigating insects—entomology. While scientific knowledge continued to develop, certain individuals became successful in studying its various component parts. Thus, Charles **Darwin**, besides examining general biological problems, was also engaged in such disciplines of zoology as ecology and zoography, as he studied insects, birds, mammals, and so on.

And still, the **trend** to subdivide persisted. Gradually, entomology was subdivided into other disciplines, including odonatology (the study of dragonflies and damselflies); orthopterology (the study of grasshoppers and crickets); coleopterology (the study of beetles); lepidopterology (the study of moths and butterflies); dipterology (the study of flies); and trichopterology (the study of caddisflies).

The **following subdisciplines** were separated within hymenopterology (the study of hymenopterans [sawflies, wasps, bees, and ants]): myrmecology (the study of ants), vespology (the study of social wasps), melittology (the study of bees), and apiology (the study of honey bees). At the same time, the following **subdisciplines** were separated within arachnology (the study of arachnids): acarology (the study of ticks and mites), araneology (the study of spiders), and scorpiology (the study of scorpions).

It is evident that this process is not complete, and one can expect emergence of new specific entomological disciplines. In other words, the **vertical structuring** of knowledge (as study goes into more and more depth) will continue into the foreseeable future. This process has resulted in jokes saying that a contemporary scholar aspires to know everything about nothing.

At the same time, there are the more general entomological disciplines that examine the structure, activities, development, evolution, ecology, and diversity of insects in their various forms. To a great extent, this book falls under that category: Here, the structure of knowledge (cognition) is **horizontal**—the scope is quite wide but its depth is very thin, corresponding only to the interactions between humans and insects. So, to rephrase the abovementioned joke, the scholar aspires to know nothing about everything.

Nevertheless, books providing the most general, detailed insight into a wide range of problems are useful. **First**, the process of focused specialization sometimes goes beyond the mark, so that scholars have trouble picturing what occurs in the sidelines of science. **Second**, it may well be that books such as this one will help scholars to discover a scientific specialization.

When creating this type of book, **two approaches** are possible. The **first** consists of inviting a large group of authors, each of whom is an expert in a particular area of knowledge, to contribute. Such a method is practiced quite often. Although it may seem reasonable, this approach also has major deficiencies. Far too often, these contributing authors overestimate the significance of their specific subject.

The **second method** lies in the package treatment of all aspects of the problem by one person. Along with its disadvantages (the author's different level of competence in different areas), there is also an obvious advantage. Such an approach dramatically reduces the degree of subjectivity and allows one to estimate more exactly the relative significance of different aspects of a problem. The possible inaccuracies or actual errors are minimized by the reviewing or editing of separate sections by competent professionals.

When preparing this book, the **second** approach was used. Its intention was to outline the comprehensive concept of interactions between humans and insects. This subject is very extensive and certain aspects of this interaction may have dropped off my grid. I am not an entomologist; therefore, when I wrote this book, I became aware of much new information. I hope that each reader will also have this experience. It is also my hope that the reviews of different chapters performed by the dedicated experts have kept to a minimum the number of errors and inaccuracies.

A simple comparison of the **volume of text** addressing the "positive" and "negative" characteristics (quoted in view of the conventionality of such subdivision, as earlier mentioned) shows that the volume of the first part of the book is essentially more. The first part of the book includes only 5 maps, while the second part includes 21 maps; this can be explained by the fact that we humans typically **take all good things for granted** and turn a blind eye to them. However, when something **troubles** human life, careful attention is paid, the issue is thoroughly studied, and the solutions are developed.

References

2015 Directory of Least-Toxic Pest Control Products. 2015. IPM Practitioner 34(11/12): 1–48. http://www.birc.org/Final2015Directory.pdf.

Abarca, K. and J.A. Oteo. 2014. Clinical approach and main tick-borne rickettsiosis present in Latin America. Revista Chilena de Infectologia 31(5): 569–576.

Abrams, A. 2009. Victorian flea circuses: a lost art form. http://www.darkroastedblend.com/2009/03/victorian-flea-circuses-lost-art-form.html.

Abuladze, K.I., S.N. Nikolsky, N.A. Kolabsky, N.V. Demidov, G.S. Dzasokhov, V.I. Potemkin, A.A. Dil'inbakt and N.V. Pavlova. 1975. Parasitology and invasive diseases of livestock animals. Moscow, Kolos. 471 pp. [in Russian].

Abuladze, K.I., N.V. Demidov, A.A. Nepoklonov, S.N. Nikolsky, N.V. Pavlova and A.V. Stepanov. 1990. Parasitology and invasive diseases of livestock animals. Moscow, Agropromizdat. 464 pp. [in Russian].

Addison, J.B., N.N. Ashton, W.S. Weber, R.J. Stewart, G.P. Holland and J.L. Yarger. 2013. β-sheet nanocrystalline domains formed from phosphorylated serine-rich motifs in caddisfly larval silk: a solid state NMR and XRD study. Biomacromolecules 14(4): 1140–1148.

Ademolu, K.O., A.B. Idowu and G.O. Olatunde. 2010. Nutritional value assessment of variegated grasshopper, *Zonocerus variegatus* (L.) (Acridoidea: Pygomorphidae), during post-embryonic development. African Entomology 18(2F): 360–364.

Adhami, J. and P. Reiter. 1998. Introduction and establishment of *Aedes* (Stegomyia) *albopictus* skuse (Diptera: Culicidae) in Albania. Journal of the American Mosquito Control Association 14(3): 340–343.

Adjare, S.O. 1990. Beekeeping in Africa. FAO Agricultural Services Bulletin 68/6. Food and Agriculture Organization of the United Nations, Rome. http://www.fao.org/docrep/t0104e/t0104e00.htm.

Adler, M.I., E.J. Cassidy, C. Fricke and R. Bonduriansky. 2013. The lifespan-reproduction trade-off under dietary restriction is sex-specific and context-dependent. Experimental Gerontology 48: 539–548.

Adler, P.H. and R.W. Crosskey. 2016. World blackflies (Diptera: Simuliidae): a comprehensive revision of the taxonomic and geographical inventory. http://www.clemson.edu/cafls/biomia/pdfs/blackflyinventory.pdf.

Afanasyev, Y. 2012. Macrophotographing of insects. Lenses for macrophotography. http://www.photoru.ru/page.php?vrub=links&vparid=0&vid=158&lang=rus [in Russian].

Agampodi, S.B. and K. Wickramage. 2013. Is there a risk of yellow fever virus transmission in South Asian countries with hyperendemic dengue? BioMed. Research International. DOI: 10.1155/2013/905043.

Agea, J.G., D. Biryomumaisho, M. Buyinza and G.N. Nabanoga. 2008. Commercialization of *Ruspolia nitidula* (Nsenene grasshoppers) in Central Uganda. African Journal of Food Agriculture and Development 8(3): 319–332.

Agricultural entomology. 1955. Moscow: Selkhozgiz, 616 pp. [in Russian].

Agricultural entomology. 1976. Moscow: Kolos, 448 pp. [in Russian].

Agrios, G.N. 2002. Transmission of plant diseases by insects. 37 pp. http://entomology.ifas.ufl.edu/capinera/eny5236/pest1/content/03/3_plant_diseases.pdf.

Ahn, M.Y., S.H. Shim, H.K. Jeong and K.S. Ryu. 2008. Purification of a dimethyladenosine compound from silkworm pupae as a vasorelaxation substance. Journal of Ethnopharmacology 117(1): 115–122.

Akbayev, M.S., F.I. Vasilevich and F.R. Rossiitseva. 1992. Parasitology and individual diseases of agricultural animals. Moscow: Agropromizdat, 447 pp. [in Russian].

Akhmetshin, N.Kh. 2002. Secrets of the silk road. Notes of the historian and the traveler. Moscow: Veche, 416 pp. [in Russian].

Akimova, I.A. (ed.). 2009. Red book of Ukraine. Animal world. Kiev, Globalmarketing. 600 pp. [in Russian].

Akimushkin, I. 1975. World animals. Stories about insects. Moscow: Molodaya Gvardiya, 240 pp. [in Russian].

Akimushkin, I. 1993. World animals. V. 5. Moscow: Mysl, 462 pp. [in Russian].

Alekseev, I.A. and Y.P. Demakov. 2000. Method for assessing the viability of pine stands. Patent of Russian Federation 2154372. http://ru-patent.info/21/50-54/2154372.html [in Russian].

Aliev, G.M.A. 1986. Equipment for dust collection and purification of industrial gases. Moscow: Metallurgy. 544 pp. [in Russian].

Alkahlout, B.H., M.M. Abid, M.M. Kasim and S.M. Haneef. 2015. Epidemiological review of scorpion stings in Qatar. The need for regional management guidelines in emergency departments. Saudi Medical Journal 36(7): 851–855.

Allan, S.A. 2001. Ticks (class Arachnida: order Acarina). pp. 72–106. *In*: Samuel, W.M., M.J. Pybus and A.A. Kocan (eds.). Parasitic Diseases of Wild Mammals. 2nd ed. Ames, IA, Iowa State University Press.

Alltech. 2015. Global Feed Survey, 2015. Executive Summary. 8 pp. http://www.alltech.com/sites/default/files/global-feed-survey-2015.pdf.

Altendorfer, R., N. Moore, H. Komsuoglu, M. Buehler, H.B. Brown Jr., D. McMordie, U. Saranli, R. Full and D.E. Koditschek. 2001. RHex: A biologically inspired hexapod runner. Autonomous Robots 11: 207–213.

Amendt, J., C.P. Campobasso, E. Gaudry, C. Reiter, H.N. LeBlanc and M.J.R. Hall. 2007. Best practice in forensic entomology—standards and guidelines. International Journal of Legal Medicine 121: 90–104. DOI: 10.1007/s00414-006-0086-x.

Amendt, J., C.S. Richards, C.P. Campobasso, R. Zehner and M.J.R. Hall. 2011. Forensic entomology: applications and limitations. Forensic Science, Medicine, and Pathology 7(4): 379–392. DOI: 10.1007/s12024-010-9209-2.

Amitai, Y., Y. Mines, M. Aker and K. Goitein. 1985. Scorpion sting in children. A review of 51 cases. Clinical Pediatrics (Phila) 24(3): 136–140.

Anand, H., A. Ganguly and P. Haldar. 2008. Potential value of acrids as high protein supplement for poultry feed. International Journal of Poultry Science 7(7): 722–725.

Andreev, A.V. 2002. Assessment of biodiversity, monitoring and ecological network. Chisinau: Biotica. 168 pp. http://www.biotica-moldova.org/library/ABook.pdf [in Russian].

Angelutsa, P. 2001. Extraordinary Australia. Brisbane, Queensland: Su Dzhok Akademiia. 322 pp. [in Russian].

Anitha, R. 2011. Indian silk industry in the global scenario. Excel International Journal of Multidisciplinary Management Studies 1(3): 100–110. http://zenithresearch.org.in/.

Annila, I.T., E.S. Karjalainen, P.A. Annila and P.A. Kuusisto. 1996. Bee and wasp sting reactions in current beekeepers. Annals of Allergy, Asthma and Immunology 77: 423–427.

Antonicelli, L., M.B. Bilò, G. Napoli, B. Farabollini and F. Bonifazi. 2003. European hornet (*Vespa crabro*) sting: a new risk factor for life-threatening reaction in hymenoptera allergic patients? European Annals of Allergy Clinical Immunology 35(6): 199–203.

Appel, E. and S.N. Gorb. 2011. Resilin-bearing wing vein joints in the dragonfly *Epiophlebia superstes*. Bioinspiration & Biomimetics 6(4): 046006. DOI: 10.1088/1748-3182/6/4/046006.

Appel, E., L. Heepe, C.P. Lin and S.N. Gorb. 2015. Ultrastructure of dragonfly wing veins: composite structure of fibrous material supplemented by resilin. Journal of Anatomy 227(4): 561–582. DOI: 10.1111/joa.12362.

Arbes, S.J., Jr., P.J. Gergen, L. Elliott and D.C. Zeldin. 2005. Prevalences of positive skin test responses to 10 common allergens in the US population: results from the third National Health and Nutrition Examination Survey. Journal of Allergy Clinical Immunology 116: 377–383.

Areddy, J.T. 2009. In Shanghai, the autumn game just isn't cricket anymore. Wall Street Journal online. http://www.wsj.com/articles/SB125935631852667013.

Armstrong, N.J. 1976. Victorian Jewelry. Macmillan. 158 pp.

Arnault, I., M. Decoux, E. Meunier, T. Hebbinckuys, S. Macrez, J. Auger and D. De Reyer. 2012. Comparison *in vitro* and *in vivo* efficiencies of three attractant products against webbing clothes moth *Tineola bisselliella* (Hummel) (Lepidoptera: Tineidae). Journal of Stored Products Research 50: 15–20.

Artomasova, A.V. 2011. First aid for the sting of the Hymenoptera (bees, wasps, hornets, bumblebees). http://svatovo.ws/health_sting.html [in Russian].

Arunkumar, K.P., M. Metta and J. Nagaraju. 2006. Molecular phylogeny of silkmoths reveals the origin of domesticated silkmoth, *Bombyx mori* from Chinese *Bombyx mandarina* and paternal inheritance of Antheraea proylei mitochondrial DNA. Mol. Phyl. Evol. 40: 419–427.

Ashby, J. 2007. Giving the people what they want. Meeting zoology collections with their audiences: a case study. Natural SCA News 11: 5–8.

Ashihmina, T.Y. (ed.). 2006. Ecological Monitoring: Educational-Methodical Manual. M.: Academic Project. 416 pp. [in Russian].

Australian Government, Forest and Wood Products Research and Development Corporation. Termite risk management. 2002. 24 pp. http://www.timber.net.au/images/downloads/termites/termite_risk_management_builders.pdf. australianbutterflies.com/.

Ayieko, M.A. and V. Oriaro. 2008. Consumption, indigenous knowledge and cultural values of the lakefly species within the Lake Victoria region. African Journal of Environmental Science and Technology 2(10): 282–286.

Azam, I., S. Afsheen, A. Zia, M. Javed, R. Saeed, M.K. Sarwar and B. Munir. 2015. Evaluating insects as bioindicators of heavy metal contamination and accumulation near industrial area of Gujrat, Pakistan. BioMed. Research International. Article ID 942751, 11 pp. http://dx.doi.org/10.1155/2015/942751.

Badia, S.B. and P.F.M.J. Verschure. 2004. A collision avoidance model based on the lobula giant movement detector (LGMD) neuron of the locust. Proceedings of 2004 IEEE International Joint Conference on Neural Networks (IJCNN) 1–4: 1757–1761.

Bae, M.J., S. Lim, D.S. Lee, K.R. Ko, W. Lee and S. Kim. 2016. Water soluble extracts from *Actinidia arguta*, PG102, attenuates house dust mite-induced murine atopic dermatitis by inhibiting the mTOR pathway with Treg generation. Journal of Ethnopharmacology 193: 96–106.

Bak, A., D. Gargani, J.L. Macia, E. Malouvet, M.S. Vernerey, S. Blanc and M. Drucke. 2013. Virus factories of cauliflower mosaic virus are virion reservoirs that engage actively in vector transmission. Journal of Virology 87(22): 12207–12215. http://jvi.asm.org/content/87/22/12207.full.pdf+html.

Balashov, Y.S. 1967. Blood-sucking ticks—vectors of diseases of man and animals. Leningrad: Nauka Publisher, 319 pp. [in Russian].

Balayiannis, G. and P. Balayiannis. 2008. Bee honey as an environmental bioindicator of pesticides' occurrence in six agricultural areas of Greece. Archives of Environmental Contamination and Toxicology 55(3): 462–470.

Balee, W. 2000. Antiquity of traditional ethnobiological knowledge in Amazonia: the Tupí-Guaraní family and time. Ethnohistory 47: 399–422.

Ballart, C., G. Vidal, A. Picado, M.R. Cortez, F Torrico, M.C. Torrico, R.E. Godoy, D. Lozano and M. Gallego. 2016. Intradomiciliary and peridomiciliary captures of sand flies (Diptera: Psychodidae) in the leishmaniasis endemic area of Chapare province, tropic of Cochabamba, Bolivia. Acta Tropica 154: 121–124.

Balozet, L. 1971. Scorpionism in the old World. pp. 349–371. *In*: Bücherl, W. and E.E. Buckley (eds.). Venomous Animals and their Venoms. Vol. 3. Venomous Invertebrates. New York: Academic Press.

Balu, R., J. Whittaker, N.K. Dutta, C.M. Elvin and N.R. Choudhury. 2014. Multi-responsive biomaterials and nanobioconjugates from resilin-like protein polymers. Journal of Materials Chemistry B 2(36): 5936–5947.

Banerjee, D., V. Kumar, A. Maity, B. Ghosh, K. Tyagi, D. Singha, S. Kundu, B.A. Laskar, A. Naskar and S. Rath. 2015. Identification through DNA barcoding of Tabanidae (Diptera) vectors of surra disease in India. Acta Tropica 150: 52–58.

Banks, J.C. and A.M. Paterson. 2004. A penguin-chewing louse (Insecta: Phthiraptera) phylogeny derived from morphology. Invertebrate Systematics 18(1): 89–100. DOI: 10.1071/IS03022.

bannerghattabiologicalpark.org/butterfly_park.html.

Bao, L., J.S. Hu, Y.L. Yu, P. Cheng, B.Q. Xu and B.G. Tong. 2006. Viscoelastic constitutive model related to deformation of insect wing under loading in flapping motion. Applied Mathematics and Mechanics (English edition) 27(6): 741–748. DOI: 10.1007/s10483-006-0604-1.

Barabanshchikova, N.S. 2013. Analysis of the pattern of the elytra bug-soldier (*Pyrrhocoris apterus* L.) in populations with different degree of urbanization. *In*: Proceedings of the XVIII International Ecological Student Conference "Ecology of Russia and Adjacent Territories." Novosibirsk: Novosibirsk National Research State University, p. 67 [in Russian].

Barabanshchikova, N.S. 2014. *Pyrrhocoris apterus* as a bioindicator of the environment. *In*: Proceedings of the Conference "Field and Experimental Investigations of Biologic Systems". Ishim, p. 5–8 [in Russian].

Barbaro, K.C., M.L. Ferreira, D.F. Cardoso, V.R. Eickstedt and I. Mota. 1996. Identification and neutralization of biological activities in the venoms of *Loxosceles* spiders. Brazilian Journal of Medicine and Biological Research 29(11): 1491–1497.

Barbier, M. 1978. Introduction to chemical ecology. Moscow: Mir. 229 pp. [in Russian].

Barbosa de Oliveira-Junior, J.M., Y. Shimano, T.A. Gardner, R.M. Hughes, P. de Marco Júnior and L. Juen. 2015. Neotropical dragonflies (Insecta: Odonata) as indicators of ecological condition of small streams in the eastern Amazon. Austral Ecology 40(6): 733–744.

Barfield, A.S., J.C. Bergstrom, S. Ferreira, A.P. Covich and K.S. Delaplane. 2015. An economic valuation of biotic pollination services in Georgia. Journal of Economic Entomology 108(2): 388–398. DOI: 10.1093/jee/tou045.

Barganska, Z., M. Slebioda and J. Namiesnik. 2016. Honey bees and their products: Bioindicators of environmental contamination. Critical Reviews in Environmental Science and Technology 46(3): 235–248. DOI: 10.1080/10643389.2015.1078220.

Barman, B. and S. Gupta. 2015. Aquatic insects as bio-indicator of water quality—A study on Bakuamari stream, Chakrashila Wildlife Sanctuary, Assam, North East India. Journal of Entomology and Zoology Studies 3(3): 178–186.

Barnard, J. 1973. Studies of 400 Hymenoptera sting deaths in the United States. Journal of Allergy and Clinical Immunology 52: 259–264.

Barrett, A.D. and S. Higgs. 2007. Yellow fever: a disease that has yet to be conquered. Annual Review of Entomology 52: 209–229. DOI: 10.1146/annurev.ento.52.110405.091454.

Barrett, M.P. and S.L. Croft. 2012. Management of trypanosomiasis and leishmaniasis. British Medical Bulletin 104(1): 175–196.

Barros-Souza, A.S., R.L. Ferreira-Keppler and D.B. Agra. 2012. Development period of forensic importance Calliphoridae (Diptera: Brachycera) in urban area under natural conditions in Manaus, Amazonas, Brazil. EntomoBrasilis 5(2): 99–105.

Barth, C.C., W.G. Anderson, S.J. Peake and P. Nelson. 2013. Seasonal variation in the diet of juvenile lake sturgeon, Acipenser fulvescens, Rafinesque, 1817, in the Winnipeg River, Manitoba, Canada. Journal of Applied Ichthyology 29: 721–729.

Barton, P.S., H.J. Weaver and A.D. Manning. 2014. Contrasting diversity dynamics of phoretic mites and beetles associated with vertebrate carrion. Experimental and Applied Acarology 63(1): 1–13. DOI: 10.1007/s10493-013-9758-7.

Basista, K.M. and B. Filipek. 2012. Allergy to propolis in Polish beekeepers. Postepy Dermatologii I Alergologii 6: 440–445.

Basista-Sołtys, K. 2013. Allergy to propolis in beekeepers—a literature review. Occupational Medicine and Health Affairs 1: 1. DOI: 10.4172/2329-6879.1000105.

Battersby, S. and W.H. Bassett. 2012. Clay's Handbook of Environmental Health. Routledge, 968 pp.

Bauer, L.S., J.J. Duan, J.R. Gould and R. van Driesche. 2015. Progress in the classical biological control of *Agrilus planipennis* Fairmaire (Coleoptera: Bupresitdae) in North America. The Canadian Entomologist 147: 300–317. DOI: 10.4039/tce.2015.18.

Baxter, C.V., K.D. Fausch and W.C. Saunders. 2005. Tangled webs: reciprocal flows of invertebrate prey link streams and riparian zones. Freshwater Biology 50(2): 201–220.

Bayram, A., N. Yiğit, T. Danişman, İ. Çorak, Z. Sancak and D. Ulaşoğlu. 2007. Venomous spiders of Turkey (Araneae). Journal of Applied Biological Sciences 1(3): 33–36.

bearmountainbutterflies.com/.

Beati, L., A.G. Cáceres, J.A. Lee and L.E. Munstermann. 2004. Systematic relationships among Lutzomyia sand flies (Diptera: Psychodidae) of Peru and Colombia based on the analysis of 12S and 28S ribosomal DNA sequences. International Journal for Parasitology 34(2): 225–234. DOI: 10.1016/j.ijpara.2003.10.012.

Beatus, T. and I. Cohen. 2015. Wing-pitch modulation in maneuvering fruit flies is explained by an interplay between aerodynamics and a torsional spring. Physical Review E 92(2). DOI: 10.1103/PhysRevE.92.022712.

Bechah, Y., C. Capo, J.L. Mege and D. Raoult. 2008. Epidemic typhus. Lancet Infectious Diseases 8(7): 417–426.

Beljaev, E.A. and O.A. Velyaev. 2016. First records of subtropical noctuoid moth *Risoba Yanagitai* Nakao, Fukuda et Hayashi, 2016 (Lepidoptera: Nolidae, Risobinae) from Russia and Korea. Far Eastern Entomologist N 325: 13–17.

Bell, J.R., D.A. Bohan, E.M. Shaw and G.S. Weyman. 2005. Ballooning dispersal using silk: world fauna, phylogenies, genetics and models. Bulletin of Entomological Research 95: 69–114.

Belli, A., J. Arostegui, J. Garcia, C. Aguilar, E. Lugo, D. Lopez, S. Valle, M. Lopez and E. Harris. 2015. Introduction and establishment of *Aedes albopictus* (Diptera: Culicidae) in Managua, Nicaragua. Journal of Medical Entomology 52(4): 713–718. DOI: 10.1093/jme/tjv049.

Ben-Dov, Y. 1993. A Systematic Catalogue of the Soft Scale Insects of the World (Homoptera: Coccoidea: Coccidae). Gainesville, FL: Sandhill Crane Press, 536 pp.

Benecke, M. 2001. A brief history of forensic entomology. Forensic Science International 120(1–2): 2–14. DOI: 10.1016/S0379-0738(01)00409-1.

Bengtsson, J. 2015. Biological control as an ecosystem service: partitioning contributions of nature and human inputs to yield. Ecological Entomology 40: 45–55. DOI: 10.1111/een.12247.

Berenbaum, M. 2009. Insect biodiversity—millions and millions. pp. 575–582. *In*: Foottit, R.G. and P.H. Adler (eds.). Insect Biodiversity: Science and Society. Blackwell Publishing Ltd. http://www.lacbiosafety.org/wp-content/uploads/2011/09/insect-biodiversity-science-and-society1.pdf.

Berezhnova, O.N. and M.N. Thsurikov. 2013. To the study of Coleoptera necrobionts in the "Galichya gora" nature reserve and their role in the utilization of animal bodies decay. Vestnik MGOU 3. www.evestnik-mgou.ru [in Russian].

Berg, B. and C. McClaugherty. 2008. Plant Litter. Decomposition, Humus Formation, Carbon Sequestration. Springer, 340 pp.

Berman, D.I. and Z.A. Zhigulsky. 1996. Ants in the communities of cedar in the North-East of Russia. Readings memory Alekseya Ivanovicha Kurentsova. Vol. VI. Vladivostok: Dalnauka, pp. 21–32 [in Russian].

Bermudez, J., C. Davies, A. Simonazzi, J.P. Real and S. Palma. 2016. Current drug therapy and pharmaceutical challenges for Chagas disease. Acta Tropica 156: 1–16.

Bernadou, A. and V. Fourcassie. 2008. Does substrate coarseness matter for foraging ants? An experiment with *Lasius niger* (Hymenoptera; Formicidae). Journal of Insect Physiology 54: 534–542.

Berryman, A. 1990. Protection of forests against destructive insects. Moscow: Agropromizdat, 288 pp. [in Russian].

Berryman, A.A. (ed.). 2013. Dynamics of Forest Insect Populations: Patterns, Causes, Implications. New York: Plenum Press. 603 pp.

Bertaccini, A. and B. Duduk. 2009. Phytoplasma and phytoplasma diseases: a review of recent research. Phytopathologia Mediterranea 48: 355–378.

Bicknell, J.E., S.P. Phelps, R.G. Davies, D.J. Mann, M.J. Struebig and Z.G. Davies. 2014. Dung beetles as indicators for rapid impact assessments: Evaluating best practice forestry in the neotropics. Ecological Indicators 43: 154–161. DOI: 10.1016/j.ecolind.2014.02.030.

Biegelsen, D.K., A. Berlin, P. Cheung, P.J. Fromherz, D. Goldberg, W. Jackson, B. Preas, J. Reich and L.E. Swartz. 2000. Air-jet paper mover: an example of mesoscale MEMS. SPIE Proceedings, Micromachined Devices and Components VI; Santa Clara, CA. Bellingham, WA: SPIE 4176: 122–129.

Bilo, B.M. and F. Bonifazi. 2008. Epidemiology of insect-venom anaphylaxis. Current Opinion in Allergy and Clinical Immunology 8: 330–337.

Bilo, M.B., F. Rueff, H. Mosbech, F. Bonifazi and J.N.G. Oude-Elberink. 2005. EACCI interest group on insect venom hypersensitivity. Diagnosis of Hymenoptera venom allergy. Allergy 60: 1339–1349.

Binford, G.J., M.S. Callahan, M.R. Bodner, M.R. Rynerson, P.B. Núñez, C.E. Ellison and R.P. Duncan. 2008. Phylogenetic relationships of Loxosceles and Sicarius spiders are consistent with Western Gondwanan vicariance. Molecular Phylogenetics and Evolution 49: 538–553.

BioQuip Books Catalog. 2012. 109 pp. http://www.bioquip.com/html/catalog.htm.

Björnsson, E., C. Janson, P. Plaschke, E. Norrman and O. Sjöberg. 1995. Venom allergy in adult Swedes: a population study. Allergy 50: 800–805.

Blyummer, A.G. 2014. Some features of introduction to European countries and European part of Russia insects of Asian origin—serious pests of woody plants. The Kataev Memorial Readings—VIII. Pests and Diseases of Woody Plants in Russia. Proceedings of the International Conference. St. Petersburg, Russia: St. Petersburg State Forest Technical University, pp. 5–6 [in Russian].

Blyummer, A.G. 2016. Oriental chestnut gall wasp *Dryocosmus Kuriphilus* Yasumatsu, 1951 (Hymenoptera, Cynipidae)— dangerous invasive chestnut pest in the United States and Europe: is it possible to prevent the introduction of the phytophage in Russia? Plant Health Research and Practice 2(16): 27–39. http://vniikr.ru/files/Doc/publ/journal_16.pdf [in Russian].

Bode, H.B. 2009. Insects: true pioneers in anti-infective therapy and what we can learn from them. Angewandte Chemie (International ed. in English) 48(35): 6394–6396. DOI: 10.1002/anie.200902152.

Bogdanov, S. 2011. The bee products: the wonders of the bee hexagon. 200 pp. http://elearning1.uniss.it/moodle/file.php/173/Materiale_didattico_A.A._2010_2011/BeeProducts_prima_parte.pdf.

Bogdanov, S. 2016. Chapter 2: Beeswax: history, uses and trade. *In*: Bee Product Science. http://www.bee-hexagon.net/files/file/fileE/Wax/WaxBook2.pdf.

Bogdasarov, M. 2008. Amber deposits and finds in Belarus. Proceedings of 15th Seminar "Organic inclusions, amber and other fossil resin finds in Europe." Warsaw: Gdańsk, pp. 12–15.

Bogdasarov, M.A. 2006. Problem of amber and other fossil resins formation. Geology and Mineralogy Bulletin 2(16): 18–26 [in Russian]. http://knu.edu.ua/Files/GMV/GMV_16_06/2.pdf.

Bogoni, J.A. and M.I.M. Hernandez. 2014. Attractiveness of native mammal's feces of different trophic guilds to dung beetles (Coleoptera: Scarabaeidae). Journal of Insect Science 14(299): 1–7. DOI: 10.1093/jisesa/ieu161.

Bonacci, T., S. Greco and T.Z. Brandmayr. 2011. Insect fauna and degradation activity of Thanatophilus species on carrion in southern Italy (Coleoptera: Silphidae). Entomologia Generalis 33(1–2): 63–70.

Bonadonna, P., M. Schiappoli, A. Dama, M. Olivieri, L. Perbellini, G. Senna and G. Passalacqua. 2008. Is hymenoptera venom allergy an occupational disease? Occupational and Environmental Medicine 65(3): 217–218.

Bondarenko, N.V., S.V. Pospelov and M.P. Persov. 1991. General and agricultural entomology. Leningrad: Agropromizdat, 431 pp. [in Russian].

Bondarenko, N.V., L.A. Guskova and S.G. Pegelman. 1993. Harmful nematodes, ticks and rodents. Moscow: Kolos, 269 pp. [in Russian].

Bondarev, A.I. and V.V. Soldatov. 1999. Monitoring of Siberian silk moth population in Krasnoyarsk Krai. Ecological monitoring of the forest ecosystems. Petrozavodsk: Publ. KNTs RAS, 70 pp. [in Russian].

Bondarev, L.G. 2003. Locusta in Europe. Herald of Moscow University. Ser. 5, Geography 1: 67–70.

Bonizzoni, M., G. Gasperi, X. Chen and A.A. James. 2013. The invasive mosquito species *Aedes albopictus*: current knowledge and future perspectives. Trends in Parasitology 29(9): 460–468. DOI: 10.1016/j.pt.2013.07.003.

Boppre, M. and R.I. Vane-Wright. 2012. The butterfly house industry: Conservation risks and education opportunities. Conservation and Society 10: 285–303. http://www.conservationandsociety.org/text.asp?2012/10/3/285/101831.

Borges, M.E., R.L. Tejera, L. Díaz, P. Esparza and E. Ibáñez. 2012. Natural dyes extraction from cochineal (*Dactylopius coccus*). New extraction methods. Food Chemistry 132: 1855–1860.

bornholm.info/en/bornholm-butterfly-park.

Bosch, G., S. Zhang, D.G.A.B. Oonincx and W.H. Hendriks. 2014. Protein quality of insects as potential ingredients for dog and cat foods. Journal of Nutritional Science 3: e29. DOI: 10.1017/jns.2014.23.

Bouchemla, F. and V.A. Agoltsov. 2014. Analysis the epizootic situation of bluetongue in the Mediterranean countries. pp. 62–69. *In*: Bauman, N.Uh. (ed.). Scientific Notes of the Kazan State Academy of Veterinary Medicine. Vol. 3. http://cyberleninka.ru/article/n/analiz-epizooticheskoy-situatsii-po-blyutangu-v-stranah-sredizemnomorya [in Russian].

Bourzac, K. 2012. Electronic sensor rivals sensitivity of human skin. Devices inspired by beetle wings could give robots a more nuanced sense of touch. DOI: 10.1038/nature.2012.11081. http://www.nature.com/news/electronic-sensor-rivals-sensitivity-of-human-skin-1.11081.

Boutellis, A., L. Abi-Rached and D. Raoult. 2014. The origin and distribution of human lice in the world. Infection, Genetics and Evolution 23: 209–217.

Bownes, A. 2014. Suitability of a leaf-mining fly, *Hydrellia* sp., for biological control of the invasive aquatic weed, *Hydrilla verticillata* in South Africa. Biocontrol 59(6): 771–780.

Boz, C., S. Velioglu and M. Ozmenoglu. 2003. Acute disseminated encephalomyelitis after bee sting. Neurological Sciences 23: 313–315.

Bradbear, N. 2009. Bees and their role in forest livelihoods: a guide to the services provided by bees and the sustainable harvesting, processing and marketing of their products. Rome: Food and Agriculture Organization of the United Nations. 194 pp. ftp://ftp.fao.org/docrep/fao/012/i0842e/i0842e12.pdf.

Bretagnolle, V. and S. Gaba. 2015. Weeds for bees? A review. Agronomy for Sustainable Development 35(3): 891–909. DOI: 10.1007/s13593-015-0302.

Brewer, W. 2008. Gypsy moth, *Lymantria dispar* Linnaeus (Lepidoptera: Lymantriidae). pp. 1756–1759. *In*: Capinera, J.L. (ed.). Encyclopedia of Entomology. University of Florida, 2nd Ed. Vol. 4.

Brimblecombe, P. and P. Querner. 2014. Webbing clothes moth catch and the management of heritage environments. International Biodeterioration & Biodegradation 96: 50–57.

Brodsky, A.K. and V.D. Ivanov. 1983. Air flow visualization around a flying insect. Reports of the Academy of Sciences 271: 742.

Brodsky, A.K. and D.L. Grodnitskiy. 1986. Aerodynamics of fixed flight of Thymelicus lineola Ochs. (Lepidoptera, Hesperiidae). Entomological Review (Washington) 65(3): 60–69.

Brodsky, A.K. 1991. Vortex formation in the tethered flight of the peacock butterfly *Inachis io l.* (Lepidoptera, Nymphalidae) and some aspects of insect flight evolution. Journal of Experimental Biology 161(11): 77–95.

Brodsky, A.K. 1994. The Evolution of Insect Flight. Oxford: Oxford University Press. 243 pp.

Brown, R.E. 1988. Biological control of Tansy Ragwort (*Senecio jacobaea*) in Western Oregon, U.S.A., 1975–87. pp. 299–305. *In*: Delfosse, E.S. (ed.). Proceedings of the 7th International Symposium on Biological Control of Weeds, 6–11 March 1988, Rome, Italy,

Brown, S.G. and R.J. Heddle. 2003. Prevention of anaphylaxis with ant venom immunotherapy. Current Opinion in Allergy and Clinical Immunology 3(6): 511–516.

Brown, S.G.A. 2001. Funnel web spider envenomations. eMedicine Journal 2(9). http://www.emedicine.com/emerg/topic548.htm.

Brown, S.G.A., R.W. Franks, B.A. Baldo and R.J. Heddle. 2003. Prevalence, severity, and natural history of jack jumper ant venom allergy in Tasmania. Journal of Allergy and Clinical Immunology 111(1): 187–192.

Brown, S.G.A., P. van Eeden, M.D. Wiese, R.J. Mullins, G.O. Solley, R. Puy, R.W. Taylor and R.J. Heddle. 2011. Causes of ant sting anaphylaxis in Australia: the Australian ant venom allergy study. Medical Journal of Australia 195(2): 69–73.

Browne, R.K. 2009. Amphibian diet and nutrition. Aark Science and Research. http://portal.isis.org/partners/AARK/ResearchGuide/Amphibian%20husbandry/Amphibian%20diet%20%20and%20nutrition.pdf.

Browne-Cooper, R., B. Bush, B. Maryan and D. Robinson. 2007. Reptiles and Frogs in the Bush: Southwestern Australia. University of Western Australia Press, pp. 46, 65, 158.

Bryer, M.A.H., C.A. Chapman, D. Raubenheimer, J.E. Lambert and J.M. Rothman. 2015. Macronutrient and energy contributions of insects to the diet of a frugivorous monkey (Cercopithecus ascanius). International Journal of Primatology 36(4): 839–854. DOI: 10.1007/s10764-015-9857-x.

Bubova, T., V. Vrabec, M. Kulma and P. Nowicki. 2015. Land management impacts on European butterflies of conservation concern: a review. Journal of Insect Conservation 19: 805–821. DOI 10.1007/s10841-015-9819-9.

Bucaretchi, F., C.R. de Deus Reinaldo, S. Hyslop, P.R. Madureira, E.M. De Capitani and R.J. Vieira. 2000. A clinico-epidemiological study of bites by spiders of the genus Phoneutria. Revista do Instituto de Medicina Tropical de São Paulo 42(1): 17–21.

Bucher, C., P. Korner and B. Wüthrich. 2001. Allergy to bumblebee venom. Current Opin. Allergy Clin. Immunol. 1(4): 361–365.

Bukejs, A. and V.I. Alekseev. 2015. A second Eocene species of death-watch beetle belonging to the genus Microbregma Seidlitz (Coleoptera: Bostrichoidea) with a checklist of fossil Ptinidae. Zootaxa 3947(4): 553–562.

Burakova, O.V. 2005. Insect pests of stored food. http://initor.by/nasekomye-vrediteli-prodovolstvennyx-zapasov [in Russian].

Buren, W.F. 1982. Red imported fire ant now in Puerto Rico. The Florida Entomologist 65: 188–189. DOI: 10.2307/3494163.

Burks, C.S. and J.A. Johnson. 2012. Biology, behavior, and ecology of stored fruit and nut insects. pp. 21–32. *In*: Hagstrum, D.W., T.W. Phillips and G. Cuperus (eds.). Stored Product Protection. Kansas State University, http://www.bookstore.ksre.ksu.edu/pubs/s156.pdf.

Burns, D.A. 2010. Diseases caused by arthropods and other noxious animals. pp. 38.1–38.61. *In*: Burns, T., S. Breathnach, N. Cox and C. Griffiths (eds.). Rook's Textbook of Dermatology. 8th ed. Chichester: Wiley Blackwell.

Burns, J.N., R. Acuna-Soto and D.W. Stahle. 2014. Drought and epidemic typhus, central Mexico, 1655–1918. Emerging Infectious Diseases 20(3): 442–447.

Burrows, M. and G.P. Sutton. 2012. Locusts use a composite of resilin and hard cuticle as an energy store for jumping and kicking. Journal of Experimental Biology 215(19): 3501–3512.

Busch, D.D. and R. Sheppard. 2012. David Busch's Close-Up and Macro Photography Compact Field Guide. 112 pp.

Bush, B. 2007. Australia's Venomous Snakes: The Modern Myth or Are You a Man or Mouse? http://members.iinet.au/~bush/myth.html.

Butenko, A.M. and I.N. Trusova. 2013. Morbidity rates for Crimean hemorrhagic fever in European, African and Asian countries (1943–2012). Epidemiology and Infectious Diseases 5: 46–49 [in Russian].

Butler, T. 2013. Plague gives surprises in the first decade of the 21st century in the United States and worldwide. American Journal of Tropical Medicine and Hygiene 89(4): 788–793.

butterflywonderland.com/.

Byalt, V.V., L.V. Orlova and A.F. Potokin. 2014. Botany. Guidance on Herbarium: a training manual. SPb.: SPBGLTU. 59 pp. [in Russian].

Cabezas-Cruz, A. and J.J. Valdés. 2014. Are ticks venomous animals? Frontiers in Zoology 11: 47.

Cabrera, R.B. 2005. Downstream processing of natural products: carminic acid. Doctoral Thesis. International University Bremen, School of Engineering and Science.

Calderone, N.W. 2012. Insect pollinated crops, insect pollinators and US agriculture: trend analysis of aggregate data for the period 1992–2009. Public Library of Science One 7(5). DOI: 10.1371/journal.pone.0037235.

Cale, J.A., S.A. Teale, M.T. Johnston, G.L. Boyer, K.A. Perri and J.D. Castello. 2015. New ecological and physiological dimensions of beech bark disease development in aftermath forests. Forest Ecology and Management 336: 99–108. DOI: 10.1016/j.foreco.2014.10.019.

Campana, M.G., N.M. Robles García and N. Tuross. 2015. America's red gold: multiple lineages of cultivated cochineal in Mexico. Ecology and Evolution 5(3): 607–617. DOI: 10.1002/ece3.1398.

Campobasso, C., D. Marchetti, F. Introna and M. Colonna. 2009. Postmortem artifacts made by ants and the effect of ant activity on decompositional rates. American Journal of Forensic Medicine and Pathology 30(1): 84–87.

Cano, E.B. and J.C. Schuster. 2008. Beetles as indicators for forest conservation in Central America. Encyclopedia of Life Support Systems (UNESCO-EOLSS). http://www.eolss.net/sample-chapters/c20/e6-142-tpe-04.pdf.

Caplat, P., P. Edelaar, R.Y. Dudaniec, A.J. Green, B. Okamura, J. Cote, J. Ekroos, P.R. Jonsson, J. Löndahl, S.V.M. Tesson and E.J. Petit. 2016. Looking beyond the mountain: dispersal barriers in a changing world. Frontiers in Ecology and the Environment 14(5): 261–268. DOI: 10.1002/fee.1280.

Carrillo, D., R.E. Duncan, J.N. Ploetz, A.F. Campbell, R.C. Ploetz and J.E. Pena. 2013. Lateral transfer of a phytopathogenic symbiont among native and exotic ambrosia beetles. Plant Pathology 63: 54–62.

Carvalho, R.G., R. Lourenco-de-Oliveira and I.A. Braga. 2014. Updating the geographical distribution and frequency of Aedes albopictus in Brazil with remarks regarding its range in the Americas. Memórias do Instituto Oswaldo Cruz 109(6): 787–796.

Caterino, M.S., K. Wolf-Schwenninger and G. Bechly. 2015. *Cretonthophilus tuberculatus*, a remarkable new genus and species of hister beetle (Coleoptera: Histeridae) from Cretaceous Burmese amber. Zootaxa 4052(2): 241–245. http://dx.doi.org/10.11646/zootaxa.4052.2.10.

Cauchie, H.-M. 2002. Chitin production by arthropods in the hydrosphere. Hydrobiologia 470: 63–95.

Ceccarelli, F.S., B.D. Opell, C.R. Haddad, R.J. Raven, E.M. Soto and M.J. Ramírez. 2016. Around the world in eight million years: historical biogeography and evolution of the spray zone spider Amaurobioides (Araneae: Anyphaenidae). Public Library of Science One 11(10): e0163740. DOI: 10.1371/journal.pone.0163740.

Cerda, H., R. Martinez, N. Briceno, L. Pizzoferrato, P. Manzi, M. Tommaseo Ponzetta, O. Marin and M.G. Paoletti. 2001. Palm worm (*Rhynchophorus palmarum*): traditional food in Amazonas, Venezuela. Nutritional composition, small scale production and tourist palatability. Ecology of Food and Nutrition 40(1): 13–32.

Cerritos, R. 2009. Insects as food: an ecological, social and economical approach. CAB Reviews: Perspectives in Agriculture, Veterinary Science, Nutrition and Natural Resources 4(27): 1–10.

Cerritos, R. 2011. Grasshoppers in agrosystems: Pest or food? CAB Reviews: Perspectives in Agriculture, Veterinary Science, Nutrition and Natural Resources 6: 1–9.

Ceurstemont, S. 2015. 3D-printed bionic ants team up to get the job done. https://www.newscientist.com/article/dn27248-3d-printed-bionic-ants-team-up-to-get-the-job-done/#.VRUv5vmsV8F.

Chaboo, C. 2011. Defensive behaviors in leaf beetles: From the unusual to the weird. pp. 59–69. *In*: Weir, T. and J.M. Vivanco (eds.). Chemical Biology of the Tropics: An Interdisciplinary Approach. Signaling and Communication in Plants. Berlin: Springer Verlag.

Chaffin, J.D. and D.D. Kane. 2010. Burrowing mayfly (Ephemeroptera: Ephemeridae: Hexagenia spp.) bioturbation and bioirrigation: a source of internal phosphorus loading in Lake Erie. Journal of Great Lakes Research 36: 57–63.

Chaika, S.Y. 2003. Forensic entomology. Moscow: MAKS Press, 60 pp. [in Russian].

Chakravorty, J., S. Ghosh and V.B. Meyer-Rochow. 2011. Practices of entomophagy and entomotherapy by members of the Nyishi and Galo tribes, two ethnic groups of the state of Arunachal Pradesh (North-East India). Journal of Ethnobiology and Ethnomedicine 7(5).

Chan, J.F.W., G.K.Y. Choi, C.C.Y. Yip, V.C.C. Cheng and K.-Y. Yuen. 2016. Zika fever and congenital Zika syndrome: An unexpected emerging arboviral disease. Journal of Infection 72: 507–524.

Charpin, D., J. Birnbaum and D. Vervloet. 1994. Epidemiology of hymenoptera allergy. Clinical and Experimental Allergy 24: 1010–1015.

Chauvin, R. 1970. World of insects. [Translated from French by N.B. Kobrina.] Moscow: Mir, 240 pp. [in Russian].

Chávez-Moreno, C.K., A. Tecante and A. Casas. 2009. The Opuntia (Cactaceae) and Dactylopius (Hemiptera: Dactylopiidae) in Mexico: a historical perspective of use, interaction and distribution. Biodiversity and Conservation 18(13): 3337–3355.

Chebyshev, N.V. (ed.). 2012. Medical parasitology. Moscow, Medicina. 304 pp. [in Russian].

Chen, F., D. Porter and F. Vollrath. 2012. Structure and physical properties of silkworm cocoons. J. R. Soc. Interface 1–10. DOI: 10.1098/rsif.2011.0887.

Chen, D., J. Yin, K. Zhao, W. Zheng and T. Wang. 2011. Bionic mechanism and kinematics analysis of hopping robot inspired by locust jumping. Journal of Bionic Engineering 8: 429–439.

Chen, H., X. Shen, X. Li and Y. Jin. 2011. Bionic mosaic method of panoramic image based on compound eye of fly. Journal of Bionic Engineering 8: 440–448.

Chen, H., R. He, Z.L. Wang, S.Y. Wang, Y. Chen, Z.C. Zhu and X.M. Chen. 2015. Genetic diversity and variability in populations of the white wax insect Ericerus pela, assessed by AFLP analysis. Genetics and Molecular Research 14(4): 17820–17827. http://www.funpecrp.com.br/gmr/year2015/vol14-4/pdf/gmr6452.pdf.

Chen, X., Y. Feng and H. Zhang. 2008. Review of the nutritive value of edible insects. pp. 85–92. *In*: Durst, P.B., D.V. Johnson, R.N. Leslie and K. Shono (eds.). Forest Insects as Food: Humans Bite Back. Proceedings of a Workshop on

Asia-Pacific Resources and Their Potential for Development, 19–21 February 2008, Chiang Mai, Thailand. Bangkok: Food and Agriculture Organization of the United Nations.

Chen, X., Y. Feng and Z. Chen. 2009. Common edible insects and their utilization in China. Entomological Research 39: 299–303. DOI: 10.1111/j.1748-5967.2009.00237.x.

Chen, X., Y. Feng and Z. Chen. 2009. Common edible insects and their utilization in China. Journal of Entomological Research 39: 299–303.

Chen, X.M., Z.L. Wang, Y. Chen, S.D. Ye, S.Y. Wang and Y. Feng. 2007. The main climate factors affecting wax excretion of *Ericerus pela* Chavannes (Homopetera: Coccidae) and an analysis of its ecological adaptability. Acta Entomologica Sinica 50(02). http://en.cnki.com.cn/Article_en/CJFDTOTAL-KCXB200702006.htm.

Chen, Y.H., M. Skote, Y. Zhao and W.M. Huang. 2013. Stiffness evaluation of the leading edge of the dragonfly wing via laser vibrometer. Materials Letters 97: 166–168. DOI: 10.1016/j.matlet.2013.01.110.

Cheng, D., J.A. Dattaro and R. Yakobi. Scorpion sting. Updated November 8, 2007. http://emedicine.medscape.com/article/168230-overview.

Cheng, D., S.P. Bush, R. Yakobi, C.J. Gerardo and J.A. Dattaro. 2016. Scorpion Envenomation. http://emedicine.medscape.com/article/168230-overview.

Chenikalova, E.V., R.S. Eremenko and A.A. Mohren. 2008. Protection of rare and useful insects: a training manual. Stavropol: AGRUS, 130 pp. [in Russian].

Cherepnev, I.A., V.I. Dyakonov and A.A. Varako. 2013. Possibilities of electromagnetic technologies to counteract the emergencies in elevators and in grain storages. *In*: Collection of Scientific Works of Kharkiv Air Force University, 2(35): 209–216 [in Russian].

Cherniack, E.P. 2010. Bugs as Drugs, Part 1: Insects. The "New" Alternative Medicine for the 21st Century? Alternative Medicine Review 15(2): 124–135.

Cherniack, E.P. 2011. Bugs as drugs, Part 2: Worms, leeches, scorpions, snails, ticks, centipedes, and spiders. Alternative Medicine Review 16(1): 50–58.

Chernysh, S., S.I. Kim, G. Bekker, V.A. Pleskach, N.A. Filatova, V.B. Anikin, V.G. Platonov and P. Bulet. 2002. Antiviral and antitumor peptides from insects. Proceedings of the National Academy of Sciences of the United States of America 99(20): 12628–12632. DOI: 10.1073/pnas.192301899.

Chernysh, S. and I. Kozuharova. 2013. Antitumor activity of a peptide combining patterns of insect alloferons and mammalian immunoglobulins in naive and tumor antigen vaccinated mice. International Immunopharmacology 17: 1090–1093.

Chernysh, S., N. Gordya and T. Suborova. 2015. Insect antimicrobial peptide complexes prevent resistance development in bacteria. Public Library of Science One 10(7): e0130788. doi:10.1371/journal.pone.0130788.

Chernysh, S.I. and N. Gordia. 2011. The immune system of larvae of *Calliphora vicina* (Diptera, Calliphoridae) as a source of medicinal substances. Journal of Evolutionary Biochemistry and Physiology 47(6): 444–452 [in Russian].

Chernyshev, V.B. 1996. Ecology of insects. Moscow: MGU, 304 pp. [in Russian].

Cherpakov, V. 2014. Xylophagous insects as vectors and symbionts of pathogenic microflora of tree species. Bulletin of Saint Petersburg Forestry Academy 207: 71–83 [in Russian].

Cherry, R. 1987. History of sericulture. Bull Entomol. Soc. Am. 35: 83–84. http://www.insects.org/ced1/seric.html.

Chidami, S. and M. Amyot. 2008. Fish decomposition in boreal lakes and biogeochemical implications. Limnology and Oceanography 53(5): 1988–1996.

Chin Heo, C., M. Marwi, R. Hashim, N. Abdullah, C. Dhang, J. Jeffery, H. Kurahashi and B. Omar. 2009. Ants (Hymenoptera: Formicidae) associated with pig carcasses in Malaysia. Tropical Biomedicine 26(1): 106–109.

Chippaux, J.P. and M. Goyffon. 2008. Epidemiology of scorpionism: a global appraisal. Acta Tropica 107(2): 71–79.

Chippaux, J.-P. 2012. Emerging options for the management of scorpion stings. Drug Design, Development and Therapy 6: 165–173.

Chippaux, J.-P. 2015. Epidemiology of envenomations by terrestrial venomous animals in Brazil based on case reporting: from obvious facts to contingencies. Journal of Venomous Animals and Toxins including Tropical Diseases 21: 13. DOI: 10.1186/s40409-015-0011-1.

Chistyakov, V.A. and Y.V. Denisenko. 2010. Simulation of ageing of Drosophila *in silico*. Successes of Gerontology 23: 557–563 [in Russian].

Choo, J., E.L. Zent and B.B. Simpson. 2009. The importance of traditional ecological knowledge for palm-weevil cultivation in the Venezuelan Amazon. Journal of Ethnobiology 29(1): 113–128.

Choudhary, S., G. Sageena and M. Shakarad. 2014. Hymenopteran venom: a blessing in disguise. International Journal of Applied Engineering Research 9(9): 1111–1118.

Choufani, J., W. El-Halabi, D. Azar and A. Nel. 2015. First fossil insect from Lower Cretaceous Lebanese amber in Syria (Diptera: Ceratopogonidae). Cretaceous Research 54: 106–116. DOI: 10.1016/j.cretres.2014.12.006.

Chowell, G., P. Diaz-Duenas, R. Bustos-Saldana, A. Aleman Mireles and V. Fet. 2006. Epidemiological and clinical characteristics of scorpionism in Colima, Mexico (2000–2001). Toxicon 47: 753–758.

Chshieva, F.T. 2006. Evaluation of the genotoxic action of several drugs in the test system *Drosophila melanogaster* and mammals (Mammalia). Dissertation of candidate of biological Sciences. Vladikavkaz. 182 pp. [in Russian].

Civen, R. and V. Ngo. 2008. Murine typhus: an unrecognized suburban vectorborne disease. Clinical Infectious Diseases 46(6): 913–918. DOI: 10.1086/527443. http://cid.oxfordjournals.org/content/46/6/913.long#sec-2.

Clark, N.D.L. 2010. Amber: Tears of the Gods. 118 pp. Dunedin Academic Press, Edinburgh.

Clark, W.H. and P.E. Blom. 1991. Observations of ants (Hymenoptera: Formicidae: Myrmicinae, Formicinae, Dolichoderinae) utilizing carrion. Southwest Naturalist 36(1): 140–142.

Clayton, J. 2010. Chagas disease 101. Nature 465: S4–S5. DOI: 10.1038/nature09220.

Cloyd, R.A. 2013. Management of insect-vectored diseases (Chapter 7). pp. 37–39. *In*: Krischik, V. and J. Davidson (eds.). IPM (Integrated Pest Management) of Midwest Landscapes. http://cues.cfans.umn.edu/old/ipmbook.htm.

Cocroft, R.B. and R.L. Rodríguez. 2005. The behavioral ecology of insect vibrational communication. BioScience 55(4): 323–334.

Coelho, J.S., E.A.Y. Ishikawa, P.R.S. Garcez dos Santos and P.P. de Oliveira Pardal. 2016. Scorpionism by *Tityus silvestris* in eastern Brazilian Amazon. Journal of Venomous Animals and Toxins including Tropical Diseases 22: 24. DOI: 10.1186/s40409-016-0079-2.

Cohen, A.C. 2001. Formalizing insect rearing and artificial diet technology. American Entomologist 47: 198–206.

Cole, M. 2011. Insect macro photography hints and tips. http://mattcolephotography.blogspot.ru/2011/09/macro-photography-hints-and-tips.html.

Coleman, C.O. 2006. An amphipod of the genus Synurella Wrzesniowski, 1877 (Crustacea, Amphipoda, Crangonyctidae) found in Baltic amber. Organisms, Diversity & Evolution 6: 103–108.

Coles, B.J. and J.M. Coles. 1988. Sweet Track to Glastonbury: The Somerset Levels in Prehistory (New Aspects of Antiquity). London: Thames and Hudson. 200 pp.

Coles, D.M. 2014. The oldest road in the world. http://historicaldis.ru/blog/43611842092/Samaya-drevnyaya-doroga-v-mire [in Russian].

Collavo, A., R.H. Glew, Y.S. Huang, L.T. Chuang, R. Bosse and M.G. Paoletti. 2005. House cricket small-scale farming. pp. 519–544. *In*: Paoletti, M.G. (ed.). Ecological Implications of Minilivestock: Potential of Insects, Rodents, Frogs and Snails. New Hampshire, Science Publishers.

Combes, S.A. and R. Dudley. 2009. Turbulence-driven instabilities limit insect flight performance. Proceedings of the National Academy of Sciences of the United States of America 106(22): 9105–9108. DOI: 10.1073/pnas.0902186106.

Common Rules for Formation, Accounting, Preservation and Use of Museum Objects and Museum Collections in the Museums of the Russian Federation. 2009. Approved by Order of the Ministry of Culture of the Russian Federation. 842. http://www.studfiles.ru/preview/1721344/page:12/ [in Russian].

Concomitant of Forester: Handbook. 1990. Moscow: Agropromizdat, 416 pp. [in Russian].

Coppel, H.C. and J.W. Mertins. 1977. Biological Insect Pest Suppression. 314 pp.

Cordeiro, F.A., F.G. Amorim, F.A.P. Anjolette and E.C. Arantes. 2015. Arachnids of medical importance in Brazil: main active compounds present in scorpion and spider venoms and tick saliva. Journal of Venomous Animals and Toxins Including Tropical Diseases 21: 24. DOI: 10.1186/s40409-015-0028-5.

Corona, M., R.A. Velarde, S. Remolina, A. Moran-Lauter, Y. Wang, K.A. Hughes and G.E. Robinson. 2007. Vitellogenin, juvenile hormone, insulin signaling, and queen honey bee longevity. Proceedings of the National Academy of Sciences of the United States of America 104(17): 7128–7133.

Costa-Neto, E.M. 2002. The use of insects in folk medicine in the state of Bahia, Northeastern Brazil, with notes on insects reported elsewhere in Brazilian folk medicine. Human Ecology 30(2): 245–263.

Costa-Neto, E.M. 2005. Animal-based medicines: biological prospection and the sustainable use of zootherapeutic resources. Annals of the Brazilian Academy of Sciences 77(1): 33–43. http://www.scielo.br/pdf/aabc/v77n1/a04v77n1.pdf.

Costa-Neto, E.M. 2005. Entomotherapy, or the medicinal use of insects. Journal of Ethnobiology 25(1): 93–114.

Costanza, R., R. d'Arge, R. de Groot, S. Farber, M. Grasso, B. Hannon, K. Limburg, S. Naeem, R.V. Oneill, J. Paruelo, R.G. Raskin, P. Sutton and M. van den Belt. 1997. The value of the world's ecosystem services and natural capital. Nature 387: 253–260.

Costanza, R., R. de Groot, P. Sutton, S. van der Ploeg, S.J. Anderson, I. Kubiszewski, S. Farber and R.K. Turneret. 2014. Changes in the global value of ecosystem services. Global Environmental Change 26: 152–158.

Crailsheim, K. 2015. What's new in honey bee science? *In*: Apimondia 2015 Abstract book, p. 138. http://www.apimondia2015.com/2015/eng/file/Abstract%20book_0904_1030%E2%98%85.pdf.

Crane, E. 1990. Bees and Beekeeping: Science Practice and World Resources. Ithaca, New York: Cornell University Press, 640 pp.

Crane, E. 1999. The World History of Beekeeping and Honey Hunting. Routledge: London, 682 pp.

Cranston, P.S. 2010. Insect biodiversity and conservation in Australasia. Annual Review of Entomology 55: 55–75. DOI: 10.1146/annurev-ento-112408-085348.

Creed, R.P., R.P. Cherry, J.R. Pflaum and C.J. Wood. 2009. Dominant species can produce a negative relationship between species diversity and ecosystem function. Oikos 118: 723–732.

Cricket Fighting Contests in China. 2011. http://www.amusingplanet.com/2011/11/cricket-fighting-contests-in-china.html.

Cruz, M., I. Martinez, J. Lopez-Collado, M. Vargas-Mendoza, H. Gonzalez-Hernandez and D.E. Platas-Rosado. 2012. Degradation of cattle dung by dung beetles in tropical grassland in Veracruz, Mexico. Revista Colombiana de Entomologia 38(1): 148–155.

Cuda, J. 2009a. Introduction to biological control of aquatic weeds (Chapter 8). pp. 51–58. *In*: Gettys, L.A., W.T. Haller and D.G. Petty (eds.). Biology and Control of Aquatic Plants. A Best Management Practices Handbook. Marietta, Georgia: Aquatic Ecosystem Restoration Foundation. http://www.aquatics.org/bmp%203rd%20edition.pdf.

Cuda, J. 2009b. Insects for biocontrol of aquatic weeds (Chapter 9). pp. 59–66. *In*: Gettys, L.A., W.T. Haller and D.G. Petty (eds.). Biology and Control of Aquatic Plants. A Best Management Practices Handbook. Marietta, Georgia: Aquatic Ecosystem Restoration Foundation. http://www.aquatics.org/bmp%203rd%20edition.pdf.

Cuda, J.P., R. Charudattan, M.J. Grodowitz, R.M. Newman, J.F. Shearer, M.L. Tamayo and B. Villegas. 2008. Recent advances in biological control of submersed aquatic weeds. Journal of Aquatic Plant Management 46: 15–32.

Curt, M.D., G. Curt, P.L. Aguado and J. Fernández. 2010. Proposal for the biological control of Egeria densa in small reservoirs: a Spanish case study. Journal of Aquatic Plant Management 48: 124–127.

Cushing, P.E. 2012. Spider-ant associations: an updated review of myrmecomorphy, myrmecophily, and myrmecophagy in spiders. Psyche, Article ID 151989. DOI: 10.1155/2012/151989.

Cutler, G.C., C.D. Scott-Dupree and D.M. Drexler. 2014. Honey bees, neonicotinoids and bee incident reports: the Canadian situation. Pest Management Science 70(5): 779–783. DOI: 10.1002/ps.3613.

Dafni, A., P. Kevan, C.L. Gross and K. Goka. 2010. *Bombus terrestris*, pollinator, invasive and pest: An assessment of problems associated with its widespread introductions for commercial purposes. Applied Entomology and Zoology 45(1): 101–113.

Dai, Z., J. Sun, D. Yue, Z. Xia and Z.L. Wang. 2007. Morphology and contact mechanics influence adhesive characteristics of Dung Beetle's bristle and Gecko's setae. Progress in Natural Science 17(9): 1074–1081.

Dalla Pozza, G.L., R. Romi and C. Severini. 1994. Source and spread of *Aedes albopictus* in the Veneto region of Italy. Journal of the American Mosquito Control Association 10(4): 589–592. PMID: 7707070.

Daniel, M. 1990. Secret paths of death bearers. Progress, Moscow, 416 pp. [in Russian].

Dantart, J., C. Stefanescu, A. Àvila and M. Alarcón. 2009. Long-distance wind-borne dispersal of the moth Cornifrons ulceratalis (Lepidoptera: Crambidae: Evergestinae) into the northern Mediterranean. European Journal of Entomology 106: 225–229.

Das, P. and P.K. Singh. 2014. The Munda and their Lac Culture: A case study of Gulllu Area of Murhu Block of Khunti District. Asian Mirror-International Journal of Research 1(1). http://www.asianmirror.in/The%20Munda%20and%20their%20Lac%20Culture.pdf.

Das, P.K., S.P. Pani and K. Krishnamoorthy. 2002. Prospects of elimination of lymphatic filariasis in India. ICMR Bulletin 32(5&6). http://icmr.nic.in/bumayjun02.pdf.

Davis, C.J., E. Yoshioka and D. Kageler. 1992. Biological control of lantana, prickly pear, and Hamakua pamakani in Hawai'i: a review and update. http://www2.hawaii.edu/~theodore/Images/biocontrol_weeds_hawaii.pdf.

Davis, L.R., R.K. Vander Meer and S.D. Porter. 2001. Red imported fire ants expand their range across the West Indies. Florida Entomologist 84(4): 735–736. DOI: 10.2307/3496416.

Davis, P. and M. Grimm. 2003. Red imported fire ant (RIFA), Farmnote, Western Australia Department of Agriculture, Perth. No. 25/2003. www.soe.wa.gov.au/site/../5_WA_SOE2007_BIODIVERSITY.pdf.

de Bortolia, S.A., A.M. Vacaria, V.L. Laurentisa, C.P. De Bortolia, R.F. Santosa and A.K. Otuka. 2016. Selection of prey to improve biological parameters of the predator *Podisus nigrispinus* (Dallas, 1851) (Hemiptera: Pentatomidae) in laboratory conditions. Brazilian Journal of Biology 76(2): 307–314.

de Góes Cavalcanti, L.P., P.L. Tauil, C.H. Alencar, W. Oliveira, M.M. Teixeira and J. Heukelbach. 2016. Zika virus infection, associated microcephaly, and low yellow fever vaccination coverage in Brazil: is there any causal link? Journal of Infection in Developing Countries [Electronic Resource] 10(6): 563–566. DOI:10.3855/jidc.8575.

de Graaf, D.C., M. Aerts, E. Danneels and B. Devreese. 2009. Bee, wasp and ant venomics pave the way for a component-resolved diagnosis of sting allergy. Journal of Proteomics 72: 145–154.

de Groot, H. 2006. Allergy to bumblebees. Current Opinion in Allergy and Clinical Immunology 6(4): 294–297.

de la Giroday, H.-M.C., A.L. Carroll and B.H. Aukema. 2012. Breach of the northern Rocky Mountain geoclimatic barrier: initiation of range expansion by the mountain pine beetle. Journal of Biogeography 39: 1112–1123.

de la Torre-Morin, F., J.C. García-Robaina, C. Vázquez-Moncholí, J. Fierro and C. Bonnet-Moreno. 1995. Epidemiology of allergic reactions in beekeepers: a lower prevalence in subjects with more than 5 years exposure. Allergologia et Immunopathologia 23(3): 127–132.

de Mik, E.L., W. van Pelt, B.D. Docters-van Leeuwen, A. van der Veen, J.F. Schellekens and M.W. Borgdorff. 1997. The geographical distribution of tick bites and erythema migrans in general practice in The Netherlands. International Journal of Epidemiology 26(2): 451–457. DOI: 10.1093/ije/26.2.451.

de Regge, N., M. Madder, I. Deblauwe, B. Losson, C. Fassotte, J. Demeulemeester, F. Smeets, M. Tomme and A. Cay. 2014. Schmallenberg virus circulation in Culicoides in Belgium in 2012: field validation of a real time RT-PCR approach to assess virus replication and dissemination in midges. Public Library of Science One 9(1): e87005. DOI: 10.1371/journal.pone.0087005.

de Roodt, A.R., S.I. Garcia, O.D. Salomon, L. Segre, S.A. Dolap, R.F. Funes and E.H. de Titto. 2003. Epidemiological and clinical aspects of scorpionism by Tityus trivittatus in Argentina. Toxicon 41(8): 971–977.

Dekeirsschieter, J., F.J. Verheggen, E. Haubruge and Y. Brostaux. 2011. Carrion beetles visiting pig carcasses during early spring in urban, forest and agricultural biotopes of Western Europe. Journal of Insect Science 11.

Delil, R.K., T.K. Dileba, Y.A. Habtu, T.F. Gone and T.J. Leta. 2016. Magnitude of malaria and factors among febrile cases in low transmission areas of Hadiya zone, Ethiopia: A facility based cross sectional study. Public Library of Science One 11(5): e0154277. DOI: 10.1371/journal.pone.0154277.

Demidenko, Y. 2012. Stronger than steel, thinner than web. Fashion Theory, Clothing, Body Culture 25. http://www. nlobooks.ru/node/2652 [in Russian].

Demontis, F., R. Piccirillo, A.L. Goldberg and N. Perrimon. 2013. Mechanisms of skeletal muscle aging: insights from Drosophila and mammalian models. Disease Models and Mechanisms 6: 1339–1352.

Derry, J. 2012. Investigating Shellac: Documenting the Process, Defining the Product (PDF). Project-Based Masters Thesis, University of Oslo. 159 pp.

Desquesnes, M., P. Holzmuller, D.H. Lai, A. Dargantes, Z.R. Lun and S. Jittaplapong. 2013. Trypanosoma evansi and Surra: a review and perspectives on origin, history, distribution, taxonomy, morphology, hosts, and pathogenic effects. BioMed. Research International. Article ID 194176. http://dx.doi.org/10.1155/2013/194176.

detroitzoo.org/animal-habitat/butterfly-garden/.

Devinder, S. and M. Bharti. 2001. Ants (Hymenoptera: Formicidae) associated with decaying rabbit carcasses. Uttar Pradesh Journal of Zoology 21: 93–94.

Dhami, S., U. Nurmatov, E.M. Varga, G. Sturm, A. Muraro, C.A. Akdis, D. Antolín-Amérigo, M.B. Bilò, D. Bokanovic, M.A. Calderon, E. Cichocka-Jarosz, J.N.G. Oude Elberink, R. Gawlik, T. Jakob, M. Kosnik, J. Lange, E. Mingomataj, D.I. Mitsias, H. Mosbech, O. Pfaar, C. Pitsios, V. Pravettoni, G. Roberts, F. Ruëff, B.A. Sin and A. Sheikh. 2016. Allergen immunotherapy for insect venom allergy: protocol for a systematic review. Clinical and Translational Allergy 6. UNSP 6 http://ctajournal.biomedcentral.com/articles/10.1186/s13601-016-0095-x.

Diaz, J.H. 2004. The global epidemiology, syndromic classification, management, and prevention of spider bites. American Journal of Tropical Medicine and Hygiene 71(2): 239–250.

Díaz-García, A., L. Morier-Díaz, Y. Frión-Herrera, H. Rodríguez-Sánchez, Y. Caballero-Lorenzo, D. Mendoza-Llanes, Y. Riquenes-Garlobo and J.A. Fraga-Castro. 2013. *In vitro* anticancer effect of venom from Cuban scorpion *Rhopalurus junceus* against a panel of human cancer cell lines. Journal of Venom Research 4: 5–12. http://www.ncbi.nlm.nih.gov/pubmed/23946884.

Dicke, M. 2004. From Venice to Fabre: insects in western art. Proceedings of the Netherlands Entomological Society 15: 9–14. http://www.nev.nl/pages/publicaties/proceedings/nummers/15/9-14.pdf.

Dijkstra, K.D., M.T. Monaghan and S.U. Pauls. 2014. Freshwater biodiversity and aquatic insect diversification. Annual Review of Entomology 59: 143–163.

Doggett, S.L. and R. Russell. 2009. Bed bugs—What the GP needs to know. Australian Family Physician 38(11): 880–884.

Doszhanov T.N. 2003. Louse flies (Diptera, Hippoboscidae) in the Palaearctic. Almaty: Nauka, 277 p. (in Russian).

Dragomirescu, A. and S. Sood. 2009. Bee venom therapy. http://www.davayurvedaezine.com/ezine/feb_2009/bee_venom_therapy.php.

Drake, V.A. and A.G. Gatehouse (eds.). 1995. Insect Migration: Tracking Resources Through Space and Time. Cambridge University Press, 478 pp.

Droogendijk, H., R.A. Brookhuis, M.J. de Boer, R.G.P. Sanders and G.J.M. Krijnen. 2014. Towards a biomimetic gyroscope inspired by the fly's haltere using microelectromechanical systems technology. Journal of the Royal Society Interface 11(99). DOI: 10.1098/rsif.2014.0573.

Duan, J.J., L.S. Bauer, K.J. Abell, M.D. Ulyshen and R.G. van Driesche. 2015. Population dynamics of an invasive forest insect and associated natural enemies in the aftermath of invasion: implications for biological control. Journal of Applied Ecology 52(5): 1246–1254. DOI: 10.1111/1365-2664.12485.

Dubern, J. 1994. Transmission of African cassava mosaic geminivirus by the whitefly (Bemisia tabaci). Tropical Science 34(1): 82–91.

DuBois, C.G. 1908. The religion of the Luiseño Indians of Southern California. http://www.sacred-texts.com/nam/ca/roli/roli07.htm.

Dunayev, E.A. 1997. Ants of the Moscow region: methods for environmental studies. MosGorsYun. 96 pp. [in Russian].

Dunn, R. 2007. Insects as medicine: The ant and the grasshopper. http://www.robrdunn.com/2007/01/the-ant-and-the-grasshopper/.

Dwi, R.L. 2008. Teak caterpillars and other edible insects in Java. pp. 99–103. *In*: Durst, P.B., D.V. Johnson, R.N. Leslie and K. Shono (eds.). Forest Insects as Food: Humans Bite Back. Proceedings of a Workshop on Asia-Pacific Resources and Their Potential for Development, 19–21 February 2008, Chiang Mai, Thailand. Bangkok: Food and Agriculture Organization of the United Nations.

Early, R., B.A. Bradley, J.S. Dukes, J.J. Lawler, J.D. Olden, D.M. Blumenthal, P. Gonzalez, E.D. Grosholz, I. Ibanez, L.P. Miller, C.J.B. Sorte and A.J. Tatem. 2016. Global threats from invasive alien species in the twenty-first century and national response capacities. Nature Communications 7: 12485. DOI: 10.1038/ncomms12485.

Eaten alive. 1998. New Science 159(2143): 12.

Ebeling, W. 2002. Pests of fabrics and paper. *In*: Urban Entomology. Chapter 8. University of California, Division of Agricultural Sciences; pp. 310–322. http://www.entomology.ucr.edu/ebeling/ebeling8.html.

Ebeling, W. 2002. Wood-destroying insects and fungi. *In*: Urban Entomology. Chapter 5, part 1. University of California, Division of Agricultural Sciences; pp. 128–167.

Eberle, A.L., B.H. Dickerson and T.L. Daniel. 2015. A new twist on gyroscopic sensing: body rotations lead to torsion in flapping, flexing insect wings. Journal of the Royal Society Interface 12(104). DOI: 10.1098/rsif.2014.1088.

Edwards, R. and A.E. Mill. 1986. Termites in Buildings: Their Biology and Control. Rentokil Ltd., East Grinstead, United Kingdom. 261 pp.

Egerton, F.N. 2006. A history of the ecological sciences, Part 21: Réaumur and his history of insects. Bulletin of the Ecological Society of America: 212–224. http://esapubs.org/bulletin/current/history_list/history21.pdf.

Ehelamalpe, C. 2015. Insect, mite and nematode pests of forest species and plant protection methods in practice. http://www.slideshare.net/chandikeehelamalpe/insect-mite-and-nematode-pests-of-forest.

Elbers, A.R., R. Meiswinkel, E. van Weezep, M.M. van Oldruitenborgh-Oosterbaan and E.A. Kooi. 2013a. Schmallenberg virus in Culicoides spp. biting midges, the Netherlands, 2011. Emerging Infectious Diseases 19(1): 106–109. DOI: 10.3201/eid1901.121054.

Elbers, A.R., R. Meiswinkel, E. van Weezep, E.A. Kooi and W.H. van der Poel. 2013b. Schmallenberg virus in Culicoides biting midges in the Netherlands in 2012. Transboundary and Emerging Diseases 62(3): 339–342. DOI: 10.1111/tbed.12128.

Elias, S.A. 1994. Quaternary Insects and Their Environments. Washington, D.C.: Smithsonian Institution Press.

Elias, S.A., J.T. Andrews and K.H. Anderson. 1999. Insights on the climatic constraints on the beetle fauna of coastal Alaska, U.S.A., derived from the Mutual Climatic Range method of paleoclimate reconstruction. Arctic, Antarctic, and Alpine Research 31: 94–98.

Elias, S.A. 2000. Climatic tolerances and zoogeography of the late Pleistocene beetle fauna of Beringia. Géographie physique et Quaternair 54(2): 143–155.

Ellis, J. 2014. Lecture 9: The History of Beekeeping. http://leon.ifas.ufl.edu/agr/files/2014/04/History-of-Beekeeping.pdf.

Ellison, A.M. and N.J. Gotelli. 2009. Energetics and the evolution of carnivorous plants. Journal of Experimental Botany 60: 19–42. DOI: 10.1093/jxb/ern179.

encenter.org/visit-us/butterfly-house/.

Endicott, S. and E. Hagerman. 1999. The United States and Biological Warfare. Secrets from the Early Cold War and Korea. Indiana University Press. 304 pp.

Engel, M.S., R.C. McKellar, S. Gibb and B.D.E. Chatterton. 2012. A new Cenomanian-Turonian (Late Cretaceous) insect assemblage from southeastern Morocco. Cretaceous Research 35: 88–93. DOI: 10.1016/j.cretres.2011.11.022.

Enserink, M. 2015. An obscure mosquito-borne disease goes global. Science 350(6264): 1012–1013. DOI: 10.1126/science.350.6264.1012.

Ermakov, A.I. 2013. Change necrofile complex invertebrates under the action of pollution with emissions from the Sredneural copper smelter. Ecology 6: 463–470 [in Russian].

Ernst, L., F. Zlochevskiy and G. Erastus. 2007. Waste management of livestock and poultry. http://webpticeprom.ru/ru/articles-processing-waste.html?pageID=1177395301 [in Russian].

Eroshenko, E. 2013. Five problems of the world beekeeping. Belarusian Agriculture 5(133): 30–33. http://agriculture.by/articles/agrarnaja-politika/pjat-problem-mirovogo-pchelovodstva [in Russian].

Eshetu, E. and B. Begejo. 2015. The current situation and diagnostic approach of Nagana in Africa: A review. Journal of Natural Sciences Research 5(17): 117–124.

Esipenko, L.P. 2012. A new method of the ragweed inhibition (*Ambrosia artemi-siifolia l.*) of south Russia. The Scientific Journal of the Kuban State Agrarian University 79(05). http://ej.kubagro.ru/2012/05/pdf/51.pdf [in Russian].

Esipenko, L.P. 2012. Biological invasions as a global environmental problem of the South of Russia. South of Russia: Ecology, Development 7(4): 21–25 [in Russian].

Esipenko, L.P. and A.S. Zamotajlov. 2014. Introduction of phytophagous insects for biological supression of common ragweed (*Ambrosia artemisiifolia l.*). Russia: Retrospective Overview in Plant Protection News 2: 43–46 [in Russian].

Europe preaches entomophagy. 2010. http://survincity.com/2010/11/europe-preaches-entomophagy/.

European Environmental Agency. The European Grassland Butterfly Indicator: 1990–2011. 2013. EEA Technical report. No. 11/2013. Luxembourg, Publications Office of the European Union. 36 pp. DOI: 10.2800/89760. https://www.ufz.de/export/data/24/56765_European_Grassland_Butterfly_Indicator_1990-2011.pdf.

European Food Safety Authority Scientific Committee. 2015. Risk profile related to production and consumption of insects as food and feed. EFSA Journal 13(10): 4257. DOI: 10.2903/j.efsa.2015.4257.

Evenhuis, N.L. 2016. The insect and spider collections of the world. http://hbs.bishopmuseum.org/codens/.

Faccoli, M. and P. Gatto. 2016. Analysis of costs and benefits of Asian longhorned beetle eradication in Italy. Forestry 89(3): 301–309.

Fact Sheet. 2014. The Economic Challenge Posed by Declining Pollinator Populations. https://www.whitehouse.gov/the-press-office/2014/06/20/fact-sheet-economic-challenge-posed-declining-pollinator-populations.

Farag, Y. 2010. Characterization of different shellac types and development of shellac-coated dosage forms. Dissertation. Hamburg, University of Hamburg, 143 pp. https://www.chemie.uni-hamburg.de/bibliothek/2010/DissertationFarag.pdf.

Farajollahi, A., W.J. Crans, D. Nickerson, P. Bryant, B. Wolf, A. Glaser and T.G. Andreadis. 2005. Detection of West Nile virus RNA from the louse fly *Icosta americana* (Diptera : Hippoboscidae). Journal of the American Mosquito Control Association 21(4): 474–476.

Fassakhov, R.S., I.D. Reshetnikova, G.S. Voitsekhovich, L.V. Makarova and N. Gorshunova. 2009. Anaphylactic shock: causes, symptoms, and emergency treatment, prevention. Practical Medicine 3: 25–31 [in Russian].

Federle, W., M. Riehle, A.S.G. Curtis and R.J. Full. 2002. An integrative study of insect adhesion: mechanics and wet adhesion of pretarsal pads in ants. Integrative and Comparative Biology 42(6): 1100–1106.

FEDIAF. 2014. The European Pet Food Industry. Facts and Figures 2014. http://www.fediaf.org/facts-figures/.

Fedorova, O.A. 2009. Blood-sucking midges (Diptera, Simuliidae) south of the Tyumen region (biological bases of protection of cattle). Dis. kand.biol.nauk. Tyumen, 185 pp. [in Russian].

Fedoskova, T.G. 2007. Allergy to insects. Modern principles of diagnostics and treatment. Russian Medical Journal 2: 65–73 [in Russian].

Fedyaeva, N. 2011. Achievements of bionics: what people can learn from cockroaches, lizards and sea shells? http://theoryandpractice.ru/posts/2745-dostizheniya-bioniki-chemu-lyudi-mogut-nauchitsya-u-tarakanov-yashcherits-i-morskikh-rakovin [in Russian].

Feng, H., X. Wu, B. Wu and K. Wu. 2009. Seasonal migration of *Helicoverpa armigera* (Lepidoptera: Noctuidae) over the Bohai Sea. Journal of Economic Entomology 102(1): 95–104.

Feng, Y., X. Chen, L. Sun and Z. Chen. 2008. Common edible wasps in Yunnan Province, China and their nutritional value. pp. 93–98. *In*: Durst, P.B., D.V. Johnson, R.N. Leslie and K. Shono (eds.). Forest Insects as Food: Humans Bite Back. Proceedings of a Workshop on Asia-Pacific Resources and Their Potential for Development, 19–21 February 2008, Chiang Mai, Thailand. Bangkok, Food and Agriculture Organization of the United Nations.

Feng, Y., M. Zhao, Z. He, Z. Chen and L. Sun. 2009. Research and utilization of medicinal insects in China. Entomological Research 39: 313–316.

Fenoglio, S., T. Bo., M. Cammarata, G. Malacarne and G. Del Frate. 2010. Contribution of macro- and micro-consumers to the decomposition of fish carcasses in low order streams: an experimental study. Hydrobiologia 637: 219–228.

Fenoglio, S., R.W. Merritt and K.W. Cummins. 2014. Why do no specialized necrophagous species exist among aquatic insects? Freshwater Science 33(3): 711–715. DOI: 10.1086/677038.

Fernandez, J., M. Blanca, P. Soriano, F. Sanchez and C. Juarez. 1999. Epidemiological study of the prevalence of allergic reactions to Hymenoptera in a rural population in the Mediterranean area. Clinical and Experimental Allergy 29: 1069–1074.

Fet, V.Y. 2006. Travels under Scorpio constellation. Science at First Hand 5: 110–121.

Feyereisen, R. and M. Jindra. 2012. The silkworm coming of age early. Public Library of Science Genetics 8: e1002591.

Figueiredo, A.C., D. de Sanctis, R. Gutiérrez-Gallego, T.B. Cereija, S. Macedo-Ribeiro, P. Fuentes-Prior and P.J.B. Pereira. 2012. Unique thrombin-inhibition mechanism by anophelin, an anticoagulant from the malaria vector. Proceedings of the National Academy of Sciences of the United States of America 109(52): E3649–E3658.

Fitzgerald, K.T. and A.A. Flood. 2006. Hymenoptera stings. Clinical Techniques in Small Animal Practice 21(4): 194–204.

Fleire-Maia, L., J.A. Campos and C.F. Amaral. 1994. Approaches to the treatment of scorpion envenoming. Toxicon 32(9): 1009–1014.

Flindt, R. 1992. Biology in the Figures. M.: Mir. 303 pp. [in Russian].

Flinn, A. 2011. Shellac and food glaze. http://gentleworld.org/shellac-food-glaze/.

Food and Agriculture Organization of the United Nations. 2007. Overview of forest pests—Kyrgyz republic. Working Paper FBS/21E. Rome, FAO. http://www.fao.org/docrep/012/al010e/al010e00.pdf.

Food and Agriculture Organization of the United Nations. 2012. FAO Statistics. Rome, FAO.

Food and Agriculture Organization of the United Nations. 2014. The State of World Fisheries and Aquaculture Opportunities and challenges. Rome, FAO. 223 pp.

Ford, R.P. and P.J. Barry. 2004. Spider mites. http://www.forestpests.org/nursery/spidermites.html.

Forensic entomology. Use of insects to help solve crime. 2007. 8 pp. http://www.clt.uwa.edu.au/__data/assets/pdf_file/0015/2301612/fse07_forensic_entomology.pdf.

Forest Encyclopedia, Vol. 2. Vorobyov G.I. (ed.). Moscow, Sov. Encyclopedia, 1986. 631 pp. [in Russian].

Francis, T.B. and D.E. Schindler. 2009. Shoreline urbanization reduces terrestrial insect subsidies to fishes in North American lakes. Oikos 118(12): 1872–1882. DOI: 10.1111/j.1600-0706.2009.17723.x.

Frank, K.D. 2015. Ecology of Center City, Philadelphia. Philadelphia, Pennsylvania: Fitler Square Press, 404 pp.

Frelik, A. 2014. Predation of adult large diving beetles Dytiscus marginalis (Linnaeus, 1758), Dytiscus circumcinctus (Ahrens, 1811) and Cybister lateralimarginalis (De Geer, 1774) (Coleoptera: Dytiscidae) on fish fry. Oceanological and Hydrobiological Studies 43(4): 360–365.

Freymann, B.P., R. Buitenwerf, O. Desouza and H. Olff. 2008. The importance of termites (Isoptera) for the recycling of herbivore dung in tropical ecosystems: A review. European Journal of Entomology 105: 165–173.

Frost, S.A., R. Streeter, C.H.G. Wright and S.F. Barrett. 2013. Bio-mimetic optical sensor for structural deflection measurement. http://nari.arc.nasa.gov/sites/default/files/Frost_FinalReportFrost2013.pdf.

Gaju, M., M.J. Notario, R. Moral, E. Alcaide, T. Moreno, R. Molero and C.B. de Roca. 2002. Termite damage to buildings in the Province of Cordoba. Spain Sociobiology 40: 75–85.

Galperin, M.V. 2007. General ecology. St. Petersburg: Forum, Infra-M. 336 pp. [in Russian].

Galvan, J.A. (ed.). 2014. They do what? A cultural encyclopedia of extraordinary and exotic customs from around the world. Santa Barbara, California: ABC-CLIO. 374 pp.

Ganez, A.Y., I.K. Baker, R. Lindsay, A. Dibernardo, K. McKeever and B. Hunter. 2002. West Nile virus outbreak in North American owls, Ontario. Emerg. Infect. Dis. 10(12): 2135–42.

Ganguli, R., S. Gorb, F.O. Lehmann and S. Mukherjee. 2010. An experimental and numerical study of calliphora wing structure. Experimental Mechanics 50(8): 1183–1197. DOI: 10.1007/s11340-009-9316-8.

Ganie, N.A., A.S. Kamili, M.F. Baqual, R.K. Sharma, K.A. Dar and I.L. Khan. 2012. Indian sericulture industry with particular reference to Jammu and Kashmir. I.J.A.B.R. 2(2): 194–202. ISSN: 2250-3579.

Ganin, G.N. 2011. Structural and functional organization of mezopedobiont communities of the Southern Russian Far East. Vladivostok, Dalnauka. 380 pp. [in Russian].

Ganta, S. and S. Pallamparthi. 2010. Indian sericulture industry—future prospects and challenges. pp. 252–255. *In*: Proceedings of 2010 International Conference on Agricultural and Animal Science, Singapore, 26–28 February.

Gao, X., X. Yan, X. Yao, L. Xu, K. Zhang, J. Zhang, B. Yang and L. Jiang. 2007. The dry-style antifogging properties of mosquito compound eyes and artificial analogues prepared by soft lithography. Advanced Materials 19: 2213–2217. DOI: 10.1002/adma.200601946.

Garratt, M.P.D., T.D. Breeze, N. Jenner, C. Polce, J.C. Biesmeijer and S.G. Potts. 2014. Avoiding a bad apple: Insect pollination enhances fruit quality and economic value. Agriculture Ecosystems and Environment 184: 34–40. DOI: 10.1016/j.agee.2013.10.032.

Gelashvili, D.B., V.N. Krylov and E.B. Romanova. 2015. Zootoxinology: bioecological and biomedical aspects. Manual. Nizhni Novgorod: Nizhni Novgorod State University Press, 770 pp. [in Russian].

Gennard, D.E. 2007. Forensic Entomology. An Introduction. Chichester, England: John Wiley & Sons Ltd., 254 pp.

George, N.T., T.C. Irving, C.D. Williams and T.L. Daniel. 2013. The cross-bridge spring: can cool muscles store elastic energy? Science 340(6137): 1217–1220. DOI: 10.1126/science.1229573.

Gerlach, J., M. Samways and J. Pryke. 2013. Terrestrial invertebrates as bioindicators: an overview of available taxonomic groups. Journal of Insect Conservation 17: 831–850. DOI 10.1007/s10841-013-9565-9.

Giannini, T.C., G.D. Cordeiro, B.M. Freitas, A.M. Saraiva and V.L. Imperatriz-Fonseca. 2015. The dependence of crops for pollinators and the economic value of pollination in Brazil. Journal of Economic Entomology 108(3): 849–857. DOI: 10.1093/jee/tov093.

Gibson, A.J. 1942. The story of Lac. Journal of the Royal Society of Arts 90(4611): 318.

Giller, P. and L. Greenberg. 2015. The relationship between individual habitat use and diet in brown trout. Freshwater Biology 60(2): 256–266. DOI: 10.1111/fwb.12472.

Gillespie, R.G., B.G. Baldwin, J.M. Waters, C.I. Fraser, R. Nikula and G.K. Roderick. 2012. Long-distance dispersal: a framework for hypothesis testing. Trends in Ecology and Evolution 27(1): 47–56.

Gilyarov, M.S. (ed.). 1986. Biological encyclopaedic dictionary. Moscow, 831 pp. [in Russian].

Glatz, R. and K. Bailey-Hill. 2010. Mimicking nature's noses: From receptor deorphaning to olfactory biosensing. Progress in Neurobiology 93(2011): 270–296. DOI: 10.1016/j.pneurobio.2010.11.004.

Glick, M., Y. Levy and Y. Gafni. 2009. The viral etiology of tomato yellow leaf curl disease—a review. Plant Protection Sciences 3: 81–97.

Global Invasive Species Database. 2017. 100 of the world's worst invasive alien species. http://www.iucngisd.org/gisd/100_worst.php.

Global Invasive Species Database. 2017. Species profile: *Lymantria dispar*. http://www.iucngisd.org/gisd/species.php?sc=96 on 28-01-2017.

Goddard, J. and R. de Shazo. 2009. Bed bugs (*Cimex lectularius*) and clinical consequences of their bites. Journal of the American Medical Association 301(13): 1358–1366.

Goddard, J. 2012. Public Health Entomology. CRC Press, 230 pp.

Goetz, D.W. 2008. *Harmonia axyridis* ladybug invasion and allergy. Allergy and Asthma Proceedings 29: 123–129. DOI: 10.2500/aap.2008.29.3092.

Goetz, D.W. 2009. Seasonal inhalant insect allergy: *Harmonia axyridis* ladybug. Current Opinion in Allergy and Clinical Immunology 9(4): 329–333.

Goka, K. 2010. Introduction to the special feature for ecological risk assessment of introduced bumblebees: Status of the European bumblebee, *Bombus terrestris*, in Japan as a beneficial pollinator and an invasive alien species. Applied Entomology and Zoology 45(1): 1–6.

Golden, D.B. 2015. Anaphylaxis to insect stings. Immunology and Allergy Clinics of North America 35(2): 287–302. doi:10.1016/j.iac.2015.01.007.

Golden, D.B.K., K.R.N. Denise, R.G. Hamilton and T.J. Craig. 2009. Venom immunotherapy reduces large local reactions to insect stings. Journal of Allergy and Clinical Immunology 123(6): 1371–1375.

Goldfinch, P.D. 2014. The cochineal insect: a bug that changed history. http://www.pinebrookhills.org/Press/Articles/cochineal_insect.htm.

Golub, V.B., M.N. Tsurikov and A.A. Prokin. 2012. Collections of insects: collection, processing and storage of the material. Moscow, KMK Publishing, 339 pp. [in Russian].

Golubev, A.G. 2003. Biochemistry of life extension. Successes of Gerontology 12: 57–76 [in Russian].

Gomes, C., M.J. Pons, J. del Valle Mendoza and J. Ruiz1. 2016. Carrion's disease: an eradicable illness? Infectious Diseases of Poverty 5(105). DOI: 10.1186/s40249-016-0197-7.

Goodman, L.F. and B.A. Hungate. 2006. Managing forests infested by spruce beetles in south-central Alaska: Effects on nitrogen availability, understory biomass, and spruce regeneration. Forest Ecology and Management 227: 267–274.

Gorb, S.N. and A.E. Filippov. 2014. Fibrillar adhesion with no clusterisation: Functional significance of material gradient along adhesive setae of insects. Beilstein Journal of Nanotechnology 5: 837–845.

Gorbunov, O.G. and V.S. Murzin. 2009. Butterflies. http://gordon0030.narod.ru/archive/19977/2009_Butterflies.pdf [in Russian].

Gordon, D.G. 2013. The Eat-a-Bug Cookbook, Revised—40 Ways to Cook Crickets, Grasshoppers, Ants, Water Bugs, Spiders, Centipedes, and Their Kin. Ten Speed Press, 136 pp.

Gornostaev, G.N. 1986. Problems of protection of endangered insects. The Results of Science and Technology. Ser. Entomology. Vol. 6. Moscow, VINITI, pp. 116–204 [in Russian].

Gorshkov, S.P. 2001. Basics of geoecology. Zeldorizdat, Moscow, 592 pp. [in Russian].

Gouache, T., Y. Gao, Y. Gourinat and P. Coste. 2010. Wood wasp inspired planetary and earth drill. *In*: Mukherjee, A. (ed.). Biomimetics Learning from Nature. InTech. http://www.intechopen.com/books/biomimetics-learning-from-nature/wood-wasp-inspiredplanetary-and-earth-drill.

Gough, A. 2008. The bee: Part 1—Beedazzled. http://andrewgough.co.uk/articles_bee1/.

Goulson, D. 2010. Impacts of non-native bumblebees in Western Europe and North America. Applied Entomology and Zoology 45: 7–12.

Govorushko, S.M. 2009. Geoecological Designing and Expertise., Vladivostok: Far-Eastern State University. 388 pp. [in Russian].

Govorushko, S.M. 2012. Natural processes and Human impacts: Interaction between Humanity and the Environment. Dordrecht: Springer. 678 pp.

Govorushko, S.M. 2016. Human Impact on the Environment. An Illustrated World Atlas. Cham: Springer International Publishing AG Switzerland. 367 pp.

Goyffon, M., M. Vachon and N. Broglio. 1982. Epidemiological and clinical characteristics of the scorpion envenomation in Tunisia. Toxicon 20(1): 337–344.

Graif, Y., O. Romano-Zelekha, I. Livne, M.S. Green and T. Shohat. 2006. Allergic reactions to insect stings: Results from a national survey of 10,000 junior high school children in Israel. Journal of Allergy and Clinical Immunology 117(6): 1435–1439.

Grasshopper—new chassis for robots. 2015. http://vefnews.com/en/news/innovations/grasshopper-new-chassis-robots.

Great Medicinal Encyclopaedia. 1975–1988. Soviet encyclopaedia, Moscow, V. 3, 1976, V. 5, 1977, V. 6, 1977, V. 8, 1978, V. 10, 1979, V. 13, 1980, V. 15, 1981, V. 17, 1981, V. 18, 1982, V. 20, 1983, V. 22, 1984, V. 23, 1984, V. 24, 1985, V. 25, 1985, V. 27, 1986, V. 28, 1986, V. 28, 1987.

Grebennikov, V.S. 2014. In the country insects. Notes and sketches of an entomologist and artist. http://litresp.ru/chitat/ru/%D0%93/grebennikov-viktor-stepanovich/v-strane-nasekomih-zapiski-i-zarisovki-entomologa-i-hudozhnika/7 [in Russian].

Green, A. 2012. A brief history of the chainsaw. http://www.popularmechanics.com/home/tools/reviews/a8162/a-brief-history-of-the-chain-saw-13626055/.

Green, P.W.C. 2008. Fungal isolates involved in biodeterioration of book-paper and their effects on substrate selection by *Liposcelis bostrychophila* (Badonnel) (Psocoptera: Liposcelididae). Journal of Stored Products Research 44(3): 258–263. DOI: 10.1016/j.jspr.2008.01.003.

Green, P.W.C. and D.I. Farman. 2015. Can paper and adhesive alone sustain damaging populations of booklice? Journal of Conservation and Museum Studies 13(1): 3. DOI: http://doi.org/10.5334/jcms.1021222.

Greenberg, B. and J.C. Kunich. 2002. Entomology and law: flies as forensic indicators. Cambridge, Cambridge University Press.

Greenwood, B.M., K. Bojang, C.J. Whitty and G.A. Targett. 2005. Malaria. Lancet 365(9469): 1487–1498. DOI: 10.1016/S0140-6736(05)66420-3.

Gregory, L.G. and C.M. Lloyd. 2011. Orchestrating house dust mite-associated allergy in the lung. Trends in Immunology 32: 402–411.

Grenier, S. 2012. Artificial rearing of entomophagous insects, with emphasis on nutrition and parasitoids—general outlines from personal experience. Karaelmas Science and Engineering Journal 2(2): 1–12.

Griffin, J.M., M. Simard and M.G. Turner. 2013. Salvage harvest effects on advance tree regeneration, soil nitrogen, and fuels following mountain pine beetle outbreak in lodgepole pine. Forest Ecology and Management 291: 228–239.

Griffiths, S. 2015. Bionic ants, butterfly drones and chameleon grippers: The robo-insect workers coming to a factory near you. http://www.technocrazed.com/discover-the-impressive-bionic-insects-from-insect-labs-photo-gallery.

Grigoreas, C., I.D. Galatas, C. Kiamouris and D. Papaioannou. 1997. Insect-venom allergy in Greek adults. Allergy 52: 51–57.

Grigoriev, Y.S., N.V. Pojarkova, S.V. Prudnikova and O.E. Kruchkova. 2008. Biological control of the environment. Krasnoyarsk, 2008, 117 pp. [in Russian].

Grodnitsky, D.L. 1989. Structure and possible functions of the scaly cover on the wings of diurnal butterflies (Lepidoptera: Hesperioidea, Papilionoidea). Entomological Review (Washington) 68(2): 11–17.

Grodnitsky, D.L. and P.P. Morozov. 1993. Vortex formation during tethered flight of functionally and morphologically two-winged insects, including evolutionary considerations on insect flight. Journal of Experimental Biology 182: 11–40. http://jeb.biologists.org/content/jexbio/182/1/11.full.pdf.

Grodnitsky, D.L. 1996. Adaptation to flapping flight in different insects with complete transformation. Zoological Journal 75(5): 699–700.

Grodnitsky, D.L. 1996. Functional morphology of the wings of insects with complete metamorphosis. The dissertation of a scientific degree of the doctor of biological sciences. Saint Petersburg, 29 pp. [in Russian].

Grodnitsky, D.L. 1999. Form and Function of Insect Wings: The Evolution of Biological Structures. Baltimore, MD, Johns Hopkins University Press, 261 pp.

Gu, G., C. Zhang and F. Hu. 2002. Analysis on the structure of honey production and trade in the world. Apiacta 2: 1–5. http://www.apimondiafoundation.org/foundation/files/2002/GU%20G.%20ZHANG%20CH.pdf.

Guerreiro, O., P. Cardoso, J.M. Ferreira, M.T. Ferreira and P.A.V. Borges. 2014. Potential distribution and cost estimation of the damage caused by *Cryptotermes brevis* (Isoptera: Kalotermitidae) in the Azores. Journal of Economic Entomology 107(4): 1554–1562. DOI: 10.1603/EC13501.

Guglielmone, A.A., R.G. Robbing, D.A. Apanaskevich, T.N. Petney, A. Estrada-Peña, I.G. Horak, R. Shao and S.C. Barker. 2010. The Argasidae, Ixodidae and Nuttalliellidae (Acari: Ixodida) of the world: a list of valid species names. Zootaxa 2528: 1–28.

Guillot, A. and J.A. Meyer. 2013. Bionics: when science imitates nature. Moscow: Technosphera, 280 pp. [in Russian].

Gurr, G. and S.D. Wratten (eds.). 2000. Biological Control: Measures of Success. Springer, 432 pp. DOI: 10.1007/978-94-011-4014-0.

Ha, N.S., Q.T. Truong, N.S. Goo and H.C. Park. 2013. Relationship between wingbeat frequency and resonant frequency of the wing in insects. Bioinspiration and Biomimetics 8(4). DOI: 10.1088/1748-3182/8/4/046008.

Haack, R.A., K.R. Law, V.C. Mastro, S. Ossenbruggen and B.J. Raimo. 1997. New York's battle with the Asian long-horned beetle. Journal of Forestry 95(12): 11–15.

Haavik, L.J., K.J. Dodds and J.D. Allison. 2015. Do native insects and associated fungi limit non-native woodwasp, *Sirex noctilio*, survival in a newly invaded environment? Public Library of Science One 10(10): e0138516. DOI: 10.1371/journal.pone.0138516.

Haddad, L.S., L. Kelbert and A.J. Hulbert. 2007. Extended longevity of queen honey bees compared to workers is associated with peroxidation-resistant membranes. Experimental Gerontology (42)7: 601–609.

Haddad, V. Jr., J.L.C. Cardoso, O. Lupi and S.K. Tyring. 2012. Tropical dermatology: Venomous arthropods and human skin. Part I. Insecta. Journal of the American Academy of Dermatology 67(339): e1–e14.

Haddad, V., Jr., P.C. Haddad de Amorim, W.T. Haddad, Jr. and J.L.C. Cardoso. 2015. Venomous and poisonous arthropods: identification, clinical manifestations of envenomation, and treatments used in human injuries. Revista da Sociedade Brasileira de Medicina Tropical 48(6): 650–657. http://dx.doi.org/10.1590/0037-8682-0242-2015.

Hagstrum, D.W., T.W. Phillips and G. Cuperus. 2012. Stored Product Protection. 358 pp. Kansas State University. http://www.bookstore.ksre.ksu.edu/pubs/s156.pdf.

Hahn, M.B., R.J. Eisen, L. Eisen, K.A. Boegler, C.G. Moore, J. McAllister, H.M. Savage and J.P. Mutebi. 2016. Reported distribution of *Aedes (Stegomyia) aegypti* and *Aedes (Stegomyia) albopictus* in the United States, 1995–2016 (Diptera: Culicidae). Journal of Medical Entomology 53(5): 1169–1175.

Hajek, A.E., B.P. Hurley, J.R. Garnas, S.J. Bush, M.J. Wingfield, A.E. Hajek, M. Kenis, J.C. Van Lenteren and M.J.W. Cock. 2016. Exotic biological control agents: A solution or contribution to arthropod invasions? Biological Invasions 18(4). DOI: 10.1007/s10530-016-1075-8.

Haller, W.T. 2009. Hydrilla (Chapter 13.1). pp. 89–94. *In*: Gettys, L.A., W.T. Haller and D.G. Petty (eds.). Biology and Control of Aquatic Plants. A Best Management Practices Handbook. http://www.fs.usda.gov/Internet/FSE_DOCUMENTS/stelprdb5373509.pdf.

Halliday, R.B., B.M. O'Connor and A.S. Baker. 2000. Global diversity of mites. pp. 192–212. *In*: Raven, P.H. and T. Williams (eds.). Nature and Human Society: the Quest for a Sustainable World: Proceedings of the 1997 Forum on Biodiversity. National Academies.

Hall-Mendelin, S., P. O'Donoghue, R.B. Atwell, R. Lee and R.A. Hall. 2011. An ELISA to detect serum antibodies to the salivary gland toxin of *Ixodes holocyclus* Neumann in dogs and rodents. Journal of Parasitology Research 2011: 283416. http://dx.doi.org/10.1155/2011/283416.

Halvorson, H.M., C. Fuller, S.A. Entrekin and M.A. Evans-White. 2015. Dietary influences on production, stoichiometry and decomposition of particulate wastes from shredders. Freshwater Biology 60(3): 466–478. DOI: 10.1111/fwb.12462.

Hamdin, M.S., M. Mohamed and L. Tokiman. 2015. Potential of entomotourism at Taman Negara Johor Endau Rompin. International Journal of Administration and Governance 1(4): 92–97.

Hamidoghli, A., B. Falahatkar, M. Khoshkholgh and A. Sahragard. 2014. Production and enrichment of chironomid larva with different levels of vitamin C and effects on performance of Persian sturgeon larvae. North American Journal of Aquaculture 76: 289–295.

Hanboonsong, Y., T. Jamjanya and P.B. Durst. 2013. Six-legged livestock: edible insect farming, collection and marketing in Thailand. FAO 69 pp. http://www.fao.org/docrep/017/i3246e/i3246e00.htm.

Handbook of Agronomist on Plant Protection. 1990. Agropromizdat, Moscow, 366 pp. [in Russian].

Hangay, G. and P. Zborowski. 2010. A Guide to the Beetles of Australia. Csiro Publishing, 248 pp.

Hansen, L.S., H. Skovgard and K. Hell. 2004. Life table study of *Sitotroga cerealella* (Lepidoptera: Gelechiidae), a strain from West Africa. Journal of Economy and Entomology 97: 1484–1490.

Hartland-Rowe, R. 1992. The biology of the wild silkmoth *Gonometa rufobrunnea* Aurivillius (Lasiocampidae) in northeastern Botswana, with some comments on its potential as a source of wild silk. Botswana Notes and Records 24: 123–133.

Harves, A. and L. Millikan. 2008. Current concepts of therapy and pathophysiology in arthropod bites and stings. International Journal of Dermatology 14(9): 621–634.

Harvey, B.J., D.C. Donato and M.G. Turnera. 2014. Recent mountain pine beetle outbreaks, wildfire severity, and postfire tree regeneration in the US Northern Rockies. Proceedings of National Academy of Science of the United States of America 111(42): 15120–15125. DOI: 10.1073/pnas.1411346111.

Harvey, P., S. Sperber, F. Kette, R.J. Heddle and P.J. Roberts-Thomson. 1984. Bee-sting mortality in Australia. Medical Journal of Australia 140(4): 209–211.

Haseman, L. 2007. An Elementary Study of Insects. 145 pp. http://www.gutenberg.org/files/23434/23434-h/23434-h.htm.

Haskell, N.H. 2006. The science of forensic entomology. pp. 431–440. *In*: Wecht, C.H. and J.T. Rago (eds.). Forensic Science and Law: Investigative Applications in Criminal, Civil, and Family Justice. Boca Raton, CRC Press.

Hayashi, M., M. Bakkali, A. Hyde and S.L. Goodacre. 2015. Sail or sink: novel behavioural adaptations on water in aerially dispersing species. BioMed. Central Evolutionary Biology. 15: 118. http://bmcevolbiol.biomedcentral.com/articles/10.1186/s12862-015-0402-5. DOI: 10.1186/s12862-015-0402-5.

Heinen, T.E. and A.B. Gorini da Veiga. 2011. Arthropod venoms and cancer. Toxicon 57: 497–511.

Hellqvist, M. 2004. Local environment and human impact at Gamla Uppsala, SE Sweden, during the Iron Age, as inferred from fossil beetle remains. Journal of Nordic Archaeological Science 14: 89–99.

Helson, J.E., T.L. Capson, T. Johns, A. Aiello and D.M. Windsor. 2009. Ecological and evolutionary bioprospecting: Using aposematic insects as guides to rain-forest plants active against disease. Frontiers in Ecology and the Environment 7: 130–134.

Henshaw, M.T., N. Kunzmann, C. Vanderwoude, M. Sanetra and R.H. Crozier. 2005. Population genetics and history of the introduced fire ant, *Solenopsis invicta* Buren (Hymenoptera: Formicidae), in Australia. Australian Journal of Entomology 44(1): 37–44. DOI: 10.1111/j.1440-6055.2005.00421.x.

Herrero, L.C., C.R.M. Pena, M.S. Garcia and G.J. Barciela. 2013. A fast chemometric procedure based on NIR data for authentication of honey with protected geographical indication. Food Chemistry 141(4): 3559–3565. DOI:10.1016/j.foodchem.2013.06.022.

Herzig, V., R. Ward and W. Ferreira dos Santos. 2002. Intersexual variations in the venom of the Brazilian 'armed' spider *Phoneutria nigriventer* (Keyserling, 1891). Toxicon 40(10): 1399–1406.

Hieber, M. and M.O. Gessner. 2002. Contribution of stream detrivores, fungi, and bacteria to leaf breakdown based on biomass estimates. Ecology 83: 1026–1038.

Hile, D.C., T.P. Coon, C.G. Skinner, L.M. Hile, P. Levy, M.M. Patel and M.A. Miller. 2006. Treatment of imported fire ant stings with mitigator sting and bite treatment—a randomized control study. Wilderness and Environmental Medicine 17(1): 21–25.

Hodson, H. 2014. Six legs tasty: First edible insect farm opens in US. New Scientist 2970. https://www.newscientist.com/article/mg22229702-900-six-legs-tasty-first-edible-insect-farm-opens-in-us/.

Hoebeke, E.R. 2003. Invasive insects as major pests in the United States. pp. 1–4. *In*: Pimentel, D. (ed.). Encyclopaedia of Pest Management. Marcel Dekker, Inc. http://www.informaworld.com/smpp/content~content=a713568446~db=all~order=title.

Hoerauf, A.M. 2011. Onchocerciasis. pp. 741–749. *In*: Guerrant, R.L., D.H. Walker and P.F. Weller (eds.). Tropical Infectious Diseases. Philadelphia, Elsevier Saunders.

Hoffmann, J., R. Romey, C. Fink and T. Roeder. 2013. Drosophila as a model to study metabolic disorders. Advances in Biochemical Engineering – Biotechnology 135: 41–61.

Holker, F., M.J. Vanni, J.J. Kuiper, C. Meile, H.-P. Grossart, P. Stief, R. Adrian, A. Lorke, O. Dellwig, A. Brand, M. Hupfer, W.M. Mooij, G. Nützmann and J. Lewandowski. 2015. Tube-dwelling invertebrates: tiny ecosystem engineers have large effects in lake ecosystems. Ecological Monographs 85(3): 333–351. DOI: 10.1890/014-1160.1.

Holter, P. and C.H. Scholtz. 2007. What do dung beetles eat? Ecological Entomology 32(6): 690–697. DOI: 10.1111/j.1365-2311.2007.00915.x.

Holyoak, M., T.S. Talley and S.E. Hogle. 2010. The effectiveness of US mitigation and monitoring practices for the threatened valley elderberry longhorn beetle. Journal of Insect Conservation 14: 43–52. DOI: 10.1007/s10841-009-9223-4.

Honga, K.-J., J.-H. Lee, G.-S. Lee and S. Lee. 2012. The status quo of invasive alien insect species and plant quarantine in Korea. Journal of Asia-Pacific Entomology 15(4): 521–532.

Hopkins, J., A. Bourdain and M. Freeman. 2004. Extreme Cuisine: The Weird and Wonderful Foods That People Eat. Periplus Editions. 352 pp.

Hoppenheit, A. 2013. Tsetse (Diptera: Glossinidae) bloodmeal analysis by pcr and species differentiation by MALDI TOF MS as contributions to rational vector control. Biomedical Sciences PhD Dissertation, Freie University, Berlin, Germany.

Hoppenheit, A., B. Bauer, S. Steuber, W. Terhalle, O. Diall, K.-H. Zessin and P.-H. Clausen. 2013. Multiple host feeding in *Glossina palpalis gambiensis* and *Glossina tachinoides* in southeast Mali. Medical and Veterinary Entomology 27: 222–225. DOI: 10.1111/j.1365-2915.2012.01046.x.

Hormiga, G. 2002. Orsonwells, a new genus of giant linyphild spiders (Araneae) from the Hawaiian Islands. Invertebrate Systematics 16(3): 369–448. DOI: 10.1071/IT01026. https://www2.gwu.edu/~spiders/content/publications/Hormiga%202002.pdf.

Horn, H. and G. Leybold. 2006. Medicines from the hive: honey, pollen, royal jelly, beeswax, propolis, bee venom. Moscow, AST: ASTREL. 238 pp. [in Russian].

Hosoda, N. and S.N. Gorb. 2012. Underwater locomotion in a terrestrial beetle: combination of surface de-wetting and capillary forces. Proceeedings Biological Sciences 279(1745): 4236–4242. DOI: 10.1098/rspb.2012.1297.
Houston, D.R. 1994. Major new tree disease epidemics: beech bark disease. Annual Review of Phytopathology 32: 75–87.
http://4108.ru/u/karminovaya_kislota.
http://4ento.com/2015/05/07/pet-food-challenges-why-need-change/.
http://ag.arizona.edu/pubs/insects/ahb/inf13.html.
http://agro-archive.ru/zaschita-rasteniy/141-fitopatogennye-mikoplazmy.html.
http://agro-portal24.ru/bolezni-rasteniy/5078-istochniki-infekcii-i-sposoby-rasprostraneniya-bakteriozov-chast-5.html.
http://allthingsbugs.com/.
http://amber-trade.com/mining/.
http://amccorona.com/wp-content/uploads/2014/11/Feeding-Insect-eating-reptiles.pdf.
http://anbp.org/index.php/members-products.
http://andy321.proboards.com/thread/46825.
http://animalreader.ru/termitnik-shedevr-inzhenernogo-iskusstva.html.
http://animals.about.com/od/reptiles/a/what-reptiles-eat.htm.
http://animals.mom.me/list-insecteating-mammals-8065.html.
http://antclub.ru/lib/brian-m-v/obshchestvennye-nasekomye/pishcha/osy-i-muravi-kak-khishchniki.
http://anysite.ru/publication/silk.
http://apismelliferamellifera.0pk.ru/viewtopic.php?id=222.
http://architectuul.com/architecture/honeycomb-apartments.
http://aviation.cours-de-math.eu/ATPL-081-POF/flutter.php.
http://babochkindom.ru/articles/286623.
http://baob.wikidot.com/aerial-insectivores.
http://belboh.com/magflori/115-mag45.html.
http://beneficialbugs.org/bugs/Diving_Beetle/diving_beetle.htm.
http://bezvrediteley.ru/kleshhy/chem-opasen-krasnyj-kleshh.
http://bigcricketfarms.com/.
http://bijlmakers.com/insects/insect-proverbs-and-quotes/.
http://bijlmakers.com/proverbs-and-quotes/harry-potter/.
http://biotechlearn.org.nz/news_and_events/news/2011_archive/dung_beetles_imported_to_aid_soil_fertility.
http://borboletario.aguiasdaserra.com.br/.
http://bortnic.ru/?p=464.
http://botan0.ru/?cat=3&id=200.
http://boutique.ald-entomologie.fr/.
http://campfiresmusic.com/Helvetica%20Condensed/?p=libelle-g-suit.
http://carnivoraforum.com/topic/9754556/1/.
http://classifieds.insectnet.com/hsx/classifieds.hsx?session_key=&search_and_display_db_button=on&db_id=32262&query=retrieval.
http://cogmtl.net/Articles/138.htm.
http://dic.academic.ru/dic.nsf/ruwiki/1525388.
http://dic.academic.ru/dic.nsf/ruwiki/88741.
http://diseases.academic.ru/1070/ТИФ_СЫПНОЙ_ЭНДЕМИЧЕСКИЙ.
http://dnevniki.ykt.ru/_Emerald_/522481.
http://dommedika.com/laboratoria/474.html.
http://dom-sad-og.ru/nasekomye-protiv-nasekomyx-vreditelej/.
http://ec.europa.eu/agriculture/evaluation/market-and-income-reports/2013/apiculture/chap3_en.pdf.
http://ecdc.europa.eu/en/healthtopics/emerging_and_vector-borne_diseases/louse-borne-diseases/pages/louse-borne-relapsing-fever.aspx.
http://en.wikipedia.org/wiki/Hornet.
http://en.wikipedia.org/wiki/Polistes.
http://en.wikipedia.org/wiki/Scorpiones.
http://en.wikipedia.org/wiki/Scorpiones.
http://en.wikipedia.org/wiki/Scorpiones.
http://en.wikipedia.org/wiki/Vespula.
http://encyclopedia2.thefreedictionary.com/Honey+Extractor.
http://entnemdept.ufl.edu/creatures/livestock/deer_fly.htm.
http://entnemdept.ufl.edu/creatures/URBAN/MEDICAL/Stomoxys_calcitrans.htm.
http://entnemdept.ufl.edu/creatures/URBAN/MEDICAL/Stomoxys_calcitrans.htm.
http://entnemdept.ufl.edu/pubs/proverbs.htm.
http://entomofarms.com/.
http://entomolog.narod.ru/news_ent_yarmarki.html.
http://entomology.com.ua/.

http://entomology.com.ua/travels/?route=list-region.
http://entomologytoday.org/tag/apocryta-westwoodi/.
http://espacepourlavie.ca/en/insectarium.
http://finesell.ru/vsjo-pro-jantarj/raznovidnosti-jantarja.html.
http://forest.geoman.ru/forest/item/f00/s00/e0000403/index.shtml.
http://g.janecraft.net/virusy-rastenijj/.
http://gambledor.com/ru/articles/cockroach-racing.
http://gems.minsoc.ru/eng/articles/amber.
http://geolike.ru/page/gl_1921.htm.
http://givoyles.ru/articles/uhod/tainye-puti-infekcii/.
http://gothic.com.ua/ukrrus/forum/index.php?showtopic=6460.
http://hotlix.com/candy/index.php?route=common/home.
http://http-wikipediya.ru/wiki/Оленья_кровососка.
http://hundredways.ru/babochki-i-nasekomye-v-ramkah.html.
http://iabes.org/.
http://insectalib.ru/news/item/f00/s02/n0000209/index.shtml.
http://insectalib.ru/news/item/f00/s02/n0000221/index.shtml.
http://insectalib.ru/news/item/f00/s02/n0000224/index.shtml.
http://insect-collections.ru/.
http://insecticea.ru/yekologiya_nasekomyh_i_biologicheskie_ritmy/vliyanie_cheloveka_na_nasekomyh.
http://insecticea.ru/yekologiya_nasekomyh_i_biologicheskie_ritmy/vliyanie_cheloveka_na_nasekomyh.
http://insects.org/ced1/numis.html.
http://inserco.org/en/?q=statistics.
http://intranet.puhinui.school.nz/Topics/Insects/InsectAnatomyDutch/bijlmakers/entomology/citaten_engels.htm.
http://ipm.ucanr.edu/PMG/PESTNOTES/pn7436.html.
http://izhchara.ru/gromnichnye-svechi/.
http://johnjhalseth.tumblr.com/post/93847655335/swami-bills-flea-circus-one-of-the-better-shows.
http://kartravel.ru/fotomastera.html.
http://klbutterflypark.com/.
http://klop911.ru/tarakany/o-tarakanax/tarakani-bega.html.
http://kodomo.fbb.msu.ru/~partyhard/term2/pr6/nobelprize1995.html.
http://kungrad.com/nature/fauna/zlatki/.
http://labs.russell.wisc.edu/insectsasfood/about-dr-defoliart/.
http://ladybirdconsulting.co.in/past-events.html#BFFR.
http://lefortvet.ru/bolezn-shmallenberga.
http://lib4all.ru/base/B3161/B3161Part11-46.php.
http://lib7.com/aziatyy/244-shelkovodstvo-kitai.html.
http://lsdinfo.org/gribnye-bolezni-lesnyx-nasazhdenij-i-borba-s-nimi/.
http://medbiol.ru/medbiol/env_fact/000688e2.htm.
http://mednadom.com/statiy/pcheliniy-vosk/.
http://medvuz.info/load/allergologija/insektnaja_allergija/48-1-0-566.
http://mentalfloss.com/article/23038/9-insane-torture-techniques.
http://mnzoo.org/.
http://montre24.com/brand/Vulcain/Vulcain/.
http://mrunal.org/2013/07/geography-location-factors-silk-industry-china-india-karnataka-kanchipuram.html.
http://mydoctorhouse.ru/ticks/ticks-nature/vidy-kleshhej.html.
http://mylektsii.ru/9-1665.html.
http://myreptile.ru/forum/index.php?topic=5237.0.
http://mywatch.ru/watch-art/art_1109.html.
http://nacekomoe.ru/publ/hiwnye-nasekomye/.
http://naperekate.narod.ru/alesfishi224.html.
http://naturalhistory.si.edu/butterflies/.
http://nazeb.ru/1454-vosk.html.
http://nlinsectarium.com/.
http://noev-kovcheg.ru/mag/2011-20/2880.html#ixzz424QwfykC.
http://onbird.ru/terminy-i-opredelenija-ornitologii/e/entomofagi.
http://onkologia.maxbb.ru/topic239.html.
http://os-nauka.ru/scorpions/.
http://otagomuseum.nz.
http://ours-nature.ru/b/book/15/page/5-navozniki-i-bronzovki/102-nora-i-navoznaya-kolbasa.
http://paukoobraznye.ru/books/item/f00/s00/z0000010/st016.shtml.
http://penelope.uchicago.edu/Thayer/e/roman/texts/historia_augusta/aurelian/2*.html

http://pestcontrol.lk/termites/.

http://phytopath.ca/wp-content/uploads/2015/03/DPVCC-Chapter-16-potato.pdf.

http://planeta.by/article/752.

http://portaleco.ru/ekologija-ryb/nasekomye-igrajut-znachitelno-bolshuju-rol-v-zhizni-ryb-v-presnyh-vodah-chem-v-morskih.html.

http://portaleco.ru/ekologija-ryb/nasekomye-igrajut-znachitelno-bolshuju-rol-v-zhizni-ryb-v-presnyh-vodah-chem-v-morskih.html.

http://pubs.acs.org/action/doSearch?text1=spider+silk&field1=Title.

http://roypchel.ru/pchely/vidy-pchel.html.

http://ru.encydia.com/en/Antheraea_polyphemus.

http://ru.focus.lv/life/zhivotnye/v-germanii-pogibla-truppa-bloshinogo-cirka.

http://ru.wikipedia.org/wiki/Dolichovespula.

http://schmetterlinghaus.at/en/.

http://schools-wikipedia.org/wp/a/Amber.htm.

http://schools-wikipedia.org/wp/a/Amber.htm.

http://screw-wholesale.myweb.hinet.net/.

http://showhistory.com/acts/flea-circus.

http://slovariki.org/ekologiceskij-slovar/2667.

http://sputnik-rybolova.org.ua/zhivotnie-nazhivki/nasekomie-i-ich-lichinki-v-kachestve-nazhivki.

http://static1.squarespace.com/static/54dae2b6e4b011d9d8f95d9e/t/54f90909e4b0d13f30b89ea7/1425606921096/BugPoems.pdf.

http://stopvreditel.ru/parazity/perenoschiki/krasnotelkovyje-kleshhi.html.

http://stores.ebay.co.uk/Norfolk-Pet-Supplies/Fishing-Bait-/_i.html?_fsub=1276540016.

http://substance-en.etsmtl.ca/festo-unveils-its-new-bionic-insects/.

http://survinat.ru/2010/02/mozhzhevelnik/#ixzz4HxYsVy2v.

http://survinat.ru/2010/02/mozhzhevelnik/#ixzz4HxYsVy2v.

http://therivardreport.com/members-night-at-bracken-bat-cave/.

http://tinref.ru/000_uchebniki/04800selskoe_hozaistvo/002_tehhnolog_muku_krupi_kombikorm_1_3/038.htm.

http://tolweb.org/Phthiraptera.

http://travee.co/1098.

http://ucrazy.ru/interesting/1327197690-pytki-nasekomymi.html.

http://vespabellicosus2008.narod.ru/polistes.html.

http://vsebabochki.ru/salut/.

http://what-when-how.com/insects/apis-species-honey-bees-insects/.

http://what-when-how.com/insects/apis-species-honey-bees-insects/.

http://wine.historic.ru/books/item/f00/s00/z0000001/st008.shtml.

http://wol.jw.org/ru/wol/d/r2/lp-u-ru/102011093#h=1.

http://www.8lap.ru/section/interesnye-fakty/pochemu-shmel-letaet/.

http://www.8lap.ru/section/interesnye-fakty/vsye-o-paukakh-kak-letayut-pauki/.

http://www.afrol.com/articles/10447.

http://www.agroatlas.ru/en/content/pests/Bruchus_pisorum/.

http://www.alaris-schmetterlingspark.de/.

http://www.amazonanimalorphanage.org/.

http://www.amazonianbutterflies.com/.

http://www.amazoninsect.com/collecting_trips.html.

http://www.animalsandenglish.com/animal--insect-eating-plants.html.

http://www.animalsglobe.ru/ehidnyi/.

http://www.animalsglobe.ru/muravyedi/.

http://www.antwiki.org/wiki/Centromyrmex_gigas.

http://www.antwiki.org/wiki/Hypoponera.

http://www.apus.ru/site.xp/049052056055124054050050056124.html.

http://www.apus.ru/site.xp/050050055048124.html.

http://www.asahi-net.or.jp/~ch2m-nitu/indexe.htm.

http://www.asiahouse24.de.

http://www.azerbaijanrugs.com/arfp-natural_dyes_insect_dyes.htm.

http://www.badgersett.com/sites/default/files/info/publications/Bulletin4v1_0.pdf.

http://www.bcogoa.org/.

http://www.bees-products.com/?p=articles&j=27.

http://www.beetlebreeding.ch/beetle-shops-of-japan-insect-shop-global-in-osaka/.

http://www.beetlebreeding.ch/beetle-shops-of-japan-insect-shop-global-in-osaka/.

http://www.biochemi.ru/chems-1103-2.html.

http://www.biochemi.ru/chems-1103-2.html.

http://www.biocontrol.entomology.cornell.edu/success.php.
http://www.bioquip.com/.
http://www.brocgaus.ru/text/054/838.htm.
http://www.brost.se/eng/education/find.html.
http://www.bterfoundation.org/bvt.
http://www.bugdesign.com.ua/roach_race/.
http://www.butterfliesandmoths.org/species/Anisota-senatoria.
http://www.butterfly.co.nz/index.html.
http://www.butterflyfarm.co.cr/en/the-butterfly-farm/butterfly-farm-costa-rica-tour-information/index.html.
http://www.butterflygardens.com.
http://www.butterflyhouse.com.au/.
http://www.butterfly-insect.com/whoweare.php.
http://www.butterflyreleasecompany.com/.
http://www.butterflyworkx.com.
http://www.calacademy.org/.
http://www.callawaygardens.com/things-to-do/attractions/day-butterfly-center.
http://www.cals.ncsu.edu/course/ent425/text01/impact1.html.
http://www.carrboro.com/butterflyhouse.html.
http://www.catgut.ru/info/16.html.
http://www.city-data.com.
http://www.city-data.com.
http://www.cockroachraces.com.au/.
http://www.collector-secret.com/index.php/top-auctions/.
http://www.cricketflours.com/about/.
http://www.cyberlipid.org/wax/wax0001.htm.
http://www.danaida.ru/obsh/sad.htm.
http://www.denvercountyfair.org/#!special-events/c457.
http://www.dltk-kids.com/crafts/insects/songs.htm.
http://www.duhzemli.ru/animal/bird-protection/insect.html.
http://www.eafe.org/.
http://www.earthlife.net/insects/conservation.html.
http://www.ebay.co.uk/sch/Live-Fishing-Bait/179963/bn_10030959/i.html.
http://www.ecosystema.ru/08nature/birds/168s.php.
http://www.ecosystema.ru/08nature/birds/morf/morf2.htm.
http://www.egir.ru/bird/46.html.
http://www.enterrafeed.com/.
http://www.entomolog.ru/.
http://www.entomology.msstate.edu/resources/croplosses/2009loss.asp.
http://www.entomology.org.uk/.
http://www.entoproducts.com.
http://www.entosupplies.com.au/equipment/field.
http://www.entsoc.org/resources/faq.
http://www.enviroflight.net/.
http://www.etawau.com/Life/Gallery/Insects.htm.
http://www.eurolab.ua/diseases/508/.
http://www.evrey.com/sitep/talm/print.php?trkt=hullin&menu=65.
http://www.exotic-insects.com.
http://www.fao.org/3/a-v8879e/v8879e08.htm.
http://www.fao.org/docrep/005/x9895e/x9895e03.htm.
http://www.fao.org/docrep/009/p5178e/P5178E05.htm.
http://www.fao.org/docrep/u8480e/U8480E0j.htm.
http://www.fastcompany.com/3037716/inside-the-edible-insect-industrial-complex.
http://www.fauna-toxin.ru/15.html.
http://www.fda.gov/ForIndustry/ColorAdditives/GuidanceComplianceRegulatoryInformation/ucm153038.htm.
http://www.fish-uk.com/online_fishing_tackle_and_bait_s.htm.
http://www.fleacircus.co.uk/History.htm.
http://www.flmnh.ufl.edu/.
http://www.f-mx.ru/biologiya/nasekomye_v_zhizni_cheloveka.html.
http://www.focussingapore.com/singapore-tourism/butterfly-park-insect-kingdom.html.
http://www.gersociety.ru/information/magg2009/.
http://www.giradis-insect.com.
http://www.globalosaka.co.jp.

http://www.godofinsects.com/.
http://www.hcmealworm.com/.
http://www.hiroshima-navi.or.jp/en/sightseeing/shizen_koen/dobutsu_shokubutsu/21754.php.
http://www.hlscientific.in/entomologicals-equipments.html.
http://www.honeyo.com/org-International.shtml.
http://www.hyaenidae.org/the-hyaenidae/aardwolf-proteles-cristatus/cristatus-diet-and-foraging.html.
http://www.indiamart.com/biolinx-labsystems/search.html?ss=entomological%20equipment.
http://www.indiamart.com/industrial-commercial-services/entamology-equipments.html.
http://www.insect-collections.ru/6_0.html.
http://www.insectnet.eu/contact.php.
http://www.insectopia.ru/.
http://www.insects.de/anzeigen.php?art=CCE.
http://www.insects.org/ced1/numismatic-entomology.html.
http://www.insects4sale.com/.
http://www.insect-sale.com/.
http://www.insect-supply.com.
http://www.insect-supply.eu.
http://www.insect-trade.eu.
http://www.interinsects.com.
http://www.iucngisd.org/gisd/.
http://www.iucngisd.org/gisd/100_worst.php on 21-01-2017.
http://www.kerala-traveller.com/2008/02/kerala-butterfly-safari-park-thenmala.html.
http://www.korolevpharm.ru/dokumentatsiya/syrevye-komponenty/264-beeswax-pchelinyj-vosk.html.
http://www.kreca.com.
http://www.kupi-uley.ru/rasteniya_medonosy.php.
http://www.labcarescientific.in/entomology-product.html.
http://www.liquisearch.com/wild_silk/list_of_some_wild_silk_moths_and_their_silk.
http://www.livescience.com/45539-what-do-turtles-eat.htm.
http://www.livescience.com/46117-ticks-lyme-disease.html.
http://www.livingrichlyonabudget.com/insects-used-as-medicine-around-the-world.
http://www.mariposario.org.mx/.
http://www.meddiscover.ru/newmed-520.html.
http://www.medicinform.net/infec/infec62.htm.
http://www.medoviy.ru/stat-mean-sort-1825-Pad_kak_syre_dlja_lesnogo_meda.html.
http://www.medoviy.ru/stat-pasek-medosbor-1868-Falsifikacija_meda.html.
http://www.michigan.org/property/michigan-state-university-horticultural-gardens-and-butterfly-house/.
http://www.micromegamondo.com/casadellefarfalle/index.php?centro=2.
http://www.milanoperibambini.it/in-citta/musei/364-loasi-delle-farfalle.html.
http://www.missouribotanicalgarden.org/visit/family-of-attractions/butterfly-house.aspx.
http://www.montgomeryparks.org/brookside/.
http://www.mosi.org/.
http://www.nafea.net/.
http://www.naturalhandyman.com/iip/infpai/shellac.html.
http://www.nature-et-passion.com.
http://www.nhm.org/site/explore-exhibits/special-exhibits/butterfly-pavilion.
http://www.niagaraparks.com/niagara-falls-attractions/butterfly-conservatory.html.
http://www.oceanpark.com.hk/en/park-experience/attractions.
http://www.odonata.su/content-view-152.html.
http://www.orkin.com/other/mosquitoes/mosquito-predators/.
http://www.ozinsects.com/.
http://www.passiflorahoeve.nl/home-en/.
http://www.pchela.com.ua/?page_id=8.
http://www.pitersad.ru/galli.html.
http://www.priroda.su/item/1376.
http://www.psciences.net/main/sciences/biology/articles/shuky.html.
http://www.reimangardens.com/.
http://www.rnsp.su/istoriya-pchelovodstva/.
http://www.ronl.ru/lektsii/biologiya/847004/.
http://www.roseentomology.com/.
http://www.ru-expo.ru/novost-kozha_prochnee_bronezhileta.
http://www.rumbur.ru/nature/1182-trubkozub.
http://www.ryadi.ru/tekstil/436.html.

http://www.sad-sevzap.ru/vse-o-zashchite-rastenii/ptitsy-pomogayut-sadu.
http://www.salkova.ru/Product_bee/Beewax/description.php.
http://www.scud.ru/pencil/ink_watery_powder/01.php.
http://www.seq.qc.ca/english/activites/leaflets/f2a.htm.
http://www.si.edu/encyclopedia_si/nmnh/buginfo/benefits.htm.
http://www.sonnik.ru/articles/art1398.html.
http://www.swallowtailfarms.com/.
http://www.thaibugs.com/?page_id=690.
http://www.thebugmaniac.com/.
http://www.thefullwiki.org/Halyomorpha_halys.
http://www.theinsectcollector.com.
http://www.thornesinsects.com.
http://www.tinlib.ru/biologija/priklyuchenija_s_nasekomymi/p37.php.
http://www.trainedfleas.com/.
http://www.treehugger.com/culture/qa-is-silk-green.html.
http://www.ucmp.berkeley.edu/history/leeuwenhoek.html.
http://www.udec.ru/vrediteli/zernovka.php.
http://www.vegparadise.com/news13.html.
http://www.vinograd7.ru/docs/vrediteli/filloksera.htm.
http://www.vokrugsveta.ru/vs/article/49/.
http://www.waitingman.net/BugLit.html.
http://www.wardsci.com/.
http://www.watdon.co.uk/acatalog/entomological-equipment.html.
http://www.watdon.co.uk/acatalog/entomological-equipment.html.
http://www.webring.org/l/rd?ring=famousquotesandf;id=1;url=http%3A%2F%2Fwebspace%2Ewebring%2Ecom%2Fpeop
   le%2Fif%2Ffiqabil%2Finsectpro%2Ehtml.
http://www.wellness.com/ reference/allergies/insect-sting-allergy.
http://www.wetlandpark.gov.hk/en/exhibition/reserve_butterfly.asp.
http://www.wheelerfarms.com/.
http://www.woodyman.ru/publ/173-1-0-4640.
http://www.wormsdirectuk.co.uk/acatalog/maggots.html.
http://www.wpwa.org/documents/education/Biological%20sampling.pdf.
http://www.zin.ru/animalia/Coleoptera/rus/incoel.htm.
http://www.zoo.wroclaw.pl/index.php?lang=2.
http://www.zooeco.com/0-plant/0-plant22-8-5.html.
http://www.zoonewengland.org/franklin-park-zoo/exhibits/butterfly-landing.
http://xlegio.ru/sources/aeneas-tacticus/de-obsidione-toleranda.html.
http://ylik.ru/pcheliniy-vosk/izgotovlenie-svechej/#ixzz4500tbckK.
http://zipcodezoo.com/Animals/P/Polistes_annularis/.
http://zoogalaktika.ru/photos/invertebrata/arthropoda/arachnida/solifugae.
http://zooresurs.ru/horse/horse-zb/748-afrikanskaya-chuma-loshadej.html.
http://zooschool.ru/amfib/33_2.shtml.
https://accessuar.ru/articles/wild_silk/.
https://bioweb.uwlax.edu/bio203/s2012/grosshue_crai/diet.htm.
https://bronxzoo.com/exhibits/butterfly-garden.
https://butterflyhousemarinecove.org/.
https://butterflyplace-ma.com/.
https://butterflyworld.com/.
https://d3.ru/comments/329344/.
https://en.wikipedia.org/wiki/1989_California_medfly_attack.
https://en.wikipedia.org/wiki/2015%E2%80%9316_Zika_virus_epidemic.
https://en.wikipedia.org/wiki/African_horse_sickness.
https://en.wikipedia.org/wiki/Africanized_bee.
https://en.wikipedia.org/wiki/Al-Biruni.
https://en.wikipedia.org/wiki/Alexander_von_Humboldt.
https://en.wikipedia.org/wiki/Alternanthera_philoxeroides.
https://en.wikipedia.org/wiki/American_black_bear.
https://en.wikipedia.org/wiki/Amphibian#Feeding_and_diet.
https://en.wikipedia.org/wiki/Amphibian.
https://en.wikipedia.org/wiki/Anisota_senatoria.
https://en.wikipedia.org/wiki/Antbird.
https://en.wikipedia.org/wiki/Anteater.

https://en.wikipedia.org/wiki/Antheraea_pernyi.
https://en.wikipedia.org/wiki/Aristotle.
https://en.wikipedia.org/wiki/Armadillo.
https://en.wikipedia.org/wiki/Asian_black_bear.
https://en.wikipedia.org/wiki/Automeris_io.
https://en.wikipedia.org/wiki/Ballooning_(spider).
https://en.wikipedia.org/wiki/Ballooning_(spider).
https://en.wikipedia.org/wiki/Balsam_woolly_adelgid.
https://en.wikipedia.org/wiki/Baltic_amber.
https://en.wikipedia.org/wiki/Beekeeping.
https://en.wikipedia.org/wiki/Beeswax.
https://en.wikipedia.org/wiki/Bombardier_beetle.
https://en.wikipedia.org/wiki/Bombyx_mandarina.
https://en.wikipedia.org/wiki/Bumblebee.
https://en.wikipedia.org/wiki/Carpenter_ant.
https://en.wikipedia.org/wiki/Centaurea_diffusa.
https://en.wikipedia.org/wiki/Ceratitis_capitata.
https://en.wikipedia.org/wiki/Ceratopogonidae.
https://en.wikipedia.org/wiki/Chagas_disease.
https://en.wikipedia.org/wiki/Chikungunya.
https://en.wikipedia.org/wiki/Cockroach_racing.
https://en.wikipedia.org/wiki/Coprolite.
https://en.wikipedia.org/wiki/Cricket_fighting.
https://en.wikipedia.org/wiki/Crimean–Congo_hemorrhagic_fever.
https://en.wikipedia.org/wiki/Cucumber_mosaic_virus.
https://en.wikipedia.org/wiki/Cuevas_de_la_Ara%C3%B1a_en_Bicorp.
https://en.wikipedia.org/wiki/Cultural_entomology.
https://en.wikipedia.org/wiki/Cyrtobagous_salviniae.
https://en.wikipedia.org/wiki/Dead_Sea_Scrolls.
https://en.wikipedia.org/wiki/Dengue_fever.
https://en.wikipedia.org/wiki/Diamphotoxin.
https://en.wikipedia.org/wiki/Dominican_amber.
https://en.wikipedia.org/wiki/Dragonfly.
https://en.wikipedia.org/wiki/Elizabeth_B%C3%A1thory.
https://en.wikipedia.org/wiki/Elm_yellows.
https://en.wikipedia.org/wiki/Embalming.
https://en.wikipedia.org/wiki/Emerald_ash_borer.
https://en.wikipedia.org/wiki/Entomophagy.
https://en.wikipedia.org/wiki/Epidemic_typhus.
https://en.wikipedia.org/wiki/European_hedgehog.
https://en.wikipedia.org/wiki/Eye#Types_of_eye.
https://en.wikipedia.org/wiki/Filariasis.
https://en.wikipedia.org/wiki/Fire_blight.
https://en.wikipedia.org/wiki/Flea_circus.
https://en.wikipedia.org/wiki/Formica_rufa.
https://en.wikipedia.org/wiki/Francesco_Redi.
https://en.wikipedia.org/wiki/Gonimbrasia_belina.
https://en.wikipedia.org/wiki/Gonometa_postica.
https://en.wikipedia.org/wiki/G-suit.
https://en.wikipedia.org/wiki/History_of_silk.
https://en.wikipedia.org/wiki/Honey_hunting.
https://en.wikipedia.org/wiki/Honeymoon.
https://en.wikipedia.org/wiki/Horse-fly.
https://en.wikipedia.org/wiki/House_dust_mite).
https://en.wikipedia.org/wiki/Iberian_worm_lizard.
https://en.wikipedia.org/wiki/Initiation.
https://en.wikipedia.org/wiki/Insect_biodiversity.
https://en.wikipedia.org/wiki/Insect_collecting.
https://en.wikipedia.org/wiki/IUCN_Red_List.
https://en.wikipedia.org/wiki/Japanese_beetle.
https://en.wikipedia.org/wiki/Jared_Gold.
https://en.wikipedia.org/wiki/Joseph_Buford_Cox.

https://en.wikipedia.org/wiki/Kajiki,_Kagoshima.
https://en.wikipedia.org/wiki/Kashrut.
https://en.wikipedia.org/wiki/L._L._Langstroth.
https://en.wikipedia.org/wiki/La_Brea_Tar_Pits.
https://en.wikipedia.org/wiki/La_Crosse_encephalitis.
https://en.wikipedia.org/wiki/Lantana_camara.
https://en.wikipedia.org/wiki/Leishmaniasis.
https://en.wikipedia.org/wiki/Leo_Africanus.
https://en.wikipedia.org/wiki/Lepidopterism.
https://en.wikipedia.org/wiki/Li_Shizhen.
https://en.wikipedia.org/wiki/List_of_extinct_and_endangered_species_of_Lithuania.
https://en.wikipedia.org/wiki/List_of_honeydew_sources.
https://en.wikipedia.org/wiki/List_of_insects_in_the_Red_Data_Book_of_Russia.
https://en.wikipedia.org/wiki/List_of_medically_significant_spider_bites.
https://en.wikipedia.org/wiki/List_of_national_parks_of_Armenia.
https://en.wikipedia.org/wiki/List_of_protected_areas_of_Armenia.
https://en.wikipedia.org/wiki/List_of_types_of_amber.
https://en.wikipedia.org/wiki/London_Butterfly_House.
https://en.wikipedia.org/wiki/Louse.
https://en.wikipedia.org/wiki/Lyme_disease.
https://en.wikipedia.org/wiki/Mar%C3%ADa_Fernanda_Cardoso.
https://en.wikipedia.org/wiki/Maw%C3%A9_people.
https://en.wikipedia.org/wiki/Mead.
https://en.wikipedia.org/wiki/Mikhail_Lomonosov.
https://en.wikipedia.org/wiki/Mites_of_livestock.
https://en.wikipedia.org/wiki/Monarch_butterfly.
https://en.wikipedia.org/wiki/Monarch_Butterfly_Biosphere_Reserve.
https://en.wikipedia.org/wiki/Mongol_military_tactics_and_organization.
https://en.wikipedia.org/wiki/Montreal_Insectarium.
https://en.wikipedia.org/wiki/Mosquito.
https://en.wikipedia.org/wiki/Mound-building_termites.
https://en.wikipedia.org/wiki/Mountain_pine_beetle.
https://en.wikipedia.org/wiki/Murine_typhus.
https://en.wikipedia.org/wiki/Museum_Witt.
https://en.wikipedia.org/wiki/Myrmecophagy.
https://en.wikipedia.org/wiki/Nectar.
https://en.wikipedia.org/wiki/New_Delhi_metallo-beta-lactamase_1.
https://en.wikipedia.org/wiki/Numbat.
https://en.wikipedia.org/wiki/Oligonychus_ununguis.
https://en.wikipedia.org/wiki/Onchocerciasis.
https://en.wikipedia.org/wiki/Otto_Staudinger.
https://en.wikipedia.org/wiki/Paraponera_clavata.
https://en.wikipedia.org/wiki/Parnassius_autocrator.
https://en.wikipedia.org/wiki/Petro_Prokopovych.
https://en.wikipedia.org/wiki/Phylloxera.
https://en.wikipedia.org/wiki/Plague_(disease).
https://en.wikipedia.org/wiki/Polish_cochineal.
https://en.wikipedia.org/wiki/Pollination.
https://en.wikipedia.org/wiki/Polposipus_herculeanus.
https://en.wikipedia.org/wiki/Potato_virus_X.
https://en.wikipedia.org/wiki/Pterostigma.
https://en.wikipedia.org/wiki/Red_Data_Book_of_the_Russian_Federation.
https://en.wikipedia.org/wiki/Red_imported_fire_ant.
https://en.wikipedia.org/wiki/Resilin.
https://en.wikipedia.org/wiki/Rhex.
https://en.wikipedia.org/wiki/Saga_pedo.
https://en.wikipedia.org/wiki/Salvinia_molesta.
https://en.wikipedia.org/wiki/Samia_cynthia.
https://en.wikipedia.org/wiki/Sandfly.
https://en.wikipedia.org/wiki/Scaphism.
https://en.wikipedia.org/wiki/Schmallenberg_virus.
https://en.wikipedia.org/wiki/Second_plague_pandemic.

https://en.wikipedia.org/wiki/Shellac.
https://en.wikipedia.org/wiki/Shrew.
https://en.wikipedia.org/wiki/Shrilk.
https://en.wikipedia.org/wiki/Silphidae.
https://en.wikipedia.org/wiki/Sloth_bear.
https://en.wikipedia.org/wiki/Smuggling_of_silkworm_eggs_into_the_Byzantine_Empire.
https://en.wikipedia.org/wiki/Spider_silk.
https://en.wikipedia.org/wiki/Spider_web.
https://en.wikipedia.org/wiki/Spinneret_(spider).
https://en.wikipedia.org/wiki/Sucking_louse.
https://en.wikipedia.org/wiki/Sun_bear.
https://en.wikipedia.org/wiki/Termite.
https://en.wikipedia.org/wiki/The_Food_Defect_Action_Levels.
https://en.wikipedia.org/wiki/Thorny_dragon#Diet.
https://en.wikipedia.org/wiki/Tick-borne_disease.
https://en.wikipedia.org/wiki/Timeline_of_the_evolutionary_history_of_life.
https://en.wikipedia.org/wiki/Trench_fever.
https://en.wikipedia.org/wiki/Tsetse_fly.
https://en.wikipedia.org/wiki/Tunga_penetrans.
https://en.wikipedia.org/wiki/Unit_731.
https://en.wikipedia.org/wiki/Walter_Rothschild,_2nd_Baron_Rothschild.
https://en.wikipedia.org/wiki/Wax_foundation.
https://en.wikipedia.org/wiki/Westminster_Abbey.
https://en.wikipedia.org/wiki/Wilbur_R._Franks.
https://en.wikipedia.org/wiki/Wild_silk.
https://en.wikipedia.org/wiki/Yellow_fever.
https://otvet.mail.ru/question/14358814
https://ru.wikipedia.org/wiki/Hymenaea_protera.
https://ru.wikipedia.org/wiki/Аргасовые_клещи.
https://ru.wikipedia.org/wiki/Иксодовые_клещи.
https://ru.wikipedia.org/wiki/Инсектная_аллергия.
https://ru.wikipedia.org/wiki/История_шёлка.
https://ru.wikipedia.org/wiki/Мошки.
https://ru.wikipedia.org/wiki/Народная_медицина.
https://ru.wikipedia.org/wiki/Слепни.
https://ru.wikipedia.org/wiki/Сыпной_тиф.
https://ru.wikipedia.org/wiki/Энтомологическое_оборудование.
https://ru.wikipedia.org/wiki/Японская_дубовая_павлиноглазка.
https://tidcf.nrcan.gc.ca/en/insects/factsheet/42.
https://www.amentsoc.org/events/exhibitions.html.
https://www.amentsoc.org/insects/conservation/habitat.html.
https://www.birminghamzoo.com/.
https://www.butterflies.org/.
https://www.butterfliesandthings.com.
https://www.catseyepest.com/pest-library/pantry-pests/beetles/hide-beetle.
https://www.catseyepest.com/pest-library/pantry-pests/beetles/hide-beetle.
https://www.cites.org/eng.
https://www.cites.org/sites/default/files/eng/disc/species_02.10.2013.pdf.
https://www.dbg.org/butterfly-exhibit.
https://www.diergaardeblijdorp.nl/en/.
https://www.fatsecret.com/calories-nutrition/usda/honey?portionid=45776&portionamount=1.000.
https://www.flickr.com/photos/squatbetty/6244800448/.
https://www.insectframe.co.uk/.
https://www.k-state.edu/butterfly/.
https://www.mindat.org/show.php?id=188&ld=1#themap.
https://www.mpm.edu/plan-visit/exhibitions/permanent-exhibits/first-floor-exhibits/puelicher-butterfly-wing.
https://www.nwf.org/Wildlife/Wildlife-Library/Mammals/Nine-Banded-Armadillo.aspx.
https://www.pacificsciencecenter.org/exhibits/tropical-butterfly-house/.
https://www.sixflags.com/discoverykingdom/attractions/butterfly-habitat.
https://www.stlzoo.org/visit/thingstoseeanddo/discoverycorner/insectarium/.
https://www.tripadvisor.ru/ShowUserReviews-g186338-d211599-r318858741-Golders_Hill_Park-London_England.html.
https://www.zin.ru/animalia/coleoptera/rus/anoglamd.htm.

https://www.zin.ru/animalia/coleoptera/rus/incobu_.htm.

https://www.zin.ru/Animalia/Siphonaptera/distr.htm.

Huarcaya, E., C. Maguiña, R. Torres, J. Rupay and L. Fuentes. 2004. Bartonelosis (Carrion's Disease) in the pediatric population of Peru: an overview and update. Brazilian Journal of Infectious Diseases 8(5). http://dx.doi.org/10.1590/S1413-86702004000500001.

Huelsman, M.F., J. Kovach, J. Jasinski, C. Young and B. Eisley. 2002. Multicolored Asian lady beetle (*Harmonia axyridis*) as a nuisance pest in households in Ohio. pp. 243–250. *In*: Jones, S.C., J. Zhai and W.H. Robinson (eds.). Proceedings of 4th International Conference on Urban Pests.

Huis, A. van. 2013. Potential of insects as food and feed in assuring food security. Annual Review of Entomology 58: 563–583.

Huis, A. van, J. van Itterbeeck, H. Klunder, E. Mertens, A. Halloran, G. Muir and P. Vantomme. 2013. Edible insects: future prospects for food and feed security. FAO Forestry Paper 171. Rome, Food and Agriculture Organization of the United Nations. 201 pp.

Huis, A. van, H. van Gurp and M. Dicke. 2015. The Insect Cookbook. Food for a Sustainable Planet. 216 pp.

Humphries, C.J. and L.R. Parenti. 1999. Cladistic Biogeography: Interpreting Patterns of Plant and Animal Distributions, ed. 2. Oxford, Oxford University Press.

Hung, S.-W. and T.-L. Wong. 2004. Arachnid envenomation in Taiwan. Annals of Disaster Medicine 1: 12–17.

Hunt, R.S. 2002. Forest losses to pest insects/mites and plant pathogens. pp. 298–300. *In*: Pimentel, D. (ed.). Encyclopedia of Pest Management. New York, Marcel Dekker, Inc.

Hunter, C.D. 1997. Suppliers of beneficial organisms in North America. California Environmental Protection Agency. http://www.cdpr.ca.gov/docs/pestmgt/ipminov/bensup.pdf.

Hussey, S. 2010. The complete guide for photographing live insects at home. http://www.diyphotography.net/the-comlete-guide-for-photographing-live-insects-at-home/.

Hutson, A.M. 1984. Keds, flat-flies and bat-flies. Diptera, Hippoboscidae and Nycteribiidae. *In*: Handbooks for the Identification of British Insects. Vol. 10, Part 7. Royal Entomological Society of London. 40 pp.

Hvenegaard, G., T.A. Delamere, R.H. Lemelin, K. Brager and A. Auger. 2012. Insect Festivals. Cambridge Books Online. Cambridge University Press. http://dx.doi.org/10.1017/CBO9781139003339.016.

Hwangbo, J., E.C. Hong, A. Jang, H.K. Kang, J.S. Oh, B.W. Kim and B.S. Park. 2009. Utilization of house fly-maggots, a feed supplement in the production of broiler chickens. Journal of Environmental Biology 30(4): 609–614.

Ibrahimkhalilova, I.V. 2006. Forensic entomology—its history and development. Pest management. No. 4. pp. 47–50. http://molbiol.ru/forums/index.php?act=Attach&type=post&id=71555 [in Russian].

Idge, M., M. Acici, F.A. Igde and S. Umur. 2009. Carpet Beetle Anthrenus Verbasci, Linnaeus 1767: A New Seasonal Indoor Allergen: Case Report in Turkiye Klinikleri Tip Bilimleri Dergisi 29(6): 1729–1731.

Ilyichev, V.D., B.V. Bocharov, A.A. Anisimov, M.V. Gorlenko and Nyuksha. 1987. Biodeterioration. Moscow: Vysshaya shkola. 352 pp. [in Russian].

Imamura, M., J. Nakai, S. Inoue, G.X. Quan, T. Kanda and T. Tamura. 2013. Targeted gene expression using the GAL4/UAS system in the silkworm *Bombyx mori*. Genetics 165: 1329–1340.

Ings, T.C., N.L. Ings, L. Chittka and P. Rasmont. 2010. A failed invasion? Commercially introduced pollinators in Southern France. Apidologie 41: 1–13.

Insects and Ticks of the Far East Being of Medico-veterinary Importance. 1987. Leningrad, Nauka. 311 pp. [in Russian].

Insects and Ticks—Crop Pests. Volume 3, Part 2. 1999. Leningrad, Nauka. 407 pp. [in Russian].

Inzhuvatova, M.V., N.O. Novikova, T.E. Vlasova and Y.B. Vasilyeva. 2016. Analysis of outbreaks of bluetongue in Russia. 9 pp. www.scienceforum.ru/2016/pdf/24134.pdf [in Russian].

Ioyrish, N.P. 1974. Bees for humanity. Moscow, Nauka, 1974. 182 pp. [in Russian].

Ioyrish, N.P. 1976. Bee products and their use. Moscow, Rosselkhozizdat. 175 pp. [in Russian].

Irwin, M.E. and G.E. Kampmeier. 2003. Commercial products, from insects. pp. 251–260. *In*: Resh, V.H. and R. Carde (eds.). Encyclopedia of Insects. San Diego, Academic Press.

Isbister, G.K. and M.R. Gray. 2003. White-tail spider bite: a prospective study of 130 definite bites by Lampona species. Medical Journal of Australia 179: 199–202.

Iwai, T., K. Ito, T. Ohta, T. Mizushige, T. Kishida, C. Miura and T. Miura. 2015. Dietary effects of housefly (*Musca domestica*) (Diptera: Muscidae) pupae on the growth performance and the resistance against bacterial pathogen in red sea bream (*Pagrus major*) (Perciformes: Sparidae). Applied Entomology and Zoology 50: 213–221.

Izhevsky, S.S. 1990. Alien pests of plants in the Soviet Union. Plant Protection 8: 30–32 [in Russian].

Izhevsky, S.S., A.L. Lobanov and A.Y. Sosnin. 2014. Life of wonderful bugs. Moscow, Kodeks. 368 pp. [in Russian].

Izmailov, I. 2005. Venomous evolution. Round the World 10: 63–65 [in Russian].

Jacquemin, S.J., M. Pyron, M. Allen and L. Etchison. 2014. Wabash River freshwater drum Aplodinotus grunniens diet: effects of body size, sex and river gradient. Journal of Fish and Wildlife Management 5: 133–140.

Jahan, S., A. Mohammed Al Saigul and S. Abdul Rahim Hamed. 2007. Scorpion stings in Qassim, Saudi Arabia—a 5-year surveillance report. Toxicon 50(2): 302–305.

Jakubas-Zawalska, J., M. Asman, M. Klys and K. Solarza. 2016. Sensitization to *Sitophilus granarius* in selected suburban population of South Poland. Journal of Stored Products Research 69: 1–6.

James, W.D., T. Berger and D. Elston. 2016. Andrews' Diseases of the Skin: Clinical Dermatology. 12th ed. Saunders Elsevier. 968 pp.

Jarisch, R. and W. Hemmer. 2010. Insect allergy: house fly, mosquito, horsefly. Allergologie 33(2): 86–92 [in German].

Jaroszewicz, A., J. Sasiadek and K. Sibilski. 2013. Modeling and simulation of flapping wings entomopter in martian atmosphere. pp. 143–162. *In*: Aerospace Robotics. Springer-verlag, Berlin.

Jarrar, B.M. and M.A. Al-Rowaily. 2008. Histology and histochemistry of the venom apparatus of the black scorpion *Androctonus crassicauda* (Olivier, 1807) (scorpiones: Buthidae). Journal of Venomous Animals and Toxins including Tropical Diseases 14(3): 514–526.

Jelinek, G.A. 1997. Widow spider envenomation (latrodectism): A worldwide problem. Wilderness and Environmental Medicine 8: 226–231.

Jijakli, M.H. 2010. European market of biological control agents: actual situation and perspectives. https://orbi.ulg.ac.be/bitstream/2268/59851/1/N%C2%B0416%20-%20jijakli%202010.doc.

Jin, X.-B. 2011. Chinese cricket culture. http://www.insects.org/ced3/chinese_crcul.html.

Johansson, B., A. Eriksson and L. Örnehult. 1991. Human fatalities caused by wasp and bee stings in Sweden. International Journal of Legal Medicine 104(2): 99–103.

Johansson, L.C., S. Engel, E. Baird, M. Dacke, F.T. Muijres and A. Hedenstrom. 2012. Elytra boost lift, but reduce aerodynamic efficiency in flying beetles. Journal of the Royal Society Interface 9: 2745–2748. DOI: 10.1098/rsif.2012.0053.

Johnson, M.W. 2000a. History of biological control. History and development of biological control. *In*: Biological Control of Pests. Lecture Notes for ENTO 675. University of Hawaii at Manoa, Fall 2000. pp. 6–11. http://ucanr.edu/sites/W2185/files/109429.pdf.

Johnson, M.W. 2000b. Biological control case histories. Analysis of BC successes and case histories. *In*: Biological Control of Pests. Lecture Notes for ENTO 675. University of Hawaii at Manoa, Fall 2000. pp. 123–128. http://nature.berkeley.edu/biocon/BC%20Class%20Notes/123-128%20Case%20Histories.pdf.

Johny, S., C.E. Lange, L.F. Solter, A. Merisko and D.W. Whitman. 2007. New insect system for testing antibiotics. Journal of Parasitology 93: 1505–1511.

Jones, J., H. Rothfuss, H. Steinkraus and R. Lewis. 2010. Transgenic goats producing spider silk protein in their milk; behavior, protein purification and obstacles. Transgenic Research 19(1): 135.

Jongema, Y. 2012. List of edible insect species of the world. Wageningen, Laboratory of Entomology, Wageningen University. (available at www.ent.wur.nl/UK/Edible+insects/Worldwide+species+list/).

Jongema, Y. 2014. List of edible insects of the world. Wageningen University, Wageningen, the Netherlands. http://tinyurl.com/mestm6p.

Jongema, Y. 2015. List of edible insects of the world (June 1, 2015). http://www.wageningenur.nl/en/Expertise-Services/Chair-groups/Plant-Sciences/Laboratory-of-Entomology/Edible-insects/Worldwide-species-list.htm.

Jorgensen, S.E. 2010. Ecosystem services, sustainability, and thermodynamic indicators. Ecological Complexity 7: 311–313.

Joseph, I., D.G. Mathew, P. Sathyan and G. Vargheese. 2011. The use of insects in forensic investigations: An overview on the scope of forensic entomology. Journal of Forensic Dental Sciences 3(2): 89–91. DOI: 10.4103/0975-1475.92154.

Jouquet, P., S. Traoré, C. Choosai, C. Hartmann and D. Bignell. 2011. Influence of termites on ecosystem functioning. Ecosystem services provided by termites. European Journal of Soil Biology 47: 215–222.

Jun, M. 2008. The future use of insects as human food. pp. 115–122. *In*: Durst, P.B., D.V. Johnson, R.N. Leslie and K. Shono (eds.). Forest Insects as Food: Humans Bite Back. Proceedings of a Workshop on Asia-Pacific Resources and Their Potential for Development, 19–21 February 2008, Chiang Mai, Thailand. Bangkok: Food and Agriculture Organization of the United Nations.

Jung, C., S. Aryal and R. Thapa. 2015. An overview of beekeeping economy and its constraints in Nepal. Journal of Apiculture 30(3): 135–142.

Jurga, F. and S. Morrison. 2009. Maggot debridement therapy. alternative therapy for hoof infection and necrosis. Hoofcare and Lameness Magazine 78: 28–31. http://www.hoofcare.com/article_pdf/hoofcaremaggotsmorrison.pdf.

Kaabak, L.V. and A.V. Sochivko. 2012. Butterflies of the world. Moscow, Avanta. 184 pp. [in Russian].

Kadoya, T., S. Suda and I. Washitani. 2004. Dragonfly species richness on man-made ponds: Effects on ponds size and pond age on newly established assemblages. Ecological Research 19(5): 461–467.

Kagen, S.L. 1990. Inhalant allergy to arthropods. Insects, arachnids, and crustaceans. Clinical Review of Allergy 8: 99–125.

Kakharov, K.K. 2008. Bioecological features of the Colorado potato beetle (*Leptinotarsa decemlineata*, Say) and measures of combating it in the context of Tajikistan. Abstract of dissertation of Doctor of Agricultural Sciences. St. Petersburg. 40 pp. [in Russian].

Kalugina, N.S. 1980. Insects in aquatic ecosystems of the past. pp. 224–240. *In*: Historical Development of Insects. Moscow, Nauka (Proceedings of Paleontological Institute, Vol. 175) [in Russian].

Kappico, J.T., A. Suzuki and N. Hongu. 2012. Is Honey the Same as Sugar? University of Arizona. 4 pp. http://extension.arizona.edu/sites/extension.arizona.edu/files/pubs/az1577.pdf.

Karakaya, G., E. Celebioglu, A.U. Demir and A.F. Kalyoncu. 2011. The analysis of Hymenoptera hypersensitive patients in Ankara, Turkey. Allergol Immunopathol (Madr). DOI: 10.1016/j.aller.2010.11.002.

Karakaya, G., E. Celebioglu, A.U. Demir and A.F. Kalyoncu. 2012. The analysis of Hymenoptera hypersensitive patients in Ankara, Turkey. Allergologia et Immunopathologia 40(1): 9–13. DOI: 10.1016/j.aller.2010.11.002.

Kearns, C.A. and D.M. Oliveras. 2009. Environmental factors affecting bee diversity in urban and remote grassland plots in Boulder, Colorado. Journal of Insect Conservation 13: 655–665. DOI: 10.1007/s10841-009-9215-4.

Keeling, C.I., H. Henderson, M. Li, M. Yuen, E.L. Clark, J.D. Fraser, D.P. Huber, N.Y. Liao, T.R. Docking, I. Birol, S.K. Chan, G.A. Taylor, D. Palmquist, S.J. Jones and J. Bohlmann. 2012. Transcriptome and full-length cDNA resources for the mountain pine beetle, *Dendroctonus ponderosae* Hopkins, a major insect pest of pine forests. Insect Biochemistry and Molecular Biology 42(8): 525–536. DOI: 10.1016/j.ibmb.2012.03.010.

Kegler, H. and W. Hartman. 1998. Present status of controlling conventional strains of plum pox virus. pp. 616–628. *In*: Hadidi, A., R.K. Khetarpal and H. Koganezawa (eds.). Plant Virus Disease Control. St. Paul, MN, APS Press.

Kelemu, S., S. Niassy, B. Torto, K. Fiaboe, H. Affognon, H. Tonnang, N.K. Maniania and S. Ekesi. 2015. African edible insects for food and feed: inventory, diversity, commonalities and contribution to food security. Journal of Insects as Food and Feed 1(2): 103–119. DOI: 10.3920/JIFF2014.0016 103.

Kelly, C.D. and B.T. Tawes. 2013. Sex-specific effect of juvenile diet on adult disease resistance in a field cricket. Public Library of Science One 8: e61301.

Kelly, R. 2007. Spider bites: Fact or fiction? Georgia Epidemiology Report 23(3): 17–22.

Kenis, M., M.A. Auger-Rozenberg, A. Roques, L. Timms, C. Pere, M.J.W. Cock, J. Settele, S. Augustin and C. Lopez-Vaamonde. 2009. Ecological effects of invasive alien insects. Biol. Invasions 11: 21–45.

Khatukhov, A.M. 2015. Grasshopper *Bradyporus multituberculatus* (Fischer – Waldheim, 1833) in the Kabardino-Balkar Republic. Modern Problems of Science and Education 2 (part 1). http://www.science-education.ru/ru/article/view?id=19036 [in Russian].

Khismatullina, N.Z. 2005. Apitherapy. Perm, Mobile, 296 pp. [in Russian].

Khodorych, A. 2002. The insect hunters. http://www.kommersant.ru/doc/313577 [in Russian].

Khorasani, S. 2013. Mid-late holocene environmental change in northern Sweden: an investigation using fossil insect remains. Thesis submitted in fulfilment of the degree of Doctor of Philosophy. School of GeoSciences, University of Edinburgh, 320 pp.

Khoruzhik, L.I., L.M. Susana and V.I. Parfenov. 2005. The Red Book of Belarus: Rare and threatened species of wild plants. Minsk, Belarussian Encyclopedia, 456 pp. [in Russian].

Kinyuru, J.N., G.M. Kenji, S.M. Njoroge and M. Ayieko. 2010. Effect of processing methods on the *in vitro* protein digestibility and vitamin content of edible winged termite (*Macrotermes subhylanus*) and grasshopper (*Ruspolia differens*). Food and Bioprocess Technology 3(5): 778–782.

Kireeva, N.A., T.R. Kabirov and T.C. Onegova. 2007. Bioremediation of oil-contaminated soils using cellulose-containing substrates. *In*: Biological Recultivation and Monitoring of Disturbed Lands. Yekaterinburg, Ural University Press, pp. 346–358 [in Russian].

Kiselev, S.V. 2005. Ecological aspects of the insect fauna of industrial zones the city of Tula. Dissertation of candidate of Doctor of Biological Sciences. Tula, 178 pp. [in Russian].

Klein, A.M., B.E. Vaissiere, J.H. Cane, I. Steffan-Dewenter, S.A. Cunningham, C. Kremen and T. Tscharntke. 2007. Importance of pollinators in changing landscapes for world crops. Proceedings of the Royal Society B Biological Sciences 274(1608): 303–313.

Klein, D. 2008. Identifying museum insect pest damage. Conserve O Gram 3(11): 1–7. https://www.nps.gov/museum/publications/conserveogram/03-11.pdf.

Klotz, J.H., R.D. de Shazo, J.L. Pinnas, A.M. Frishman, J.O. Schmidt, D.R. Suiter, G.W. Price and S.A. Klotz. 2005. Adverse reactions to ants other than imported fire ants. Annals of Allergy, Asthma and Immunology 95: 418–425.

Klotz, J.H., S.A. Klotz and J.L. Pinnas. 2009. Animal bites and stings with anaphylactic potential. The Journal of Emergency Medicine 36(2): 148–156.

Klotz, J.H., P.L. Dorn, J.L. Logan, L. Stevens, J.L. Pinnas, J.O. Schmidt and S.A. Klotz. 2010. "Kissing bugs": potential disease vectors and cause of anaphylaxis. Clinical Infectious Diseases 50(12): 1629–1634.

Knight, K. 2014. Parasitic fig wasps bore with zinc-tipped drill bit. Journal of Experimental Biology 217: 1833. DOI: 10.1242/jeb.107920.

Kocarek, P. 2003. Decomposition and coleoptera succession on exposed carrion of small mammal in Opava, the Czech Republic. European Journal of Soil Biology 39(1): 31–45.

Koch, R.L. 2003. The multicolored Asian lady beetle, *Harmonia axyridis*: A review of its biology, uses in biological control, and non-target impacts. Journal of Insect Science 3: 32. http://www.ncbi.nlm.nih.gov/pmc/articles/PMC524671/.

Kolle, M., P.M. Salgard-Cunha, M.R.J. Scherer, F. Huang, P. Vukusic, S. Mahajan, J.J. Baumberg and U Steiner. 2010. Mimicking the colourful wing scale structure of the *Papilio blumei* butterfly. Nature Nanotechnology 5: 511–515. DOI: 10.1038/nnano.2010.101.

Komarova, L.A. 2008. General entomology. Biysk: Biysk Pedagogical University. 60 pp. [in Russian].

Kondo, T. and P.J. Gullan. 2007. Taxonomic review of the lac insect genus Paratachardina Balachowsky (Hemiptera: Coccoidea: Kerriidae), with a revised key to genera of Kerriidae and description of two new species. Zootaxa 1617: 1–41. http://www.mapress.com/zootaxa/2007f/z01617p041f.pdf.

Korasaki, V., J. Lopes, G.G. Brown and J. Louzada. 2013. Using dung beetles to evaluate the effects of urbanization on Atlantic Forest biodiversity. Insect Science 20(3): 393–406.

Koryakov, D.E. and I.F. Zhimulev. 2009. Chromosomes: Structure and function. Novosibirsk: SB RAN. 258 pp. [in Russian].

Korzeniewski, K., D. Juszczak and E. Zwolinska. 2016. Zika—another threat on the epidemiological map of the world. International Maritime Health 67(1): 31–37. DOI: 10.5603/imh.2016.0007.

Korzh, V.N. 2010. A complete directory of the beekeeper. Kharkov, 416 pp. [in Russian].

Korzunovich, P.V. 2005. Wasp Ephialtes (Hymenoptera: Ichneumonidae)—parasite of longhorn and jewel beetles. http://www.zin.ru/ANIMALIA/coleoptera/rus/ephialt1.htm [in Russian].

Koshkin, E.S. and V.G. Bezborodov. 2013. First records of hawkmoth *Ambulyx tobii* (Inoue, 1976) (Lepidoptera, Sphingidae) from the southern part of Primorsky Krai, Russia. Euroasian Entomological Journal 12(4): 415–419 [in Russian with English summary].

Koshkin, E.S., V.G. Bezborodov, A.A. Voronkov, A.V. Korshunov, A.E. Kostyunin and K.M. Prokopenko. 2015. Distribution of the hawk moths *Clanis undulosa* Moore, 1879 and *Ambulyx tobii* (Inoue, 1976) (Lepidoptera, Sphingidae) in Russia. Far Eastern Entomologist 302: 14–17.

Kovalev, I.S. 2008. The functional role of the hollow region of the butterfly *Pyrameis atalanta* (L.) scale. Journal of Bionic Engineering 5(3): 224–230.

Kozak, M.F. 2007. Drosophila—the Model Object for Genetics: Educational-Methodical Manual. Astrakhan, Publishing House Astrakhan University, 87 pp. [in Russian].

Kozin, E.K. 2013. Massive decays of coniferous forests as a natural stage of age development. Bulletin of the Botanical Garden Institute FEB RAS 10: 4–14 [in Russian].

Kozlov, M.V. 1990. Influence of anthropogenic factors on populations of terrestrial insects. *In*: Results of Science and Technology: VINITI. Entomology Series. Vol. 13, 166 pp. [in Russian].

Krichevsky, G.E. 2011. Nano-, bio- and chemical technologies in the production of a new generation of fibers, textiles and clothing. Moscow, Publishing House Izvestia, 528 pp. [in Russian].

Krikken, J. and J. Huijbregts. 2001. Insects as forensic informants: the Dutch experience and procedure. Proceedings of the Section Experimental and Applied Entomology of the Netherlands Entomological Society (N.E.V.) Amsterdam 12: 159–164.

Kritsky, G. and D. Mader. 2010. Leonardo's insects. American Entomologist 56(3): 178–184.

Kritsky, G. and D. Mader. 2011. The insects of Pieter Bruegel the Elder. American Entomologist 57: 245–251.

Kritsky, G., D. Mader and J.J. Smith. 2013. Surreal entomology: The insect imagery of Salvador Dali. American Entomologist 59(1): 28–37.

Krivolutsky, D.A. and E.A. Sidorchuk. 2005. The oribatid mites as bioindicators of the holocene paleogeographical conditions of the Northern European Russia. Bulletin of the Russian Academy of Sciences, Geographical Series, No. 1: 60–67 [in Russian].

Krivtsov, N.I. and V.I. Lebedev. 1993. The receipt and use of bee products. Moscow, Niva Russia, 288 pp. [in Russian].

Kruglikova, A.A. and S.I. Chernysh. 2013. Surgical maggots and the history of their medical use. Entomological Review 93(6): 667–674. ISSN 0013-8738.

Krutov, V.I. and I.I. Minkevich. 2002. Fungal Disease of the Wood Species. Karelian Scientific Center of Russian Academy of Sciences, Petrozavodsk, 196 pp. [in Russian].

Kukharenko, E.A. 2013. Beetles—living jewels [in Russian]. http://www.zin.ru/animalia/coleoptera/rus/livejew.htm.

Kumschick, S., A. Devenish, M. Kenis, W. Rabitsch, D.M. Richardson and J.R.U. Wilson. 2016. Intentionally Introduced Terrestrial Invertebrates: patterns, risks, and options for management. Biological Invasions 18(4): 1077–1088. DOI: 10.1007/s10530-016-1086-5.

Kundanati, L. and N. Gundiah. 2014. Biomechanics of substrate boring by fig wasps. Journal of Experimental Biology 217: 1946–1954. DOI: 10.1242/jeb.098228.

Kuno, G. 2012. Revisiting Houston and Memphis: the background histories behind the discovery of the infestation by *Aedes albopictus* (Diptera: Culicidae) in the United States and their significance in the contemporary research. Journal of Medical Entomology 49(6): 1163–1176.

Kuz'michev, E.P., E.S. Sokolova and E.G. Mozolevskaya. 2004. Diseases of woody plants: A handbook [Pests and Diseases in the Russian Forests. Vol. 1]. Moscow, VNIILM, 120 pp. [in Russian].

Kuzmina, S.A. 2003. Quaternary insects. http://www.zin.ru/animalia/COLEOPTERA/rus/kuzmina1.htm [in Russian].

Kuznetsov, V.N. 1997. On a problem of gipsy moth (Lymantria dispar L.) and Siberian moth (Dendrolimus superans Bult.) in Promorsky Krai. Vestnik of Far-Eastern Branch of RAS 3: 24–30 [in Russian].

Kuznetsov, V.N. 2005. Invasions of insects in terrestrial ecosystems of the Russian Far East. *In*: Kurentsov's A.I. (ed.). Annual Memorial Meetings, Vol. XVI: 91–97 [in Russian].

Kuznetsov, V.N. and S.Y. Storozhenko. 2010. Invasions of insects in terrestrial ecosystems of the Russian Far East. Russian Journal of Biological Invasions 3(1): 12–18 [in Russian].

Lacour, G., L. Chanaud, G. L'Ambert and T. Hance. 2015. Seasonal synchronization of diapause phases in *Aedes albopictus* (Diptera: Culicidae). Public Library of Science One 10(12): e0145311. DOI: 10.1371/journal.pone.0145311.

Łagowska, B. and K. Golan. 2009. Scale insects (Hemiptera, Coccoidea) as a source of natural dye and other useful substances. Aphids and Other Hemipterous Insects 15: 151–167. http://www.kul.pl/files/658/aphids15/11lagowskagolan.pdf.

Langley, R. 1999. Physical hazards of animal handlers. Occup. Med. 14: 181–193.

Langley, R.L. 2005. Animal-related fatalities in the United States—an update. Wilderness Environmental Medicine 16(2): 67–74.

Lambrinos, D., R. Möller, T. Labhart, R. Pfeifer and R. Wehner. 2000. A mobile robot employing insect strategies for navigation. Robotics and Autonomous Systems 30: 39–64.

Lattorff, H.M.G., J. Buchholz, I. Fries and R.F.A. Moritz. 2015. A selective sweep in a *Varroa destructor* resistant honeybee (*Apis mellifera*) population. Infection Genetics and Evolution 31: 169–176. DOI: 10.1016/j.meegid.2015.01.025.

Laustsen, A.H., M. Solà, E.C. Jappe, S. Oscoz, L.P. Lauridsen and M. Engmark. 2016. Biotechnological trends in spider and scorpion antivenom development. Toxins 8: 226. DOI: 10.3390/toxins8080226.

Le Roux, P. 2009. Butterflies. An Innovative Guide to the Use of Butterfly Remedies in Homeopathy. Narayan Verlag, 138 pp.

Lee, C.-Y. 2002. Subterranean termite pests and their control in the urban environment in Malaysia. Sociobiology 40: 3–9.

Lee, C.-Y., C. Vongkaluang and M. Lenz. 2007. Challenges to subterranean termite management of multi-genera faunas in Southeast Asia and Australia. Sociobiology 50: 213–221.

Lee, H., S. Halverson and R. Mackey. 2016. Insect allergy. Primary Care 43(3): 417–431.

Lee, H.-Y., S.-H. Lee and K.-J. Min. 2015. Insects as a model system for aging studies. Entomological Research 45: 1–8.

Lee, J.-D., H.-J. Park, Y. Chae and S. Lim. 2005. An overview of bee venom acupuncture in the treatment of arthritis. Evidence-Based Complementary and Alternative Medicine 2(1): 79–84. DOI: 10.1093/ecam/neh070.

Lee, S.H. and K.J. Min. 2013. Caloric restriction and its mimetics. Biochemistry and Molecular Biology Reports 46: 181–187.

Lee, Y.-H., W. San-an and S.-J. Suh. 2012. Notes on the Indian wax scale, *Ceroplastes ceriferus* (Fabricius), from Korea (Hemiptera: Coccidae). Korean Journal of Applied Entomology 51(2): 157–162. DOI: http://dx.doi.org/10.5656/KSAE.2012.04.0.24.

Leeflang, M.M.G., C.W. Ang, J. Berkhout, H.A. Bijlmer, W. Van Bortel, A.H. Brandenburg, N.D. Van Burgel, A.P. Van Dam, R.B. Dessau, V. Fingerle, J.W.R. Hovius, B. Jaulhac, B. Meijer, W. Van Pelt, J.F.P. Schellekens, R. Spijker, F.F. Stelma, G. Stanek, F. Verduyn-Lunel, H. Zeller and H. Sprong. 2016. The diagnostic accuracy of serological tests for Lyme borreliosis in Europe: a systematic review and meta-analysis. BMC Infectious Diseases 16: 140. DOI 10.1186/s12879-016-1468-4.

Leite, N.B., A. Aufderhorst-Roberts, M.S. Palma, S.D. Connell, J.R. Neto and P.A. Beales. 2015. PE and ps lipids synergistically enhance membrane poration by a peptide with anticancer properties. Biophysical Journal 109(5): 936–947. DOI: http://dx.doi.org/10.1016/j.bpj.2015.07.033.

Lemdahl, G. 1990. Insect assemblages from an Iron Age settlement in the clay district of Butjadingen, NW Germany. International Quaternary Union Subcommission for the Study of the Holocene, Cultural Landscapes Meetings Abstracts, pp. 12–22.

Lemelin, R.H. 2009. Goodwill hunting: dragon hunters, dragonflies and leisure. Current Issues in Tourism 12(5): 553–571.

Lemelin, R.H. (ed.). 2012a. The Management of Insects in Recreation and Tourism. Cambridge University Press. Online. Hardback. DOI: http://dx.doi.org/10.1017/CBO9781139003339.

Lemelin, R.H. 2012b. Six-legged sojourns: insect-based recreation and tourism. Wings. Essays of Invertebrate Conservation, pp. 11–15.

Lemelin, R.H. and G.A. Fine. 2013. Leisure on the recreational fringe. Naturework and the place of amateur mycology and entomology. PAN: Philosophy, Activism, Nature 10: 77–86.

Lenteren, J.C. van, M.M. Roskam and R. Timmer. 1997. Commercial mass production and pricing of organisms for biological control of pests in Europe. Biological Control 10: 143–149. http://www.nhm.ac.uk/resources/research-curation/projects/chalcidoids/pdf_Y/LenterRoTi997.pdf.

Lenteren, J.C. van and V.H.P. Bueno. 2003. Augmentative biological control of arthropods in Latin America. BioControl 48: 123–139.

Lenteren, J.C. van, J. Bale, F. Bigler, H.M.T. Hokkanen and A.J.M. Loomans. 2006. Assessing risks of releasing exotic biological control agents of arthropod pests. Annual Review of Entomology 51: 609–634.

Lenteren, J.C. van (ed.). 2012. IOBC Internet Book of Biological Control, version 6. 182 pp. http://www.iobc-global.org/download/IOBC_InternetBookBiCoVersion6Spring2012.pdf.

Leonhardt, S.D., N. Gallai, L.A. Garibaldi, M. Kuhlmann and A.M. Klein. 2013. Economic gain, stability of pollination and bee diversity decrease from southern to northern Europe. Basic and Applied Ecology 14(6): 461–471. DOI: 10.1016/j.baae.2013.06.003.

Li, C., P.B. Umbanhowar, H. Komsuoglu and D.I. Goldman. 2011. Erratum to: The effect of limb kinematics on the speed of a legged robot on granular media. Experimental Mechanics 51: 1017. DOI 10.1007/s11340-011-9497-9.

Li, L. and K.L. Kiick. 2013. Resilin-based materials for biomedical applications. ACS Macro Letters 2(8): 635–640.

Liabzina, S.N. 2011a. Species composition and the structure of the complex of necrophagous arthropods of South Karelia. Scientific notes of Petrozavodsk State University. Biology 4: 10–18 [in Russian].

Liabzina, S.N. 2011b. Participation of ants (Formicidae, Hymenoptera) in the destruction of dead animals. Proceedings of the PSPU named V.G. Belinsky. No. 25, pp. 383–385 [in Russian].

Liabzina, S.N. 2013. Invertebrate necrobiont in the littoral zone of different lakes in Karelia. Biologia Vhutrennikh Vod. 2: 51–59 [in Russian].

Liabzina, S.N. and S.D. Uzenbaev. 2013. Ecology of the burying beetles (Coleoptera, Silphidae) in Karelia. Scientific Notes of Petrozavodsk State University. No. 2. Biology, pp. 27–32 [in Russian].

Liang, S.P., D.Y. Zhang, X. Pan, Q. Chen and P.A. Zhou. 1993. Properties and amino acid sequence of huwentoxin–I, a neurotoxin purified from the venom of the Chinese bird spider *Selenocosmia huwena*. Toxicon 31(8): 969–978.

Liao, L.-Y., M.-C. Chiu, Y.-S. Huang and M.-H. Kuo. 2012. Size-dependent foraging on aquatic and terrestrial prey by the endangered Taiwan salmon Oncorhynchus masou formosanus. Zoological Studies 51(5): 671–678.

Liebhold, A.M., K.W. Gottschalk, D.A. Mason and R.R. Bush. 1997. Forest susceptibility to the gypsy moth. Journal of Forestry 95(5): 20–24.

Liew, W.K., E. Williamson and M.L.K. Tang. 2009. Anaphylaxis fatalities and admissions in Australia. Journal of Allergy and Clinical Immunology 123(2): 434–442.

Ligi, M., P.P. Sengupta, G.R. Rudramurthy and H. Rahman. 2016. Flagellar antigen based CI-ELISA for sero-surveillance of surra. Veterinary Parasitology 219: 17–23. http://dx.doi.org/10.1016/j.vetpar.2016.01.019.

Likhanova, N.V. 2014. The role of tree waste in the litter layer formation in cutting areas of middle Taiga spruce forests. Lesnoy Journal 3: 52–66 [in Russian].

Lim, S.M. and S.H. Lee. 2015. Effectiveness of bee venom acupuncture in alleviating post-stroke shoulder pain: a systematic review and meta-analysis. Journal of Integrative Medicine 13(4): 241–247. DOI: 10.1016/S2095-4964(15)60178-9.

Lindeman, G.V. 1993. Relationship of xylophagous insects and deciduous trees in Arid conditions. Moscow, Nauka. 207 pp. [in Russian].

Lindgren, N., S. Bucheli, A. Archambeault and J. Bytheway. 2011. Exclusion of forensically important flies due to burying behavior by the red imported fire ant (Solenopsis invicta) in southeast Texas. Forensic Science International 204: 1–3.

Linnie, M.J. and M.J. Keatinge. 2000. Pest control in museums: toxicity of para-dichlorobenzene, 'Vapona', and naphthalene against all stages in the life-cycle of museum pests, *Dermestes maculatus* Degeer, and *Anthrenus verbasci* (L.) (Coleostidae). International Biodeterioration and Biodegradation 45 (1–2): 1–13. DOI: 10.1016/S0964-8305(00)00034-2.

Lipnitsky, S.S. and A.F. Piluy. 1991. Medicinal Poisons in Veterinary Medicine. Minsk, Uradzhay. 303 pp. [in Russian].

Litovitz, T.L., W. Klein-Schwartz, K.S. Dyer, M. Shannon, S. Lee and M. Powers. 1998. 1997 annual report of the American association of poison control centers toxic exposure surveillance system. American Journal of Emergency Medicine 6(5): 443–497.

Liu, C., J. Liu, L. Xu and W. Xiang. 2014. Recent achievements in bionic implementations of insect structure and functions. Kybernetes 43(2): 307–324. DOI: 10.1108/K-09-2013-0192.

Liu, N., Y.C. Li and R.Z. Zhang. 2012. Invasion of Colorado potato beetle, *Leptinotarsa decemlineata*, in China: dispersal, occurrence, and economic impact. Entomologia Experimentalis et Applicata 143(3): 207–217. DOI: 10.1111/j.1570-7458.2012.01259.x.

Liu, Q., J.K. Tomberlin, J.A. Brady, M.R. Sanford and Z. Yu. 2008. Black soldier fly (Diptera: Stratiomyidae) larvae reduce *Escherichia coli* in dairy manure. Eberle 37(6): 1525–1530.

Livingston, R.L. and L. Pederson. 2010. Management Guide for Balsam Woolly Adelgid *Adelges piceae* (Ratzeburg). 4 pp. http://www.fs.usda.gov/Internet/FSE_DOCUMENTS/stelprdb5187218.pdf.

Ljamin, M.J. and N.M. Pakhorukov. 2009. Biodiversity and Ecology of Invertebrates. Terrestrial fauna. Perm State University, Perm, Russia. 176 pp. [in Russian].

Lo Vecchio, F. and T.V. Tran. 2004. Allergic reactions from insect bites. American Journal of Emergency Medicine 22(7): 631.

Lochynska, M. 2010. History of sericulture in Poland. Journal of Natural Fibers 7(4): 334–337. DOI: 10.1080/15440478.2010.529320.

Łochyńska, M. 2015. The mulberry silkworm—a new source of bioactive proteins. Journal of Agricultural Science and Technology A 5: 639–645. DOI: 10.17265/2161-6256/2015.08.001.

Lockey, T.C. 2007. Imported fire ants. http://ipmworld.umn.edu/chapters/lockley.htm.

Lockwood, J.A. 2008. Six-legged Soldiers: Using Insects as Weapons of War. Oxford University Press, USA. 400 pp.

Lockwood, J.A. 2012. Insects as weapons of war, terror, and torture. Annual Review of Entomology 57: 205–227. DOI: 10.1146/annurev-ento-120710-100618.

Löser, Z. 2001. Exotic insects. Moscow, Aquarium LTD.192 pp. [in Russian].

Losey, J.E. and M. Vaughan. 2006. The economic value of ecological services provided by insects. Bioscience 56(4): 311–323. DOI: 10.1641/0006-3568(2006)56[311:TEVOES]2.0.CO;2.

Lotfy, W.M. 2015. Plague in Egypt: Disease biology, history and contemporary analysis: A minireview. Journal of Advanced Research 6(4): 549–554.

Lourenço, W.R., J.L. Cloudsley-Thompson, O. Cuellar, V.R.D. Eickstedt, B. Barraviera and M.B. Knox. 1996. The evolution of scorpionism in Brazil in recent years. The Journal of Venomous Animals and Toxins 2(2): 121–134.

Low, I. and S. Stables. 2006. Anaphylactic deaths in Auckland, New Zealand: a review of coronial autopsies from 1985 to 2005. Pathology 38: 328–332.

Lowe, S., M. Browne, S. Boudjelas and M. De Poorter. 2000. 100 of the World's Worst Invasive Alien Species A selection from the Global Invasive Species Database. 11 pp. http://www.issg.org/pdf/publications/worst_100/english_100_worst.pdf.

Lozano, R. 2012. Global and regional mortality from 235 causes of death for 20 age groups in 1990 and 2010: a systematic analysis for the Global Burden of Disease Study 2010. Lancet 380(9859): 2095–2128. DOI: 10.1016/S0140-6736(12)61728-0.

Lucia, E. 1981. A lesson from nature: Joe Cox and his revolutionary saw chain. Journal of Forest History 159–165.

Lucientes-Curdi, J., R. Molina-Moreno, C. Amela-Heras, F. Simon-Soria, S. Santos-Sanz, A. Sanchez-Gomez, B. Suarez-Rodriguez and M.J. Sierra-Moros. 2014. Dispersion of *Aedes albopictus* in the Spanish Mediterranean Area. European Journal of Public Health 24: 637–640.

Lundin, O., M. Rundlöf, H.G. Smith, I. Fries and R. Bommarco. 2015. Neonicotinoid insecticides and their impacts on bees: A systematic review of research approaches and identification of knowledge gaps. Public Library of Science One 10(8): e0136928. DOI: 10.1371/journal.pone.0136928.

Lung, T.-N. 2004. The Culturing of *Ericerus pela* (Chavannes) in China and the Transmission of Chinese wax into Europe. http://agri-history.ihns.ac.cn/agrobiology/wax.htm.

Lychkovskaya, I.Y. 2006. Composition and structure of complexes of Hemiptera insects in calciphyte habitats of Central Russian forest-steppe. The dissertation of candidate of biological sciences. Voronezh, Voronezh State University. 20 pp. [in Russian].

Lyn, J.C., N. Wida, A. Vadim and C.D. Rollo. 2011. Influence of two methods of dietary restriction on life history features and aging of the cricket Acheta domesticus. Age 33: 509–522.

Lyukhin, A.M. 2014. Amber—the mystery of the origin of the solar mineral. 8 pp. http://lyukhin.ru/wp-content/uploads/201 4/04/%D0%9F%D1%80%D0%BE%20%D1%8F%D0%BD%D1%82%D0%B0%D1%80%D1%8C.pdf [in Russian].

Ma, Y., J.G. Ning, H.L. Ren, P.F. Zhang and H.Y. Zhao. 2015. The function of resilin in honeybee wings. Journal of Experimental Biology 218(13): 2136–2142. DOI: 10.1242/jeb.117325.

Macadam, C.R. and J.A. Stockan. 2015. More than just fish food: ecosystem services provided by freshwater insects. Ecological Entomology 40(Suppl. 1): 113–123. DOI: 10.1111/een.12245.

MacEvilly, C. 2000. Bugs in the system. Nutrition Bulletin 25: 267–268.

Maguina, C., P.J. Garcia, E. Gotuzzo, L. Cordero and D.H. Spach. 2001. Bartonellosis (Carrion's disease) in the modern era. Clinical Infectious Diseases 33(6): 772–779. DOI: 10.1086/322614.

Majewski, J. 2014. Economic value of pollination of major crops in poland in 2012. Economic Science for Rural Development: Production and Co-Operation in Agriculture 34: 14–21.

Makore, T.A., P. Garamumhango, T. Chirikure and S.D. Chikambi. 2015. Determination of nutritional composition of *Encosternum delegorguei* caught in Nerumedzo community of Bikita, Zimbabwe. International Journal of Biology 7(4): 13–19. DOI: 10.5539/ijb.v7n4p13.

Malaisse, F. 2005. Human consumption of Lepidoptera, termites, Orthoptera, and ants in Africa. pp. 175–230. *In*: Paoletti, M.G. (ed.). Ecological Implications of Minilivestock. Enfield, New Hampshire, Science Pub.

Malkhazova, S.M. (ed.). 2015. Medico-geographical Atlas of Russia. Natural Focal Diseases. Moscow, Moscow State University. 208 pp. [in Russian].

Mamaev, B.M. and E.A. Bordakova. 1985. Entomology for teachers. Moscow, Prosveshchenie, 114 pp. [in Russian].

Mangodt, E.A., A.L. Van Gasse, C.H. Bridts, V. Sabato and D.G. Ebo. 2015. Simultaneous oral mite anaphylaxis (pancake syndrome) in a father and daughter and a review of the literature. Journal of Investigational Allergology and Clinical Immunology 25(1): 75–76.

Mani, M., C. Shivaraju and M. Srinivasa Rao. 2014. Pests of grapevine: A worldwide list. Pest Management in Horticultural Ecosystems 20(2): 170–216.

Mans, B.J., C.M.L. Steinmann, J.D. Venter, A.I. Louw and A.W.H. Neitz. 2002. Pathogenic mechanisms of sand tampan toxicoses induced by the tick, *Ornithodoros savignyi*. Toxicon 40: 1007–1016.

Mans, B.J. and A.W.H. Neitz. 2004. The sand tampan, *Ornithodoros savignyi*, as a model for tick–host interactions. South African Journal of Science 100: 283–288.

Mans, B.J., R. Gothe and A.W. Neitz. 2004. Biochemical perspectives on paralysis and other forms of toxicoses caused by ticks. Parasitology 129: S95–S111.

Manukyan, A. and V. Vaishat. 2010. Dominican amber—window to the past of the Earth. Science and Life 3: 76–80 [in Russian].

Manukyan, A. 2011. Color chronicle of evolution. Science and Life 8: 53–58 [in Russian].

Manukyan, A.R. 2013. The Kaliningrad market of inclusion imitations. Techniques of non-invasive identification. Amber and its imitations. Proceedings of the International Scientific and Practical Conference, Kaliningrad, pp. 40–48 [in Russian].

Manzoli-Palma, M.F., N. Gobbi and M.S. Palma. 2003. Insects as biological models to assay spider venom toxicity. Journal of Venomous Animals and Toxins 9: 64–79.

Manzoor, F., T. Rafique and B.T. Forschler. 2015. Survivorship and tunneling activity of *Heterotermes indicola* (Wassman) (Isoptera: Rhinotermitidae) in response to soil treated with four different insecticides. Journal of Innovative Sciences 1(1): 11–14.

Marchenko, M.I. and V.I. Kononenko. 1991. Practical guide to forensic entomology. Kharkov. 69 pp. [in Russian].

Mariani, R., R. Garcia-Mancuso, G.L. Varela and A.M. Inda. 2014. Entomofauna of a buried body: Study of the exhumation of a human cadaver in Buenos Aires, Argentina. Forensic Science International 237: 19–26. DOI: 10.1016/j.forsciint.2013.12.029.

Marikovskii, P.I. 1947. Venomous steppe spider. Alma-Ata, 32 pp. [in Russian].

Marikovsky, P.I. 2012. Wonderful world of insects (entertaining entomology). Vol. 2. Almaty, 453 pp. [in Russian].

Marka, A., A. Diamantidis, A. Papa, G. Valiakos, S.C. Chaintoutis, D. Doukas, P. Tserkezou, A. Giannakopoulos, K. Papaspyropoulos, E. Patsoula, E. Badieritakis, A. Baka, M. Tseroni, D. Pervanidou, N.T. Papadopoulos, G. Koliopoulos, D. Tontis, C.I. Dovas, C. Billinis, A. Tsakris, J. Kremastinou, C. Hadjichristodoulou and the MALWEST project. 2013. West Nile virus state of the art report of MALWEST project. International Journal of Environmental Research of Public Health 10: 6534–6610. DOI: 10.3390/ijerph10126534.

Markin, G.P. and J.L. Littlefield. 2008. Biological control of tansy ragwort (*Senecio jacobaeae* L.) by the cinnabar moth, *Tyria jacobaeae* (CL) (*Lepidoptera: Arctiidae*), in the northern Rocky Mountains. Proceedings of XII International Symposium on Biological Control of Weeds, pp. 583–588.

Marques, L. and A.T. de Almeida. 2000. Electronic nose-based odour source localization. Proceedings of 6th International Workshop on Advanced Motion Control, Nagoya Institute of Technology, Nagoya, Japan, pp. 36–40.

Marsh, L. 2016. Vladimir Nabokov, a scientific genius. http://inosmi.ru/culture/20160424/236277705.html [in Russian].

Martib, E., X. Wanga, N.N. Jambaria, C. Rhynerc, J. Olzhausenc, J.J. Pérez-Bareaa, G.P. Figueredod and M.J.C. Alcocera. 2015. Novel *in vitro* diagnosis of equine allergies using a protein array and mathematical modelling approach: A proof of concept using insect bite hypersensitivity. Veterinary Immunology and Immunopathology 167(3–4): 171–177.

Maslov, A.D. (ed.). Protection of forest against pests and diseases: reference book 1988. Moscow: Agropromizdat, 414 pp. [in Russian].

Maslyakov, V.Y. and S.S. Izhevsky. 2011. Alien phytophagous insects invasions in the European part of Russia. Moscow, Igras. 272 pp. [in Russian].

Mason, L.J. and M. McDonough. 2012. Biology, Behavior, and Ecology of Stored Grain and Legume Insects. pp. 7–20. *In*: Hagstrum, D.W., T.W. Phillips and G. Cuperus (eds.). Stored Product Protection. Kansas State University, http://www.bookstore.ksre.ksu.edu/pubs/s156.pdf.

Matosevic, D., N. Lackovic, G. Melika, K. Kos, I. Franic, E. Kriston, M. Bozso, G. Seljak and M. Rot. 2016. Biological control of invasive *Dryocosmus kuriphilus* with introduced parasitoid *Torymus sinensis* in Croatia, Slovenia and Hungary. Periodicum Biologorum 117(4): 471–477.

Matsui, V.M. 2010. From gallipot—resin upto amber. Proceedings of the National Museum of Natural History, 2010, Nr 8. pp. 135–142 [in Russian].

Matthews, K.A., T.C. Kaufman and W.M. Gelbart. 2005. Research resources for Drosophila: the expanding universe. Nature Reviews Genetics 6: 179–193.

Matyukhin, A.V., A.V. Zabashta and M.V. Zabashta. 2012. Louse Flies (Hippoboscidae) of Falconiformes and Strigiformes in Palearctic. Birds of prey in the dynamic environment of the third Millennium: status and prospects. Kryvyi Rig, pp. 530–533 [in Russian].

Matyukhin, A.V., A.N. Matrosov, A.M. Porshakov and A.A. Kuznetsov. 2013. Blood-sucking fly *Icosta ardeae*—its distribution and probable role in circulation of West-Nile fever virus. Problems of Particularly Dangerous Infections 4: 110–111 [in Russian].

Mavrischev, V.V. 2000. Fundamentals of General Ecology. Minsk, Russia, Vysheishaya Shkola. 317 pp. [in Russian].

Maximova, Y.V. 2014. Biological Methods of Forest Protection: A Training Manual. Tomsk, Tomsk State University Publishing. 172 pp. [in Russian].

Mazur, N., M. Nagel, U. Leppin, G. Bierbaum and J. Rust. 2014. The extraction of fossil arthropods from Lower Eocene Cambay amber. Acta Palaeontologica Polonica 59(2): 455–459. DOI: 10.4202/app.2012.0018.

Mazzuca, M., C. Heurteaux, A. Alloui, S. Diochot, A. Baron, N. Voilley, N. Blondeau, P. Escoubas, A. Gélot, A. Cupo, A. Zimmer, A.M. Zimmer, A. Eschalier and M. Lazdunski. 2007. A tarantula peptide against pain via ASIC1a channels and opioid mechanisms. Nature Neuroscience 10: 943–945.

McCubbin, K. and J. Weiner. 2002. Fire ants in Australia: a new medical and ecological hazard. Medical Journal of Australia 176(11): 518–519.

McFadyen, R.E.C. 1998. Biological control of weeds. Annual Review of Entomology 43: 369–393.

McFadyen, R.E.C. 2000. Successes in biological control. pp. 3–14. *In*: Spencer, N.R. (ed.). Proceedings of the 10th International Symposium of the Biological Control of Weeds, Montana State University, Bozeman, MT.

McFadyen, R.E.C. 2003. Biological control of weeds using exotic insects. pp. 163–183. *In*: Koul, O. and G.S. Dhaliwal (eds.). Predators and Parasitoids. London, Taylor & Francis.

McGain, F. and K.D. Winkel. 2002. Ant sting mortality in Australia. Toxicon 40(8): 1095–1100.

McGeoch, M.A., S.H.M. Butchart, D. Spear, E. Marais, E.J. Kleynhans, A. Symes, J. Chanson and M. Hoffmann. 2010. Global indicators of biological invasion: species numbers, biodiversity impact and policy responses. Diversity and Distributions 16: 95–108.

McKellar, R.C. and M.S. Engel. 2014. The first Mesozoic Leptopodidae (Hemiptera: Heteroptera: Leptopodomorpha), from Canadian Late Cretaceous amber. Historical Biology 26(6): 702–709. DOI: 10.1080/08912963.2013.838753.

Mehner, T., J. Ihlau, H. Domer, M. Hupfer and F. Holker. 2005. Can feeding of fish on terrestrial insects subsidize the nutrient pool of lakes? Limnology and Oceanography 50: 2022–2031.

Mehravaran, A., M. Moradi, Z. Telmadarraiy, E. Mostafavi, A.R. Moradi, S. Khakifirouz, N. Shah-Hosseini, F.S. Varaie, T. Jalali, S. Hekmat, S.M. Ghiasi and S. Chinikar. 2012. Molecular detection of Crimean–Congo haemorrhagic fever (CCHF) virus in ticks from southeastern Iran. Ticks and Tick-Borne Diseases 4(1–2): 35–38. DOI: 10.1016/j.ttbdis.2012.06.006.

Melathopoulos, A.P., G.C. Cutler and P. Tyedmers. 2015. Where is the value in valuing pollination ecosystem services to agriculture? Ecological Economics 109: 59–70. DOI: 10.1016/j.ecolecon.2014.11.007.

Mele van, P. 2008. A historical review of research on the weaver ant Oecophylla in biological control. Agricultural and Forest Entomology 10(1): 13–22. DOI: 10.1111/j.1461-9563.2007.00350.x.

Melhorn, H. 2007. First occurrence of Culicoides obsoletus-transmitted Bluetongue virus epidemic in Central Europe. Parasitology Research 101(1): 219–228. DOI: 10.1007/s00436-007-0519-6.

Meng, P.S., K. Hoover and M.A. Keena. 2015. Asian longhorned beetle (Coleoptera: Cerambycidae), an introduced pest of maple and other hardwood trees in North America and Europe. Journal of Integrative Pest Management 2015 6(1): 4. DOI: 10.1093/jipm/pmv003.

Menzel, P. and F. D'Aluisio. 2004. Man Eating Bugs: The Art and Science of Eating Insects. Random House. 192 pp.

Merritt, R.W. and J.R. Wallace. 2009. Chapter 7: The role of aquatic insects in forensic investigations. pp. 271–319. *In*: Byrd, J.H. and J.L. Castner (eds.). Forensic Entomology: The Utility of Arthropods in Legal Investigations. CRC Press, DOI: 10.1201/NOE0849392153.ch7.

Mertz, T.L., S.B. Jacobs, T.J. Craig and F.T. Ishmael. 2012. The brown marmorated stinkbug as a new aeroallergen. Journal of Allergy and Clinical Immunology 130(4): 999–1001.

Messenger, S. 2011. Trillions of insects killed by cars every year, says study. http://www.treehugger.com/cars/trillions-of-insects-killed-by-cars-every-year-says-study.html.

Meyer, J.R. 2007. Sericulture. http://www.cals.ncsu.edu/course/ent425/text01/sericulture.html.

Micke, T. 1996. Xylotrupes – rhinoceros beetle: The Sumo wrestlers of the animal world. http://www.articlesextra.com/beetle-fight-xylotrupes-thailand.htm.

Mikkola, K. 1986. Direction of insect migrations in relation to the wind. pp. 152–171. *In*: Danthanarayana, W. (ed.). Insect Flight: Dispersal and Migration. Springer-Verlag, Berlin.

Milano, S. and L.R. Fontes. 2002. Termite pests and their control in urban Brazil. Sociobiology 40: 163–177.

Millennium Ecosystem Assessment (MA). 2005. Ecosystems and Human Well-Being: Synthesis. Island Press, Washington. 155 pp.

Miller, T. 1992. Latrodectism: bite of the black widow spider. American Family Physician 45: 181.

Mironova, M.K. and S.S. Izhevsky. 2007. Paths of invasions of the strange insects-phytophags (by example of quarantine species). http://83.149.228.85/invasive/publications/mir_izh_02.html.

Mirzoyan, S.A., I.D. Batiashvili, V.N. Gramma, S.A. Vardikyan, Z.F. Klyuchko, I.G. Kritskaya, R.E. Effendi, I.A. Khalifman, E.N. Vasilyeva, N.G. Samedov, I. Novak and Z. Tsapetsky. 1982. Rare insects. Moscow, Lesnaya Promyshlennost. 165 pp. [in Russian].

Mitsuhashi, J. 2010. The future use of insects as human food. pp. 115–122. *In*: Durst, P.B., D.V. Johnson, R.N. Leslie and K. Shono (eds.). Forest Insects as Food: Humans Bite Back. Proceedings of a Workshop on Asia-Pacific Resources and Their Potential for Development, 19–21 February 2008, Chiang Mai, Thailand. Bangkok: Food and Agriculture Organization of the United Nations.

Mlcek, J., O. Rop, M. Borkovcova and M. Bednarova. 2014. A comprehensive look at the possibilities of edible insects as food in Europe—a review. Polish Journal of Food and Nutritional Science 64(3): 147–157. DOI: 10.2478/v10222-012-0099-8 http://journal.pan.olsztyn.pl.

Modern Medical Encyclopaedia. 2002. St. Petersburg, Norint. 1236 pp. [in Russian].

Mordkovich, I.B. and E.A. Sokolov (eds.). 1999. The Directory determinant of quarantine and other dangerous pests of raw materials, products stock and seed. Moscow, Kolos. 384 pp. [in Russian].

Mordkovich, V.G. 1997. Catena approach for monitoring of biodiversity. Monitoring of Biodiversity. Moscow, pp. 226–231 [in Russian].

Moreva, L.Y., A.V. Ambramchuk, A.A. Efimenko, E.V. Shakhmatova and M.S. Izmerlaync. 2010. Maintenance of heavy metals and radioactive substances in the products of beekeeping received close highways in Krasnodar region. Science of Kuban 2: 29–32 [in Russian].

Moro, G., P. Charvet and R.S. Rosa. 2012. Insectivory in *Potamotrygon signata* (Chondrichthyes: Potamotrygonidae), an endemic freshwater stingray from the Parnaiba River basin, northeastern Brazil. Brazilian Journal of Biology 72: 885–891.

Mosbech, H. 1983. Deaths resulting from bee and wasp stings in Denmark 1960–1980. Ugeskr, Laeger. Vol. 145, pp. 1757–1760.

Mosov, A.V. 2015. There are cockroaches in every bar of chocolate. https://roscontrol.com/community/article/shokolad-kakao-bobi-vperemeshku-s-tarakanami/ [in Russian].

Motoyashiki, T., A.T. Tu, D.A. Azimov and K. Ibragim. 2003. Isolation of anticoagulant from the venom of tick, *Boophilus calcaratus*, from Uzbekistan. Thrombosis Research 110(4): 235–241.

Mountcastle, A.M. and S.A. Combes. 2013. Wing flexibility enhances load-lifting capacity in bumblebees. Proceedings of the Royal Society B Biological Sciences 280(1759). DOI: 10.1098/rspb.2013.0531.

Mowatt, T. 2011. Inspired by insect cuticle, Wyss researchers develop low-cost material with exceptional strength and toughness. http://wyss.harvard.edu/viewpressrelease/72.

Muafor, F.J., P. Levang, T.E. Angwafo and P. Le Gall. 2012. Making a living with forest insects: beetles as an income source in Southwest Cameroon. International Forestry Review 14(4).

Müeller, U.R. 1990. Insect Sting Allergy. Clinical picture, diagnosis and treatment. 183 pp.

Mukhamedzhanova, E.J. and M.D. Kon'shin. 2015. Habitat Monitoring. Omsk, Publishing House of OMGTU. 110 pp. [in Russian].

Mukherjee, S. and R. Ganguli. 2012. A comparative study of dragonfly inspired flapping wings actuated by single crystal piezoceramic. Smart Structures and Systems 10(1): 67–87.

Mukhina, A.A. 2007. Insects as a matter of bioindication. http://www.zoology.dp.ua/z_07_157.html [in Russian].

Mullen, G.R. and L.A. Durden (eds.). 2009. Medical and Veterinary Entomology. New York, Academic Press. 637 pp.

Mullens, B.A. 2009. Temporal changes in distribution, prevalence and intensity of northern fowl mite (*Ornithonyssus sylviarium*) parasitism in commercial caged laying hens, with a comprehensive economic analysis of parasite impact. Veterinary Parasitology 160: 116–133. DOI: 10.1016/j.vetpar.2008.10.076.

Müller, U., H. Mosbech, P. Blaauw, S. Dreborg, H.J. Malling, B. Przybilla, R. Urbanek, E. Pastorello, M. Blanca, J. Bousquet, R. Jarisch and L. Youlten. 1990. Emergency treatment of allergy reactions to Hymenoptera stings. Clinical and Experimental Allergy 21: 281–288.

Müller, U.R., G. Haeberli and A. Helbling. 2008. Chapter 43. Allergic reactions to stinging and biting insects. pp. 657–666. *In*: Rich, R.R., T. Fleisher, W. Shearer, H. Schroeder, A. Frew and C. Weyand (eds.). Clinical Immunology. Principles and Practice. St. Louis, MO, Mosby-Elsevier.

Müller, U.R. 2010. Insect venoms. Chemical Immunology and Allergy 95: 141–156. DOI: 10.1159/000315948.

Muniaraj, M. 2014. Fading chikungunya fever from India: beginning of the end of another episode? Indian Journal of Medical Research 139(3): 468–470.

Münstedt, K. and M. Kalder. 2009. Contact allergy to propolis in beekeepers. Allergologia et Immunopathologia 37: 298–301.

Murray, T.E., M.F. Coffey, E. Kehoe and F.G. Horgan. 2013. Pathogen prevalence in commercially reared bumble bees and evidence of spillover in conspecific populations. Biological Conservation 159: 269–276.

Mutsamba, E.F., I. Nyagumbo and P. Mafongoya. 2016. Termite prevalence and crop lodging under conservation agriculture in sub-humid Zimbabwe. Crop Protection 82: 60–64.

Myers, J.H. 2000. What can we learn from biological control failures? Proceedings of the 10th International Symposium on Biological Control of Weeds. 4–14 July 1999, Montana State University, Bozeman, MT, USA, pp. 151–154.

Na, J.P.S. and C.Y. Lee. 2001. Identification key to common urban pest ants in Malaysia. Tropical Biomedicine 18(1): 1–17.

Naconieczny, M., A. Kedziorski and K. Michalczyk. 2007. Apollo butterfly (*Parnassius Apollo* L.) in Europe—its history, decline and perspectives of conservation. Functional Ecosystems and Communities. Global Science Books 1(1): 56–79. http://www.globalsciencebooks.info/Online/GSBOnline/images/0706/FEC_1(1)/FEC_1(1)56-79o.pdf.

Nadeau, L., I. Nadeau, F. Franklin and F. Dunkel. 2014. The potential for entomophagy to address undernutrition. Ecology of Food and Nutrition 00: 1–9. DOI: 10.1080/03670244.2014.930032.

Nall, T.M. 1985. Analysis of 677 death certificates and 168 of autopsies of stinging insect deaths. Journal of Allergy and Clinical Immunology 75: 207.

Nandasena, M.R.M.P., D.M.S.K. Disanayake and L. Weeratunga. 2008. Sri Lanka as a potential gene pool of edible insects. pp. 161–164. *In*: Durst, P.B., D.V. Johnson, R.N. Leslie and K. Shono (eds.). Forest Insects as Food: Humans Bite Back. Proceedings of a Workshop on Asia-Pacific Resources and Their Potential for Development, 19–21 February 2008, Chiang Mai, Thailand. Bangkok: Food and Agriculture Organization of the United Nations.

Nartshuk, E.P. 2003. Key to families of Diptera (Insecta) of the fauna of Russia and adjacent countries. Proceedings of the Zoological institute, Vol. 294. 253 pp. [in Russian].

Nash, D., F. Mostashari, A. Fine, J. Miller, D. O'Leary, K. Murray, A. Huang, A. Rosenberg, A. Greenberg, M. Sherman, S. Wong, M. Layton; 1999; West Nile Outbreak Response Working Group. 2001. The outbreak of West Nile virus infection in the New York City area in 1999. New England Journal of Medicine 344: 1807–1814.

Nasirian, H., B. Vazirianzadeh, S. Mohammad, T. Sadeghi and S. Nazmara. 2014. *Culiseta subochrea* as a bioindicator of metal contamination in Shadegan International Wetland, Iran (Diptera: Culicidae). J. Insect Sci. 14(258): ii. DOI: 10.1093/jisesa/ieu120.

naturalsciences.org/living-collections/living-conservatory.

Neoh, K.-B. 2013. Termites and human society in Southeast Asia. The Focus. The Newsletter 66: 30–31. http://iias.asia/sites/default/files/IIAS_NL66_3031.pdf.

Neumann, D. and A. Kureck. 2013. Composite structure of silken threads and a proteinaceous hydrogel which form the diving bell wall of the water spider *Agyroneta aquatica*. Springer Plus 2(223). DOI: 10.1186/2193-1801-2-223.

Newman, A. 1989. Lungs of our planet. Moscow, Mir. 335 pp. [in Russian].

Nicholls, C.I. and M.A. Altieri. 2013. Plant biodiversity enhances bees and other insect pollinators in agroecosystems. A review. Agronomy for Sustainable Development 33(2): 257–274. DOI: 10.1007/s13593-012-0092-y.

Nieuwenhuys, E. 2008. The demystification of the toxicity of spiders. 16 pp. https://ednieuw.home.xs4all.nl/Spiders/Nasty-Spiders/The%20demystification%20of%20the%20toxicity%20of%20spiders.pdf.

Nikitsky, N.B. and A.V. Sviridov. 1987. Insects of the Red data book of the USSR. Moscow, Pedagogika. 176 pp. [in Russian].

Nikolsky, G.V. 1961. Ecology of fishes. Moscow, Vyshaya shkola 1961. 336 pp. [in Russian].

Nikuze, A. 2013. Silk sector profile. 22 pp. http://www.rdb.rw/fileadmin/user_upload/Documents/Manufacturing/4_Silk_sector_Profile.pdf.

Niu, S., B. Li, Z. Mu, M. Yang, J. Zhang, Z. Han and L. Ren. 2015. Excellent structure-based multifunction of morpho butterfly wings: a review. Journal of Bionic Engineering 12: 170–189.

Nogaro, G., F. Mermillod-Blondin, B. Montuelle, J.-C. Boisson and J. Gibert. 2008. Chironomid larvae stimulate biogeochemical and microbial processes in a riverbed covered with fine sediment. Aquatic Sciences 70: 156–168.

Noli, C. and W. Beck. 2007. Flea bite allergy and the control of fleas. Kleintierpraxis 52(7): 438–450 [in German].

Noordijk, J. 2010. A risk analysis for fire ants in the Netherlands. Leiden, Stichting European Invertebrate Survey, 37 pp.

Noordwijk van, C.G.E., D.E. Flierman, E. Remke, M.F. Wallis de Vries and M.P. Berg. 2012. Impact of grazing management on hibernating caterpillars of the butterfly *Melitaea cinxia* in calcareous grasslands. Journal of Insect Conservation 16: 909–920. DOI: 10.1007/s10841-012-9478-z.

Noormahomed, E.V., K. Akrami and C. Mascaro-Lazcano. 2016. Onchocerciasis, an undiagnosed disease in Mozambique: identifying research opportunities. Parasites and Vectors 9: 180. DOI: 10.1186/s13071-016-1468-7.

Novikov, Y.V. 1999. Ecology, environment and humanity. Fair-Press, Moscow, 320 pp. [in Russian].

Nowicki, P., V. Vrabec, B. Binzenhoer, J. Feil, B. Zaksek, T. Hovestadt and J. Settele. 2014. Butterfly dispersal in inhospitable matrix: rare, risky, but long-distance. Landscape Ecology 29: 401–412.

Ntonifor, N.H. and S. Veyufambom. 2016. Assessing the effective use of mosquito nets in the prevention of malaria in some parts of Mezam division, Northwest Region Cameroon. Malaria Journal 15(1): 390. DOI: 10.1186/s12936-016-1419-y.

Nuorteva, M. and K.A. Kinnunen. 2008. Insect frass in Baltic amber. Bulletin of the Geological Society of Finland 80: 105–124.

Nyffeler, M. and B.J. Pusey. 2014. Fish predation by semi-aquatic spiders: a global pattern. Public Library of Science One 9(6): e99459. http://journals.plos.org/plosone/article/asset?id=10.1371%2Fjournal.pone.0099459.PDF.

Nzonzi, R. 2003. Biologic features of Tomato Russet Mite (*Aculops lycopersici* Massee) and measures of combating it. Abstract of dissertation of candidate of biological Sciences. Moscow, 18 pp. [in Russian].

Obopile, M. and T.G. Seeletso. 2013. Eat or not eat: an analysis of the status of entomophagy in Botswana. Food Security 5: 817–824.

O'Brien, M. and M. Walton. 2010. Got Silk? http://phys.org/news/2010-05-scientists-goats-spider-silk.html.

Odum, E.P. 1983. Basic Ecology: Fundamentals of Ecology. Holt-Saunders (Japan), 325 pp.

Oerke, E.-C. 2006. Crop losses to pests. Journal of Agricultural Science 144: 31–43. DOI: 10.1017/S0021859605005708. http://journals.cambridge.org/download.php?file=%2FAGS%2FAGS144_01%2FS0021859605005708a.pdf&code=ce 99ed5b3a21eba4a69d566e33aeaf5e.

O'Hehir, R.E. and J.A. Douglass. 1999. Stinging insect allergy. The Medical Journal of Australia 171: 649–650.

Oliveira, C.M., A.M. Auad, S.M. Mendes and M.R. Frizzas. 2014. Crop losses and the economic impact of insect pests on Brazilian agriculture. Crop Protection 56: 50–54.

Omarov, Sh.M. 2009. Apitherapy: bee products in the world of medicine. Rostov na Donu, Feniks. 351 pp. [in Russian].

Omkar, P. 2016. Ecofriendly Pest Management for Food Security. Elsevier Science Publishing Co. Inc. 762 pp.

Ooninсх, D.G.A.B. and I.J.M. de Boer. 2012. Environmental impact of the production of mealworms as a protein source for humans—a life cycle assessment. Public Library of Science One 7(12): e51145. DOI: 10.1371/journal.pone.0051145.

Opatova, V. and M.A. Arnedo. 2014. Spiders on a hot volcanic roof: colonisation pathways and phylogeography of the Canary Islands endemic trap-door spider *Titanidiops canariensis* (Araneae, Idiopidae). Public Library of Science One 9: e115078. DOI: 10.1371/journal.pone.0115078.

Opell, B.D., S.G. Helweg and K.M. Kiser. 2016. Phylogeography of Australian and New Zealand spray zone spiders (Anyphaenidae: Amaurobioides): Moa's Ark loses a few more passengers. Biological Journal of the Linn Society 118: 959–969.

Ori, M. and H. Ikeda. 1998. Spider venoms and spider toxins. Journal of Toxicology Toxin Reviews 17(3): 405–426.

Orlov, B.N. and D.B. Gelashvili. 1985. Zootoxinology (venomous animals and their poisons). Moscow, Vysshaya shkola. 280 pp. [in Russian].

Orlov, B.N., D.B. Gelashvili and A.K. Ibragimov. 1990. Poisonous animals and plants of the USSR. Moscow, Vysshaya shkola. 272 pp. [in Russian].

Orlov, V.M. 1990. Insects and the electric field. Tomsk. 111 pp. [in Russian].

Orr, D. 2009. Biological control and integrated pest management. Chapter 9. pp. 207–240. *In*: Peshin, R. and A.K. Dhawan (eds.). Integrated Pest Management: Innovation-Development Process. Vol. 1. Springer, DOI: 10.1007/978-1-4020-8992-3.

Oršolić, N. 2012. Bee venom in cancer therapy. Cancer Metastasis Rev. 31: 173–194. DOI 10.1007/s10555-011-9339-3.

Otero, R., E. Navio, F.A. Cespedes, M.J. Núñez, L. Lozano, E.R. Moscoso, C. Matallana, N.B. Arsuza, J. García, D. Fernández, J.H. Rodas, O.J. Rodríguez, J.E. Zuleta, J.P. Gómez, M. Saldarriaga, J.C. Quintana, V. Núñez, S. Cárdenas, J. Barona, R. Valderrama, N. Paz, A. Diaz, O.L. Rodriguez, M.D. Martinez, R. Maturana, L.E. Beltran, M.B. Mesa, J. Paniagua, E. Florez and W.R. Lourenco. 2004. Scorpion envenoming in two regions of Colombia: clinical, epidemiological and therapeutic aspects. Transactions of the Royal Society of Tropical Medicine and Hygiene 98(12): 742–750.

Overton, M. 1998. Spider fighting. http://arachnophiliac.info/burrow/news/spider_fighting.htm.

Ozkan, O. and A. Filazi. 2004. The determination of acute lethal dose-50 (LD50) levels of venom in mice, obtained by different methods from scorpions, Androctonus crassicauda (Oliver 1807). T Parazitol. Derg. 28(1): 50–53.

Paini, D.R. and D. Yemshano. 2012. Modelling the arrival of invasive organisms via the international marine shipping network: a Khapra beetle study. Public Library of Science One 7(9): e44589. DOI: 10.1371/journal.pone.0044589.

Painter, R. 1953. Plant resistance to insects. Moscow, Izdat Inostran Liter. 443 pp.

Panfilov, D.V. 1977. In the world of insects. Moscow, Lesnaya Promyshlennost. 128 pp. [in Russian].

Pang, C., D. Kang, T.I. Kim and K.Y. Suh. 2012. Analysis of preload-dependent reversible mechanical interlocking using beetle-inspired wing locking device. Langmuir 28(4): 2181–2186. DOI: 10.1021/la203853r. panhandlebutterflyhouse.org/.

Panteleev, S.V. 2012. Assessment of the role of insects in the spread of pathogens of black mold and blackspot in forest nurseries on the basis of the use of DNA analysis techniques. Proceedings of BGTU 1: 253–257. ISSN 1683-0377 [in Russian].

Paoletti, M.G. and D.L. Dufour. 2005. Edible invertebrates among Amazonian Indians: a critical review of disappearing knowledge. pp. 293–342. *In*: Paoletti, M.G. (ed.). Ecological Implications of Minilivestock: Role of Rodents, Frogs, Snails, and Insects for Sustainable Development. Enfield, NH, Science Publishers.

Pape, T., V. Blagoderov and M.B. Mostovski. 2011. Order Diptera Linnaeus, 1758. *In*: Zhang, Z.-Q. (ed.). Animal Biodiversity: An Outline of Higher-level Classification and Survey of Taxonomic Richness. Zootaxa 3148: 222–229. http://www.mapress.com/zootaxa/2011/f/zt03148p229.pdf.

Pappagallo, L. 2012. Termite mounds inspire energy neutral buildings. http://www.greenprophet.com/2012/01/termite-mounds-middle-east/.

Park, I.G., Y.I. Mah, H.J. Yoon and S.Y. Yang. 1998. Studies on the regional distribution and some ecological characteristics of Chinese white-wax scale in Korea. Korean Journal of Applied Entomology 37(2): 137–142.

Park, S.P., B.M. Kim, J.Y. Koo, H. Cho, C.H. Lee, M. Kim, H.S. Na and U. Oh. 2008. A tarantula spider toxin, GsMTx4, reduces mechanical and neuropathic pain. Pain 137(1): 208–217. DOI: 10.1016/j.pain.2008.02.013.

Parker, A.R. and C.R. Lawrence. 2001. Water capture by a desert beetle. Nature 414(6859): 33–34. DOI: 10.1038/35102108.

Parrish, H.M. 1963. Analysis of 460 fatalities from venomous animals in the United States. American Journal of Medical Science 245: 129–141.

Parsons, J.R. 2010. The pastoral niche in pre-hispanic Mesoamerica. pp. 109–136. *In*: Staller, J.E. and M.D. Carrasco (eds.). Pre-Columbian Foodways: Interdisciplinary Approaches to Food, Culture and Markets in Ancient Mesoamerica. New York, Springer.

Patel, S. and B.R. Meher. 2016. A review on emerging frontiers of house dust mite and cockroach allergy research. Allergologia et immunopathologia 44(6): 580–593. doi: 10.1016/j.aller.2015.11.001.

Patiny, S., P. Rasmont and D. Michez. 2009. Survey and review of the status of wild bees in the West-Palaearctic region. Apidologie 40(3): 313–331. DOI: 10.1051/apido/2009028.

Patrick, C.J. 2013. The effect of shredder community composition on the production and quality of fine particulate organic matter. Freshwater Science 32(3): 1026–1035. DOI: 10.1899/12-090.1.

Pavlov, V.N. and A.V. Barsukova. 1976. Herbarium: Guidance on the collection, processing and storage of plant collections. Moscow, Publishing House of Moscow University. 32 pp. [in Russian].

Pavlovsky, E.N. 1931. Poisonous Animals of the USSR. Moscow, Gos. Medits. Publishing House. 202 pp. [in Russian].

Pavlovsky, E.N. 1950. Venomous Animals of Central Asia. Stalinabad, 1950. 108 pp. [in Russian].

Paynter, Q. 2013. Biocontrol for aquatic weeds, a step closer. Weed Biocontrol 64. http://www.landcareresearch.co.nz/publications/newsletters/biological-control-of-weeds/issue-64/biocontrol-for-aquatic-weeds.

Peattie, A.M. 2009. Attachment forces of single tarantula adhesive setae. Comparative Biochemistry and Physiology A—Molecular and Integrative Physiology 153A(2): S123–S123.

Peisker, H., J. Michels and S.N. Gorb. 2013. Evidence for a material gradient in the adhesive tarsal setae of the ladybird beetle *Coccinella septempunctata*. Nature Communications 4(1661).

Pekhtasheva, E.L. 2012. Biodeterioration of non-food goods. Moscow, Dashkov and Ko. 332 p. [in Russian].

Penney, D. and R.F. Preziosi. 2014. Estimating fossil ant species richness in Eocene Baltic amber. Acta Palaeontologica Polonica 59(4): 927–929.

Pérez-Pimiento, A.J., L.A. González-Sánchez, L. Prieto-Lastra, M.I. Rodríguez-Cabreros, A. Iglesias-Cadarso and M. Rodríguez-Mosquera. 2005. Anaphylaxis to hymenoptera sting: study of 113 patients. Med. Clin. (Barc) 125(11): 417–420.

Peris, D., M.M. Solorzano Kraemer, E. Penalver and X. Delclos. 2015. New ambrosia beetles (Coleoptera: Curculionidae: Platypodinae) from Miocene Mexican and Dominican ambers and their paleobiogeographical implications. Organisms, Diversity, and Evolution 15(3): 527–542.

Perkovsky, E.E., A.P. Rasnitsyn, A.P. Vlaskin and M.V. Taraschuk. 2007. A comparative analysis of the Baltic and Rovno amber arthropod faunas: representative samples. African Invertebrates 48(1): 229–245.

Perkovsky, E.E., V.Y. Zosimovich and A.P. Vlaskin. 2010. Rovno Amber. pp. 116–136. *In*: Penney, D. (ed.). Biodiversity of Fossils in Amber from the Major World Deposits. Manchester, Siri Scientific Press.

Perotti, M.A. and H.R. Braig. 2009. Phoretic mites associated with animal and human decomposition. Experimental and Applied Acarology 49(1–2): 85–124. DOI: 10.1007/s10493-009-9280-0.

Perotti, M.A., M.L. Goff, A.S. Baker, B.D. Turner and H.R. Braig. 2009. Forensic acarology: an introduction. Experimental and Applied Acarology 49(1–2): 3–13. DOI: 10.1007/s10493-009-9285-8.

Perrichot, V. 2015. A new species of Baikuris (Hymenoptera: Formicidae: Sphecomyrminae) in mid-Cretaceous amber from France. Cretaceous Research 52: 585–590. DOI: 10.1016/j.cretres.2014.03.005.

Perry, S. and C. Randall (eds.). 2000. Forest Pest Management. A Guide for Commercial Applicators Category 2. Michigan State University. http://www.ipm.msu.edu/uploads/files/TrainingManuals_Forest/Forest_WholeManual.pdf.

Pesek, R.D. and R.F. Lockey. 2014. Treatment of Hymenoptera venom allergy: an update. Current Opinion in Allergy and Clinical Immunology 14(4): 340–346.

Peshin, R. and A.K. Dhawan (eds.). 2009. Integrated Pest Management: Innovation-Development Process. Vol. 1. Springer. 689 pp.

Peshin, R., R.S. Bandral, W.J. Zhang, L. Wilson and A.K. Dhawan. 2009. Integrated pest management: a global overview of history, programs and adoption. Chapter 1. pp. 1–50. *In*: Peshin, R. and A.K. Dhawan (eds.). Integrated Pest Management: Innovation-Development Process. Vol. 1. Springer, DOI: 10.1007/978-1-4020-8992-3.

Pests of stored products. pp. 129–149. http://cals.arizona.edu/apmc/docs/8%20Stored%20products-food,%20fabric%20F. pdf.

Peterson, I. 2004. Flight of the Bumblebee. Science News. https://www.sciencenews.org/article/flight-bumblebee.

Petney, T.N. and L.J. Fourie. 1990. The dispersion of the Karoo Paralysis Tick, *Ixodes rubicundu*s, within a naturally infested population of sheep in South Africa. Veterinary Parasitology 34: 345–352.

Petrashova, D.A. 2010. *Monotarsobius curtipes* as nonspecific bioindicator of soils contamination by pollutant emission from industrial enterprises. Proceedings of the Samara Scientific Center, Russian Academy of Sciences. Vol. 12. Number: 1–8. 2010. pp. 1947–1950 [in Russian].

Phillips, F.H. 2004. How to Do Macro Insect Photography. http://frankphillips.com/beautifulbugs/howto.htm.

Phipps, E. Cochineal red: the art history of a color. New York: Metropolitan Museum of Art; 2010.

Physico-geographical Atlas of the World. 1964. Moscow, GUGK. 298 pp. [in Russian].

Pigulevsky, S.V. 1975. Venomous animals. Toxicology of invertebrates. Leningrad, Meditsina, 1975. 184 pp. [in Russian].

Pilgrim, R.L.C. 1992. An historic collection of fleas (Siphonaptera) in the Macleay Museum, Sydney, Australia. Proceedings of the Linnean Society of New South Wales 113: 77–86.

Pimentel, D., L. McLaughlin, A. Zepp, B. Lakitan, T. Kraus, P. Kleinman, F. Vancini, W.J. Roach, E. Graap, W.S. Keeton and G. Selig. 1993. Environmental and economic effects of reducing pesticide use in agriculture. Agriculture, Ecosystems and Environment 46(1–4): 273–288.

Pimentel, D., C. Wilson, C. McCullum, R. Huang, P. Dwen, J. Flack, Q. Tran, T. Saltman and B. Cliff. 1997. Economic and environmental benefits of biodiversity. BioScience 47(11): 747–755.

Pimentel, D. (ed.). 2002. Biological Invasions: Economic and Environmental Costs of Alien Plant, Animal, and Microbe Species. CRC Press, 384 pp.

Pimentel, D. and M. Pimentel. 2003. Sustainability of meat-based and plant-based diets and the environment. American Journal of Clinical Nutrition 78: 660S–663S.

Pimentel, D. 2009. Pesticides and pest control. pp. 83–88. *In*: Peshin, R. and A.K. Dhawan (eds.). Integrated Pest Management: Innovation-Development Process. DOI: 10.1007/978-1-4020-8992-35.

Pinniger, D. 2015. Integrated Pest Management for Cultural Heritage. Archetype Publications, London. 156 pp.

Pinniger, D.B. and J.D. Harmon. 1999. Pest management, prevention and control. pp. 152–176. *In*: Carter, D. and A. Walker (eds.). Care and Conservation of Natural History Collections. Oxford: Butterwoth Heinemann, http://www.natsca.org/care-and-conservation.

Plantinga, E.A., G. Bosch and W.H. Hendriks. 2011. Estimation of the dietary nutrient profile of free-roaming feral cats: possible implications for nutrition of domestic cats. British Journal of Nutrition 106: S35–S48.

Plonsky, M. 2002. Bug Pictures (Insect Macro Photography). http://www.mplonsky.com/photo/article.htm.

Poinar, G. and A.A. Legalov. 2015. Two new species of the genus Rhynchitobius Sharp, 1889 (Coleoptera: Rhynchitidae) in Dominican amber. Annales de la societe entomologique de France 51(1): 70–77. DOI: 10.1080/00379271.2015.1059996.

Poinar, G., Jr. 2015. A new genus of fleas with associated microorganisms in Dominican Amber. Journal of Medical Entomology 52(6): 1234–1240.

Poinar, G.O. 1992. Life in Amber. Stanford, CA, Stanford University Press. 368 pp.

Poinar, G.O. and R. Poinar. 1999. The amber forest: a reconstruction of a vanished world. Princeton, NJ, Princeton University Press.

Polevod, V.A. 2016. The objects of museum entomology of Kemerovo region: dermestid beetles (Coleoptera, Dermestidae). Bulletin of the Kemerovo State University of Culture and Arts 34: 169–174 [in Russian].

Poltavsky, A.N. 2012. Entomological refugia and their significance in the maintenance of the red book of the Rostov region. Rostov-na-Donu: Izd-vo Cubesh. 184 pp. [in Russian].

Polyakov, A.N. and N.M. Nabatov. 1992. Forestry and forest inventory. Ekologiya, Moscow, 335 pp. (in Russian).

Ponomarenko, A.G. and A.A. Prokin. 2014. Insects in ancient lakes of Mongolia. Limnology and paleolimnology of Mongolia. Proceedings of the Joint Russian-Mongolian Complex Biological Expedition 60: 285–315. http://www.bio.vsu.ru/heteroptera/profile_prokin/prokin_pdf/Ponomarenko_Prokin2014.pdf [in Russian].

Popescu, A. 2013. Trends in world silk cocoons and silk production and trade, 2007–2010. Animal Science and Biotechnologies 46(2): 418–423.

Popov, I.A., V.V. Yasyukevich and E.N. Popova. 2016. Modeling of change of climatic areal of the European tick *Ixodes ricinus*. Proceedings of the Stavropol Department of Russian Entomological Society. Materials of IX International Scientific-Practical Internet-Xonference 30–32 [in Russian].

Popova, E.N. and I.O. Popov. 2013. Climatic factors determining ranges of agricultural pests and agents of plant diseases and model methodology for assessment of change in ranges. Moscow, IGCE. Problems of Ecological Monitoring and Ecosystem Modeling 25: 175–204 [in Russian].

Popova, E.N. and S.M. Semenov. 2013. Current and expected changes in Colorado beetle climatic habitat in Russia and neighboring countries. Russian Meteorology and Hydrology 38(7): 509–514.

Popova, E.N., S.M. Semenov and I.O. Popov. 2016. Assessment of possible expansion of the climatic range of Italian locust (*Calliptamus italicus* L.) in Russia in the 21st century at simulated climate changes. Russian Meteorology and Hydrology 41(3): 213–217.

Porrini, C., E. Caprio, D. Tesoriero and G. Di Prisco. 2014. Using honey bee as bioindicator of chemicals in Campanian agroecosystems (South Italy). Bulletin of Insectology 67(1): 137–146.

Poulsen, M., M.V.W. Kofoed, L.H. Larsen, A. Schramm and P. Stief. 2014. Chironomus plumosus larvae increase fluxes of denitrification products and diversity of nitrate-reducing bacteria in freshwater sediment. Systematic and Applied Microbiology 37(1): 51–59. DOI: 10.1016/j.syapm.2013.07.006.

Prahlow, J.A. and J.J. Barnard. 1998. Fatal anaphylaxis due to fire ant stings. American Journal of Forensic Medicine and Pathology 19: 137–142.

Premke, K., P. Fischer, M. Hempel and K. Rothhaupt. 2010. Ecological studies on the decomposition rate of fish carcasses by benthic organisms in the littoral zone of Lake Constance, Germany. Annales de Limnologie – International Journal of Limnology 46: 157–168.

Presman, A.S. 2003. Electromagnetic field and life. Moscow, Science. 215 pp. [in Russian].

Primack, R., H. Kobori and S. Mori. 2000. Dragonfly pond restoration promotes conservation awareness in Japan. Conservation Biology 14(5): 1553–1554.

Primack, R.B. 2002. Essentials of conservation biology. Moscow, Publishing House of the Scientific and Educational Center. 256 pp. [in Russian].

Pukhlik, V.M. 2002. Basic Immunology. Vinnitsa, Veles. 148 pp. [in Russian].

Pumphrey, R. 2004. Anaphylaxis: can we tell who is at risk of a fatal reaction? Current Opinion in Allergy and Clinical Immunology 4: 285.

Purse, B.V., P.S. Mellor, D.J. Rogers, A.R. Samuel, P.P. Mertens and M. Baylis. 2005. Climate change and the recent emergence of bluetongue in Europe. National Review of Microbiology 3(2): 171–181. DOI: 10.1038/nrmicro1090.

Pushkin, S.V. 2004. Necrobiotic entomocomplexes the highlands of North-West Caucasus. Eurasian Entomological Journal 3(3): 195–202 [in Russian].

Pushkin, S.V. 2012. Necrobiont beetles (Insecta: Coleoptera) of North Caucasus and adjacent areas (fauna, ecology, biocenotic and practical importance): avtoreferat of dissertation of doctor of Biological Sciences. Stavropol. 32 pp. [in Russian].

Pushkin, S.V. 2015. Rare and disappearing types of insects of the Central Caucasus: insects. Moscow-Berlin, Direct Media. 105 pp. [in Russian].

Qi, D. and R. Gordnier. 2015. Effects of deformation on lift and power efficiency in a hovering motion of a chord-wise flexible wing. Journal of Fluids and Structures 54: 142–170.

Qi, J. and J.-F. Zhao. 2004. Polycosanols from *Ericerus pela* wax. US 20040152787 A1. http://www.google.ch/patents/US20040152787.

Qin, G.K., X. Hu, P. Cebe and D.L. Kaplan. 2012. Mechanism of resilin elasticity. Nature Communications 3(1003). DOI: 10.1038/ncomms2004.

Qin, T.-K. 1997. The pela wax scale and commercial wax production. World Crop Pests. Soft Scale Insects their Biology, Natural Enemies and Control 7(Part A): 303–321.

Querner, P., S. Simon, M. Morelli and S. Fürnkranz. 2013. Results from the insect pest monitoring 2010 and implementing an integrated pest management concept in two large Museum collections in Berlin and Vienna. International Biodeterioration & Biodegradation 84: 275–280. http://dx.doi.org/10.1016/j.ibiod.2012.04.024.

Querner, P. 2015. Insect pests and integrated pest management in museums, libraries and historic buildings. Insects 6(2): 595–607. DOI: 10.3390/insects6020595.

Querner, P. 2016. Linking webbing clothes moths to infested object or other food source in museums. Studies in Conservation 61: 111–117. http://www.tandfonline.com/doi/pdf/10.1179/2047058414Y.0000000153.

Querner, P., K. Sterflinger, D. Piombino-Mascali, J.J. Morrow, R. Pospischil and G. Piñar. 2017. Insect pests, microorganisms, and integrated pest management in the Capuchin Catacombs of Palermo, Italy. International Biodeterioration & Biodegradation pp. 1–8.

Quicke, D.L.J. and M.G. Fitton. 1995. Ovipositor steering mechanisms in parasitic wasps of the families Gasteruptiidae and Aulacidae (Hymenoptera). Proceedings of the Royal Society of London B 261: 99–103.

Quicke, D.L.J., P. Wyeth, J.D. Fawke, H.H. Basibuyuk and J.F.V. Vincent. 1998. Manganese and zinc in the ovipositors and mandibles of hymenopterous insects. Zoological Journal of the Linnean Society 124(4): 387–396.

Quinn, T. 2011. Military's Strangest Campaigns & Characters: Extraordinary but True Stories from Over Two Thousand Years of Military History. Anova Books. 241 pp.

Radmanesh, M. 1990. *Androctonus crassicauda* sting and its clinical study in Iran. Journal of Tropical Medicine and Hygiene 93(5): 323–326.

Ragazzi, E. 2008. Fossil resin deposits or finds in Italy. Proceedings of 15th Seminar, Organic Inclusions, Amber and Other Fossil Resin Finds in Europe. Warsaw, Gdańsk, pp. 5–11.

Ragulin, V.V. 1981. Technology of tire production. Moscow, Chemistry. 264 pp. [in Russian].

Rajagopal, T., S.P. Sevarkodiyone and M. Sekar. 2005. Ant species richness, diversity and similarity index at five selected localities of Sattur Taluk. Indian Journal of Environmental Education 5: 7–12.

Raksakantong, P., N. Meeso, J. Kubola and S. Siriamornpun. 2010. Fatty acids and proximate composition of eight Thai edible terricolous insects. Food Research International 43: 350–355.

Ramaiah, K.D., P.K. Das, E. Michael and H. Guyatt. 2000. The Economic burden of lymphatic filariasis in India. Parasitol. Today 16: 251.

Ramirez, R. and K. Zeoli. 2015. Skeeter syndrome—an uncommon insect bite reaction. Journal of the American Academy of Dermatology 72(5): AB80–AB80.

Ramos-Elorduy de Concini, J. and J.M. Pino Moreno. 1988. The utilization of insects in the empirical medicine of ancient Mexicans. Journal of Ethnobiology 8(2): 195–202.

Ramos-Elorduy, J. 1998. Creepy crawly cuisine: the gourmet guide to edible insects. Rochester, VT, Park Street Press.

Ramos-Elorduy, J. 2005. Insects: a hopeful food source. pp. 263–291. *In*: Paoletti, M.G. (ed.). Ecological Implications of Minilivestock; Role of Rodents, Frogs, Snails, and Insects for Sustainable Development. Enfield, NH, Science Publishers.

Ramos-Elorduy, J. 2009. Anthropo-entomophagy: Cultures, evolution and sustainability. Entomological Research 39(5): 271–288.

Ranjit, S. and N. Kissoon. 2011. Dengue hemorrhagic fever and shock syndromes. Pediatric Critical Care Medicine 12(1): 90–100. DOI: 10.1097/PCC.0b013e3181e911a7.

Raposeiro, P.M., G.M. Martins, I. Moniz, A. Cunha, A.C. Costa and V. Goncalves. 2014. Leaf litter decomposition in remote oceanic islands: the role of macroinvertebrates vs. microbial decomposition of native vs. exotic plant species. Limnologica 45: 80–87.

Rasnitsyn, A.P. and D.L.J. Quicke. 2002. History of Insects. Dordrecht, Kluwer Academic Publishers. 517 pp.

Rassi, A., A. Rassi and J.A. Marin-Neto. 2010. Chagas disease. Lancet 375(9723): 1388–1402. DOI: 10.1016/S0140-6736(10)60061-X.

Ratcliffe, N., P. Azambuja and C.B. Mello. 2014. Recent Advances in Developing Insect Natural Products as Potential Modern Day Medicines. Evidence-Based Complementary and Alternative Medicine 2014, Article ID 904958. 21 pp. http://dx.doi.org/10.1155/2014/904958.

Ratcliffe, N.A., C.B. Mello, E.S. Garcia, T.M. Butt and P. Azambuja. 2011. Insect natural products and processes: new treatments for human disease. Insect Biochemistry and Molecular Biology 41(10): 747–769.

Raw silk manufacture. Terms and definitions. 1998. GOST 3398-74. Moscow, Publishing House of Standards. 17 pp. http://standartgost.ru/g/ГОСТ_3398-74 [in Russian].

Ready, P.D. 2013. Biology of phlebotomine sand flies as vectors of disease agents. Annual Review of Entomology 58: 227–250. DOI: 10.1146/annurev-ento-120811-153557.

Red book of the Russian Federation. Vol. 1: Animals. 2001. Moscow, Astrel. [in Russian].

Regan, J.J., M.S. Traeger, D. Humpherys, D.L. Mahoney, M. Martinez, G.L. Emerson, D.M. Tack, A. Geissler, S. Yasmin, R. Lawson, V. Williams, C. Hamilton, C. Levy, K. Komatsu, D.A. Yost and J.H. McQuiston. 2015. Risk factors for fatal outcome from Rocky Mountain Spotted Fever in a highly endemic area–Arizona, 2002–2011. Clinical Infectious Diseases 60(11): 1659–1666.

Reimer, N. 2000. Biological Control of Weeds. Biological Control of Pests. Lecture Notes for ENTO 675. University of Hawaii at Manoa, Fall 2000, 52–59. http://nature.berkeley.edu/biocon/BC%20Class%20Notes/52-59%20%20Weed%20BC.pdf.

Reis, L.R.G.D. and A.D.D.A. Santos. 2014. Dieta de duas especies de peixes da familia Cichlidae (Astronotus ocellatus e Cichla pinima) introduzidos no rio Paraguacu, Bahia. Biotemas 27: 83–91 [in Portuguese].

Renner, J.N., K.M. Cherry and R.S.-C. Su. 2012. Characterization of resilin-based materials for tissue engineering applications. Biomacromolecules 13(11): 3678–3685.

Revzen, S., S.A. Burden, T.Y. Moore, J.-M. Mongeau and R.J. Full. 2013. Instantaneous kinematic phase reflects neuromechanical response to lateral perturbations of running cockroaches. Biology and Cybernetics 107: 179–200. DOI: 10.1007/s00422-012-0545-z.

Rezende, D., J.W. Melo, J.E. Oliveira and M.G. Gondim Jr. 2016. Estimated crop loss due to coconut mite and financial analysis of controlling the pest using the acaricide abamectin. Experimental and Applied Acarology 69(3): 297–310. DOI: 10.1007/s10493-016-0039-0.

Ribeiro, J.M.C. and B. Arca. 2009. From sialomes to the sialoverse: an insight into salivary potion of blood-feeding insects. Advances in Insect Physiology 37: 59e118.

Rice, E.D.L. 1986. Natural agents of plant protection against pests. Moscow, Mir. 194 pp.

Richardson, M. 2004. The management of allergic reaction to venomous insect stings. Nursing Practice, Clinical Research 100(32): 48.

Richter, A.G., P. Nightingale, A.P. Huissoon and M.T. Krishna. 2011. Risk factors for systemic reactions to bee venom in British beekeepers. Annals of Allergy, Asthma and Immunology 106(2): 159–163.

Riklefs, R. 1979. Fundamentals of general ecology. Mir, Moscow, 424 pp.

Ristic, M. and I. McIntyre (eds.). 1981. Diseases of Cattle in the Tropics: Economic and Zoonotic Relevance. Springer, 662 pages.

Rivers, V.Z., L. Mills and J. Revoir. 2011. Beetles in textiles. http://www.insects.org/ced2/beetles_tex.html.

Roach, D.A. and J.R. Carey. 2014. Population biology of aging in the wild. Annual Review of Ecology, Evolution, and Systematics 45: 421–443.

Rodríguez, H., A. Montoya, I. Miranda, Y. Rodríguez, T.L. Depestre, M. Ramos and M.H. Badii-Zabeh. 2015. Biological control of *Polyphagotarsonemus latus* (Banks) by the predatory mite *Amblyseius largoensis* (Muma) on sheltered pepper production in Cuba. Review of Protección Veg. 30(1): 70–76. ISSN: 2224-4697.

Rodríguez, L.C., M.A. Méndez and H.M. Niemeyer. 2001. Direction of dispersion of cochineal (*Dactylopius coccus* Costa) within the Americas. Antiquity 75: 73–77.

Rodriguez-Lainz, A., C.L. Fritz and W.R. McKenna. 1999. Animal and human health risks associated with Africanized honeybees. Journal of the American Veterinary Medicine Association 215: 1799–1804.

Rogina, B. 2011. For the special issue: aging studies in *Drosophila melanogaster*. Experimental Gerontology 46: 317–319.

Rolain, J.M., P. Brouqui, J.E. Koehler, C. Maguina, M.J. Dolan and D. Raoult. 2004. Recommendations for treatment of human infections caused by Bartonella species. Antimicrobial Agents and Chemotherapy 48(6): 1921–1933. DOI: 10.1128/AAC.48.6.1921-1933.2004.

Roques, A., M.-A. Auger-Rozenberg, T.M. Blackburn, J. Garnas, P. Pysek, W. Rabitsch, D.M. Richardson, M.J. Wingfield, A.M. Liebhold and R.P. Duncan. 2016. Temporal and interspecific variation in rates of spread for insect species invading Europe during the last 200 years. Biological Invasions 18(4): 907–920.

Rosin, Z.M., Ł. Myczko, P. Skorka, M. Lenda, D. Moron, T.H. Sparks and P. Tryjanowski. 2012. Butterfly responses to environmental factors in fragmented calcareous grasslands. Journal of Insect Conservation 16: 321–329. DOI 10.1007/s10841-011-9416-5.

Ross, G., H. Ross and J. Ross. 1985. Entomology. Moscow, Mir. 576 pp. [in Russian].

Roth, A., D. Hoy, P.F. Horwood, B. Ropa, T. Hancock, L. Guillaumot, K. Rickart, P. Frison, B. Pavlin and Y. Souares. 2014. Preparedness for threat of chikungunya in the Pacific. Emerging Infectious Diseases 20(8). DOI: 10.3201/eid2008.130696.

Rothman, J.M., D. Raubenheimer, M.A.H. Bryer, M. Takahashi and C.C. Gilbert. 2014. Nutritional contributions of insects to primate diets: Implications for primate evolution. Journal of Human Evolution 71: 59–69. DOI: 10.1016/j.jhevol.2014.02.016.

Roubieu, F.L., F. Expert, M. Boyron, B.-J. Fuschlock, S. Viollet and F. Ruffier. 2012. A novel 1-gram insect based device measuring visual motion along 5 optical directions. IEEE Sensors 2011 Conference, Limerick, Ireland, pp. 687–690.

Rousseau, M. 2013. Hexapod menu: a guide to culinary entomology. Pro Science 20(13). polit.ru/article/2013/05/23/ps_edible_insects/ [in Russian].

Rousseau, M. 2014. Chitin army. Pro Science 17(2). http://polit.ru/article/2014/02/22/ps_insect/ [in Russian].

Roy, H.E., T. Adriaens, N.B. Isaac, M. Kenis, G. San Martin y Gomez, T. Onkelinx, P.M.J. Brown, P. Poland, H.P. Ravn, D.B. Roy, R. Comont, J.C. Grégoire, J.C. de Biseau, L. Hautier, R. Eschen, R. Frost, R. Zindel, J. Van Vlaenderen, O. Nedved and D. Maes. 2012. Invasive alien predator causes rapid declines of native European ladybirds. Diversity and Distribution 18: 717–725.

Rozhkov, A.S. 1965. The mass reproduction of Siberian moth and measures to control them. Moscow, Nauka. 180 pp. [in Russian].

Ruchin, A.B. and S.K. Alekseev. 2012. To the study of feeding spectrum of three living together species of amphibians in the pine forest (Kaluga oblast). Bulletin of Penza State Pedagogical University named V.G. Belinsky 29: 82–85 [in Russian].

Ruchin, A.B. 2014. The spectrum of food of the grassy frog (*Rana temporaria*) in Mordovia. World of Science, Culture and Education 1(44): 387–391 [in Russian].

Rueff, F., B. Przybilla, M.B. Bilo, U. Muller, F. Scheipl, W. Aberer, J. Birnbaum, A. Bodzenta-Lukaszyk, F. Bonifazi, C. Bucher, P. Campi, U. Darsow, C. Egger, G. Haeberli, T. Hawranek, M. Korner, I. Kucharewicz, H. Kuchenhoff, R. Lang, O. Quercia, N. Reider, M. Severino, M. Sticherling, G.J. Sturm and B. Wuthrich. 2009. Predictors of severe systemic anaphylactic reactions in Hymenoptera venom allergy: the importance of baseline serum tryptase concentration and concurrent clinical variables. Journal of Allergy and Clinical Immunology 124: 1047–1054.

Rueppell, O., C. Bachelier, M.K. Fondrk and R.E. Page, Jr. 2007. Regulation of life history determines lifespan of worker honey bees (*Apis mellifera*). Experimental Gerontology 42: 1020–1032.

Rumpold, B.A. and O.K. Schlüter. 2013. Nutritional composition and safety aspects of edible insects. Molecular Nutrition and Food Research 57(5): 802–823. DOI: 10.1002/mnfr.201200735.

Rust, J., H. Singh, R.S. Rana, T. McCann, L. Singh, K. Anderson, N. Sarkar, P.C. Nascimbene, F. Stebner, J.C. Thomas, M. Solórzano Kraemer, C.J. Williams, M.S. Engel, A. Sahni and D. Grimaldi. 2010. Biogeographic and evolutionary implications of a diverse paleobiota in amber from the early Eocene of India. Proceedings of the National Academy of Sciences of the United States of America 107(43): 18360–18365. DOI: 10.1073/pnas.1007407107.

Rybicki, E.P. 2015. A top ten list for economically important plant viruses. Archives of Virology 160(1): 17–20.

Sadovsky, V.V. and N.M. Nesmelov. 2012. Commodity research and examination of textiles. Minsk, BSEU. 544 pp. [in Russian].

Saikim, F.H., A.M.R.A. Matusin, N.M. Suki and M.M. Dawood. 2015. Tourists perspective: Inclusion of entotourism concept in ecotourism activity. Journal of Tropical Biology and Conservation 12: 55–74. http://www.ums.edu.my/ibtpv2/images/publication/JTBC/JTBC-VOL-12/05_JTBC12_016_11.pdf.

Sailer, R.I. 1978. Our immigrant insect fauna. Bulletin of the Entomological Society of America 24(1): 3–11.

Sako, A., A.J. Mills and A.N. Roychoudhury. 2009. Rare earth and trace element geochemistry of termite mounds in central and northeastern Namibia: mechanisms for micro-nutrient accumulation. Geoderma. 153: 1–2.

Sallam, M.N. 2000. Insect damage: Damage on Post-harvest. FAO, Rome. 38 pp. www.fao.org/3/a-av013e.pdf.

Sampaio, M.V., V.H.P. Bueno, L.C.P. Silveira and A.M. Auad. 2009. Biological control of insect pests in the tropics. pp. 71–106. *In*: Del-Claro, K., P.S. Oliveira and V. Rico-Gray (eds.). Tropical Biology and Conservation Management—Volume III: Agriculture.

Sample, C.S., A.K. Xu, S.M. Swartz and L.J. Gibson. 2015. Nanomechanical properties of wing membrane layers in the house cricket (*Acheta domesticus* Linnaeus). Journal of Insect Physiology 74: 10–15. DOI: 10.1016/j.jinsphys.2015.01.013.

Samways, M.J. 2005. Insect Diversity Conservation. Cambridge University Press. 342 pp.

Samways, M.J. 2009. Insect conservation. Tropical biology and conservation management, pp. 142–180. http://www.eolss.net/sample-chapters/c20/e6-142-tpe-28.pdf.

Sanchez-Bayo, F. and K. Goka. 2014. Pesticide residues and bees—a risk assessment. Public Library of Science One 9(4). DOI: 10.1371/journal.pone.0094482.

Sánchez-Piñeroa, F. and F.C. Bolívar. 2004. Indirect effects of a non-target species, *Pyrrhalta luteola* (Chrysomelidae) on the biodeterioration of Brussels tapestries. International Biodeterioration and Biodegradation 54(4): 297–302. DOI: 10.1016/j.ibiod.2003.12.005.

Sane, S.P. 2003. The aerodynamics of insect flight. Journal of Experimental Biology 206: 4191–4208. DOI: 10.1242/jeb.00663. http://faculty.washington.edu/danielt/sane_review.pdf.

Sang Eon Shin, Min Suk Jang, Ji Hye Park and Seong Hwan Park. 2015. A forensic entomology case estimating the minimum postmortem interval using the distribution of fly pupae in fallow ground and maggots with freezing injury. Korean Journal of Legal Medicine 39: 17–21.

Sankarana, S., L.R. Khota and S. Panigrahi. 2012. Biology and applications of olfactory sensing system: A review. Sensors and Actuators B 171–172: 1–17. DOI: 10.1016/j.snb.2012.03.029.

Saranli, U., M. Buehler and D.E. Koditschek. 2001. RHex: A simple and highly mobile hexapod robot. The International Journal of Robotics Research 20(7): 616. DOI: 10.1177/02783640122067570.

Saravanan, D. 2006. Spider silk-structure, properties and spinning. Journal of Textile and Apparel, Technology and Management 5(1): 1–20.

Sarsour, J., T. Stegmaier, M. Linke and H. Planck. 2010. Bionic development of textile materials for harvesting water from fog. Proceedings of 5th International Conference on Fog, Fog Collection and Dew. Münster, Germany, 25–30 July 2010.

Sathe, T.V., A. Sathe and N.T. Sathe. 2013. Diversity of dipterous forensic insects from Western Maharashtra, India. International Journal of Pharma and Bio Sciences 4(2): 173–179.

Saulich, A.K. and D.L. Mussolin. 2013. Biology and ecology of parasitic Hymenoptera (Hymenoptera: Apocrita: Parasitica). St. Petersburg, St. Petersburg State University. 94 pp. [in Russian].

Schabel, H.G. 2010. Forest insects as food: a global review. pp. 37–64. *In*: Durst, P.B., D.V. Johnson, R.N. Leslie and K. Shono (eds.). Forest Insects as Food: Humans Bite Back. Proceedings of a Workshop on Asia-Pacific Resources and Their Potential for Development, 19–21 February 2008, Chiang Mai, Thailand. Bangkok: Food and Agriculture Organization of the United Nations.

Schauff, M.E. (ed.). 1986. Collecting and preserving insects and mites: techniques and tools. National Museum of Natural History, Washington, DC. 69 pp. http://www.ars.usda.gov/SP2UserFiles/ad_hoc/12754100CollectingandPreservingInsectsandMites/collpres.pdf.

Schenone, H., T. Saavedra, A. Rojas and F. Villarroel. 1989. Loxoscelism in Chile. Epidemiologic, clinical and experimental studies. Revista do Instituto de Medicina Tropical de São Paulo 31: 403–415.

Schiesari, L., J. Zuanon, C. Azevedo-Ramos, M. Garcia, M. Gordo, M. Messias and E.M. Vieira. 2003. Macrophyte rafts as dispersal vectors for fishes and amphibians in the lower Solimões River, Central Amazon. Journal of Tropical Ecology 19: 333–336.

Schmidt, A.R., V. Perrichot, M. Svojtka, K.B. Anderson, K.H. Belete, R. Bussert, H. Dörfelt, S. Jancke, B. Mohr, E. Mohrmann, P.C. Nascimbene, A. Nel, P. Nel, E. Ragazzi, G. Roghi, E.E. Saupe, K. Schmidt, H. Schneider, P.A. Selden and N. Vávra. 2010. Cretaceous African life captured in amber. Proceedings from the National Academy of Sciences 107(16): 7329–7334.

Schmidtmann, E.T., M.V. Herrero, A.L. Green, D.A. Dargatz, J.M. Rodriquez and T.E. Walton. 2011. Distribution of *Culicoides sonorensis* (Diptera: Ceratopogonidae) in Nebraska, South Dakota, and North Dakota: Clarifying the Epidemiology of Bluetongue Disease in the Northern Great Plains Region of the United States. In Other Publications in Zoonotics and Wildlife Disease. Paper 161. http://digitalcommons.unl.edu/zoonoticspub/161.

Schneider, J.C. (ed.). 2009. Principles and procedures for rearing high quality insects. Mississippi State University. 352 pp.

Schofield, C.J. and C. Galvno. 2009. Classification, evolution, and species groups within the Triatominae. Acta Tropica 110(2–3): 88–100.

Schowalter, T.D. 2013. Insects and Sustainability of Ecosystem Services. CRC Press. 362 pp.

Schuch, S., J. Bock, C. Leuschner, M. Schaefer and K. Wesche. 2011. Minor changes in orthopteran assemblages of Central European protected dry grasslands during the last 40 years. Journal of Insect Conservation 15: 811–822. DOI 10.1007/s10841-011-9379-6.

Schuetze, G.E., J. Forster, P.J. Hauk, K. Friedl and J. Kuehr. 2002. Bee-venom allergy in children: long-term predictive value of standardized challenge tests. Pediatric Allergy and Immunology 13: 18–23 [in Spanish].

Segal, S. 1990. Flowers and nature: Netherlandish flower painting of four centuries. The Hague: SDU Publishers, 302 pp.

Sekine, M. 2002. Not a cockfight but a spider-fight: Kajiki. http://www.natureoz.net/spfight.htm.

Selikhovkin, A.V. 2013. The response of dendrofagous insects on industrial air pollution. Biosphere 5(1): 47–76 [in Russian].

Sengupta, N., P. Akoury and R.S. Gautam. 2011. Report of the Study on Lac Sub Sector. Raipur, Chhattisgarh State Institute of Rural Development. 68 pp. http://www.cgsird.gov.in/Lac%20-Version%20-2.pdf.

Seong, K.M., C.S. Kim, S.W. Seo, and Y.W. Jin. 2013. Chronic low-dose radiation inhibits cisplatin-induced formation of tumorous clones in *Drosophila melanogaster* wts/+ heterozygotes. Entomological Research 43: 79–83.

Sergeev, V.A. and A.S. Blagodatskaya. 2015. Insects and bionics: mysteries of the visual system. Priroda 1: 22–27 [in Russian].

Sericulture in the countries of the world. Moscow, 1968. 24 pp. [in Russian].

Serrano, A., A. van den Doel, M. van Bommel, J. Hallett, I. Joosten and K.J. van den Berg. 2015. Investigation of crimson-dyed fibres for a new approach on the characterization of cochineal and kermes dyes in historical textiles. Analytica Chimica Acta 897: 116–127. http://dx.doi.org/10.1016/j.aca.2015.09.046.

Seymour, R.S. and S.K. Hetz. 2011. The diving bell and the spider: the physical gill of *Argyroneta aquatica*. Journal of Experimental Biology 214(13): 2175–2181.

Sezerino, U.M., M. Zannin, L.K. Coelho, J. Gonçalves Júnior, M. Grando, S.G. Mattosinho, J.L. Cardoso, V.R. von Eickstedt, F.O. França, K.C. Barbaro and H.W. Fan. 1998. A clinical and epidemiological study of Loxosceles spider envenoming in Santa Catarina, Brazil. Transactions of the Royal Society of Tropical Medicine and Hygiene 92: 546–548.

Shahbazzadeh, D., A. Amirkhani, N.D. Djadid, S. Bigdeli, A. Akbari and H. Ahari. 2009. Epidemiological and clinical survey of scorpionism in Khuzestan province, Iran. Toxicon 53(4): 454–459.

Shapiro, E.D. 2014. Clinical practice. Lyme disease. New England Journal of Medicine 370(18): 1724–1731.

Shatza, A.J., J. Rogana, F. Sangermanoaa, Y. Ogneva-Himmelbergerb and H. Chen. 2013. Characterizing the potential distribution of the invasive Asian longhorned beetle (Anoplophora glabripennis) in Worcester County, Massachusetts. Applied Geography 45: 259–268.

Shek, L.P.-C., N.S.P. Ngiam and B.-W. Lee. 2004. Ant allergy in Asia and Australia. Current Opinion in Allergy and Clinical Immunology 4(4): 325–328.

Sherman, R.A. 2014. Mechanisms of maggot-induced wound healing: what do we know, and where do we go from here? Evidence-Based Complementary and Alternative Medicine 2014. Article ID 592419. 13 pp. http://dx.doi.org/10.1155/2014/592419.

Shetty, P.K. and M. Sabitha. 2009. Economic and ecological externalities of pesticide use in India. pp. 113–130. *In*: Peshin, R. and A.K. Dhawan (eds.). Integrated Pest Management: Innovation-Development Process. DOI: 10.1007/978-1-4020-8992-35.

Sheumack, D.D., B.A. Baldo, P.R. Carroll, F. Hampson, M.E. Howden and A. Skorulis. 1984. A comparative study of properties and toxic constituents of funnel web spider (Atrax) venoms. Comparative Biochemistry and Physiology 78(1): 55–68.

Shin, E.H., J.Y. Roh, W.I. Park, B.G. Song, K.-S. Chang, W.-G. Lee, H.I. Lee, C. Park, M.-Y. Park and E-H. Shin. 2014. Transovarial transmission of *Orientia tsutsugamushi* in *Leptotrombidium palpale* (Acari: *Trombiculidae*). Public Library of Science One 9(4): e88453. DOI: 10.1371/journal.pone.0088453.

Shin, S.E., M.S. Jang, J.H. Park and S.H. Park. 2015. A forensic entomology case estimating the minimum postmortem interval using the distribution of fly pupae in fallow ground and maggots with freezing injury. Korean Journal of Legal Medicine 39(1): 17–21.

Shkenderov, S. and T.S. Ivanov. 1985. Bees products. Sofia, Zemizdat. 227 pp. [in Russian].

Shteinmann, G. 1984. Weapons of animals. Moscow, Lesnaya Promyshlennost. 144 pp. [in Russian].

Sidorchuk, E.A. 2004. Subfossil oribatid mites as the bioindicators of the deep environmental change during the holocene. Reports of Academy of Sciences 396(5): 710–713 [in Russian].

Sikka, V., V.K. Chattu, R.K. Popli, S. Galwankar, D. Kelkar, S.G. Sawicki, S.P. Stawicki and T.J. Papadimos. 2016. The emergence of zika virus as a global health security threat: A review and a consensus statement of the INDUSEM Joint Working Group (JWG). Journal of Global Infectious Diseases 8(1): 3–15.

Silva, J.C., F.N. Oliveira, K.G. Moreira, A.B. Mayer, D.O. Freire, M.D. Cherobim, N. Gomes de Oliveira Junior, C.A. Schwartz, E.F. Schwartz and M.R. Mortari. 2016. Pathophysiological effects caused by the venom of the social wasp *Synoeca surinama*. Toxicon 113: 41–48.

Simarro, P.P., G. Cecchi, J.R. Franco, M. Paone, A. Diarra, J.A. Ruiz-Postigo, E.M. Fèvre, R.C. Mattioli and J.G. Jannin. 2012. Estimating and mapping the population at risk of sleeping sickness. PLOS Neglected Tropical Diseases 6(10): e1859. DOI: 10.1371/journal.pntd.0001859.

Simics, M. 1999. Basic bee venom in products. http://www.beevenom.com/basicBVinProducts.htm.

Simmons, C.P., J.J. Farrar, N. van Vinh Chau and B. Wills. 2012. Dengue. New England Journal of Medicine 366(15): 1423–1432. DOI: 10.1056/NEJMra1110265.

Simó, M. and D.A. Brescovit. 2001. Revision and cladistic analysis of the Neotropical spider genus *Phoneutria* Perty, 1833 (*Araneae, Ctenidae*), with notes on related *Cteninae*. Bulletin British Arachnology Society 12(2): 67–82.

Simon, M.R. and Z.D. Mulla. 2008. A population-based epidemiologic analysis of deaths from anaphylaxis in Florida. Allergy 63: 1077–1083.

Simon-Delso, N., G. San Martin, E. Bruneau, L.-A. Minsart, C. Mouret and L. Hautier. 2014. Honeybee colony disorder in crop areas: The role of pesticides and viruses. Public Library of Science One 9(7): e103073. DOI: 10.1371/journal.pone.0103073.

simplybutterfliesproject.com/conservation_center.html.

Sin, S.-H., B.C. McNulty, G.G. Kennedy and J.W. Moyer. 2005. Viral genetic determinants for thrips transmission of Tomato spotted wilt virus. Proceedings of the National Academy of Sciences of the United States of America 102: 5168–5173.

Singh, J. and B.R. Sharma. 2008. Forensic entomology: A supplement to forensic death investigation. JPAFMAT 8(1): 26–33. ISSN 0972-5687.

Singh, R. 2006. Lac culture. Varanasi: Udai Pratap Autonomous College. 18 pp. http://nsdl.niscair.res.in/jspui/bitstream/123456789/219/1/LAC%20CULTURE.pdf.

Singh, S. and B.K. Mann. 2013. Insect bite reactions. Indian J. Dermatol. Venereol. Leprol. 79: 151–64.

Sirimungkararat, S., W. Saksirirat, T. Nopparat and A. Natongkham. 2010. Edible products from eri and mulberry silkworms in Thailand. pp. 189–200. *In*: Durst, P.B., D.V. Johnson, R.N. Leslie and K. Shono (eds.). Forest Insects as Food: Humans Bite Back. Proceedings of a Workshop on Asia-Pacific Resources and Their Potential for Development, 19–21 February 2008, Chiang Mai, Thailand. Bangkok: Food and Agriculture Organization of the United Nations.

Skorer, R. 1980. Aerohydrodynamics of environment. Moscow, Mir. 549 pp. [in Russian].

Skurlatova, M.V. 2015. Bionics as the relationship of nature and technology. Young Scientist 10: 1283–1289 [in Russian].

Slepchenko, L.G. 2006. Course of lectures on entomology. Textbook. Grodno. 189 pp. [in Russian].

Small, G.E., P.J. Torres, L.M. Schweizer, J.H. Duff and C.M. Pringle. 2013. Importance of terrestrial arthropods as subsidies in lowland neotropical rain forest stream ecosystems. Biotropica 45(1): 80–87. DOI: 10.1111/j.1744-7429.2012.00896.x.

Smil, V. 2002. Worldwide transformation of diets, burdens of meat production and opportunities for novel food proteins. Enzyme and Microbial Technology 30: 305–311.

Smith, K. 1978. Principles of applied meteorology. Leningrad, Gidrometeoizdat. 422 pp. [in Russian].

Smith, K.G.V. 1986. A manual of forensic entomology. London, British Museum (Natural History), Cornell University Press. 205 pp.

Sokolov, V.I. 2008. Natural resources of USA: peculiarities of location, evaluations, application. Moscow, ISL RAS, 148 pp. [in Russian].

Soper, A.L., C.J.K. MacQuarrie and R. Van Driesche. 2015. Introduction, establishment, and impact of *Lathrolestes thomsoni* (Hymenoptera: Ichneumonidae) for biological control of the ambermarked birch leafminer, *Profenusa thomsoni* (Hymenoptera: Tenthredinidae), in Alaska. Biological Control 83: 13–19. DOI: 10.1016/j.biocontrol.2014.12.018.

Sorte, J.B. 2013. Predicting persistence in a changing climate: flow direction and limitations to redistribution. Oikos 122: 161–170. DOI: 10.1111/j.1600-0706.2012.00066.x.

Soszynska-Maj, A. and W. Krzeminski. 2015. New representative of the family Panorpodidae (Insecta, Mecoptera) from Eocene Baltic Amber with a key to fossil species of genus Panorpodes. Palaeontologia Electronica 18(2): 33A.

Souza, A.S.B., F.D. Kirst and R.F. Krüger. 2008. Insects of forensic importance from Rio Grande do Sul state in southern Brazil. Revista Brasileira de Entomologia 52(4): 641–646.

Speranskaya, K.S. and A.S. Zaytsev. 2011. Rare insects species conservation in the protected areas of European part of Russia. Geographical fundamentals of ecological nets creation in Russia and Eastern Europe. Moscow, KMK, pp. 259–263. http://www.econet2011.narod.ru/Speranskaya_Zaitsev.htm [in Russian].

Spider bites. 2009. http://www.infectiousdisease.dhh.louisiana.gov.

Sponberg, A., T. Daniel and A. Fairhall. 2015. Decoding synergistic and independent motor features in insect flight muscles in *Manduca sexta*. Public Library of Science Computational Biology. DOI: 10.1371/journal.pcbi.1004168.

Sprygin, A.V., O.A. Fedorova, Y.Y. Babin, A.V. Kononov and A.K. Karaulov. 2015. Blood-sucking midges from the genus Culicoides (Diptera: Ceratopogonidae) act as filed vectors of human and animal diseases (review). Sel'skokhozyaistvennaya Biologiya 50(2): 183–197. DOI: 10.15389/agrobiology.2015.2.183rus [in Russian].

Srebrodolsky, B.I. 1984. Amber. Moscow, Nuauka. 112 pp. [in Russian].

Srivastava, S.K., N. Babu and H. Pandey. 2009. Traditional insect bioprospecting—As human food and medicine. Indian Journal of Traditional Knowledge 8(4): 485–494.

Stanaway, J.D., D.S. Shepard, E.A. Undurraga, Y.A. Halasa, L.E. Coffeng, O.J. Brady, S.I. Hay, N. Bedi, I.M. Bensenor, C.A. Castañeda-Orjuela, T.W. Chuang, K.B. Gibney, Z.A. Memish, A. Rafay, K.N. Ukwaja, N. Yonemoto and C.J. Murray. 2016. The global burden of dengue: an analysis from the Global Burden of Disease Study 2013. Lancet Infectious Diseases 16(6): 712–723. DOI: 10.1016/S1473-3099(16)00026-8.

Starukhina, A.O. and V.S. Bondarchuk. 2014. Tobacco mosaic tobamovirus. http://www.scienceforum.ru/2014/pdf/3744.pdf [in Russian].

Steele, W.K., R.L.C. Pilgrim and R.L. Palma. 1997. Occurrence of the flea *Glaciopsyllus antarcticus* and avian lice in central Dronning Maud Land. Polar Biology 18: 292–294. DOI: 10.1007/s003000050190.

Stegmaier, T., V. von Arnim, M. Linke, M. Milwich, J. Sarsour, A. Scherrieble, P. Schneider and H. Planck. 2008. Bionic developments based on textile materials for technical applications. *In*: Abbott, A. and M. Ellison (eds.). Biologically Inspired Textiles. Cambridge, Woodhead Publishing Limited.

Stenesen, D., J.M. Suh, J. Seo, K. Yu, K.S. Lee, J.S. Kim, K.J. Min and J.M. Graff. 2013. Adenosine nucleotide biosynthesis and AMPK regulate adult life span and mediate the longevity benefit of caloric restriction in flies. Cell Metabolism 17: 101–112.

Stenroth, K., L.E. Polvi, E. Faltstrom and M. Jonsson. 2015. Land-use effects on terrestrial consumers through changed size structure of aquatic insects. Freshwater Biology 60(1): 136–149. DOI: 10.1111/fwb.12476.

Stevens, L., P.L. Dorn, J.O. Schmidt, J.H. Klotz, D. Lucero and S.A. Klotz. 2011. Kissing Bugs. The Vectors of Chagas. Chapter 8. Advances in Parasitology 75: 169–192. Elsevier Ltd. ISSN 0065-308X. DOI: 10.1016/B978-0-12-385863-4.00008-3.

Stewart, A.D. and R.R. Anand. 2012. Source of anomalous gold concentrations in termite nests, Moolart Well, Western Australia: implications for exploration. Geochemistry: Exploration, Environment, Analysis 12(4): 327–337. DOI: 10.1144/geochem2012-126.

Stewart, A.D. and R.R. Anand. 2014. Anomalies in insect nest structures at the Garden Well gold deposit: Investigation of mound-forming termites, subterranean termites and ants. Journal of Geochemical Exploration 140: 77–86. DOI: 10.1016/j.gexplo.2014.02.011.

St-Hilaire, S., K. Cranfill, M.A. Mcguire, E.E. Mosley, J.K. Tomberlin, L. Newton, W. Sealey, C. Sheppard and S. Irving. 2007. Fish offal recycling by the Black Soldier Fly produces a foodstuff high in omega-3 fatty acids. Journal of the World Aquaculture Society 38(2): 309–313.

Straw, N.A., N.J. Fielding, C. Tilbury, D.T. Williams and T. Cull. 2016. History and development of an isolated outbreak of Asian longhorn beetle *Anoplophora glabripennis* (Coleoptera: Cerambycidae) in southern England. Agricultural and Forest Entomology. DOI: 10.1111/afe.12160.

Su, N.Y. and R.H. Scheffrahn. 2000. Termites as pests of buildings. *In*: Termites: Evolution, Sociality, Symbioses, Ecology. Springer Netherlands, pp. 437–453. DOI: 10.1007/978-94-017-3223-9_20.

Su, N.Y. 2002. Novel technologies for subterranean termite control. Sociobiology 40: 95–101.

Su, R.S.-C., Y. Kim and J.C. Liu. 2014. Resilin: Protein-based elastomeric biomaterials. Acta Biomaterialia 10(4): 1601–1611.

Suesirisawad, S., N. Malainual, A. Tungtrongchitr, P. Chatchatee, N. Suratannon and J. Ngamphaiboon. 2015. Dust mite infestation in cooking flour: experimental observations and practical recommendations. Asian Pacific Journal of Allergy and Immunology 33: 123–128.

Sun, J., Z. Miao, Z. Zhang, Z. Zhang and N.E. Gillette. 2004. Red turpentine beetle, *Dendroctonus valens* LeConte (Coleoptera: Scolytidae), response to host semiochemicals in China. Environmental Entomology 33(2): 206–212. http://naldc.nal.usda.gov/download/9017/PDF.

Sun, J.-M., L.-M. Zhang and X. Fang. 2009. A numerical taxonomic study of Phlebotominae (Diptera: Psychodidae) from China. Acta Entomologica Sinica 52(12): 1356–1365.

Sun, S.S.M. 2008. Application of agricultural biotechnology to improve food nutrition and healthcare products. Asia Pacific Journal of Clinical Nutrition 17: 87–90.

Sun, X.D., X.N. Fan, M.Q. Ling, H.H. Zhuang, Z. Zhang, M. Li and X.W. Zhang. 2014. Simulation of the mechanism of insect bionic compound eye super-resolution restoration technology of copper strip surface defect images. Progress in Mechatronics and Information Technology, pp. 302–307. DOI: 10.4028/www.scientific.net/AMM.462-463.302.

Sun, Y. and R. Xu. 2003. Ability of Ixodes persulcatus, *Haemaphysalis concinna* and *Dermacentor silvarum* ticks to acquire and transstadially transmit *Borrelia garinii*. Experimental and Applied Acarology 31(1–2): 151–160. DOI: 10.1023/B:APPA.0000005119.30172.43.

Supotnitsky, M.V. and N.S. Supotnitsky. 2006. Essays on the history of plague. Vol. I: Plague before bacteriaceae period. Moscow, University Book. 468 pp. [in Russian].

Surya Prakash Rao, K. and S.V. Raju. 1984. Geochemical analyses of termite mounds as a prospecting tool for tin deposits in Bastar, MP—a preliminary study. Proceedings of the Indian Academy of Sciences—Earth and Planetary Sciences 93(2): 141–148. DOI: 10.1007/BF02871994.

Süss, J. 2008. Tick-borne encephalitis in Europe and beyond—the epidemiological situation as of 2007. Eurosurveillance 13(4–6): 8. http://www.eurosurveillance.org/images/dynamic/EE/V13N26/art18916.pdf.

Suter, G.W. and S.M. Cormier. 2015. Why care about aquatic insects: Uses, benefits, and services. Integrated Environmental Assessment and Management 11(2): 188–194. DOI: 10.1002/ieam.1600.

Suter, R.B. 1999. Cheap transport for fishing spiders (Araneae, Pisauridae): The physics of sailing on the water surface. Journal of Arachnology 27: 489–496.

Sutherland, S.K., A.W. Duncan and J. Tibballs. 1980. Local inactivation of funnel-web spider (Atrax robustus) venom by first-aid measures: potentially lifesaving part of treatment. Medical Journal of Australia 2(8): 435–437.

Sviridov, A.V. 2011. Principles of insects protection (after the example of Lepidoptera): history and perspectives. Bulletin of Moscow Society of Naturalists, Biology Department 116(6): 3–18 [in Russian].

Swaay van, C.A.M., P. Nowicki, J. Settele and A.J. van Strien. 2008. Butterfly monitoring in Europe: methods, applications and perspectives. Biodiversity Conservation 17: 3455–3469.

Swe, P.M., S.L. Reynolds and K. Fischer. 2014. Parasitic scabies mites and associated bacteria joining forces against host complement defence. Parasite Immunology 36(11): 583–591.

Swengel, S.R., D. Schlicht, F. Olsen and A.B. Swengel. 2011. Declines of prairie butterflies in the midwestern US. J. Insect. Conserv. 15: 327–339. DOI: 10.1007/s10841-010-9323-1.

Syrjanen, J., K. Korsu, P. Louhi, R. Paavola and T. Muotka. 2011. Stream salmonids as opportunistic foragers: the importance of terrestrial invertebrates along a stream-size gradient. Canadian Journal of Fisheries and Aquatic Sciences 68(12): 2146–2156. DOI: 10.1139/F2011-118.

Tabor, K.L. 2004. Succession and Development Studies on Carrion Insects of Forensic Importance. Dissertation submitted to the Faculty of Virginia Polytechnic Institute and State University in partial fulfillment of the requirements for the degree of Doctor of Philosophy in Entomology. 129 pp.

Takeda, J.T. 2014. Global insects: silkworms, sericulture, and statecraft in Napoleonic France and Tokugawa Japan. French History 28(2): 207–225. DOI: 10.1093/fh/cru044.

Talyzin, F.F. 1970. Venomous sea and land animals. Moscow, Znanie, 1970. 96 pp. [in Russian].

Tankersley, M.S. and D.K. Ledford. 2015. Stinging insect allergy: State of the art 2015. The Journal of Allergy and Clinical Immunology in Practice 3(3): 315–322.

Tansky, V.I. 1988. Biological principles of insect harmfulness. Agropromizdat, Moscow, 182 pp. [in Russian].

Tarasov, V.V. 1996. Medical entomology. Moscow State University, Moscow, 352 pp. [in Russian].

Tarasov, V.V. 2002. The epidemiology of vector-borne diseases. Moscow, MGU, 2002. 336 pp. [in Russian].

Tarnani, I.K. 1907. Our venomous animals (biology, damage, and control measures). St. Petersburg, 127 pp. [in Russian].

Tas, E., U. Jappe, H. Beltraminelli, M. Worm, J. Kleine-Tebbe and T. Werfel. 2007. Occupational inhalant allergy to the common housefly (*Musca domestica*). Hautarzt 58(2): 156–160.

Taylor, A.J., D.H. Paris and P.N. Newton. 2015. A systematic review of mortality from untreated scrub Typhus (*Orientia tsutsugamushi*). Public Library of Science Negl. Trop. Dis. 9(8): e0003971. DOI: 10.1371/journal.pntd.0003971.

Taylor, E. 2013. What insects are used as indicators for measuring stream pollution? http://www.ehow.com/info_12297774_insects-used-indicators-measuring-stream-pollution.html.

Termite risk management. 2002. 24 pp. http://www.timber.net.au/images/downloads/termites/termite_risk_management_builders.pdf.

The Global Silk Industry: Perception of European Operators toward Thai Natural and Organic Silk Fabric and Final Products. 2011. http://www.fibre2fashion.com/industry-article/6015/the-global-silk-industry.

Thomas, J. 2006. Biomimetic building uses termite mound as model. http://www.treehugger.com/sustainable-product-design/biomimetic-building-uses-termite-mound-as-model.html.

Thomas, M.L. 2001. Dung beetle benefits in the pasture ecosystem. 12 pp. https://attra.ncat.org/attra-pub/download.php?

Thomas, S.E. and C.A. Ellison. 2000. A century of classical biological control of *Lantana camara*: Can pathogens make a significant difference? Proceedings of the X International Symposium on Biological Control of Weeds 4–14 July 1999, Bozeman, MT, Montana State University, pp. 97–104. http://www.invasive.org/publications/xsymposium/proceed/01pg97.pdf.

Thomas, W.R., W.-A. Smith and B.J. Hales. 2010. Geography of house dust mite allergens. Asian Pac. J. Allergy Immunol. 28: 211–224.

Thompson, R. 2003. Close-Up on Insects: A Photographers Guide. Guild of Master Craftsmen Publications.

Thompson, R. 2007. Close Up and Macro: A Photographers Guide. David & Charles Plc. 160 pp.

Tikhonov, A. 2002. Red book of Russia. The animals and plants. Moscow: Rosmen. 414 pp. [in Russian].

Timkina, M.V. and O.A. Odintsev. 2014. Biotopical changes of batrachians nutricion spectrum of south west taiga of Sibiria (based on the Tyumen region). http://www.scienceforum.ru/2014/539/1717 [in Russian].

Tisdell, C. and C. Wilson. 2012. Nature-based Tourism and Conservation: New Economic Insights and Case Studies. Cheltenham, U.K. and Northampton, MA, U.S.A.: Edward Elgar. 506 pp.

Tisdell, C.A. 1990. Economic impact of biological control of weeds and insects. pp. 301–316. *In*: Mackauer, M., L.E. Ehler and J. Roland (eds.). Critical Issues in Biological Control. Andover, MA, Intercept.

Tiu, L.G. 2012. Enhancing Sustainability of Freshwater Prawn Production in Ohio. Ohio State University South Centers Newsletter, Fall 2012.

Tixier, T., J.-P. Lumaret and G.T. Sullivan. 2015. Contribution of the timing of the successive waves of insect colonisation to dung removal in a grazed agro-ecosystem. European Journal of Soil Biology 69: 88–93.

Togola, A., P.A. Seck, I.A. Glitho, A. Diagne, C. Adda, A. Toure and F.E. Nwilene. 2013. Economic losses from insect pest infestation on rice stored on-farm in Benin. Journal of Applied Sciences 13: 278–285. DOI: 10.3923/jas.2013.278.285. http://scialert.net/abstract/?doi=jas.2013.278.285.

Tolkanitz, V.I. and E.E. Perkovsky. 2015. A new species of the genus Paxylommites (Hymenoptera, Ichneumonidae, Hybrizoninae) from baltic amber. Paleontological Journal 49(4): 391–393.

Tolle, M.A. 2009. Mosquito-borne diseases. Current Problems in Pediatric and Adolescent Health Care 39(4): 97–140. DOI: 10.1016/j.cppeds.2009.01.001.

Tomberlin, J.K. and M.E. Benbow (eds.). 2015. Forensic Entomology: International Dimensions and Frontiers. CRC Press. 468 pp.

Toskina, I.N. 1998. Insects—pests of art treasures. Moscow, State Research Institute of Restoration. 43 pp. [in Russian].

Toskina, I.N. and I.N. Provorova. 2007. Insects in museums (Biology, Prevention of infection, Control measures). Moscow, KMK. 220 pp. [in Russian].

Truc, P., P. Buscher, G. Cuny, M.I. Gonzatti, J. Jannin, P. Joshi, P. Juyal, Z.R. Lun and P. Mattioli. 2013. Atypical human infections by animal Trypanosomes. PLOS Neglected Tropical Diseases 7(9): e2256. DOI: 10.1371/journal.pntd.0002256.

Truong, T.V., D. Byun, L.C. Lavine, D.J. Emlen, H.C. Park and M.J. Kim. 2012. Flight behavior of the rhinoceros beetle *Trypoxylus dichotomus* during electrical nerve stimulation. Bioinspiration and Biomimetics 7(3). DOI: 10.1088/1748-3182/7/3/036021.

382 Human–Insect Interactions

Tsai, C.-Y. and C.C. Chang. 2013. Auto-adhesive transdermal drug delivery patches using beetle inspired micropillar structures. J. Mater Chem. B 1: 5963–5970. DOI: 10.1039/C3TB20735H.
Tsatsenko, L.V. 2015. Using examples of paintings in teaching cytology as a discipline. The Scientific Journal of the KubSAU, 111(07). http://ej.kubagro.ru/2015/07/pdf/14.pdf [in Russian].
Tsurikov, M.N. 2005. Invertebrates: should we be afraid of them? http://humane.evol.nw.ru/popbp3.html [in Russian].
Tuskes, P.M., J.P. Tuttle and M.M. Collins. 1996. The Wild Silk Moths of North America: The Natural History of the Saturniidae of the United States and Canada. Ithaca, NY, Cornell University Press. 250 pp.
Tüzün, A., F. Dabiri and S. Yüksel. 2010. Preliminary study and identification of insects' species of forensic importance in Urmia, Iran. African Journal of Biotechnology 9(24): 3649–3658.
Tyumaseva, Z.I. and E.V. Guskova. 2013. Study of local lore. Invertebrate animals of the southern Urals. Chelyabinsk, Abris. 128 pp. [in Russian].
U.S. Food and Drug Administration. Defect Levels Handbook. 2014. The Food Defect Action Levels. Levels of natural or unavoidable defects in foods that present no health hazards for humans. Silver spring, MD: U.S. Food and Drug Administration. https://www.fda.gov/RegulatoryInformation/Guidances/ucm056174.htm.
Uawonggul, N., A. Chaveerach, S. Thammasirirak, T. Arkaravichien, C. Chuachan and S. Daduang. 2006. Screening of plants acting against Heterometrus laoticus scorpion venom activity on fibroblast cell lysis. Journal of Ethnopharmacology 103(2): 201–207.
Underwood, D.L.A. 2008. Insects as bioindicators. California State University: Long Beach. 9 pp. http://web.csulb.edu/~dlunderw/entomology/20-InsectsBioindicators.pdf.
Upadhyay, R.K., K.G. Mukerji and B.P. Chamola (eds.). 2001. Biocontrol potential and its exploitation in sustainable agriculture: Insect pests Vol. 2.: New York, Springer Science and Business Media. 421 pp.
Uzhegov, G.N. 2007. Large family encyclopedia of folk medicine from the doctor Uzhegova. Moscow, Olma Press. 1200 pp. [in Russian].
Vail, P.V., J.R. Coulson, W.C. Kauffman and M.E. Dix. 2001. History of biological control programs in the United States Department of Agriculture. American Entomologist 47(1).
Valerio, C.E. 1977. Population structure in the spider Achaearanea tepidariorum (Araneae, Theridridae). Journal of Arachnology 3: 185–190.
Vance, J.T. and S.P. Roberts. 2014. The effects of artificial wing wear on the flight capacity of the honey bee Apis mellifera. Journal of Insect Physiology 65: 27–36.
Vanlandingham, D.L., S. Higgs and Y.J.S. Huang. 2016. Aedes albopictus (Diptera: Culicidae) and mosquito-borne viruses in the United States. Journal of Medical Entomology 53(5): 1024–1028. DOI: 10.1093/jme/tjw025.
Vantaux, A., O. Roux, A. Magro, N.T. Ghomsi, R.D. Gordon and A. Dejean. 2010. Host-specific myrmecophily and myrmecophagy in the tropical coccinellid Diomus thoracicus in French Guiana. Biotropica 42(5): 622–629.
Vavra, N. 2009. Amber, fossil resins, and copal—contributions to the terminology of fossil plant resins. Denisia 26, zugleich Kataloge der oberösterreichischen Landesmuseen Neue Serie 86: 213–222. http://www.zobodat.at/pdf/DENISIA_0026_0213-0222.pdf.
Vazhov, V.M. 2013.Buckwheat in the fields of Altai: Moscow, Publishing House of the Academy of Natural Sciences. 188 pp. [in Russian].
Vega, F.E., F. Infante, A. Castillo and J. Jaramillo. 2009. The coffee berry borer, Hypothenemus hampei (Ferrari) (Coleoptera: Curculionidae): a short review, with recent findings and future research directions. Terrestrial Arthropod Reviews 2: 129–147. DOI 10.1163/187498209X12525675906031.
Vega, F.E. and R.W. Hofstetter (eds.). 2015. Bark beetles. Biology and Ecology of Native and Invasive Species. Elsevier. 641 pp.
Veldkamp, T., G. van Duinkerken, A. van Huis, C.M.M. Lakemond, E. Ottevanger, G. Bosch and M.A.J.S. van Boekel. 2012. Insects as a sustainable feed ingredient in pig and poultry diets—a feasibility study. Wageningen UR Livestock Research. Report 638. 62 pp. https://www.wageningenur.nl/upload_mm/2/8/0/f26765b9-98b2-49a7-ae43-5251c5b694f6_234247%5B1%5D.
Veldtman, R. 2005. The ecology of southern African wild silk moths (Gonometa species, Lepidoptera: Lasiocampidae): consequences for their sustainable use. Dissertation for the degree Philosophiae Doctor (Entomology). Pretoria, University of Pretoria. 230 pp.
Veltman, K. 2012. Butterfly conservatories, butterfly ranches and insectariums. pp. 189–197. In: Lemelin, R.H. (ed.). The Management of Insects in Recreation and Tourism [online]. Cambridge: Cambridge University Press. http://dx.doi.org/10.1017/CBO9781139003339.015.
Vendrely, C. and T. Scheibel. 2007. Biotechnological production of spider-silk proteins enables new applications. Macromolecular Bioscience 7: 401–409. DOI: 10.1002/mabi.200600255.
Verma, M., S. Sharma and R. Prasad. 2009. Biological alternatives for termite control: A review. International Biodeterioration and Biodegradation 63: 959–972.
Vetter, R.S. and P.K. Visscher. 1998. Bites and stings of medically important venomous arthropods. International Journal of Dermatology 37: 481–496.
Viegas, G., C. Stener, U.H. Schulz and L. Maltchik. 2014. Dung beetle communities as biological indicators of riparian forest widths in southern Brazil. Ecological Indicators 36: 703–710. DOI: 10.1016/j.ecolind.2013.09.036.
Vincent, J.F.V. and M.J. King. 1996. The mechanism of drilling by wood wasp ovipositors. Biomimetics 3: 187–201.

Vlasov, Y.I. and T.N. Teploukhova. 1993. Brome mosaic virus and ways of its distribution in nature. pp. 7–8. *In*: Vlasov, Y.I. (ed.). Proceedings of Research Conference "Problems of Viral Diseases on Cereals and Ways of their Solving". Tver': Vniimz [in Russian].

Vlckova, J., V. Rupes, D. Horakova, H. Kollarova and O. Holy. 2015. West Nile virus transmission risk in the Czech Republic. Epidemiologie Mikrobiologie Imunologie 64(2): 80–86.

Voisin, R. 2015. Festo unveils their new bionic insects. http://substance-en.etsmtl.ca/festo-unveils-its-new-bionic-insects/.

Volovnik, S.V. 1990. Our familiar strangers. Dnepropetrovsk, Promin. 189 pp. [in Russian].

von Eickstedt, V.R.D., L.A. Ribeiro, D.M. Candido, M.J. Albuquerque and M.T. Jorge. 1996. Evolution of scorpionism by *Tityus bahiensis* (perty) and *Tityus serrulatus* lutz and mello and geographical distribution of the two species in the state of São Paulo—Brazil. Journal of Venomous Animal Toxins 2(2): 123–131.

Vorontsov, A.I. 1981. Insects—destroyers of wood. Moscow, Lesnaya Promyshlennost. 176 pp. [in Russian].

Vorontsov, A.I. 1982. Forest entomology. Moscow, Vysshaya Shkola. 384 pp. [in Russian].

Vtorova, M.A. 2012. Insects in biological research. Fourth International Student Electronic Scientific Conference: Student Scientific Forum, 15 February–31 March 2012. http://www.rae.ru/forum2012/pdf/1500.pdf [in Russian].

Wairimu, W.P. 2010. Effect of pineapple (*Ananas comosus* L. Merrill) and papaya (*Carica papaya* L.) fruit extracts on sericin removal from silk moths cocoons in Kenya. A thesis submitted in partial fulfilment of the requirements for the award of the degree of master of science (agricultural entomology) in the school of pure and applied sciences of Kenyatta University. Nairobi, Kenya. 63 pp.

Walsh, G.C., Y.M. Dalto, F.M. Mattioli, R.I. Carruthers and L.W. Anderson. 2013. Biology and ecology of Brazilian elodea (*Egeria densa*) and its specific herbivore, *Hydrellia* sp., in Argentina. BioControl 58: 133–147. DOI: 10.1007/s10526-012-9475-x.

Walter, J., E. Fletcher, R. Moussaoui, K. Gandhi and C. Weirauch. 2012. Do bites of kissing bugs cause unexplained allergies? Results from a survey in triatomine-exposed and unexposed areas in Southern California. Public Library of Science One 7(8): e44016. DOI: 10.1371/journal.pone.0044016.

Wamwiri, F.N. and R.E. Changasi. 2016. Tsetse flies (Glossina) as vectors of human african Trypanosomiasis: A review. Biomed. Research International. Number: 6201350. DOI: 10.1155/2016/6201350.

Wang, L., L. Yongyue, X. Yijuan and Z. Ling. 2013. The current status of research on *Solenopsis invicta* Buren (Hymenoptera: Formicidae) in mainland China. Asian Myrmecology 5: 125–137. ISSN 1985-1944.

Wang, P., W. Du and M. Li. 2011. Global warming and adaptive changes of forest pests. Journal of the Northwest Forestry University 26: 124–128.

Wang, Z., Y. Zhang, J. Zhang, L. Huang, J. Liu, Y. Li, G. Zhang, S.C. Kundu and L. Wang. 2014. Exploring natural silk protein sericin for regenerative medicine: an injectable, photoluminescent, cell-adhesive 3D hydrogel. Scientific Reports 4, Article number: 7064. DOI: 10.1038/srep07064. http://www.nature.com/articles/srep07064.

Waterman, J.A. 1938. Some notes on scorpion poisoning in Trinidad. Transactions of the Royal Society of Tropical Medicine and Hygiene 31: 607–624.

Waterman, J.A. 1960. Scorpions in the West Indies with special reference to *Tityus trinitatis*. Caribbean Medical Journal 12: 167–177.

Wayland, H., R. Manderino, T.O. Crist and K.J. Haynes. 2015. Microbial pesticide application during defoliator outbreaks may reduce loss of regional forest beetle richness. Ecosphere 6(6). Article 93. 14 pp. http://onlinelibrary.wiley.com/doi/10.1890/ES14-00252.1/pdf.

Weed, A.S., B.J. Bentz, M.P. Ayres and T.P. Holmes. 2015. Geographically variable response of *Dendroctonus ponderosae* to winter warming in the western United States. Landscape Ecology. 19 pp. DOI: 10.1007/s10980-015-0170-z.

Weisman, A.L. V.V. Gorbatovsky, Y.N. Gorbunov, A.D. Poyarkov, A.G. Sorokin, P.V. Fomenko and A. Yu. Tsellarius. 1999. Wild animals and plants in commerce in Russia and CIS countries. Moscow, NIA-Priroda. 156 pp. [in Russian].

Weitschat, W. and W. Wichard. 2002. Atlas of Plants and Animals in Baltic Amber. Munchen. 256 pp.

Wendt, H., A. Hillmer, K. Reimers, J.W. Kuhbier, F. Schafer-Nolte, C. Allmeling, C. Kasper and P.M. Vogt. 2011. Artificial skin – culturing of different skin cell lines for generating an artificial skin substitute on cross-weaved spider silk fibres. Public Library of Science One 6(7): e21833. DOI: 10.1371/journal.pone.0021833.

West, L. and J. Ridl. 1994. How to Photograph Insects and Spiders. Mechanicsburg, PA: Stackpole. 118 pp.

Wetterer, J.K. and R.R. Snelling. 2006. The red imported fire ant, *Solenopsis invicta*, in the Virgin Islands (Hymenoptera: Formicidae). The Florida Entomologist 89(4): 431–434.

Wetterer, J.K. 2013. Exotic spread of *Solenopsis invicta* (Hymenoptera: Formicidae) beyond North America. Sociobiology. 60: 53–63. DOI: 10.13102/sociobiology.v60i1.50-55.

Whatley, M.H., E.E. van Loon, C. Cerli, J.A.Vonk, H.G. van der Geest and W. Admiraal. 2014. Linkages between benthic microbial and freshwater insect communities in degraded peatland ditches. Ecological Indicators 46: 415–424.

Whitaker, I.S., C. Twine, M.J. Whitaker, M. Welck, C.S. Brown and A. Shandall. 2007. Larval therapy from antiquity to present day: mechanisms of action, clinical application and future potential. Postgraduate Medical Journal 83: 409–413.

Whitford, W.G., Y. Steinberger and G. Ettershank. 1982. Contributions of subterranean termites to the "economy" of Chihuahuan Desert ecosystems. Oecologia 55: 298–302.

Why are insects not allowed in animal feed? 2014. http://www.allaboutfeed.net/Global/Whitepapers/Whitepaper_Insect_meal.pdf.

Wild, A. 2013. Darwin didn't kill Bigfoot. http://blogs.scientificamerican.com/compound-eye/darwine28099s-freak-show-or-why-darwin-didnt-kill-bigfoot/.

Wildlife Tourism challenges, opportunities and managing the future. 2009. Sustainable Tourism Cooperative Research Centre. 68 pp.

Williams, P.H. and J.L. Osborne. 2009. Bumblebee vulnerability and conservation world-wide. Apidologie 40: 367–387.

Williamson, S.M., D.D. Baker and G.A. Wright. 2013. Acute exposure to a sublethal dose of imidacloprid and coumaphos enhances olfactory learning and memory in the honeybee Apis mellifera. Invertebrate Neuroscience 13(1): 63–70.

Wilson, R.T. 2012. Small animals for small farms. Diversification booklet number 14. FAO, Rome. 92 pp. http://www.fao.org/3/a-i2469e.pdf.

Wilts, B.D., A. Matsushita, K. Arikawa and D.G. Stavenga. 2015. Spectrally tuned structural and pigmentary coloration of birdwing butterfly wing scales. Journal of the Royal Society Interface 12(111). DOI: 10.1098/rsif.2015.0717.

Winfree, R. 2010. The conservation and restoration of wild bees. Year in Ecology and Conservation Biology. Annals of the New York Academy of Sciences 1195: 169–197. DOI: 10.1111/j.1749-6632.2010.05449.x.

Winston, R.L., M. Schwarzländer, H.L. Hinz, M.D. Day, M.J.W. Cock and M.H. Julien (eds.). 2014. Biological Control of Weeds: A World Catalogue of Agents and Their Target Weeds, 5th ed. USDA Forest Service, Forest Health Technology Enterprise Team, Morgantown, WV. FHTET-2014-04. 838 pp.

Wolf, M.J., H. Amrein, J.A. Izatt, M.A. Choma, M.C. Reedy and H.A. Rockman. 2006. Drosophila as a model for the identification of genes causing adult human heart disease. Proceedings of the National Academy of Sciences of the United States of America 103(5): 1394–1399.

Wolfe, A.P., R. Tappert, K. Muehlenbachs, M. Boudreau, R.C. McKellar, J.F. Basinger and A. Garrett. 2009. A new proposal concerning the botanical origin of Baltic amber. Proceedings of the Royal Society B 276: 3403–3412. DOI: 10.1098/rspb.2009.0806.

Worischka, S., S.I. Schmidt, C. Hellmann and C. Winkelmann. 2015. Selective predation by benthivorous fish on stream macroinvertebrates. The role of prey traits and prey abundance. Limnologica 52: 41–50. DOI: 10.1016/j.limno.2015.03.004.

World Health Organization. 2010. Guidelines for the treatment of malaria, 2nd ed. Geneva, World Health Organization.

World Health Organization. 2011. Bedbugs, fleas, lice, ticks and mites (Chapter 4). pp. 237–261. http://www.who.int/water_sanitation_health/resources/vector237to261.pdf.

www.amazingbutterflies.com/wedding.htm.
www.ansp.org/.
www.bangladeshbutterflypark.com.bd/.
www.biobest.be.
www.bluewillowgarden.com/conservatory.html.
www.botanicgardens.fi/.
www.butterflyfarm.co.uk/.
www.butterflyhouse.com.au/.
www.butterflyreptile.com/.
www.cambridgebutterfly.com/.
www.chapul.com/.
www.delawarenaturesociety.org/AshlandDirections.
www.entomopraxis.com.
www.entosphinx.cz/.
www.floranimal.ru/orders/2709.html.
www.hmns.org/cockrell-butterfly-center/.
www.holidaytuscany.it/eng/visit-tuscany/butterfly-house.php.
www.insectnet.eu/.
www.jardinbotanicoquindio.org/.
www.keywestbutterfly.com/.
www.kfbg.org/.
www.koppert.nl.
www.lepkehaz.ro/en/the-butterfly-house-in-praid.
www.magicoflife.org/.
www.magicwings.com/.
www.mariposariodebenalmadena.com/.
www.mariposasdemindo.com/.
www.meijergardens.org/.
www.metroparks.org/butterfly-house/.
www.montgomeryzoo.com/.
www.mos.org/exhibits/butterfly-garden.
www.nsbutterflyhouse.com/.
www.omahazoo.com/exhibits/butterfly-insect-pavilion/.
www.originalbutterflyhouse.com/.

www.papillons.lu/en/.

www.penang.ws/penang-attractions/butterfly-farm.htm.

www.phuket.com/attractions/phuket-butterfly-garden.htm.

www.puertoricodaytrips.com/la-marquesa-forest-park/.

www.sazoo-aq.org/attractions/butterflies/.

www.seafordegardens.com/.

www.straffanbutterflyfarm.com/.

www.thebutterflyfarm.com/.

www.thebutterflypalace.com/.

www.turismofvg.it/Museums/House-of-butterflies-of-Bordano.

www.zoo.com.sg/exhibits-zones/fragile-forest.html.

www.zoo.org.au/melbourne/highlights/butterfly-house.

www.zoologicodecali.com.co/.

Xerces Society. 2014. Petition to protect the Monarch butterfly (*Danaus plexippus plexippus*) under the endangered species act. 159 pp. http://www.biologicaldiversity.org/species/invertebrates/pdfs/Monarch_ESA_Petition.pdf.

Xia, Z., S. Wu, S. Pan and J.M. Kim. 2012. Nutritional evaluation of protein from *Clanis bilineata (Lepidoptera)*, an edible insect. Journal of the Science of Food and Agriculture 92: 1479–1482.

Xiaoming, C., F. Ying, Z. Hong and C. Zhiyong. 2010. Review of the nuritive value of edible insects. pp. 85–92. *In*: Durst, P.B., D.V. Johnson, R.N. Leslie and K. Shono (eds.). Forest Insects as Food: Humans Bite Back. Proceedings of a Workshop on Asia-Pacific Resources and Their Potential for Development, 19–21 February 2008, Chiang Mai, Thailand. Bangkok: Food and Agriculture Organization of the United Nations.

Xu, M. and R.V. Lewis. 1990. Structure of a protein superfiber: spider dragline silk. Proceedings of the National Academy of Sciences of the United States of America 87(18): 7120–7124.

Yang, P., J. Zhu, M. Li, J. Li and X. Chen. 2011. Soluble proteome analysis of male *Ericerus pela* Chavannes cuticle at the stage of the second instar larva. African Journal of Microbiology Research 5(9): 1108–1118. DOI: 10.5897/AJMR11.175 http://www.academicjournals.org/journal/AJMR/article-full-text-pdf/4CE0F6A31620.

Yang, P., J.-Y. Zhu, Z.-J. Gong, D.-L. Xu, X.-M. Chen, W.-W. Liu, X.D. Lin and Y.F. Li. 2012. Transcriptome analysis of the chinese white wax scale *Ericerus pela* with focus on genes involved in wax biosynthesis. Public Library of Science One 7(4): e35719. DOI: 10.1371/journal.pone.0035719.

Yang, Z.-Q., X.-Y. Wand and Y.-N. Zhang. 2014. Recent advances in biological control of important native and invasive forest pests in China. Biological Control 68: 117–128. DOI: 10.1016/j.biocontrol.2013.06.010.

Yaro, M., K. Munyard, A., M.J. Stear and D. Groth. 2016. Combatting African Animal Trypanosomiasis (AAT) in livestock: The potential role of trypanotolerance. Veterinary Parasitology 225: 43–52. DOI: 10.1016/j.vetpar.2016.05.003.

Yashchenko, R.V. and I.D. Mityaev. 2005. Red book of animals of Kazakhstan is completed. Steppe Bulletin 17: 32–33 [in Russian].

Yashina, T.V. 2011. Indicators for evaluation of biodiversity on especially protected natural territories of the Altai-Sayan Ecoregion. Instructions for use. Krasnoyarsk. 56 pp. [in Russian].

Yasyukevich, V.V., E.A. Davidovich, S.N. Litkina and N.V. Yasukevich. 2011. Climate change observed in the second half of XX-beginning of XXI century and possible changes in potential habitat of the Colorado potato beetle (*Leptinotarsa decemlineata* Say). Applied Entomology 2(5): 30–39 [in Russian].

Yates, J.R., J.K. Grace and M. Tamashiro. 1997. The Formosan subterranean termite: a review of new management methods in Hawaii. Proceedings of 1997 FAOPMA Convention, Hong Kong 59–68. https://www.researchgate.net/publication/259103701_The_Formosan_subterranean_termite_a_review_of_new_management_methods_in_Hawaii.

Yavorskaya, N.M. 2008. Chironomid larvae (diptera, chironomidae) of the Kadi river (the lower Amur river basin). Freshwater Ecosystems of the Amur River Basin. Vladivostok, Dalnauka, pp. 209–217 [in Russian].

Yen, A.L. 2008. Edible insects and other invertebrates in Australia: future prospects. pp. 65–84. *In*: Durst, P.B., D.V. Johnson, R.N. Leslie and K. Shono (eds.). Forest Insects as Food: Humans Bite Back. Proceedings of a Workshop on Asia-Pacific Resources and Their Potential for Development, 19–21 February 2008, Chiang Mai, Thailand. Bangkok: Food and Agriculture Organization of the United Nations.

Yen, A.L. 2015. Insects as food and feed in the Asia Pacific region: current perspectives and future directions. Journal of Insects as Food and Feed 1(1): 33–55.

Yhoung-Aree, J. and K. Viwatpanich. 2005. Edible insects in the Laos PDR, Myanmar, Thailand, and Vietnam. pp. 415–440. *In*: Paoletti, M.G. (ed.). Ecological Implications of Minilivestock. Enfield, NH, Science Publishers.

Yhoung-Aree, J. 2010. Edible insects in Thailand: nutritional values and health concerns. *In*: Durst, P.B., D.V. Johnson, R.N. Leslie and K. Shono (eds.). Forest Insects as Food: Humans Bite Back. Proceedings of a Workshop on Asia-Pacific Resources and Their Potential for Development, 1921 February 2008, Chiang Mai, Thailand. Bangkok: Food and Agriculture Organization of the United Nations, p. 201.

Yi, C., Q. He, L. Wang and R. Kuang. 2010. The utilization of insect resources in Chinese rural area. Journal of Agricultural Science 2(3): 146–154.

Yogi, R.K., A. Bhattacharya, A.K. Jaiswal and A. Kumar. 2015. Lac, Plant Resins and Gums Statistics 2014: At a Glance. ICAR-Indian Institute of Natural Resins and Gums, Ranchi (Jharkhand), India. Bulletin (Technical) No. 07/2015. 68 pp. http://ilri.ernet.in/~iinrg/Lac%20Statistics.pdf.

Yong, D., M.A. Toleman, C.G. Giske, H.S. Cho, K. Sundman, K. Lee and T.R. Walsh. 2009. Characterization of a new metallo-beta-lactamase gene, bla(NDM-1), and a novel erythromycin esterase gene carried on a unique genetic structure in Klebsiella pneumoniae sequence type 14 from India. Antimicrobial Agents and Chemotherapy 53(12): 5046–5054. DOI: 10.1128/AAC.00774-09. PMID 19770275.

Young, J., S.M. Walker, R.J. Bomphrey, G.K. Taylor and A.L.R. Thomas. 2009. Details of insect wing design and deformation enhance aerodynamic function and flight efficiency. Science 325(5947): 1549–1552. DOI: 10.1126/science.1175928.

Yurchenko, N.N., A.V. Ivannikov and I.K. Zakharov. 2015. History of the discoveries in Drosophila—stages of development of genetics. Vavilov Journal of Genetics and Plant Breeding 19(1): 39–49 [in Russian].

Yushkin, N.P., D.A. Bushnev and S.N. Shanina. 2006. Fossil resins of North Eurasia. Bulletin 11: 2–5. http://cyberleninka.ru/article/n/iskopaemye-smoly-severnoy-evrazii [in Russian].

Zaitsev, V.F. and S.Y. Reznik. 2004. Biocontrol and biodiversity: two views on the problem of invasions. pp. 44–53. *In*: Alimov, A.F. and N.G. Bogutskaya (eds.). Biological Invasions in Aquatic and Terrestrial Ecosystems. Moscow, KMK [in Russian].

Zaitseva, L.P. 2007. Historical tradition of military banners production in Russia (XVII–beginning of XX centuries). Gerboved 95: 107–133 [in Russian].

Zakharov, A.A. 2015. Ants of forest communities, their life and their role in forest. Moscow, KMK. 404 pp. [in Russian].

Zakhvatkin, J.A. 2003. Fundamentals of general and agricultural ecology: St. Petersburg, Mir, 360 pp. [in Russian].

Zamotajlov, A.S. 2012. History and methods of biological protection of plants. Electronic course of lectures. Krasnodar. 237 pp. [in Russian].

Zamotajlov, A.S., A.M. Devyatkin and I.V. Bedlovskaya. 2015. Entomology. Krasnodar, KubGAU. 215 pp. [in Russian].

Zanolli, L. 2014. Feed rises. http://www.technologyreview.com/news/529756/insect-farming-is-taking-shape-as-demand-for-animal-feed-rises/.

Zaslavsky, M.A. 1966. Taxidermy of birds. Manufacture of stuffed birds, skeletons and museums preparations. Moscow, Nauka. 302 pp. [in Russian].

Zeng, L., Y.Y. Lu, X.F. He, W.Q. Zhang and G.W. Liang. 2005. Identification of red imported fire ant Solenopsis invicta to invade mainland China and infestation in Wuchuan, Guangdong. Chinese Bulletin of Entomology 42(2): 144–148. ISSN 0452-8255.

Zenni, R.D. and M.A. Nuñez. 2013. The elephant in the room: the role of failed invasions in understanding invasion biology. Oikos 122: 801–815. DOI: 10.1111/j.1600-0706.2012.00254.x.

Zhang, R., Y. Li, N. Liu and S.D. Porter. 2007. An overview of the red imported fire ant (Hymenoptera: Formicidae) in Mainland China. The Florida Entomologist 90(4): 723–731. DOI: 10.1653/0015-4040(2007)90[723:aootri]2.0.co;2.

Zhang, Z., J. Lin and Y. Zhang. 1993. Feeding position of wax insect (*Ericerus pela*) on Ligustrum lucidum and the influence of parasitism on host tissue. Acta Botanica Sinica 35(Suppl.): 19–23. http://www.jipb.net/pubsoft/content/2/3489/35-13-4.pdf.

Zhang, Z., X. Teng, M. Chen and F. Li. 2014. Orthologs of human disease associated genes and RNAi analysis of silencing insulin receptor gene in *Bombyx mori*. International Journal of Molecular Sciences 15: 18102–18116.

Zhang, Z.-Q. 2013. Phylum Athropoda. Zootaxa 3703(1): 17–26. http://www.mapress.com/zootaxa/2013/f/zt03703p026.pdf.

Zhao, J., X. Zhang, N. Chen and Q. Pan. 2012. Why superhydrophobicity is crucial for a water-jumping microrobot? Experimental and Theoretical Investigations. ACS Applied Material and Interfaces 4(7): 3706–3711. DOI: 10.1021/am300794z.

Zhdanova, T.D. 2012. Role of insects in ecological balance. http://www.portal-slovo.ru/impressionism/36352.php [in Russian].

Zherikhin, V.V., A.G. Ponomarenko and A.P. Rasnitsyn. 2008. Introduction to paleoentomology. Moscow, KMK. 371 pp. [in Russian].

Zhigalski, O.A. 2011. Biological diversity estimation of forest ecosystems in the Ural Mountains. Bulletin of Udmurt University. Series 6: Biology. Earth sciences 3: 13–22 [in Russian].

Zhokhov, P.I. 1975. Forest protection manual. Lesnaya promyshlennost, Moscow, 295 pp. [in Russian].

Zhong, J.H. and L.L. Liug. 2002. Termite fauna in China and their economic importance. Sociobiology 40: 25–32.

Zhu, B.F., W. Li, R.V. Lewis, C.U. Segre and R. Wang. 2015. E-Spun composite fibers of collagen and drag line silk protein: fiber mechanics, biocompatibility, and application in stem cell differentiation. Biomacromolecules 16(1): 202–213. DOI: 10.1021/bm501403f.

Zhuzhikov, D.P. 1983. Termites of the USSR. Moscow, Moscow University. 224 pp. [in Russian].

Zillen, P. 2008. Bug Fighting History. http://www.bugfighting.com/history.htm.

Zimmerman, M.D., D.R. Murdoch, P.J. Rosmaszl, B. Basnyat, C.W. Woods, A.R. Richards, R.H. Belbase, D.A. Hammer, T.P. Anderson and L.B. Reller. 2008. Murine typhus and febrile illness, Nepal. Emerging Infectious Diseases 14(10). http://www.biomedsearch.com/article/Murine-typhus-febrile-illness-Nepal/188062795.html.

Zinovyev, E.V. 2010. An overview of the locations of the Holocene insects of Northern and Middle Urals. Dynamics of the ecosystems in the Holocene. Proceedings of Second Russian Scientific Conference, October 12–14, 2010, Ekaterinburg, Chelyabinsk, Rifei, pp. 72–76 [in Russian].

Zlotin, A.Z. 1989. Technical entomology. Kiev, Naukova Dumka. 184 pp. [in Russian].

Zotova, N.Y. 2003. Wonderful cochineal, or the story of carmine. Biology 34: 2–4 [in Russian].

APPENDIXES

Index of Scientific Names for Species

Index of Common Names for Species

Index of Geographic Names

China 22, 32, 34, 38, 39, 43–45, 63–66, 70–73, 78, 80, 83,
84, 86, 96–98, 102, 103, 109, 111, 115, 117, 118, 122,
123, 125, 129, 133, 134, 143, 144, 147, 154, 182, 206,
207, 227, 236, 243, 258, 265, 267, 307–310, 312, 314,
316–319
Chittagong 162
Chřič village 191
Christophe de Thou 189
Coahuila state 258
Coconut Creek 165
Cocos Islands 308
Cocumont commune 189
Coff Harbour 128
Colima 299
Colombia 32, 50, 71, 165, 186, 188, 189, 258, 259, 298,
299, 301, 309, 311
Colonia department 189
Colorado Desert 51
Colroy-la-Grande commune 189
Columbia, district 316, 318
Commune Bourg-de-Péage 187
Condé-sur-Iton commune 189
Congo Democratic Republic 197, 309
Constantinople 64, 75, 179, 265
Cook Islands 183, 194–197, 199, 200, 311
Cormenon commune 189
Costa Rica 16, 158, 165, 311
Cote d'Ivoire 106, 297
Coucy-lès-Eppes commune 189
Courtisols commune 189
Cracow 157
Crimea 143, 189, 225, 227, 257, 258, 266, 272, 289, 290
Cruz Machado municipality 187
Cuba 45, 78, 116, 184, 308, 309, 311
Cundinamarca department 189, 258
Currumbin 182
Cyprus 263, 289
Cyrenaica 226
Czech Republic 8, 157, 158, 191, 192

D

Daejeon 183
Daigny commune 191
Dalby 120, 122, 182
Danzig 75
Daubeuf-près-Vatteville commune 189
Dayton 164
Deer Lake 163
Delhi 40, 147
Democratic Republic of the Congo 39, 110
Denmark 31, 32, 77, 109, 154, 162, 263, 279, 327
Denver County 130
Department of Risaralda 189
Dessau 160
Detroit 164
D'Huison-Longueville commune 189
Dnieper River 31
Dnipropetrovsk 180, 181
Dobrino settlement 184
Dois Irmãos municipality 187
Dominican Republic 31, 32, 311

Donetsk 182, 183, 190
Dubai 181
Dublin 89
Dunedin 165
Dunnellon 158
Durango 258, 296, 299, 300
Durango state 258
Durham 165
Dzhankoy town 189

E

East Africa 6, 260, 287, 296, 303
East Lansing 164
East Siberia 27
Easter Island 256
eastern Canada 66, 289, 290
eastern China 64, 66, 314
Ecuador 165, 259, 290, 311
Egypt 30, 71, 86, 91, 130, 131, 175, 182, 185, 195, 205,
228, 231, 263, 265, 296, 297, 300, 302
El Dorado Hills 166
El Salvador 308
Elba island 186
Elbeuf commune 189
Elbrus Mount 214
Emmen 161, 163
Endau-Rompin National Park 127
England 24, 30, 32, 55, 64, 73, 89, 118, 128, 130, 138,
143, 160, 163, 164, 186–188, 206, 216, 217, 319, 327
English Channel 160
Ennery commune 189
Enterprise 88, 99, 100, 149, 152, 153, 160, 180, 181, 185,
241, 247
Estissac commune 189
Estonia 186, 190, 191, 206, 263
Ethiopia 32–34, 84, 97, 146, 267, 271
Eugénie-les-Bains commune 189
Eurasia 8, 120, 257, 303, 315
Europe 3, 4, 6, 30, 33, 43, 65, 66, 72, 73, 75, 76, 79, 83,
86, 90, 96–98, 101, 109, 112, 115, 116, 118, 120, 130,
147, 151, 152, 154, 159, 161–163, 165, 173, 178, 180,
181, 207, 213, 215, 218, 225, 227, 249, 251, 253,
255–257, 262, 265–267, 271, 272, 283, 287, 289, 292,
303, 305–307, 309, 310, 316–320, 322, 324, 327, 331
European part of Russia 27, 290, 305
Eyragues city 191

F

Far East 9, 66, 72, 206, 274, 277–280, 307, 314
Federated States of Micronesia 256
Feodosia 143
Fergana province 191
Fiji 83, 195, 196, 394, 311
Finland 16, 33, 77, 162, 186, 191, 215, 263, 289, 325
Fixin city 189
Floirac commune 190
Florence 109, 128
Florida 50, 92, 96, 142, 164, 166, 317, 323, 326, 327
Formiga town 186
Foucrainville commune 191
Fourchambault commune 189

Index of Personalities

Subject Index

Milton Keynes UK
Ingram Content Group UK Ltd.
UKHW050455071024
449327UK00015B/396